Nanotechnology in the Life Sciences

Series Editor

Ram Prasad
School of Environmental Science and Engineering,
Sun Yat-Sen University, Guangzhou, China

Amity Institute of Microbial Technology,
Amity University, Noida, Uttar Pradesh, India

Nano and biotechnology are two of the 21st century's most promising technologies. Nanotechnology is demarcated as the design, development, and application of materials and devices whose least functional make up is on a nanometer scale (1 to 100 nm). Meanwhile, biotechnology deals with metabolic and other physiological developments of biological subjects including microorganisms. These microbial processes have opened up new opportunities to explore novel applications, for example, the biosynthesis of metal nanomaterials, with the implication that these two technologies (i.e., thus nanobiotechnology) can play a vital role in developing and executing many valuable tools in the study of life. Nanotechnology is very diverse, ranging from extensions of conventional device physics to completely new approaches based upon molecular self-assembly, from developing new materials with dimensions on the nanoscale, to investigating whether we can directly control matters on/in the atomic scale level. This idea entails its application to diverse fields of science such as plant biology, organic chemistry, agriculture, the food industry, and more.

Nanobiotechnology offers a wide range of uses in medicine, agriculture, and the environment. Many diseases that do not have cures today may be cured by nanotechnology in the future. Use of nanotechnology in medical therapeutics needs adequate evaluation of its risk and safety factors. Scientists who are against the use of nanotechnology also agree that advancement in nanotechnology should continue because this field promises great benefits, but testing should be carried out to ensure its safety in people. It is possible that nanomedicine in the future will play a crucial role in the treatment of human and plant diseases, and also in the enhancement of normal human physiology and plant systems, respectively. If everything proceeds as expected, nanobiotechnology will, one day, become an inevitable part of our everyday life and will help save many lives.

More information about this series at http://www.springer.com/series/15921

Ram Prasad • Vivek Kumar • Manoj Kumar
Devendra Choudhary
Editors

Nanobiotechnology in Bioformulations

Springer

Editors
Ram Prasad
School of Environmental Science and Engineering
Sun Yat-Sen University
Guangzhou, China

Amity Institute of Microbial Technology
Amity University
Noida, Uttar Pradesh, India

Manoj Kumar
Department of Life Science
Central University of Jharkhand
Ranchi, Jharkhand, India

Vivek Kumar
Himalayan School of Biosciences
Swami Rama Himalayan University
Dehradun, Uttarakhand, India

Devendra Choudhary
Amity Institute of Microbial Technology
Amity University
Noida, Uttar Pradesh, India

ISSN 2523-8027 ISSN 2523-8035 (electronic)
Nanotechnology in the Life Sciences
ISBN 978-3-030-17060-8 ISBN 978-3-030-17061-5 (eBook)
https://doi.org/10.1007/978-3-030-17061-5

This Springer imprint is published by the registered company Springer Nature Switzerland AG.
The registered company address is: Gewerbestrasse 11, 6330 Cham, Switzerland

Preface

The concept of "Nanobiotechnology" represents a new-fangled leading edge in the modern agriculture and is projected to become a foremost thrust in immediate forthcoming time by propounding its prospective uses and practical applications under field conditions. This amalgamating tactic, i.e., nanobiotechnology in agriculture sector, has great prospective to cope up with universal confronts of sustainable food security and healthy food production, overall sustainability, and climate transformation. Nevertheless, in spite of the possible and prospective benefits of nanobiotechnology in agriculture sector so far, their significance and applicability has not touched the field situations. The raising worries and apprehensions about outcome, carriage, bioavailability, the toxicity of nanoparticles, and incompatibility of regulatory framework limit the widespread preference and acceptance to adopt wholeheartedly the nanobiotechnologies in agriculture field. Furthermore, the existing research inclinations lack authentic and pragmatic attitude that fail to achieve inclusive acquaintance of factors related to risk assessment and further nanoparticles toxicity toward whole cropping system components such as plants, water, soil, micro and macro microbiome into the ecosystem. Henceforth, in the present book we put our sincere efforts to propose some imperative opinions which need to be addressed. These points are in reference to the existing and forthcoming research on nanobiotechnology in the field of cropping system.

Recent transferal in agricultural traditions from widespread use of chemical fertilizers and pesticides to organic agriculture has brought into spotlight the employment of potential microbes which performs several significant functions for the benefit of plant. Microbial formulations accessible all over the globe range from powdered or charcoal based and solution and secondary metabolites based bioformulations. The perfect specifications needed for sophisticated efficient bioformulations development of biofertilizers and biopesticides comprise of effective strain development, higher shelf life, long storage at various climatic conditions, wide and easy application technology, quality control, and above all the biosafety of the product. The prospective advantages and uses of nanobiotechnology in agricultural area have generated huge interest, since it could augment production of agriculture produce while being cost-effective on energy and pocket. Notably, nanobiotechnology by virtue of use of nanoparticles has proposed gigantic potential applications in agricultural sector which includes nanobiosensor, nanobiofertilizer, nanobiopesticide, nanoherbicide, and smart transport systems for meticulous and regulated release of agrobiochemicals. Moreover, formulation of consortium of secondary metabolites against individual phytopathogens could be used irrespective of geographical position having higher disease incidence. This bioformulation approach would be incomparable by existing technology, as the bioformulation would explicitly target a particular pathogen without harming the natural microbiome of ecosystem.

The present book *Nanobiotechnology in Bioformulations* comprises constraints associated with large-scale development and commercialization of formulation of bioinoculants. Furthermore, exclusive emphasis will be on next generation efficient bioinoculants of having secondary metabolite formulations with longer shelf-life but along with advanced competence against several phytopathogens. Here valuable chapters deal with bioformulation strategies using divergent groups of microbiome along with detailed diagrammatic and pictorial representation. This book will be of great help both for experts and novices in the field of microbial bioformulation, nanotechnology, or nano/micro-biotechnology. We hope to infuse and introduce the prevailing status, realistically, and inferences of microbial researchers and scientists, agronomists, students, environmentalists, agriculturists, and agribusiness professionals, as well as other passionate people who are whole heartedly devoted to sustaining the ecosystem. We wholeheartedly thank all the contributors who have proficiency in this field of research for their cutting edge, timely chapters and their help in making this book a successful attempt.

This book should be immensely useful to bioscience and nanoscience specially microbiologists, nanotechnologists, researchers, technocrats, and scientists of microbial nanobiotechnology. I have honored that the leading researchers who have extensive, in-depth experience and expertise in microbial system and nanobiotechnology took the time and effort to develop these outstanding chapters. Each chapter is written by internationally recognized researchers/professors so the reader is given an up-to-date and detailed account of our knowledge of the nanobiotechnology and innumerable applications of microbes.

I wish to thank Eric Stannard, Senior Editor, Springer; Antony Dunlap, Springer Nature, USA; Rahul Sharma, Project Coordinator, Springer Nature; and Santhamurthy Ramamoorthy, Project Manager, SPi Global, for generous assistance, constant support, and patience in initializing the volume. Ram Prasad gives special thanks to exquisite wife Dr. Avita for her constant support and motivations in putting everything together. Dr. Prasad in particular is very thankful to Professor

Ajit Varma, Amity University, for constant encouragement. And, also gives special thanks to his esteemed friends and well wishers and all faculty colleagues of School of Environmental Science and Engineering, Sun Yat-Sen University, China, and AIMT, Amity University, India.

Guangzhou, China; Noida, Uttar Pradesh, India — Ram Prasad
Dehradun, Uttarakhand, India — Vivek Kumar
Ranchi, Jharkhand, India — Manoj Kumar
Noida, Uttar Pradesh, India — Devendra Choudhary

Contents

Contributors

M. Anju Department of Energy and Environment Sciences, Chaudhary Devi Lal University, Sirsa, Haryana, India

V. Anshid Department of Food Technology and Nutrition, Lovely Professional University, Phagwara, Punjab, India

Akshay Bagal Department of Food Technology and Nutrition, Lovely Professional University, Phagwara, Punjab, India

Estefânia V. R. Campos São Paulo State University (UNESP), Laboratory of Environmental Nanotechnology, Institute of Science and Technology of Sorocaba, Sorocaba, SP, Brazil

Prince Chawla School of Bioengineering and Food Technology, Shoolini University, Solan, Himachal Pradesh, India

Navnidhi Chhikara Department of Food Technology and Nutrition, Lovely Professional University, Phagwara, Punjab, India

Argus Cezar da Rocha Neto Department of Agronomic Engineering, Adventist University of São Paulo-UNASP, Engenheiro Coelho, SP, Brazil

César Rodrigues da Silva Department of Agronomic Engineering, Adventist University of São Paulo-UNASP, Engenheiro Coelho, SP, Brazil

Tiago Alves Jorge de Souza Department of Agronomic Engineering, Adventist University of São Paulo-UNASP, Engenheiro Coelho, SP, Brazil

Department of Genetics, FMRP-USP, São Paulo University-USP, Ribeirão Preto, SP, Brazil

Margarita del Rosario Salazar-Sánchez Facultad de Ciencias Agrarias, Universidad del Cauca, Popayán, Colombia

Sanju Bala Dhull Department of Food Science and Technology, Chaudhary Devi Lal University, Sirsa, Haryana, India

Lorena Farías-Cepeda Facultad de Ciencias Químicas, Universidad Autónoma de Coahuila, Saltillo, Coahuila, Mexico

Leonardo F. Fraceto São Paulo State University (UNESP), Laboratory of Environmental Nanotechnology, Institute of Science and Technology of Sorocaba, Sorocaba, SP, Brazil

Leonardo Pereira Franchi Department of Genetics, FMRP-USP, São Paulo University-USP, Ribeirão Preto, SP, Brazil

Suresh Kumar Gahlawat Department of Biotechnology, Chaudhary Devi Lal University, Sirsa, Haryana, India

Vishwajeet Gaikwad Department of Food Technology and Nutrition, Lovely Professional University, Phagwara, Punjab, India

Siddra Ijaz Centre of Agricultural Biochemistry and Biotechnology (CABB), Faculty of Agriculture, University of Agriculture, Faisalabad, Pakistan

Imran Ul Haq Department of Plant Pathology, Faculty of Agriculture, University of Agriculture, Faisalabad, Pakistan

Akshay Jadhav Department of Food Technology and Nutrition, Lovely Professional University, Phagwara, Punjab, India

Shubham Jadhav Department of Food Technology and Nutrition, Lovely Professional University, Phagwara, Punjab, India

Sundeep Jaglan Division of Microbial Biotechnology, CSIR-Indian Institute of Integrative Medicine, Jammu, India

Josef Jampílek Division of Biologically Active Complexes and Molecular Magnets, Regional Centre of Advanced Technologies and Materials, Palacky University, Olomouc, Czech Republic

Department of Analytical Chemistry, Faculty of Natural Sciences, Comenius University, Ilkovičova 6, Bratislava, Slovakia

Maninder Kaur Department of Food Science and Technology, Guru Nanak Dev University, Amritsar, India

Pawan Kaur Department of Biotechnology, Chaudhary Devi Lal University, Sirsa, Haryana, India

Ravinder Kaushik School of Bioengineering and Food Technology, Shoolini University, Solan, Himachal Pradesh, India

Madhuprasad Kigga Centre for Nano and Material Sciences, JAIN (Deemed-to-be University) Jain Global Campus, Bengaluru, Karnataka, India

Katarína Kráľová Institute of Chemistry, Faculty of Natural Sciences, Comenius University, Bratislava, Slovakia

Vinita Kumari School of Engineering Sciences and Technology, Jamia Hamdard, New Delhi, India

Vijay Kumar Regional Ayurveda Research Institute for Drug Development, Gwalior, Madhya Pradesh, India

Vivek Kumar Himalayan School of Biosciences, Swami Rama Himalayan University, Dehradun, Uttarakhand, India

I. K. Kurdish Department of Microbiological Processes on Solid Surface, Zabolotny Institute of Microbiology and Virology, National Academy Sciences of Ukraine, Kiev, Ukraine

Mahaveer D. Kurkuri Centre for Nano and Material Sciences, JAIN (Deemed-to-be University) Jain Global Campus, Bengaluru, Karnataka, India

Anju Malik Department of Energy and Environmental Sciences, Chaudhary Devi Lal University, Sirsa, Haryana, India

Bisma Malik Department of Bioresources, University of Kashmir, Srinagar, India

Tariq Maqbool Department of Nanotechnology, University of Kashmir, Srinagar, India

Cesar Martinez-Ledezma Universidad Autónoma de Nuevo León, Facultad de Ciencias Químicas, San Nicolás de los Garza, NL, México

Abhishek Kumar Mishra Himalayan School of Biosciences, Swami Rama Himalayan University, Jollygrant, Dehradun, Uttarakhand, India

Mahmoud Nasr Sanitary Engineering Department, Faculty of Engineering, Alexandria University, Alexandria, Egypt

Anil Panghal Department of Food Technology and Nutrition, Lovely Professional University, Phagwara, Punjab, India

AICRP-PHET, Department of Processing and Food Engineering, Chaudhary Charan Singh Haryana Agricultural University, Haryana, India

Tanveer Bilal Pirzadah Department of Bioresources, University of Kashmir, Srinagar, India

Sneh Punia Department of Food Science and Technology, Chaudhary Devi Lal University, Sirsa, Haryana, India

Usha Rani Himalayan School of Biosciences, Swami Rama Himalayan University, Dehradun, Uttarakhand, India

Sarushi Rastogi Department of Food Technology, School of Interdisciplinary Sciences, Jamia Hamdard, New Delhi, India

Reiaz Ul Rehman Department of Bioresources, University of Kashmir, Srinagar, India

Yadira Karina Reyes-Acosta Facultad de Ciencias Químicas, Universidad Autónoma de Coahuila, Saltillo, Coahuila, Mexico

Melissa Marlene Rodríguez-Delgado Universidad Autónoma de Nuevo León, Facultad de Ciencias Químicas, San Nicolás de los Garza, NL, México

Raúl Rodríguez-Herrera Facultad de Ciencias Químicas, Universidad Autónoma de Coahuila, Saltillo, Coahuila, Mexico

Romina Romero Technological Development Unit (UDT), Universidad de Concepcion, Coronel, Chile

Lucero Rosales-Marines Facultad de Ciencias Químicas, Universidad Autónoma de Coahuila, Saltillo, Coahuila, Mexico

Anilú Rubio-Ríos Facultad de Ciencias Químicas, Universidad Autónoma de Coahuila, Saltillo, Coahuila, Mexico

Manga Veera Sai Charan Department of Food Technology and Nutrition, Lovely Professional University, Phagwara, Punjab, India

Nidhi Saini Department of Biotechnology, Chaudhary Devi Lal University, Sirsa, Haryana, India

Kawaljit Singh Sandhu Department of Food Science and Technology, Chaudhary Devi Lal University, Sirsa, Haryana, India

Department of Food Science and Technology, Maharaja Ranjit Singh Punjab Technical University, Bathinda, India

Kankan Sharma Department of Biotechnology, Lovely Professional University, Phagwara, Punjab, India

Poorva Sharma Department of Food Technology and Nutrition, Lovely Professional University, Phagwara, Punjab, India

Vasudha Sharma Department of Food Technology, School of Interdisciplinary Sciences, Jamia Hamdard, New Delhi, India

Neelesh Sindhu Department of Veterinary Clinical Complex, Lala Lajpat Rai University of Veterinary and Animal Science, Hisar, India

Joginder Singh Department of Biotechnology, Lovely Professional University, Phagwara, Punjab, India

Simranjeet Singh Department of Biotechnology, Lovely Professional University, Phagwara, Punjab, India

Punjab Biotechnology Incubators, Mohali, Punjab, India

Regional Advanced Water Testing Laboratory, Mohali, Punjab, India

Anil Kumar Siroha Department of Food Science and Technology, Chaudhary Devi Lal University, Sirsa, Haryana, India

José Fernando Solanilla-Duque Facultad de Ciencias Agrarias, Universidad del Cauca, Popayán, Colombia

Lilian Rodrigues Rosa Souza Department of Chemistry, FFCLRP-USP, University of São Paulo-USP, Ribeirão Preto, SP, Brazil

Vinod Surendran Department of Food Technology and Nutrition, Lovely Professional University, Phagwara, Punjab, India

Abhilasha Thakur Department of Biotechnology, Chaudhary Devi Lal University, Sirsa, Haryana, India

U. T. Uthappa Centre for Nano and Material Sciences, JAIN (Deemed-to-be University) Jain Global Campus, Bengaluru, Karnataka, India

Juan Francisco Villarreal-Chiu Universidad Autónoma de Nuevo León, Facultad de Ciencias Químicas, San Nicolás de los Garza, NL, México

About the Authors

Ram Prasad is associated with Amity Institute of Microbial Technology, Amity University, Uttar Pradesh, India, since 2005. His research interest includes applied microbiology, plant–microbe interactions, sustainable agriculture, and nanobiotechnology. Dr. Prasad has more than 150 publications to his credit, including research papers, review articles, and book chapters and five patents issued or pending, and edited or authored several books. Dr. Prasad has 12 years of teaching experience and has been awarded the Young Scientist Award (2007) and Prof. J.S. Datta Munshi Gold Medal (2009) by the International Society for Ecological Communications; FSAB fellowship (2010) by the Society for Applied Biotechnology; the American Cancer Society UICC International Fellowship for Beginning Investigators, USA (2014); Outstanding Scientist Award (2015) in the field of Microbiology by Venus International Foundation; BRICPL Science Investigator Award (ICAABT-2017) and Research Excellence Award (2018). He has been serving as editorial board member of *Frontiers in Microbiology*, *Frontiers in Nutrition*, and *Academia Journal of Biotechnology* and is series editor of Nanotechnology in the Life Sciences, Springer Nature, USA. Previously, Dr. Prasad served as a Visiting Assistant Professor at Whiting School of Engineering, Department of Mechanical Engineering, Johns Hopkins University, USA, and presently working as a Research Associate Professor at School of Environmental Science and Engineering, Sun Yat-Sen University, Guangzhou, China.

Vivek Kumar is an Associate Professor working at the Himalayan School of Biosciences, Swami Rama Himalayan University, Jolly Grant, Dehradun, India. He currently serves on the editorial boards of various respected international journals, including *EnvironmentAsia, International Journal of Biological and Chemical Sciences, Environmental Sustainability, Journal of Advanced Botany and Zoology*, and *Journal of Ecobiotechnology*. He is also a reviewer for prestigious international journals such as the *Journal of Hazardous Materials, Environmental Sustainability, Science International, Acta Physiologiae Plantarum, Environment Science and Pollution Research*, and *Rhizosphere*. He has authored more than 100 publications, including research papers, review articles, and book chapters, and also edited several Springer books. Dr. Kumar has served as a microbiologist in the Department of Soil and Water Research, Public Authority of Agricultural Affairs and Fish Resources, Kuwait, for 8 years. He has been credited with first reporting and identification of Pink Rot inflorescence disease of date palms in Kuwait caused by *Serratia marcescens*. He was awarded the 2002 "Young Scientist Award" in agricultural microbiology by the Association of Microbiologists of India. His research areas include plant–microbe interactions, environmental microbiology, and bioremediation. He has also organized various outreach activities.

Manoj Kumar is currently working as Associate Professor in Life Sciences, Central University of Jharkhand, and is also the Head of the department since 2018. Dr. Kumar is having 15 years of research experience in Plant Developmental Biology with doctorate and postdoctoral research experiences from prestigious organizations (India and abroad). He has opted academic career with a passion in fundamental sciences (plant developmental biology) where he excelled in numerous collaborative research activities at national and international levels. He has published valuable research inputs in accredited journals and edited books, and has also guided several students at master, M.Phil., doctorate, and postdoctorate levels. He has been principle investigator of several projects of national funding agencies such as DBT and DRDO. He is associated as editor and reviewer with several research journals and is a consultant to several biotechnology firms at international level. At present he is principal investigator of DBT-BUILDER program and leading an elite research group at multidisciplinary level. Dr. Kumar is exploring academic world with a vision of empowering young generation with fact-finding approach destined for rural India.

Devendra Choudhary has more than 16 years of research experience in Microbiology and Microbial Biotechnology and is currently working as an Associate Professor in Amity University, Noida. Before joining Amity University, he spent several years in Mody University, Lakshmangarh, Rajasthan, as an Assistant Professor preceded by People's University and Barkatullah University, Bhopal, as an Assistant Professor cum Scientist. Dr. Choudhary received his PhD in Microbiology in 2005 from GB Pant University of Agriculture & Technology, Pantnagar, after having received his MSc in Microbiology from MDS University, Ajmer, and qualifying CSIR-UGC-NET in 2002. Dr. Choudhary has worked on GOI-sponsored major projects as Principal Investigator. Recently, he worked on two major projects for the DBT and SERB at Amity University preceded by the DST FAST-TRACK project at the Department of Biotechnology, Barkatullah University, Bhopal. As an active researcher, Dr. Choudhary has published 80 research and review articles along with several book chapters for reputed journals and edited books. He is a recipient of the Indian National Science Academy (INSA) visiting and summer research fellowship 2014. He has been selected for prestigious Membership of the National Academy of Sciences, India.

Chapter 1
Biobased Nanoemulsions: Concept, Formulation, and Applications

Anilú Rubio-Ríos, Lucero Rosales-Marines, José Fernando Solanilla-Duque, Yadira Karina Reyes-Acosta, Margarita del Rosario Salazar-Sánchez, Raúl Rodríguez-Herrera, and Lorena Farías-Cepeda

Contents

1.1 Introduction

An emulsion is a dispersion of liquid droplets in a second and different immiscible liquid. There are different types of emulsions, and they are defined in terms of their droplet size and other physical and thermodynamic properties. In general terms,

A. Rubio-Ríos · L. Rosales-Marines · Y. K. Reyes-Acosta · R. Rodríguez-Herrera
L. Farías-Cepeda (✉)
Facultad de Ciencias Químicas, Universidad Autónoma de Coahuila, Saltillo, Coahuila, Mexico
e-mail: lorenafarias@uadec.edu.mx

J. F. Solanilla-Duque · M. del Rosario Salazar-Sánchez
Facultad de Ciencias Agrarias, Universidad del Cauca, Popayán, Colombia

R. Prasad et al. (eds.), *Nanobiotechnology in Bioformulations*, Nanotechnology in the Life Sciences, https://doi.org/10.1007/978-3-030-17061-5_1

Table 1.1 Different types of emulsions and their primary characteristics

Characteristics	Macroemulsions	Nanoemulsions	Microemulsions
Droplet shape	Spherical	Spherical	Spherical, lamellar
Droplet size	1–100 μm	20–500 nm	10–100 nm
Appearance	Turbid	Transparent	Transparent
Droplet surface-to-mass ratio (m^2/g)	0.07–70	70–330	330–1300
Stability	Thermodynamically unstable Weakly kinetically stable	Thermodynamically unstable Kinetically stable	Thermodynamically stable
Preparation method	High- and low-energy methods	High- and low-energy methods	Low-energy methods
Polydispersity	Often high (>40%)	Typically low (<10–20%)	Typically low (<10%)

emulsions can be classified as classical emulsions (or macroemulsions), nanoemulsions, and microemulsions. Microemulsions are thermodynamically stable and are formed spontaneously (McClements 2012); thus, changes in temperature and composition may affect them (Gupta et al. 2016). Macroemulsions and nanoemulsions are both thermodynamically unstable; however, because they contain very small droplets, nanoemulsions can be kinetically stable. In summary, the primary differences between the different types of emulsions are based on the droplet size range and stability characteristics. These characteristics are listed in Table 1.1 (Gadhave 2014; Gupta et al. 2016).

In the particular case of nanoemulsions, some of the primary characteristics are a large interfacial area; a transparent or translucent appearance; high solubilization capacity; low viscosity; and high kinetic stability against sedimentation, flocculation, and coalescence (Salvia-Trujillo et al. 2017). Nanoemulsions may remain stable for anywhere from a few hours to years, depending upon the components and the process used for their formulation; thus, stability is an important aspect of nanoemulsion technology (Gadhave 2014).

Nanoemulsions are nonequilibrium systems that cannot be formulated spontaneously; it requires mandatory energy input from mechanical devices or from the chemical potential of compounds (Gadhave 2014). Nanoemulsions are formed from oil, water, and an emulsifier (Gupta et al. 2016). This means that they are a biphasic dispersion of two immiscible liquids: either water-in-oil (W/O) or oil-in-water (O/W) droplets stabilized by an emulsifier (commonly, an amphiphilic surfactant), or a multiple water-in-oil-in-water (W/O/W) nanoemulsion (Singh et al. 2017). The phase present in greater volume becomes the external phase. To predict the type of nanoemulsion formed under certain conditions, the interaction of the components is important; if the surfactant is water soluble it favors O/W emulsification, but if the surfactant is oil soluble it favors W/O emulsification. Figure 1.1 presents the different nanoemulsion types.

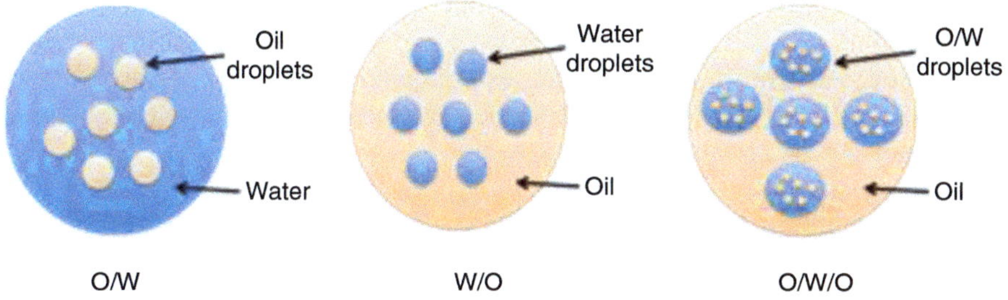

Fig. 1.1 Different types of nanoemulsion: oil-in-water (*O/W*), water-in-oil (*W/O*), and oil-in-water-in-oil (*O/W/O*)

Nanoemulsions usually contain 5–20% oil/lipid droplets in O/W emulsions, but this proportion may be significantly larger (up to 70%). For drug applications, the lipids/oils used are a function of the solubility of the drug and may be soybean oil, sesame oil, cottonseed oil, safflower oil, coconut oil, or rice bran oil, for instance (Singh et al. 2017). An appropriate emulsifier allows formation of small droplets, and it is important in the stabilization of nanoemulsions by reducing interfacial tension and the free energy required for droplet formation (Komaiko and McClements 2016).

The most common emulsifier used for preparation of nanoemulsions is a surfactant, but proteins and lipids have also been used for this purpose (Gupta et al. 2016). Some examples of the emulsifiers that are used are Tweens; Spans; egg, soy, or dairy lecithin; whey protein isolate; caseinate; gum arabic; and modified starches (Komaiko and McClements 2016).

Nanoemulsions are usually prepared in a two-step process. First, a macroemulsion is prepared, then it is converted into a nanoemulsion (Gupta et al. 2016). For the nanoemulsion preparation, there are primarily two techniques: high-energy methods and low-energy methods. These methods include high-pressure homogenization (HPH), ultrasonication, the phase inversion temperature (PIT) method, and the emulsion inversion point (EIP) method. HPH and ultrasonication are high-energy methods and consume significant energy (10^8–10^{10} W/kg), while low-energy methods, such as the PIT and EIP methods, are based on spontaneous formation of droplets; they use specific system properties to make small droplets by using low energy (10^3 W/kg) (Gupta et al. 2016).

As presented above, the methods used for nanoemulsion formation can be classified as high energy and low energy (Gadhave 2014).

1.1.1 High-Energy Methods

(i) *High-Pressure Homogenization*: In this method, the oil–water–surfactant mixture is subjected to very high pressure and pumped through a resistive valve. The shear stress produced causes fine droplets.

(ii) *Ultrasonication*: An acoustic field produces interfacial waves causing dispersion of the oil phase (as droplets) in the continuous phase. Later, ultrasound is applied, and acoustic cavitation causes formation and collapse of microbubbles by the pressure fluctuation of the sound waves. Finally, the turbulent microimplosions break up the droplets into a submicron size.
(iii) *High-Speed Devices*: In these devices, the energy provided gets dissipated, generating heat. However, the dispersion in the nanoemulsions that are produced is not good.

1.1.2 Low-Energy Methods

(i) *Phase Inversion Temperature*: This method is based on changes in the solubility of nonionic surfactants with temperature.
(ii) *Spontaneous Emulsification*: Using this method, the oil droplet diameter is a function of the ratio between the excess oil and the water-soluble solvent. The oil content that can be dispersed by this method is low.
(iii) *Membrane Emulsification*: This method requires less surfactant and produces emulsions with a narrow size distribution range.
(iv) *Emulsion Inversion Point*: In this method, the composition of the system changes at a constant temperature. The structures are formed through dilution with water or oil.

Nanoemulsions have applications in various industries, such as food, pharmaceuticals, and cosmetics (Prakash et al. 2018; Singh et al. 2017; Chellapa et al. 2016; Prasad et al. 2017). Depending on the application, they can be rendered into several forms such as liquids, creams, sprays, gels, aerosols, and foams (Singh et al. 2017).

1.1.2.1 Nanoemulsions in Cosmetics

Nanoemulsions are used in the cosmetic industry because they are easily absorbed, reducing water loss from the skin and providing rapid and effective penetration of active ingredients into the skin as a result of the large surface area of the droplets (Chellapa et al. 2016).

1.1.2.2 Nanoemulsions in Drug Delivery

Nanoemulsions are used as carriers for dermal and transdermal delivery of various kind of medicaments. They may offer more flexible therapeutics (Rai et al. 2018).

1.1.2.3 Nanoemulsions in Food

Nanoemulsions are used as delivery systems for lipids. Their optical clarity, good physical stability, and high bioavailability are important characteristics for their utilization in many foods and beverages (Komaiko and McClements 2016).

1.2 Stability of Nanoemulsions

Nanoemulsions are thermodynamically unstable systems and may break down over time through a variety of instability mechanisms, e.g., gravitational separation, coalescence, creaming, flocculation, and Ostwald ripening. This instability occurs because the free energy of the separated oil and water phases is lower than that of the nanoemulsion itself (Zeeb et al. 2015). Preparation of nanoemulsions typically requires oil, water, an emulsifier, and energy input (mechanical or physicochemical). The free energy required (ΔG) to form a nanoemulsion is given by:

$$\Delta G = \Delta A\gamma - T\Delta S \tag{1.1}$$

where A is the interfacial area, γ is the interfacial tension, T is the temperature, and S is the entropy. The term $\Delta A\gamma$ is the free energy needed to increase the oil–water interface, and $T\Delta S$ is the free energy associated with an increase in the number of possible arrangements of droplets in the nanoemulsion, in comparison with the separated phases. In both emulsions and nanoemulsions, the change in entropy is not great enough to overcome the free energy required to expand the interface; thus, the process of emulsion or nanoemulsion formation requires some free energy input. This free energy can be provided by mechanical devices or by the chemical potential of the system (Komaiko and McClements 2016).

A nanoemulsion can be kinetically stable by warranting that there is a sufficiently large energy barrier between the two states (the nanoemulsion and the separate phase). The energy barrier should be <20 kT to produce nanoemulsions with long-term stability. This metastable state is governed by the energy barriers and mass transport phenomena (McClements 2012). Physicochemical phenomena determine the energy barrier that prevents the droplets from coming into close proximity—for example, repulsive hydrodynamic and steric or electrostatic interactions operating between droplets (McClements 2012). The DLVO (Derjaguin–Landau–Verwey–Overbeek) theory predicts the energy barrier that two droplets have to overcome to come close to each other and flocculate. The DLVO theory is the combined effect of attraction forces (van der Waals) and repulsive forces (electric double layer) (see Fig. 1.2). Flocculation is when droplets come closer to each other because of attractive interactions and move as a single entity, and coalescence is when droplets merge with each other to form bigger drops. In emulsions it is difficult to distinguish between flocculation and coalescence (Gupta et al. 2016). Coalescence is generally the dominant mechanism of destabilization for microemulsions; in the

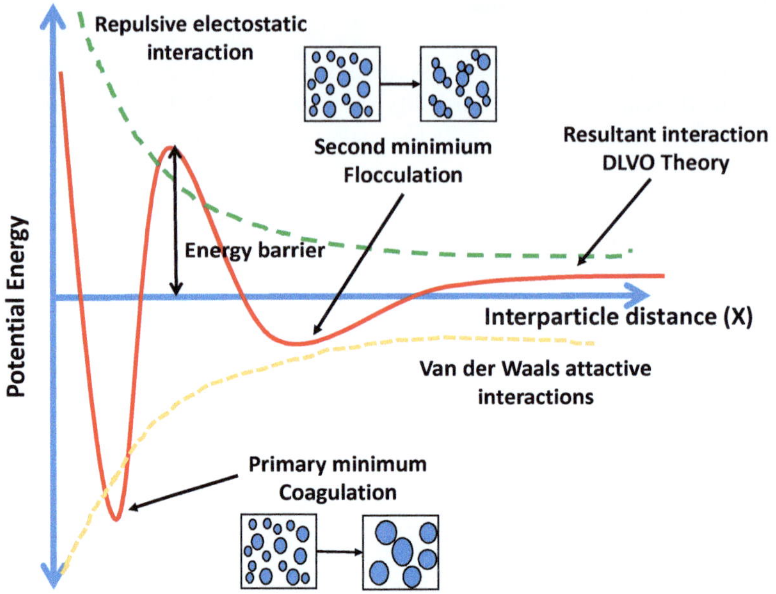

Fig. 1.2 Schematic representation of the DLVO (Derjaguin–Landau–Verwey–Overbeek) theory

case of nanoemulsions, coalescence is prevented. The small droplet size prevents nanoemulsions from undergoing reversible aggregation mechanisms that, when coupled to gravity, may accelerate coalescence because of the net increase in the film contact duration. Nanoemulsions are protected against flocculation phenomena because droplet adhesion decays with the droplet diameter. Hence, Ostwald ripening is the main nanoemulsion destabilizing process (Delmas et al. 2011).

O/W nanoemulsions consist of small spherical droplets of oil and surfactant molecules dispersed within an aqueous medium. The droplets consist of a hydrophobic core (oil) and surfactant tails, and a hydrophilic shell made up of surfactant head groups. The nanoemulsion droplets are of a small size (100 nm) with a broad particle size distribution and a large surface area. The sphere has the lowest interfacial area for a given volume of material, because of the relatively large Laplace pressure (Eq. 1.2); in other words, nanoemulsion droplets trend to be spherical with high curvature (Sharma et al. 2010) because their relatively high interfacial tension, the small droplet radius (r), and the Laplace pressure increase favor a spherical shape.

$$\Delta P_{\mathrm{L}} = \frac{2\gamma}{r} \tag{1.2}$$

Ostwald ripening occurs because of the difference in the chemical potential of the solute within droplets of different sizes. Because of the Laplace pressure, the chemical potential of the dispersed phase is higher in smaller droplets than in larger ones, providing the driving force for mass transfer from the smaller droplets to the larger droplets. Thus, the smaller droplets become smaller and the larger droplets grow. As mass transfer occurs when the dispersed phase molecule travels through the continuous phase, solubility of the dispersed phase in the continuous phase is a critical factor affecting the Ostwald ripening rate. From first principles, Lifshitz and Slyozov (1961) proposed the following rate equation for Ostwald ripening:

$$\overline{d^3} = \overline{d_0^3} + \frac{64\sigma C_\infty v^3 D}{9RT} t \tag{1.3}$$

where $\overline{d_0}$ is the initial number average diameter, σ is the interfacial tension, C_∞ is the solubility of the dispersed phase in the continuous phase, v is the molar volume of the dispersed phase, D is the diffusivity of the dispersed phase in the continuous phase, R is the ideal gas constant, and T is the system temperature. The Ostwald ripening constant (ω_0) is the second term of Eq. 1.3, and the rate of Ostwald ripening is $\frac{1}{\overline{\overline{d^2}}}$, meaning that this destabilization mechanism is more prevalent in nanoemulsions than in macroemulsions. Creaming is insignificant in nanoemulsions until the droplet size increases to the order of a few microns as a result of Ostwald ripening, flocculation, and coalescence (Delmas et al. 2011; Gupta et al. 2016).

$$\frac{d}{dt}\left(\overline{d}\right) \sim \frac{\omega_0}{\overline{\overline{d^2}}} \tag{1.4}$$

The ripening rate in O/W emulsions is directly proportional to the solubility of the oil in the aqueous phase, the system composition, and environmental conditions. Like the oil type, the surfactant type, relative concentrations, pH, ionic strength, solvent type, temperature, applied forces, etc., can modify the kinetic stability for commercial applications. For example, Delmas et al. (2011) used ultrasonication to obtain nanoemulsion systems that could present very long-term kinetic stability. They obtained a master curve by plotting the mean diameter size evolution as a function of the sonication energy and also showed that Ostwald ripening is the main destabilization mechanism, whereas coalescence can be easily prevented because of the nanometric size of the droplets. Also, they found that addition of a second component to the dispersed phase that is insoluble in the continuous phase can reduce Ostwald ripening.

It is well known that the Ostwald ripening rate is reduced with an increase in the surfactant concentration, which alters the interfacial structure and thereby creates an interface with minimal free energy. Zdrali et al. (2017) investigated the stability and molecular structure of oil nanodroplets in water stabilized with anionic, cationic, neutral, or zwitterionic surfactants. They found that dilute anionic surfactant monolayers produce more stable nanodroplets than dilute cationic and dense geometrically packed neutral surfactant monolayers. They proposed that the droplet stability depends on interactions corresponding to charge–charge, charge–dipole, and hydrogen bonding, because the difference between the extended planar interfaces and nanoscale interfaces stems from a difference in the thermally averaged total charge–charge interactions in the two systems.

Wooster et al. (2016) studied microemulsion-inspired approaches in the formation and stabilization mechanisms of homogenized triglyceride nanoemulsions. They successfully reduced the droplet diameter and reduced the destabilization process by addition of n-alcohol cosolvents and Span 80 cosurfactants. The n-alcohols change the solvent quality near the interface and the curvature, dramatically reducing the interfacial tension. Also, they suggested that this effect is magnified by n-alcohol partitioning behavior, and this tendency is associated with the head group of polyoxyethylene (POE) surfactants. However, if n-alcohol is added in excess, it leads to nanoemulsion destabilization; unusually for nanoemulsions, destabilization does not occur via Ostwald ripening; coalescence has been found to be the primary destabilization mechanism.

The development of biobased materials with smaller environmental impacts has seen increased interest in recent years; because of that, nanoceullulose can be used as a surfactant. Pickering emulsion, which is an emulsion stabilized by solid particles, offers a wide range of potential applications because it generally provides a more stable system than surfactant-stabilized emulsions. Nanocelluloses efficiently stabilize oil–water interfaces because of their amphiphilic surface nature, which originates from the hydrophobic face and hydrophilic edge of cellulose chains (Fujisawa et al. 2017). For example, Zhang et al. functionalized carboxylic acid cellulose nanocrystals (CNCs) with different hydrophobic groups used to stabilize styrene-in-water nanoemulsions. They found that the hydrophobic/hydrophilic balance of the functionalized CNCs was critical in lowering the interfacial tension between oil and water, which allowed access to stable emulsions with droplet diameters <1 μm (Zhang et al. 2017). Also, Jiménez Saelices and Capron studied the formation of O/W nanoemulsions stabilized by different types of nanocellulose. They successful produced nanosized hexadecane droplets by using CNCs and cellulose nanofibrils (CNFs) produced by 2,2,6,6-tetramethylpiperidine-1-oxyl (TEMPO)-mediated oxidation (w-TCNF), and by decreasing the drop size to 100–600 nm, using HPH (Jiménez Saelices and Capron 2018).

1.3 Formulation of Biobased Nanoemulsions

The formulation and understanding of nanoemulsions involve many scientific concepts. These colloidal dispersions allow creation of formulations of stable nanoemulsions, with the possibility of wide applications. Nanoemulsions are understood as dispersed systems with nanosized liquid droplets (a dispersed phase) in a liquid continuous phase; they are kinetically stable over long time scales and in a thermodynamically unstable system (Gupta et al. 2016; Akbas et al. 2018). Improvements in their formulation have allowed creation of advanced nanoemulsions that have desirable functionality and controlled physical properties.

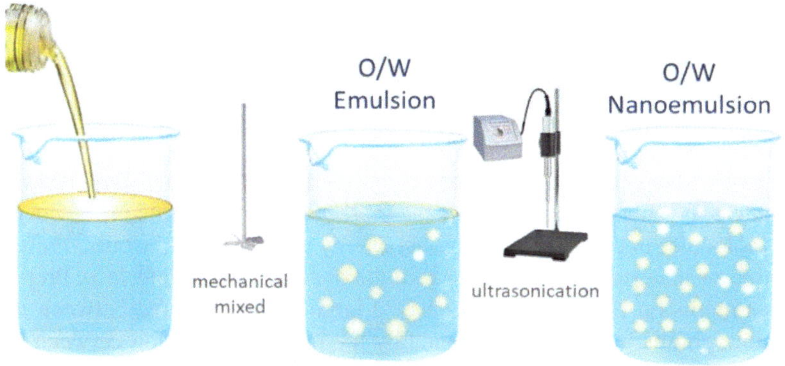

Fig. 1.3 Overview of preparation of an oil-in-water (*O/W*) nanoemulsion

A variety of techniques are used to generate nanoemulsions from an emulsion, through deformation and disruption of droplets and stabilization of the newly formed interfaces. It is necessary to review these two stages and the process of mechanical homogenization of an emulsion in order to discuss nanoemulsion formulation.

At the first stage, a process of deformation and disruption of the monomer drops is performed, which increases the specific surface area of the emulsion (Fryd and Mason 2012). Mechanical stress must therefore be applied to form and nanoemulsion to disrupt and mix the oil and water phases. Mechanical homogenization of emulsions can be achieved by different methods. Initially, it was done by simple stirring, but high-energy and low-energy methods are now the most widely used methods. At the second stage, stabilization of the new interface formed by the surfactant takes place (Fig. 1.3).

Low-energy methods are based on oil–water–surfactant systems that achieve spontaneous emulsification by mechanical mixing. Mechanical devices such as high-pressure homogenizers, microfluidizers, and sonicators are capable of generating the high-energy densities required to form small droplets (Salvia-Trujillo et al. 2017) in which the Laplace pressure is equal to or less than the osmotic pressure for all drops once the homogenization process is finished. These devices are based on applying highly disruptive forces with mechanical devices capable of producing the break-up of oil droplets and dispersing them into the water phase (Salvia-Trujillo et al. 2017). During sonication, the drop size decreases constantly until it reaches a state of equilibrium. Initially, the polydispersity of the droplets is maintained at a high level, but the constant fusion and fission processes that are taking place make the polydispersity diminish, finally reaching a stable state.

Water, oil, and an emulsifier are the elementary components used to formulate a nanoemulsion. Their characteristics and concentrations determine the final properties of the nanoemulsion. The formulation has a substantial effect on the stability of the nanoemulsion, since this is affected by the volume fraction of the dispersed phase, as well as use of an appropriate emulsifier (Salvia-Trujillo et al. 2017).

Considering that nanoemulsions are heterogeneous systems with minimal stability, it is important to specify the physical condition of each of the immiscible phases of the system. Nanoemulsions are classified into W/O and O/W types (Pengon et al. 2018), where the first phase that is mentioned represents the dispersed phase and the second represents the continuous phase.

In a two-phase system, the presence of an interface confers a difference in forces, which alters the energetic situation of the molecules at or near the interface, usually giving molecules in that region higher net energy than those in the bulk. Preferential adsorption of materials occurs because of the presence of the interface. Preparation of a nanoemulsion requires formation of a large interfacial area between the two immiscible phases, which is determined by the interfacial tension between the two liquid phases. Lowering of the interfacial tension is necessary to obtain stable systems, because small droplet production is favored.

The physicochemical properties of a nanoemulsion are largely due to the aqueous phase composition, since this phase and its adequate control are determinants of the physicochemical properties of the nanoemulsion, allowing achievement of greater stability, formation, and suitable physicochemical properties. In most cases the aqueous phase consists of water; however, other compounds can be used, such as other polar components, carbohydrates, proteins, minerals, acids, and bases (Salvia-Trujillo et al. 2017; McClements and Rao 2011).

The physicochemical characteristics of the oil phase—such as its viscosity, density, refractive index, and interfacial tension—affect the formation and stability of nanoemulsions, in which droplets must be stabilized against molecular diffusion, degradation, and coalescence by collisions. The oil droplet interactions are determined by the physicochemical characteristics of the oil phase and other factors, such as the pH and ionic strength, which have an effect on the electrostatic interactions of the droplets, varying the stability and droplet aggregation (Salvia-Trujillo et al. 2017).

Use of an appropriate emulsifier agent in the aqueous phase facilitates formation of small drops of oil, creating a stable nanoemulsion and avoiding thermodynamically unstable system characteristics of a nanoemulsion. Stabilizers (emulsifiers) plays a major role in the preparation and stabilization of a nanoemulsion, locating themselves at the interface between the two phases and the active surface material (which is soluble in one or both phases), and altering the interfacial characteristics of the system. Their presence can reduce the energetic force driving coalescence by lowering the interfacial tension and consequently helping to decrease the size of droplets resulting from flow-induced rupturing, and forming a mechanical barrier between drops. These properties protect oil droplets against breakdown due to gravitational separation, droplet aggregation, and Ostwald ripening by a short-range repulsive interaction between surfactant-coated droplet interfaces even over very long time scales, making nanoemulsions long-lived, metastable states (McClements and Rao 2011; Salvia-Trujillo et al. 2017; Fryd and Mason 2012).

To improve homogenization, use of viscosity enhancers impedes the drop movement that reduces the size of the oil droplets, as it increases the disruptive shear stresses generated inside the homogenizer, producing changes in the long-term stability by decelerating gravitational separation and droplet collisions, in addition to changing the texture of the nanoemulsion (Salvia-Trujillo et al. 2017; McClements and Rao 2011).

The type or surfactant selected for nanoemulsion formulation has an impact on the stability and properties of the nanoemulsion. Use of an emulsifier cover on the surface of oil droplets results in physical stability (Pengon et al. 2018). If one cannot get the emulsifier molecules to completely cover the surface of the drops, this will result in increased surface tension, which will cause a loss of emulsion stability.

1.3.1 Nanoemulsion Droplet Size

The design for obtaining a nanoemulsion must consider appropriate conditions to control the distribution of the drop size (the average radius and polydispersity), since these characteristics are the ones that will ultimately confer properties suitable for the intended application of the nanoemulsion. Since nanoemulsions are colloidal dispersions with nanosized droplets (Pathakoti et al. 2017; Pengon et al. 2018) and their main advantages and differences are derived from the nanosized structure of the drops that are formed, it is relevant to evaluate the influence of mechanisms and properties that have important effects on the characteristics and stability of the emulsion. The small droplet size leads to an improvement in stability, gravitational separation, and aggregation; moreover, the reduction in the droplet size and the increase in the droplet surface area may increase the functionality of bioactive compounds encapsulated in the nanoemulsion (Salvia-Trujillo et al. 2017).

The droplet size is the characteristic that gives nanoemulsions many advantages such physical stability against gravitational separation and easy handling; the effect of Brownian motion on the size of small droplets is enough to overcome their low gravitational properties, improving this trait in nanoemulsions (Pengon et al. 2018; Pathakoti et al. 2017). Having droplets in the nanosize range makes an emulsion more stable in terms of gravitational separation and particle aggregation, and can result in near transparency due to weak light scattering (Salvia-Trujillo et al. 2017).

The energy density required to break down oil droplets increases as their size decreases, which means that high-energy input is required to form nanoemulsions. The droplet size becomes roughly constant during ultrasonication but then decays exponentially with increasing ultrasonication time. The droplet size distribution can be better controlled by tuning the geometry of the homogenizer/microfluidizer (Salvia-Trujillo et al. 2017; Gupta et al. 2016)

In emulsions, the size of the drop is determined by the density and solubility of the dispersed phase, the aqueous density, and the amount of the emulsifier used. The initial drop size is a function of the mechanical force used in the formation; from this, it follows that when ultrasonication is applied for a determined standard time, conservation of the drop size and the stability of emulsion are governed by other factors. With an increase in the ultrasonication time, the size of the drops decreases and consequently the oil–water interface increases. Once that ultrasonication process ends, the emulsifier present in the medium is distributed throughout the interface to maintain a balance between the surfactant and oil–water, and the water–oil interface.

The droplet size is also determined by the Laplace pressure, which increases with a decrease in droplet size, producing a net mass flux by diffusion. If the droplets are not stabilized against diffusional degradation, small ones will disappear, increasing the average droplet size. Nanoemulsions are kinetically stable, but over long time scales they will separate into different phases. The droplet size is one of the most important factors affecting the dynamic stability of emulsions, and it is affected by the physicochemical properties of the emulsion components, as well as by the conditions of the process involving the use of such emulsions (Ricaurte et al. 2018).

The surfactant concentration also influences the droplet size, as surfactants decrease interfacial tension. The droplet size in a nanoemulsion decreases with an increase in the concentration of the emulsifier. This phenomenon is able to limit the number of emulsifier molecules emigrating from the oil phase to the aqueous phase of the emulsion; there must be sufficient emulsifier present to cover the surfaces of all oil droplets generated, preventing their coalescence (Salvia-Trujillo et al. 2017; Pengon et al. 2018).

In the same way, lower viscosity and lower interfacial tension of the oil phase facilitate droplet disruption in the homogenizer, resulting in a smaller droplet size. A lower oil viscosity results in more rapid droplet disruption in the homogenizer, facilitating the breakdown of the emulsion and creating a nanoemulsion system (Salvia-Trujillo et al. 2017).

In recent years, different methods for obtaining stable nanoemulsions have been reported. These include phase inversion composition, flow focusing, satellite droplets, membrane emulsification, and liquid–liquid nucleation. Some methods offer the advantage of producing highly concentrated nanoemulsions, which allow a greater proportion of oleic phase in the obtained nanoemulsion (Fryd and Mason 2012).

The Ostwald ripening effect can be controlled with use of an appropriate formulation. One reported method is the so-called two-component dispersed phase method or the trapped species method, which involves mixture of a dispersed phase (which is prone to Ostwald ripening after emulsification) with a second dispersed phase that is practically insoluble in the continuous phase. This causes the second dispersed phase liquid to become a species trapped inside droplets of the first dispersed phase after emulsification and gives rise to an internal osmotic pressure that depends on the concentration. Thus, as the first dispersed phase leaves the droplet, the increasing osmotic pressure ultimately surpasses the increasing Laplace pressure of the droplet, thereby stabilizing the droplet's size and inhibiting any further net transport (Fryd and Mason 2012).

Another method is evaporative maturation, which seeks to reduce the size of nanoemulsions. This involves mixing a first volatile oil and a second nonvolatile oil to form an emulsion in an aqueous solution of surfactant. After high-flux emulsification, the first oil, which has a higher vapor pressure and low solubility in the continuous phase of the nanoemulsion, is evaporated, resulting in an increase in the internal osmotic pressure caused by the second oil, which initially had a lower vapor pressure and very low solubility in the continuous phase. Finally, all the first oil can be removed by evaporation (Fryd and Mason 2012).

Emulsifier combinations have also been employed to improve the formation and stability of nanoemulsions. Likewise, amphiphilic active surface molecules in combination with low-disruption energy methods have been investigated for use as emulsion stabilizers as a function of the small size of the polar head group (McClements and Rao 2011).

1.4 Applications of Biobased Nanoemulsions

1.4.1 Industrial Applications of β-Lactoglobulin and Dextran in Food

In the food industry, proteins are of great importance for their functional properties, such as emulsifying and foaming properties. These characteristics are studied to obtain ingredients with new applications, with improvements in their functionality through physical, chemical, or enzymatic treatments. However, because of the methods used for protein modification, their use as food ingredients is limited.

1.4.2 β-Lactoglobulin and Dextran Conjugates

The structure of proteins determines their functional properties, and their functionality can be altered by formation of a covalent bond between a polysaccharide and the protein through modification of the stability and solubility of the protein—a compound called a conjugate. It has two properties: the stabilizing power of the polysaccharide and the emulsifying capacity of the protein (Dickinson and Galazka 1991; Hattori 2002; Horne and Rodríguez Patino 2003; Jiménez-Castaño et al. 2005a, b; Ganzevles et al. 2006, 2008). This protein modification can be achieved by a Maillard reaction, an effective method by which the functional characteristics of proteins are improved. This method does not require chemical catalysis, because of the interaction between proteins and polysaccharides or monosaccharides (Oliver et al. 2006).

Milk proteins are an important source of ingredients used in a wide variety of products, because of their high essential amino acid content and functional versatility (Foegeding et al. 2002). Among the globular proteins in milk are β-lactoglobulin (β-Lg), α-lactoalbumin (α-La), bovine serum albumin (BSA), and immunoglobulins. β-Lg is the majority component and is largely responsible for the functionality of the overall system. Numerous studies have reported improvements in the emulsifying and stabilizing capacity of β-Lg when it covalently binds to a carbohydrate or polysaccharide (Nagasawa et al. 1996; Chevalier et al. 2001a, b; 2002; Kalidas et al.

2001; Sanchez and Renard 2002; Foegeding et al. 2002; Zhang and Foegeding 2003; Zhang et al. 2004; Dunlap and Côté 2004; Schmitt et al. 2005a, b; Sanchez et al. 2006; Wooster and Augustin 2006; Chobert et al. 2006; Gloria 2006; Guzey and McClements 2006; Montilla et al. 2007; Schaink and Smit 2007; Broersen et al. 2007; Medrano et al. 2009).

1.4.3 Effects of Glycation on Solubility and Thermal Stability of β-Lactoglobulin

Jiménez-Castaño et al. (2007) reported that conjugates obtained by a Maillard reaction with β-Lg and dextran 10 kDa (D10) or dextran 20 kDa (D20) improve protein solubility at pH 5, but the effects are reduced at pH 4 (Fig. 1.4). It is known that glycosylation of β-Lg with monosaccharides and disaccharides changes the solubility of the protein to require a more acidic pH. This effect is observed after glycosylation of β-Lg with dextran 43 KDa (Chevalier et al. 2001a, b; Jiménez-Castaño et al. 2005a, b), resulting in a change in the isoelectric point of the conjugates to more acidic pH values (Hattori et al. 1996, 1997). Other studies have shown that glycosylated β-Lg has greater thermal stability at pH 7 and pH 5 than β-Lg under the same conditions, but without the presence of dextran (Fig. 1.4b, c) (Jiménez-Castaño et al. 2007). Finally, at pH 5 the solubility of glycosylated β-Lg is much higher than that of the native protein at temperatures above 80 °C.

Dextran-modified β-Lg has greater solubility than heated β-Lg in the reported pH ranges (Jiménez-Castaño et al. 2005a, b). The protective effect of glycation decreases the solubility due to heating by lowering its pH to 4. This behavior can be explained by the change in the isoelectric point of the modified protein, which may be due to a reduction in the number of positive charges and an increase in the negative net charge of the protein. This fact minimizes the solubility, which changes to a lower pH, explaining the increase in solubility at pH 5 in comparison with the heated β-Lg sample (Chevalier et al. 2002).

Glycosylation of β-Lg with D10 and D20 provides improved solubility of these compounds in the region of the isoelectric point of the native protein, which represents an advantage for its use as a food additive, since glycosylation improves thermal stability at pH acids that are pH neutral (Jiménez-Castaño et al. 2007).

1.4.4 Interfacial and Dilatation Properties

The increase in surface pressure as a function of the adsorption time indicates that protein is being adsorbed at the air–water interface (Graham and Phillips 1979; Damodaran and Song 1988). The magnitude of the adsorption phenomenon is dependent on the presence of dextran in the continuous phase, the molecular weight

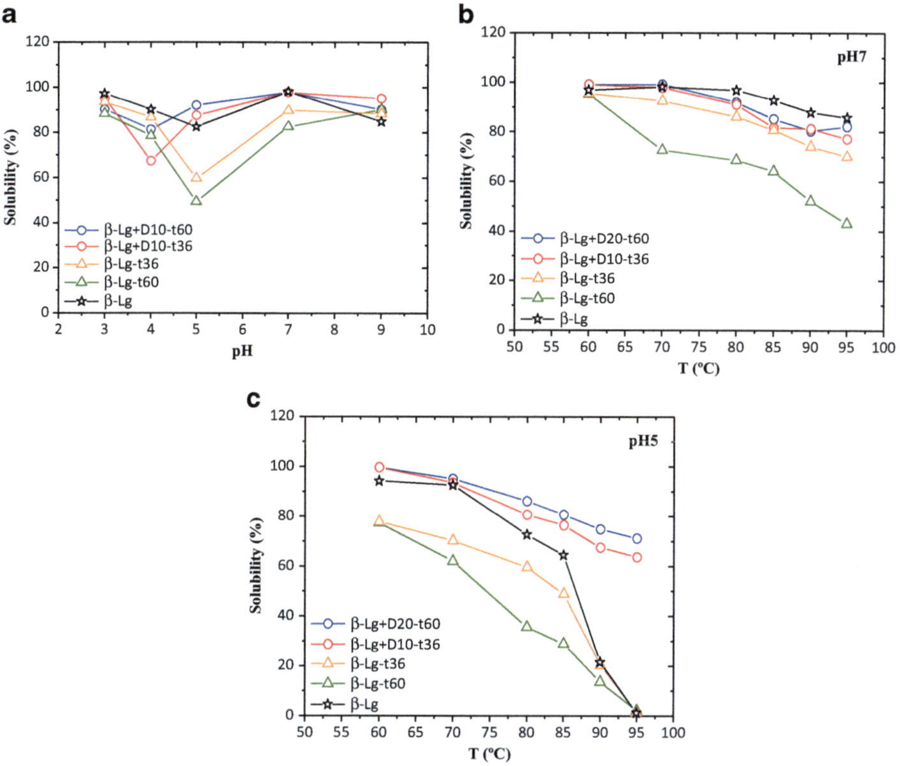

Fig. 1.4 Solubility (**a**), thermal stability at pH 7 (**b**), and thermal stability at pH 5 (**c**) of native β-lactoglobulin (*β-Lg*), dry-heated β-lactoglobulin, and glycosylated β-lactoglobulin (at 60 °C and at 0.44 aw for times corresponding to the maximum glycation level). *aw* water activity, *D10* dextran 10 kDa, *D20* dextran 20 kDa, *T* system temperature, *t36* at 36 h, *t60* at 60 h. (Reprinted from Jiménez-Castaño et al. 2007, Copyright (2007), with permission from Elsevier)

of the polysaccharide, and the glycosylation process (Dickinson and Izgi 1996; Nagasawa et al. 1996; Galazka et al. 1997; Akhtar and Dickinson 2003; Dunlap and Côté 2004; Jiménez-Castaño et al. 2005a, b; Gloria 2006; Jung et al. 2006; Wooster and Augustin 2006; Zhu et al. 2008; Vardhanabhuti and Allen Foegeding 2008).

1.4.5 Effects of the Glycation Process at pH 7 and pH 5

The presence of heat-treated protein in the mixture is a positive factor in increasing surface activity over a short time period as a result of increased diffusion, and over a long time period as a result of increased surface pressure at the "equilibrium." On the other hand, the mixture favors speed of diffusion of the protein toward the interface and an increase in surface activity at the "equilibrium." However, it is necessary

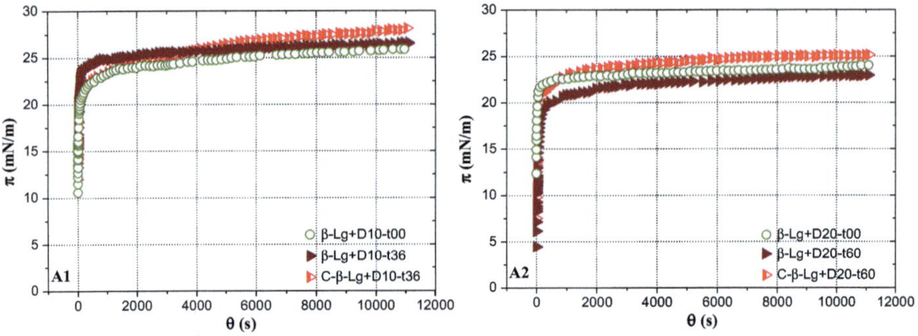

Fig. 1.5 Effects of the glycation process. Variation of the interfacial pressure (π) over time (θ) of the protein-adsorbed films β-lactoglobulin (*β-Lg*), conjugate 1 (see *panel A1*) and conjugate 2 (see *panel A2*), and their control mixtures of β-lactoglobulin and dextran 10 kDa (*C-β-Lg-D10*) and β-lactoglobulin and dextran 20 kDa (*C-β-Lg-D20*), over the air–water interface at ionic strength (*I*) 0.05 M, system temperature (*T*) 20 °C, and pH 7. *t00* at 0 h, *t36* at 36 h, *t60* at 60 h

to point out that this favorable effect is more significant for proteins conjugated to polysaccharides of a lower molecular weight (D10) (Fig. 1.5, panel A1) than those conjugated to high molecular weight polysaccharides (D20) (Fig. 1.5, panel A2). This behavior can be reaffirmed with the values of the kinetic constants of the first adsorption stage (Fig. 1.5); for conjugate 1, the glycosylation process leads to an increase in diffusion (β-Lg+D10 at 0 h (t00) versus β-Lg+D10 at 36 h (t36)), whereas for conjugate 2 this process leads to an appreciable decrease in the values of the kinetic constants of adsorption at this first stage (β-Lg+D20-t00 versus β-Lg+D20 at 60 h (t60) versus C-β-Lg+D20-t60).

The diffusion of glycosylated products is lower as a result of their degree of conjugation, the increase in molecular weight, the change in pH, and the shielding of the hydrophobic groups of the conjugated protein. Both conjugates (Fig. 1.5) have less surface activity than their respective controls where glycosylate is not present. The same trend can be seen at the first adsorption stage (Fig. 1.6).

The possible explanation for this difference in behavior between the formed glycosylates must be related to the structural characteristics of the covalent conjugates formed between β-Lg and D10 or D20. In this regard, studies by Jiménez-Castaño et al. (2007) have determined that higher glycosylation levels are achieved with use of D10 than with use of D20; that is, for each mole of β-Lg, 3.92 moles of D10 are linked, and for each mole of β-Lg, 1.4 moles of D20 are linked.

In addition, studies of fluorescence emission spectra have shown that glycosylation with dextrans decreases the fluorescence intensity of β-Lg, which has been attributed to a shielding effect of the polysaccharide chain (Hattori et al. 1996, 1997; Jiménez-Castaño et al. 2005a, b). A greater decrease in this intensity was observed in the conjugate of β-Lg with D20 than in the conjugate with D10, despite the higher degree of glycosylation in the latter. Therefore, these results could suggest that in the glycosylate formed with D20, this polysaccharide, with its larger molecular size (Wong et al. 2009), is capable of exerting a greater shielding effect on hydrophobic

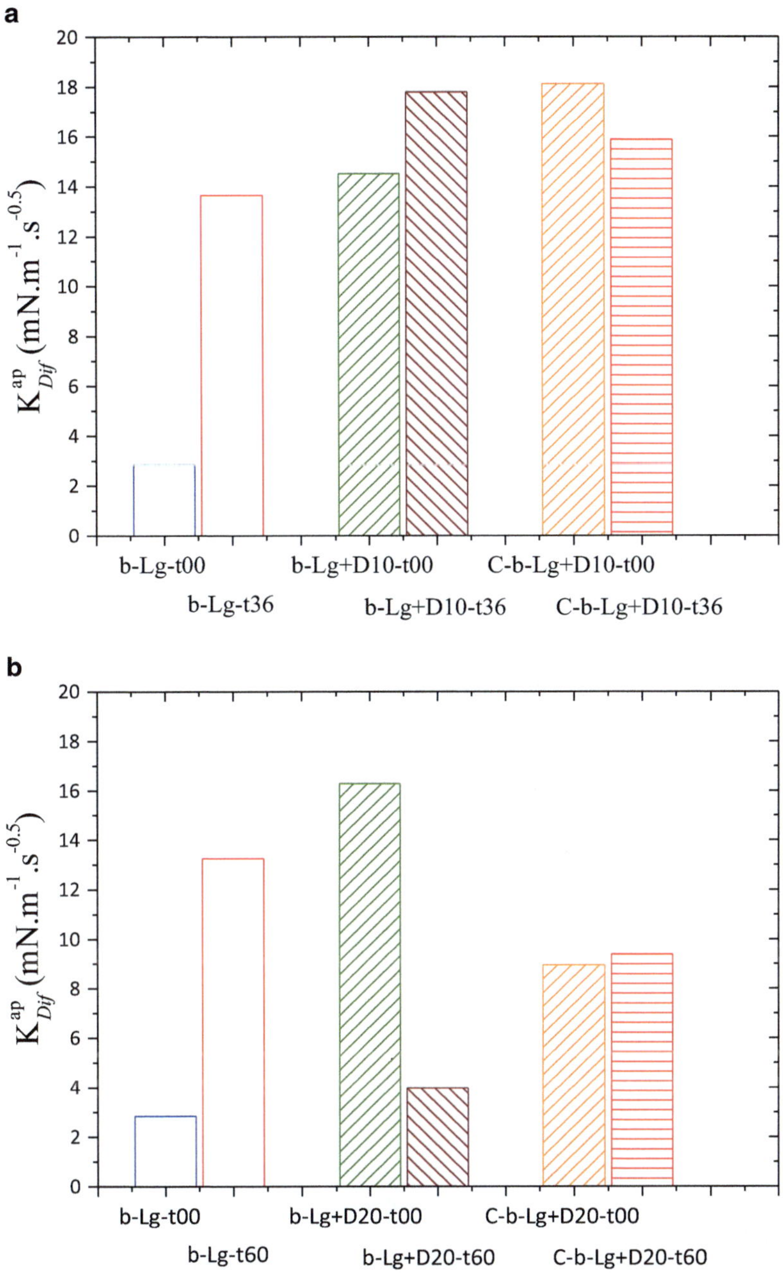

Fig. 1.6 Kinetic behavior at the first stage of absorption of β-lactoglobulin (*β-Lg*), conjugate 1 (**a**) and conjugate 2 (**b**), and their control mixtures of β-lactoglobulin and dextran 10 kDa (*C-β-Lg-D10*) and β-lactoglobulin and dextran 20 kDa (*C-β-Lg-D20*) at ionic strength (*I*) 0.05 M, system temperature (*T*) 20 °C, and pH 7. K_{Dif}^{ap} apparent diffusion constant, *t00* at 0 h, *t36* at 36 h, *t60* at 60 h

residues (Dickinson and Semenova 1992) such as tryptophan. Therefore, glycosylate has a lesser tendency to be adsorbed at the interface, as its surface hydrophobicity is partially "shielded" (Cairoli et al. 1994). Such a hypothesis would be supported by the fact that conjugate 2 has a lower diffusion velocity (over short periods) and lower surface activity (over long periods).

These results indicate that the best interfacial properties detected (increased diffusion) when the protein is previously heat treated do not significantly influence its foaming characteristics; that is, the diffusion rate and rheological characteristics of the film formed with native protein are already good enough to stabilize the system during its formation. The same behavior can be seen in Fig. 1.7, where an increase in the diffusion value does not lead to an increase in the value of the total foaming capacity (OFC).

Both conjugates are more stable at both pH 7 and pH 5 than their respective control systems. In this order of ideas, the systems in which glycosylated products are usually present, with respect to their respective control systems (a mixture of β-Lg and heat-treated dextran), lower or equal diffusion velocities, lower surface activity, and lower dilation module values at pH 7 but higher values at pH 5.

It has been widely commented in the literature that covalent complexes formed between β-Lg and dextran are capable of forming emulsified systems with excellent stability (Dickinson and Galazka 1991; Kato 2002; Akhtar and Dickinson 2003; Dunlap and Côté 2004). The explanation for this fact lies in the increased steric stabilization provided by the voluminous part corresponding to the polysaccharide, which is oriented toward the aqueous phase.

The presence of this type of adsorbed glycosylate in the air–water interface of a bubble will result in the appearance of greater steric interaction forces between adjacent bubbles than when the interface is stabilized only by protein.

1.4.6 Nanoemulsion of Essential Oils

Essential oils are highly complex mixtures of often hundreds of individual aroma compounds that are poorly soluble in water and have a pleasant odor and taste. Moreover, essential oils have been recognized as GRAS (Generally Recognized as Safe) by the US Food and Drug Administration (2018).

Plant essential oils have been used for many years in food and pharmaceutical products for their antifungal, antimycotic, and pest control properties. In fact, their antifungal effectiveness has attracted growing interest from researchers for use as food preservatives. The essential oils most widely employed as natural food preservatives are cinnamon, clove, lemongrass, oregano, thyme, nutmeg, and basil. There are very good reviews available regarding applications of biobased nanoemulsions (Donsì and Ferrari 2016; Ribes et al. 2017; Prakash et al. 2018).

Essential oils exert strong antibacterial, antiviral, and antifungal effects, stimulating their application also as natural antimicrobials in food and beverage products (Burt 2004). Essential oil nanoemulsions have been tested in vitro against different microorganisms ranging from bacterial cells to fungi, as shown in Table 1.2, which

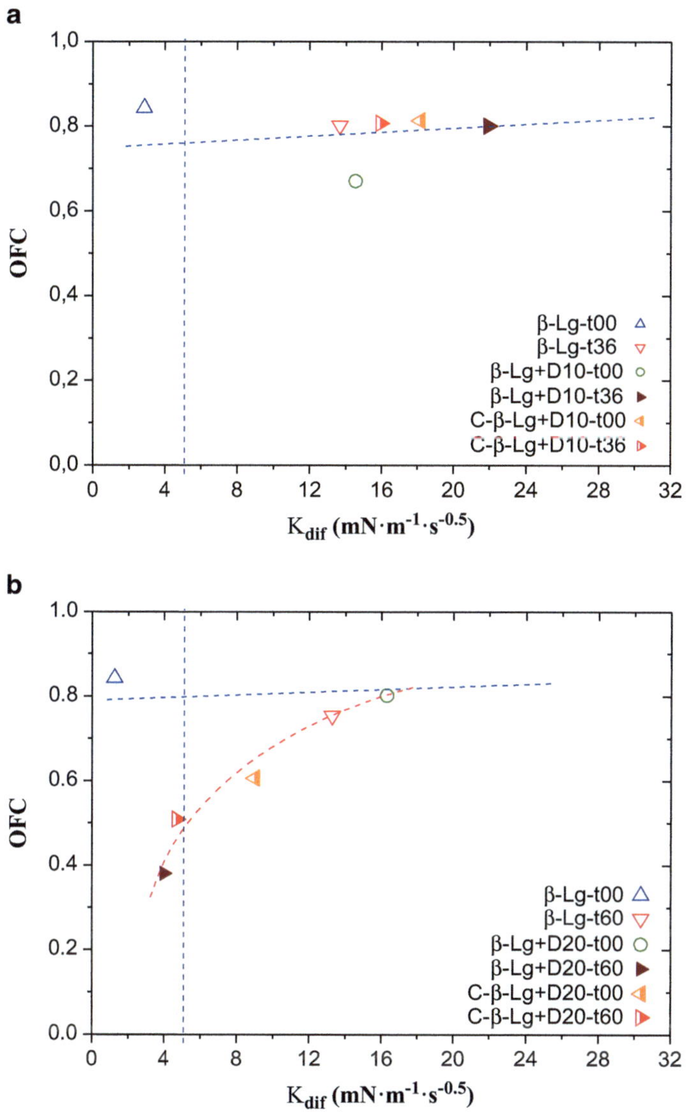

Fig. 1.7 Relationship of the effect of the diffusion velocity (K_{dif}) of conjugate 1 (**a**) and conjugate 2 (**b**) and their control mixtures of β-lactoglobulin and dextran 10 kDa (*C-β-Lg-D10*) and β-lactoglobulin and dextran 20 kDa (*C-β-Lg-D20*) on the air–water interface to the total foaming capacity (*OFC*) at ionic strength (*I*) 0.05 M, system temperature (*T*) 20 °C, pH 7, and bubbling gas (nitrogen) flow 45 mL/s. *t00* at 0 h, *t36* at 36 h, *t60* at 60 h

summarizes the main investigations reported to date, classified in terms of microorganism species, active components, and formulations used.

Tables 1.3, 1.4, and 1.5 summarize microemulsions and nanoemulsions that have been synthesized using essential oils and have demonstrated potent antibacterial,

Table 1.2 Essential oil nanoemulsion formulations tested against different microbial species and strains

Essential oils	Microorganisms	Nanoemulsion formulations	References
Carvacrol	*Escherichia coli* *Lactobacillus delbrueckii* *Saccharomyces cerevisiae*	Sunflower oil, sugar ester/Tween 20, monoolein, lecithin and pea proteins	Donsì et al. (2012)
	Listeria innocua	Medium-chain triglyceride, Tween 80	Terjun et al. (2012)
	Zygosaccharomyces	Medium-chain triglyceride, Tween 20	Chang et al. (2013)
Cinnamaldehyde	*Escherichia coli*	Tween 20	Jo et al. (2015)
		Acetem 90–50 K, Tween 60	Bilbao-Sainz et al. (2013)
		Sunflower oil, sugar ester/Tween 20, monoolein, lecithin and pea proteins	Donsì et al. (2012)
	Lactobacillus delbrueckii	Sunflower oil, sugar ester/Tween 20, monoolein, lecithin and pea proteins	Donsì et al. (2012)
	Salmonella typhimurium	Tween 20	Jo et al. (2015)
	Listeria monocytogenes	Acetem 90–50 K, Tween 60	Bilbao-Sainz et al. (2013)
Clove oil	*Escherichia coli*	Tween 80, sodium alginate	Salvia-Trujillo et al. (2015)
		Whey proteins, gum arabic, lecithin	Luo et al. (2014)
	Salmonella enteritidis *Listeria monocytogenes*	Whey proteins, gum arabic, lecithin	Luo et al. (2014)
d-Limonene	*Escherichia coli*	Tween 80	Zhang et al. (2014)
	Lactobacillus delbrueckii *Saccharomyces cerevisiae*	Palm oil, lecithin Modified starch Sunflower oil, sugar ester/Tween 20, monoolein	Donsì et al. (2011)
		Sunflower oil, sugar ester/Tween 20, monoolein, lecithin and pea proteins	Donsì et al. (2012)
	Staphylococcus aureus	Tween 80	Zhang et al. (2014)

Eugenol	*Escherichia coli* *Listeria monocytogenes*	Whey proteins, maltodextrin conjugates	Shah et al. (2013)
		Medium-chain triglyceride, Tween 80	Terjung et al. (2012)
	Staphylococcus aureus	Sesame oil, Tween 20/80	Ghosh et al. (2014)
	Listeria innocua	Medium-chain triglyceride, Tween 80	Terjung et al. (2012)
	Fusarium oxysporum	Tween 20	Abd-Elsalam and Khokhlov (2015)
Tea tree oil	*Escherichia coli* *Lactobacillus delbrueckii* *Saccharomyces cerevisiae*	Lecithin	Donsì et al. (2011)
	Escherichia coli	Tween 80, sodium alginate	Salvia-Trujillo et al. (2015)
Thyme oil	*Escherichia coli*	Tween 80, lecithin	Moghimi et al. (2016b)
		Sodium caseinate, lecithin	Xue et al. (2015)
		Tween 80, sodium alginate	Salvia-Trujillo et al. (2015)
		Soluble soybean polysaccharide	Wu et al. (2014)
	Salmonella enteritidis *Listeria monocytogenes*	Sodium caseinate, lecithin	Xue et al. (2015)
		Soluble soybean polysaccharide	Wu et al. (2014)
	Zygosaccharomyces bailii	Corn oil, Tween 80, lauric arginate	Chang et al. (2012, 2015)
Eucalyptus oil	*Staphylococcus aureus*	Tween 80	Sugumar et al. (2014)
Lemongrass oil	*Escherichia coli*	Tween 80, sodium alginate	Salvia-Trujillo et al. (2014a, b, 2015)
Peppermint oil	*Staphylococcus aureus* *Listeria monocytogenes*	Medium-chain triglyceride, modified starch	Liang et al. (2012)

Adapted from Donsì and Ferrari (2016), Copyright (2016), with permission from Elsevier

Table 1.3 Antibacterial activity of essential oil microemulsions and nanoemulsions

Essential oils	Microorganisms	Nanoemulsion formulations	References
Glycerol monolaurate	*Bacillus subtilis*, *Escherichia coli*	Tween 20 Tween 80 Pentanol Dodecane	Fu et al. (2006)
	Staphylococcus aureus	Tween 20 Tween 80 1-Pentanol 1-Dodecane	Zhang et al. (2007)
	Staphylococcus aureus, *Escherichia coli*	Tween 80 Propionic acid	Zhang et al. (2010)
	Staphylococcus aureus, *Bacillus subtilis*, *Escherichia coli*	Tween 40 Ethanol and sodium diacetate	Zhang et al. (2010)
Soybean oil, tri-*n*-butyl phosphate	*Salmonella* spp., *Escherichia coli* 0157:H7 (VT^-), *Pseudomonas aeruginosa*, *Listeria monocytogenes*	Trixton X-100	Teixeira et al. (2007)
Ethyl oleate	*Staphylococcus aureus*, *Salmonella* spp., *Escherichia coli* 0157:H7 (VT^-), *Pseudomonas aeruginosa*, *Listeria monocytogenes*	Tween 80 n-Pentanol	Al-Adham et al. (2000, 2003)
Eugenol	*Escherichia coli* O157:H7, *Listeria monocytogenes*	Surfynol 485 W	Gaysinsky et al. (2007)
Clove oil	*Staphylococcus aureus*, *Bacillus cereus*, *Escherichia coli* 0157:H7, *Pseudomonas aeruginosa*, *Salmonella typhi*, *Listeria monocytogenes*	Tween 20	Hamed et al. (2012)
Basil oil	*Escherichia coli*, *Staphylococcus aureus*	Tween 80	Ghosh et al. (2013b)
Tea tree oil	*Staphylococcus aureus*, *Staphylococcus epidermidis*, *Propionibacterium acnes*	Tween 80 Glycerol	Biju et al. (2005)
Sunflower oil	*Staphylococcus aureus*, *Bacillus cereus*, *Bacillus circulans*, *Escherichia coli*, *Salmonella typhi*, *Vibrio parahaemolyticus*	Surfactin Ethanol	Joe et al. (2012)
Cinnamon leaf oil	*Listeria monocytogenes*, *Bacillus cereus*	Tween 80	Ghosh et al. (2013a)

(continued)

Table 1.3 (continued)

Essential oils	Microorganisms	Nanoemulsion formulations	References
Black seed oil	*Staphylococcus aureus*, *Bacillus cereus*, *Salmonella typhimurium*, *Escherichia coli*, *Pseudomonas aeruginosa*, *Listeria monocytogenes*	Tween 20	Shaaban and Edris (2015)
Monocaprim	*Campylobacter jejunii*	Tween 40 Ethanol	Thormar et al. (2006)
Monolaurin	*Bacillus subtilis*, *Escherichia coli*	Tween 20 Pentanol Dodecanol	Fu et al. (2009)
1-Monoacylglycerol of capric (C:100), undecanoic (C11:0), lauric (C12:0), and myristic (C14:0) acids	*Staphylococcus aureus* subsp. *aureus*, *Enterococcus faecalis*, *Micrococcus luteus*, *Bacillus cereus*, *Bacillus subtilis* subsp. *subtilis*, *Escherichia coli*, *Pseudomonas aeruginosa*	Tween 80 Pentanol	Petra et al. (2014)
Carvacrol	*Salmonella enteritidis*, *Serratia marcescens* subsp. *marcescens*, *Citrobacter freundii*, *Staphylococcus aureus*, *Escherichia coli*, *Pseudomonas aeruginosa*, *Bacillus cereus*, *Salmonella typhimurium*, *Listeria monocytogenes*	Tween 20/ cetylpryridinium chloride Propylene glycol	Shaaban and Edris (2015)
Sage oil	*Escherichia coli*, *Salmonella typhi*, *Shigella dysenteriae*	Tween 80 Span 80	Moghimi et al. (2016a)
Thyme oil	*Escherichia coli*	Tween 80	Moghimi et al. (2016b)
Orange oil	*Saccharomyces cerevisiae*	Tween 80	Sugumar et al. (2016)

Adapted from Franklyne et al. (2016), Copyright (2016), with permission from John Wiley and Sons

Table 1.4 Antifungal activity of essential oil nanoemulsions

Essential oils	Microorganisms	Nanoemulsion formulations	References
Glycerol monolaurate	*Aspergillus niger*	Tween 20, Tween 80, pentanol, dodecane	Fu et al. (2006)
	Candida albicans, *Aspergillus niger*, *Penicillium expansum*	Tween 40, ethanol, sodium diacetate	Zhang et al. (2010)
Monolaurin	*Aspergillus niger*, *Penicillium digitatum*	Tween 20, pentanol, dodecanol	Fu et al. (2009)
Sunflower oil	*Candida* sp., *Penicillium* sp., *Aspergillus niger*, *Rhizopus nigricans*, *Aspergillus flavus*	Surfactin, ethanol	Joe et al. (2012)
Clove oil	*Penicillium digitatum*	Tween 80, ethanol	He et al. (2016)
Orange oil	*Saccharomyces cervisiae*	Tween 80	Sugumar et al. (2016)

Adapted from Franklyne et al. (2016), Copyright (2016), with permission from John Wiley and Sons

Table 1.5 Roles of microemulsions and nanoemulsions as drug delivery vehicles

Essential oils	Nanoemulsion formulations	Antimicrobial agents	Microorganisms	References
Sefsol-218	Tween 80, Tween 85	Rifampicin		Ahmed et al. (2008)
Caprylic/capric triglycerides	HCO-50, benzyl alcohol	Itraconazole		Rhee et al. (2007)
Sunflower oil	Soy lecithin, Nutralys® F85M, sucrose palmitate, Tween 20, glycerol monooleate	Carvacrol d-Limonene Cinnamaldehyde	*Escherichia coli*, *Lactobacillus delbrueckii*, *Saccharomyces cerevisiae*	Donsì et al. (2012)
Miglyol 812	Tween 80	Carvacrol	*Zygosaccharomyces bailii*, *Saccharomyces cerevisiae*, *Brettanomyces bruxellensis*, *Brettanomyces naardenensis*	Chang et al. (2013)
Soybean oil	Triton X-100	Cetylpyridinium chloride	*Acinetobacter baumannii*	Hwang et al. (2013)
Isopropyl myristate	Tween 80, propylene glycol	Amphotericin B	*Trichophyton rubrum*	Butani et al. (2014)
Cinnamon oil	Tween 80	Fluconazole		Nirmala et al. (2013a)
	Tween 20	Fluconazole		Nirmala et al. (2013b)
	Tween 20	Fluconazole		Nirmala et al. (2013b)

Adapted from Franklyne et al. (2016), Copyright (2016), with permission from John Wiley and Sons antifungal, and antiviral activities. Antimicrobial resistance is a major health concern worldwide. A narrowing of the antibiotic development pipeline and a resurgence in public enthusiasm for "natural" therapies have renewed interest in using essential oils as antimicrobial agents and their role as drug delivery vehicles (Aboalnaja et al. 2016; Franklyne et al. 2016; Salvia-Trujillo et al. 2017)

1.5 Conclusions

The field of nanoemulsion technology is still growing, since nanoemulsions offer many application opportunities because of their desirable physical properties and chemical compositions that can be achieved by their formulation constituents and processes—such as control over the drop size, polydispersity, and functionality—offering significant advantages for a wide range of applications. Nanoemulsions can be used in the cosmetic, pharmaceutical, food, and other industries because of the adaptability of their formulations as foams, creams, liquids, and sprays.

Acknowledgements The authors wish to thank the Consorcio Nacional de Recursos de Información Ceintífica y Tecnológica (CONRICYT) of the Consejo Nacional de Ciencia y Tecnología (CONACYT) for access to the information needed to write this review, Dr. Fernando Solanilla Duque for his collaboration on this chapter, and Dr. Raúl Rodriguez Herrera for his comments on this work.

Conflict of Interest The authors declare that they have no conflict of interest.

References

Abd-Elsalam KA, Khokhlov AR (2015) Eugenol oil nanoemulsion: antifungal activity against *Fusarium oxysporum* f. sp. *vasinfectum* and phytotoxicity on cottonseeds. Appl Nanosci 5(2):255–265. https://doi.org/10.1007/s13204-014-0398-y

Aboalnaja KO, Yaghmoor S, Kumosani TA et al (2016) Utilization of nanoemulsions to enhance bioactivity of pharmaceuticals, supplements, and nutraceuticals: nanoemulsion delivery systems and nanoemulsion excipient systems. Expert Opin Drug Deliv 13(9):1327–1336. https://doi.org/10.1517/17425247.2016.1162154

Ahmed M, Ramadan W, Rambhu D et al (2008) Potential of nanoemulsions for intravenous delivery of rifampicin. Pharmazie 63(11):806–811. https://doi.org/10.1691/ph.2008.8108

Akbas E, Soyler B, Oztop MH (2018) Formation of capsaicin loaded nanoemulsions with high pressure homogenization and ultrasonication. LWT 96:266–273. https://doi.org/10.1016/j.lwt.2018.05.043

Akhtar M, Dickinson E (2003) Emulsifying properties of whey protein–dextran conjugates at low pH and different salt concentrations. Colloids Surf B 31(1–4):125–132. https://doi.org/10.1016/S0927-7765(03)00049-3

Al-Adham IS, Khalil E, Al-Hmoud ND et al (2000) Microemulsions are membrane-active, antimicrobial, self-preserving systems. J Appl Microbiol 89(1):32–39. https://doi.org/10.1046/j.1365-2672.2000.01078.x

Al-Adham IS, Al-Hmoud ND, Khalil E et al (2003) Microemulsions are highly effective anti-biofilm agents. Lett Appl Microbiol 36(2):97–100. https://doi.org/10.1046/j.1472-765X.2003.01266.x

Biju SS, Ahuja A, Khar RK et al (2005) Formulation and evaluation of an effective pH balanced topical antimicrobial product containing tea tree oil. Pharmazie 60(3):208–211

Bilbao-Sainz C, Bor-Sen C, Wen-Xian D et al (2013) Influence of disperse phase characteristics on stability, physical and antimicrobial properties of emulsions containing cinnamaldehyde. JAOCS 90(2):233–241. https://doi.org/10.1007/s11746-012-2164-1

Broersen K, Elshof M, De Groot J et al (2007) Aggregation of β-lactoglobulin regulated by glucosylation. J Agric Food Chem 55(6):2431–2437. https://doi.org/10.1021/JF063178Z

Burt S (2004) Essential oils: their antibacterial properties and potential applications in foods—a review. Int J Food Microbiol 94(3):223–253. https://doi.org/10.1016/j.ijfoodmicro.2004.03.022

Butani D, Yewale C, Misra A (2014) Amphotericin B topical microemulsion: formulation, characterization and evaluation. Colloids Surf B 116:351–358. https://doi.org/10.1016/j.colsurfb.2014.01.014

Cairoli S, Iametti S, Bonomi F (1994) Reversible and irreversible modifications of beta-lactoglobulin upon exposure to heat. J Protein Chem 13(3):347–354

Chang Y, McLandsborough L, McClements DJ (2012) Physical properties and antimicrobial efficacy of thyme oil nanoemulsions: influence of ripening inhibitors. J Agric Food Chem 60(48):12056–12063. https://doi.org/10.1021/jf304045a

Chang Y, McLandsborough L, McClements DJ (2013) Physicochemical properties and antimicrobial efficacy of carvacrol nanoemulsions formed by spontaneous emulsification. J Agric Food Chem 61(37):8906–8913. doi: https://doi.org/10.1021/jf402147p

Chang Y, McLandsborough L, McClements DJ (2015) Fabrication, stability and efficacy of dual-component antimicrobial nanoemulsions: essential oil (thyme oil) and cationic surfactant (lauric arginate). Food Chem 172:298–304. https://doi.org/10.1016/j.foodchem.2014.09.081

Chellapa P, Ariffin FD, Eid AM et al (2016) Nanoemulsion for cosmetic application. EJBPS 3(7):8–11

Chevalier F, Choberta JM, Popineaub Y et al (2001a) Improvement of functional properties of β-lactoglobulin glycated through the Maillard reaction is related to the nature of the sugar. Int Dairy J 11(3):145–152. https://doi.org/10.1016/S0958-6946(01)00040-1

Chevalier F, Chobert JM, Dalgalarrondo M et al (2001b) Characterization of the Maillard reaction products of B-lactoglobulin glucosylated in mild conditions. J Food Biochem 25(1):33–55. https://doi.org/10.1111/j.1745-4514.2001.tb00723.x

Chevalier F, Chobert JM, Dalgalarrondo M et al (2002) Maillard glycation of β-lactoglobulin induces conformation changes. Nahrung 46(2):58–63. https://doi.org/10.1002/1521-3803(20020301)46:2<58::aid-food58>3.0.co;2-y

Chobert JM, Gaudin JC, Dalgalarrondo M et al (2006) Impact of Maillard type glycation on properties of beta-lactoglobulin. Biotechnol Adv 24(6):629–632. https://doi.org/10.1016/j.biotechadv.2006.07.004

Damodaran S, Song KB (1988) Kinetics of adsorption of proteins at interfaces: role of protein conformation in diffusional adsorption. Biochim Biophys Acta 954(3):253–264

Delmas T, Piraux H, Couffin AC et al (2011) How to prepare and stabilize very small nanoemulsions. Langmuir 27(5):1683–1692. https://doi.org/10.1021/la104221q

Dickinson E, Galazka VB (1991) Emulsion stabilization by ionic and covalent complexes of β-lactoglobulin with polysaccharides. Food Hydrocoll 5(3):281–296. https://doi.org/10.1016/S0268-005X(09)80114-8

Dickinson E, Izgi E (1996) Foam stabilization by protein–polysaccharide complexes. Colloids Surf A Physicochem Eng Asp 113(1–2):191–201. https://doi.org/10.1016/0927-7757(96)03647-3

Dickinson E, Semenova MG (1992) Emulsifying properties of covalent protein–dextran hybrids. Colloids Surf 64(3–4):299–310. https://doi.org/10.1016/0166-6622(92)80109-F

Donsì F, Ferrari G (2016) Essential oil nanoemulsions as antimicrobial agents in food. J Biotechnol 233:106–120. https://doi.org/10.1016/j.jbiotec.2016.07.005

Donsì F, Annunziata M, Sessaa M et al (2011) Nanoencapsulation of essential oils to enhance their antimicrobial activity in foods. LWT Food Sci Technol 44(9):1908–1914. https://doi.org/10.1016/j.lwt.2011.03.003

Donsì F, Annunziata M, Vincensi M et al (2012) Design of nanoemulsion-based delivery systems of natural antimicrobials: effect of the emulsifier. J Biotechnol 159(4):342–350. https://doi.org/10.1016/j.jbiotec.2011.07.001

Dunlap CA, Côté GL (2004) β-lactoglobulin–dextran conjugates: effect of polysaccharide size on emulsion stability. J Agric Food Chem 53(2):419–423. https://doi.org/10.1021/jf049180c

Foegeding EA, Davis JP, Doucet D et al (2002) Advances in modifying and understanding whey protein functionality. Trends Food Sci Technol 13(5):151–159. https://doi.org/10.1016/S0924-2244(02)00111-5

Franklyne JS, Mukherjee A, Chandrasekaran N (2016) Essential oil micro- and nanoemulsions: promising roles in antimicrobial therapy targeting human pathogens. Lett Appl Microbiol 63(5):322–334. https://doi.org/10.1111/lam.12631

Fryd MM, Mason TG (2012) Advanced nanoemulsions. Annu Rev Phys Chem 63(1):493–518. https://doi.org/10.1146/annurev-physchem-032210-103436

Fu X, Feng F, Huang B (2006) Physicochemical characterization and evaluation of a microemulsion system for antimicrobial activity of glycerol monolaurate. Int J Pharm 321(1–2):171–175. https://doi.org/10.1016/j.ijpharm.2006.05.019

Fu X, Zhang M, Huang B et al (2009) Enhancement of antimicrobial activities by the food-grade monolaurin microemulsion system. J Food Process Eng 32(1):104–111. https://doi.org/10.1111/j.1745-4530.2007.00209.x

Fujisawa S, Togawa E, Kuroda K (2017) Nanocellulose-stabilized Pickering emulsions and their applications. Sci Technol Adv Mater 18(1):959–971. https://doi.org/10.1080/14686996.2017.1401423

Gadhave AD (2014) Nanoemulsions: formation, stability and applications. IJRSAT 2(3):38–43

Galazka VB, Ledward DA, Sumner IG et al (1997) Influence of high pressure on bovine serum albumin and its complex with dextran sulfate. J Agric Food Chem 45(9):3465–3471. https://doi.org/10.1021/JF9700642

Ganzevles RA, Cohen M, Cohen Stuart MS, van Vliet T et al (2006) Use of polysaccharides to control protein adsorption to the air–water interface. Food Hydrocoll 20(6):872–878. https://doi.org/10.1016/j.foodhyd.2005.08.009

Ganzevles RA, Fokkink R, van Vliet T et al (2008) Structure of mixed beta-lactoglobulin/pectin adsorbed layers at air/water interfaces; a spectroscopy study. J Colloid Interface Sci 317(1):137–147. https://doi.org/10.1016/j.jcis.2007.09.030

Gaysinsky S, Taylor TM, Davidson PM et al (2007) Antimicrobial efficacy of eugenol microemulsions in milk against *Listeria monocytogenes* and *Escherichia coli* O157:H7. J Food Prot 70(11):2631–2637

Ghosh V, Saranya S, Mukherjee A, Chandrasekaran N et al (2013a) Cinnamon oil nanoemulsion formulation by ultrasonic emulsification: investigation of its bactericidal activity. J Nanosci Nanotechnol 13(1):114–122. https://doi.org/10.1166/jnn.2013.6701

Ghosh V, Mukherjee A, Chandrasekaran N (2013b) Ultrasonic emulsification of food-grade nanoemulsion formulation and evaluation of its bactericidal activity. Ultrason Sonochem 20(1):338–344. https://doi.org/10.1016/j.ultsonch.2012.08.010

Ghosh V, Mukherjee A, Chandrasekaran N (2014) Eugenol-loaded antimicrobial nanoemulsion preserves fruit juice against microbial spoilage. Colloids Surf B Biointerfaces 114:392–397. https://doi.org/10.1016/j.colsurfb.2013.10.034

Gloria HH (2006) Formation of protein–polysaccharide conjugates and study of their physico-chemical and functional properties. Dissertation, Universidad Politécnica de Valencia

Graham D, Phillips M (1979) Proteins at liquid interfaces: III. Molecular structures of adsorbed films. J Colloid Interface Sci 70(3):427–439. https://doi.org/10.1016/0021-9797(79)90050-X

Gupta A, Eral HB, Hatton TA et al (2016) Nanoemulsions: formation, properties and applications. Soft Matter 12(11):2826–2841. https://doi.org/10.1039/C5SM02958A

Guzey D, McClements DJ (2006) Characterization of β-lactoglobulin–chitosan interactions in aqueous solutions: a calorimetry, light scattering, electrophoretic mobility and solubility study. Food Hydrocoll 20(1):124–131. https://doi.org/10.1016/j.foodhyd.2005.03.009

Hamed SF, Sadek Z, Edris A (2012) Antioxidant and antimicrobial activities of clove bud essential oil and eugenol nanoparticles in alcohol-free microemulsion. J Oleo Sci 61(11):641–648. https://doi.org/10.5650/jos.61.641

Hattori M (2002) Functional improvements in food proteins in multiple aspects by conjugation with saccharides: case studies of beta-lactoglobulin–acidic polysaccharides conjugates. Food Sci Technol Res 8(4):291–299. https://doi.org/10.3136/fstr.8.291

Hattori M, Aiba Y, Nagasawa K et al (1996) Functional improvement of alginic acid by conjugating with B-lactoglobulin. J Food Sci 61(6):1171–1176. https://doi.org/10.1111/j.1365-2621.1996.tb10954.x

Hattori M, Ogino A, Nakai H et al (1997) Functional improvement of β-lactoglobulin by conjugating with alginate lyase-lysate. J Agric Food Chem 45(3):703–708. https://doi.org/10.1021/JF960431U

He S, Ren X, Lu Y et al (2016) Microemulsification of clove essential oil improves its in vitro and in vivo control of *Penicillium digitatum*. Food Control 65:106–111. https://doi.org/10.1016/j.foodcont.2016.01.020

Horne DS, Rodríguez Patino JM (2003) Adsorbed biopolymers: behavior in food applications. In: Malmsten M (ed) Biopolymers at interfaces, 2nd edn. Marcel Dekker, New York, pp 857–900

Hwang YY, Ramalingam K, Bienek DR et al (2013) Antimicrobial activity of nanoemulsion in combination with cetylpyridinium chloride in multidrug-resistant *Acinetobacter baumannii*. Antimicrob Agents Chemother 57(8):3568–3575. https://doi.org/10.1128/AAC.02109-12

Jiménez Saelices C, Capron I (2018) Design of pickering micro- and nanoemulsions based on the structural characteristics of nanocelluloses. Biomacromolecules 19(2):460–469. https://doi.org/10.1021/acs.biomac.7b01564

Jiménez-Castaño L, Villamiel M, Martín-Álvarez PJ et al (2005a) Effect of the dry-heating conditions on the glycosylation of β-lactoglobulin with dextran through the Maillard reaction. Food Hydrocoll 19(5):831–837. https://doi.org/10.1016/j.foodhyd.2004.10.033

Jiménez-Castaño L, López-Fandiño R, Olano A et al (2005b) Study on β-lactoglobulin glycosylation with dextran: effect on solubility and heat stability. Food Chem 93(4):689–695. https://doi.org/10.1016/j.foodchem.2004.09.050

Jiménez-Castaño L, Villamiel M, López-Fandiño R (2007) Glycosylation of individual whey proteins by Maillard reaction using dextran of different molecular mass. Food Hydrocoll 21(3):433–443. https://doi.org/10.1016/J.FOODHYD.2006.05.006

Jo YJ, Chun JY, Kwon YJ et al (2015) Physical and antimicrobial properties of trans-cinnamaldehyde nanoemulsions in water melon juice. LWT Food Sci Technol 60(1):444–451. https://doi.org/10.1016/J.LWT.2014.09.041

Joe MM, Bradeeba K, Parthasarathia R et al (2012) Development of surfactin based nanoemulsion formulation from selected cooking oils: evaluation for antimicrobial activity against selected food associated microorganisms. J Taiwan Inst Chem Eng 43(2):172–180. https://doi.org/10.1016/j.jtice.2011.08.008

Jung SH, Choi SJ, Kim HJ et al (2006) Molecular characteristics of bovine serum albumin–dextran conjugates. Biosci Biotechnol Biochem 70(9):2064–2070. https://doi.org/10.1271/bbb.60026

Kalidas C, Joshi L, Batt C (2001) Characterization of glycosylated variants of beta-lactoglobulin expressed in *Pichia pastoris*. Protein Eng 14(3):201–207

Kato A (2002) Industrial applications of Maillard-type protein–polysaccharide conjugates. Food Sci Technol Res 8(3):193–199. https://doi.org/10.3136/fstr.8.193

Komaiko JS, McClements DJ (2016) Formation of food-grade nanoemulsions using low-energy preparation methods: a review of available methods. Compr Rev Food Sci Food Saf 15(2):331–352. https://doi.org/10.1111/1541-4337.12189

Liang R, Xu S, Shoemaker CF et al (2012) Physical and antimicrobial properties of peppermint oil nanoemulsions. J Agric Food Chem 60(30):7548–7555. https://doi.org/10.1021/jf301129k

Lifshitz IM, Slyozov VV (1961) The kinetics of precipitation from supersaturated solid solutions. J Phys Chem Solids 19(1–2):35–50. https://doi.org/10.1016/0022-3697(61)90054-3

Luo Y, Zhang Y, Pan K et al (2014) Self-emulsification of alkaline-dissolved clove bud oil by whey protein, gum arabic, lecithin, and their combinations. J Agric Food Chem 62(19):4417–4424. https://doi.org/10.1021/jf500698k

McClements DJ (2012) Nanoemulsions versus microemulsions: terminology, differences, and similarities. Soft Matter 8(6):1719–1729. https://doi.org/10.1039/C2SM06903B

McClements DJ, Rao J (2011) Food-grade nanoemulsions: formulation, fabrication, properties, performance, biological fate, and potential toxicity. Crit Rev Food Sci Nutr 51(4):285–330. https://doi.org/10.1080/10408398.2011.559558

Medrano A, Abirached C, Panizzolo L et al (2009) The effect of glycation on foam and structural properties of β-lactoglobulin. Food Chem 113(1):127–133. https://doi.org/10.1016/j.foodchem.2008.07.036

Moghimi R, Aliahmadi A, McClements DJ et al (2016a) Investigations of the effectiveness of nanoemulsions from sage oil as antibacterial agents on some food borne pathogens. LWT Food Sci Technol 71:69–76. https://doi.org/10.1016/j.lwt.2016.03.018

Moghimi R, Ghaderi L, Rafati H et al (2016b) Superior antibacterial activity of nanoemulsion of *Thymus daenensis* essential oil against *E. coli*. Food Chem 194:410–415. https://doi.org/10.1016/j.foodchem.2015.07.139

Montilla A, Casal E, Moreno FJ et al (2007) Isolation of bovine β-lactoglobulin from complexes with chitosan. Int Dairy J 17(5):459–464. https://doi.org/10.1016/J.IDAIRYJ.2006.05.009

Nagasawa K, Ohgata K, Takahashi K, a. (1996) Role of the polysaccharide content and net charge on the emulsifying properties of β-lactoglobulin–carboxymethyldextran conjugates. J Agric Food Chem 44(9):2538–2543. https://doi.org/10.1021/JF960150M.s

Nirmala MJ, Allanki S, Mukherjee A et al (2013a) Enhancing the solubility of fluconazole using a new essential oil based microemulsion system. Int J Pharm Pharm Sci 5(3):697–699

Nirmala MJ, Shivashankar M, Mukherjee A et al (2013b) Fluconazole: a simple nanoemulsion drug delivery system. Int J Pharm Pharm Sci 5(3):716–717

Oliver CM, Melton LD, Stanley RA (2006) Creating proteins with novel functionality via the Maillard reaction: a review. Crit Rev Food Sci Nutr 46(4):337–350. https://doi.org/10.1080/10408690590957250

Pathakoti K, Manubolu M, Hwang HM (2017) Nanostructures: current uses and future applications in food science. J Food Drug Anal 25(2):245–253. https://doi.org/10.1016/j.jfda.2017.02.004

Pengon S, Chinatangkul N, Limmatvapirat C et al (2018) The effect of surfactant on the physical properties of coconut oil nanoemulsions. Asian J Pharm Sci 13(5):409–414. https://doi.org/10.1016/j.ajps.2018.02.005

Petra Š, Věra K, Iva H et al (2014) Formulation, antibacterial activity, and cytotoxicity of 1-monoacylglycerol microemulsions. Eur J Lipid Sci Technol 116(4):448–457. https://doi.org/10.1002/ejlt.201300171

Prakash A, Baskaran R, Paramasivam N et al (2018) Essential oil based nanoemulsions to improve the microbial quality of minimally processed fruits and vegetables: a review. Food Res Int 111:509–523. https://doi.org/10.1016/j.foodres.2018.05.066

Prasad R, Pandey R, Varma A, Barman I (2017) Polymer based nanoparticles for drug delivery systems and cancer therapeutics. In: Kharkwal H, Janaswamy S (eds) Natural polymers for drug delivery. CAB International, UK, pp 53–70

Rai VK, Mishra N, Yadav KS et al (2018) Nanoemulsion as pharmaceutical carrier for dermal and transdermal drug delivery: formulation development, stability issues, basic considerations and applications. J Control Release 270:203–225. https://doi.org/10.1016/j.jconrel.2017.11.049

Rhee YS, Park CW, Nam TY et al (2007) Formulation of parenteral microemulsion containing itraconazole. Arch Pharm Res 30(1):114–123

Ribes S, Fuentes A, Talens P et al (2017) Prevention of fungal spoilage in food products using natural compounds: a review. Crit Rev Food Sci Nutr 58(12):2002–2016. https://doi.org/10.1080/10408398.2017.1295017

Ricaurte L, Hernández-Carrión M, Moyano-Molano M et al (2018) Physical, thermal and thermodynamical study of high oleic palm oil nanoemulsions. Food Chem 256:62–70. https://doi.org/10.1016/j.foodchem.2018.02.102

Salvia-Trujillo L, Rojas-Graü MA, Soliva-Fortuny R et al (2014a) Formulation of antimicrobial edible nanoemulsions with pseudo-ternary phase experimental design. Food Bioprocess Technol 7(10):3022–3032. https://doi.org/10.1007/s11947-014-1314-x

Salvia-Trujillo L, Rojas-Graü MA, Soliva-Fortuny R et al (2014b) Impact of microfluidization or ultrasound processing on the antimicrobial activity against *Escherichia coli* of lemongrass oil–loaded nanoemulsions. Food Control Elsevier 37:292–297. https://doi.org/10.1016/j.foodcont.2013.09.015

Salvia-Trujillo L, Rojas-Graü A, Soliva-Fortuny R et al (2015) Physicochemical characterization and antimicrobial activity of food-grade emulsions and nanoemulsions incorporating essential oils. Food Hydrocoll 43:547–556. https://doi.org/10.1016/j.foodhyd.2014.07.012

Salvia-Trujillo L, Soliva-Fortuny R, Rojas-Graü MA et al (2017) Edible nanoemulsions as carriers of active ingredients: a review. Annu Rev Food Sci Technol 8(1):439–466. https://doi.org/10.1146/annurev-food-030216-025908

Sanchez C, Renard D (2002) Stability and structure of protein-polysaccharide coacervates in the presence of protein aggregates. Int J Pharm 242(1–2):319–324

Sanchez C, Mekhloufi G, Renard D (2006) Complex coacervation between β-lactoglobulin and acacia gum: a nucleation and growth mechanism. J Colloid Interface Sci 299(2):867–873. https://doi.org/10.1016/j.jcis.2006.02.031

Schaink HM, Smit JAM (2007) Protein–polysaccharide interactions: the determination of the osmotic second virial coefficients in aqueous solutions of β-lactoglobulin and dextran. Food Hydrocoll 21(8):1389–1396. https://doi.org/10.1016/J.FOODHYD.2006.10.019

Schmitt C, Bovay C, Frossard P (2005a) Kinetics of formation and functional properties of conjugates prepared by dry-state incubation of β-lactoglobulin/acacia gum electrostatic complexes. J Agric Food Chem 53(23):9089–9099. https://doi.org/10.1021/jf051630t

Schmitt C, da Silva TP, Bovay C et al (2005b) Effect of time on the interfacial and foaming properties of β-lactoglobulin/acacia gum electrostatic complexes and coacervates at pH 4.2. Langmuir 21(17):7786–7795. https://doi.org/10.1021/la0510984

Shaaban HA, Edris AE (2015) Factors affecting the phase behavior and antimicrobial activity of carvacrol microemulsions. J Oleo Sci 64(4):393–404. https://doi.org/10.5650/jos.ess14194

Shah B, Davidson PM, Zhong Q (2013) Nanodispersed eugenol has improved antimicrobial activity against *Escherichia coli* O157:H7 and *Listeria monocytogenes* in bovine milk. Int J Food Microbiol 161(1):53–59. https://doi.org/10.1016/j.ijfoodmicro.2012.11.020

Sharma N, Bansal M, Visht S et al (2010) Nanoemulsion: a new concept of delivery system. Chron Young Sci 1(2):2–6. http://www.cysonline.org/text.asp?2010/1/2/2/70461

Singh Y, Meher JG, Raval K et al (2017) Nanoemulsion: concepts, development and applications in drug delivery. J Control Release 252:28–49. https://doi.org/10.1016/j.jconrel.2017.03.008

Sugumar S, Ghosh V, Nirmala MJ et al (2014) Ultrasonic emulsification of eucalyptus oil nanoemulsion: antibacterial activity against *Staphylococcus aureus* and wound healing activity in Wistar rats. Ultrason Sonochem 21(3):1044–1049. https://doi.org/10.1016/j.ultsonch.2013.10.021

Sugumar S, Singh S, Mukherjee A et al (2016) Nanoemulsion of orange oil with non ionic surfactant produced emulsion using ultrasonication technique: evaluating against food spoilage yeast. Appl Nanosci 6(1):113–120. https://doi.org/10.1007/s13204-015-0412-z

Teixeira PC, Leite GM, Domingues RJ et al (2007) Antimicrobial effects of a microemulsion and a nanoemulsion on enteric and other pathogens and biofilms. Int J Food Microbiol 118(1):15–19. https://doi.org/10.1016/j.ijfoodmicro.2007.05.008

Terjung N, Löffler M, Gibis M et al (2012) Influence of droplet size on the efficacy of oil-in-water emulsions loaded with phenolic antimicrobials. Food Funct 3(3):290–301. https://doi.org/10.1039/c2fo10198j

Thormar H, Hilmarsson H, Bergsson G (2006) Stable concentrated emulsions of the 1-monoglyceride of capric acid (monocaprin) with microbicidal activities against the food-borne bacteria *Campylobacter jejuni*, *Salmonella* spp., and *Escherichia coli*. Appl Environ Microbiol 72(1):522–526. https://doi.org/10.1128/AEM.72.1.522-526.2006

US Food and Drug Administration (2018) Generally Recognized as Safe (GRAS). https://www.fda.gov/food/ingredientspackaginglabeling/gras/. Accessed 7 Aug 2018

Vardhanabhuti B, Allen Foegeding E (2008) Effects of dextran sulfate, NaCl, and initial protein concentration on thermal stability of β-lactoglobulin and α-lactalbumin at neutral pH. Food Hydrocoll 22(5):752–762. https://doi.org/10.1016/J.FOODHYD.2007.03.003

Wong BT, Day L, McNaughton D et al (2009) The effect of Maillard conjugation of deamidated wheat proteins with low molecular weight carbohydrates on the secondary structure of the protein. Food Biophys 4(1):1–12. https://doi.org/10.1007/s11483-008-9096-1

Wooster TJ, Augustin MA (2006) β-Lactoglobulin–dextran Maillard conjugates: their effect on interfacial thickness and emulsion stability. J Colloid Interface Sci 303(2):564–572. https://doi.org/10.1016/j.jcis.2006.07.081

Wooster TJ, Labbett D, Sanguansria P et al (2016) Impact of microemulsion inspired approaches on the formation and destabilisation mechanisms of triglyceride nanoemulsions. Soft Matter 12(5):1425–1435. https://doi.org/10.1039/c5sm02303c

Wu JE, Lin J, Zhong Q (2014) Physical and antimicrobial characteristics of thyme oil emulsified with soluble soybean polysaccharide. Food Hydrocoll 39:144–150. https://doi.org/10.1016/j.foodhyd.2013.12.029

Xue J, Avidson PM, Zhong Q (2015) Antimicrobial activity of thyme oil co-nanoemulsified with sodium caseinate and lecithin. Int J Food Microbiol 210:1–8. https://doi.org/10.1016/j.ijfoodmicro.2015.06.003

Zdrali E, Chen Y, Okur HI et al (2017) The molecular mechanism of nanodroplet stability. ACS Nano 11(12):12111–12120. https://doi.org/10.1021/acsnano.7b05100

Zeeb B, Herz E, McClements DJ et al (2015) Reprint of: impact of alcohols on the formation and stability of protein-stabilized nanoemulsions. J Colloid Interface Sci 449:13–20. https://doi.org/10.1016/j.jcis.2014.12.005

Zhang G, Foegeding EA (2003) Heat-induced phase behavior of β-lactoglobulin/polysaccharide mixtures. Food Hydrocoll 17(6):785–792. https://doi.org/10.1016/S0268-005X(03)00099-7

Zhang G, Foegeding EA, Hardin CC (2004) Effect of sulfated polysaccharides on heat-induced structural changes in β-lactoglobulin. J Agric Food Chem 52(12):3975–3981. https://doi.org/10.1021/jf035037s

Zhang H, Feng F, Fu X et al (2007) Antimicrobial effect of food-grade GML microemulsions against *Staphylococcus aureus*. Eur Food Res Technol 226(1–2):281–286. https://doi.org/10.1007/s00217-006-0537-0

Zhang H, Cui Y, Zhu S et al (2010) Characterization and antimicrobial activity of a pharmaceutical microemulsion. Int J Pharm 395(1–2):154–160. https://doi.org/10.1016/j.ijpharm.2010.05.022

Zhang Z, Vriesekoop F, Yuan Q et al (2014) Effects of nisin on the antimicrobial activity of d-limonene and its nanoemulsion. Food Chem 150:307–312. https://doi.org/10.1016/j.foodchem.2013.10.160

Zhang Y, Karimkhani V, Makowski BT et al (2017) Nanoemulsions and nanolatexes stabilized by hydrophobically functionalized cellulose nanocrystals. Macromolecules 50(16):6032–6042. https://doi.org/10.1021/acs.macromol.7b00982

Zhu D, Damodaran S, Lucey JA (2008) Formation of whey protein isolate (WPI)–dextran conjugates in aqueous solutions. J Agric Food Chem 56(16):7113–7118. https://doi.org/10.1021/jf800909w

Chapter 2
Bio-Based Nanoemulsion Formulations Applicable in Agriculture, Medicine, and Food Industry

Josef Jampílek, Katarína Kráľová, Estefânia V. R. Campos, and Leonardo F. Fraceto

Contents

J. Jampílek (✉)
Division of Biologically Active Complexes and Molecular Magnets, Regional Centre of Advanced Technologies and Materials, Palacky University, Olomouc, Czech Republic

Department of Analytical Chemistry, Faculty of Natural Sciences, Comenius University, Ilkovičova 6, Bratislava, Slovakia

K. Kráľová
Institute of Chemistry, Faculty of Natural Sciences, Comenius University, Bratislava, Slovakia

E. V. R. Campos · L. F. Fraceto
São Paulo State University (UNESP), Laboratory of Environmental Nanotechnology, Institute of Science and Technology of Sorocaba, Sorocaba, SP, Brazil

R. Prasad et al. (eds.), *Nanobiotechnology in Bioformulations*, Nanotechnology in the Life Sciences, https://doi.org/10.1007/978-3-030-17061-5_2

2.1 Introduction

Nanotechnology is a fast-growing field that provides for the development of materials that have new dimensions, novel properties, and a broader array of applications (e.g., Achari and Kowshik 2018; Agarwal et al. 2018; Jampílek and Kráľová 2015, 2017a, b, 2018a, b; Prasad et al. 2017a, b; Sekhon 2014; Ventola 2017). It is regarded as one of the key technologies of the twenty-first century. U.S. National Nanotechnology Initiative defines nanoparticles (NPs) in the range 1–100 nm (National Nanotechnology Initiative 2008). NPs and nanoformulations can be prepared from both inorganic and organic materials (e.g., Bhushan et al. 2014; Singh 2015; Jampílek and Kráľová 2018a; Pisarčík et al. 2018). As mentioned above, nanoscale materials change properties and behavior of all materials, and thus, a variety of industrial, agricultural, pharmaceutical, and medical products have been improved and innovated in such a way (e.g., Dolez 2015; Patra et al. 2018; Sekhon 2014).

In modern agriculture, the main form of control of diseases and agricultural pests is performed using agrochemicals. However, extensive and intensive use of these compounds has resulted in environmental contamination, development of resistance in some species, decreased food safety, and side effects in nontarget organisms (Aktar et al. 2009; Fountain and Wratten 2013; Prasad et al. 2014, 2017a). To address these issues, the development of nano-based pesticides and also the utilization of bio-based pesticides have become an important research tool (Campos et al. 2016, 2018; Hayles et al. 2017; Hemraj 2017; Kah et al. 2018; Jampílek and Kráľová 2015, 2017a, 2018c, 2019a), and the application of nanofertilizers results in increased nutrient use efficiency in crop production (Achari and Kowshik 2018; Jampílek and Kráľová 2017c; Raliya et al. 2018). Nanoemulsions (NEs) showing desirable physico-chemical characteristics have been extensively studied as carriers for pesticide delivery (Wang et al. 2007; Knowles 2008; Hayles et al. 2017; Hazra et al. 2017).

However, nanotechnology is increasingly being applied also in medicine and in theranostic and drug delivery (Prasad et al. 2016, 2017b). By encapsulation into nanoformulations, sustainable release of drugs as well as reduction of the required drug amount could be obtained, and nanomaterials represent also an alternative approach to treating and mitigating infections caused by resistant bacteria (e.g., Jampílek et al. 2015; Jampílek and Kráľová 2017b, 2018a, 2019b, c; Pentak et al. 2016; Patra et al. 2018; Prasad et al. 2018).

In the food sector, the nanotechnologies are used for food protection, including nanocomposites for protection of fruits, vegetables, cheese, dairy products, meat, or fish, in smart active packaging, responsive packaging, or edible coatings, and like that significantly contribute to enhanced food quality (Prasad et al. 2017c). Moreover, they found wide application also as nanosensors (e.g., Chellaram et al. 2014; Jampílek and Kráľová 2015, 2018b; Malhotra et al. 2014; Mihindukulasuriya and Lim 2014; Singh et al. 2017a).

For preparation of bio-based NEs, many compounds of natural origin, including encapsulated active ingredients (e.g., essential oils (EOs)), but also plant oils for NE oil phase (e.g., palm oil, rapeseed oil, sunflower oil, etc.) (e.g., Raviadaran et al. 2018; Kaci et al. 2018; Abdou et al. 2018), emulsifiers, biosurfactants, or cosurfactants (e.g., *Quillaja* saponin, phospholipids, lecithin, gum arabic, pectin, whey protein, lactoferrin, lactoferrin/alginate) (Bai et al. 2016; McClements and Gumus 2016; Ozturk et al. 2014, 2015; Liu et al. 2017; Verma et al. 2016; Artiga-Artigas et al. 2018; Zhao et al. 2018a; Pinheiro et al. 2016), or targeting ligands on the surface of NE such as folate (Liu et al. 2017; Ganta et al. 2016; Afzal et al. 2016a), increasing drug bioavailability, reducing undesirable side effects, minimizing non-specific uptake, and thus allowing specific targeting to certain target cells are used. Targeting technology usually utilizes the nanocarrier functionalization, which can be surface modification (e.g., Attia et al. 2017; Liu et al. 2017) and/or ligand grafting (e.g., Geng et al. 2016). Chitosan (CS)-based NE coatings are used to enhance mucoadhesive properties (Fachel et al. 2018; Mendes et al. 2017; Kumar et al. 2009) as well as antimicrobial activity of these nanoformulations (Marei et al. 2018; Severino et al. 2014, 2015), and also pectin, sodium caseinate, and carrageenan (Abdou et al. 2018; Qian and McClements 2011; Alarcon-Alarcon et al. 2018) are used as NE coatings.

This chapter is focused on the use of NEs in agriculture, with the main emphasis on formulations encapsulating EOs or plant extracts suitable as effective pesticide preparations as well as medicinal applications of bio-based NEs, where attention is paid to transdermal NE formulations, the use of NEs in cancer therapy and for pulmonary and ocular drug delivery. NEs formulated with natural emulsifiers, biosurfactants, and biopolymers are presented, and bio-based NEs of EOs and their constituents as well as NEs with encapsulated vitamins, fatty acids, and some bioactive compounds are discussed. Applications of NEs in edible coatings are outlined as well.

2.2 Nanoemulsions and Methods of Their Preparations

Nanoemulsions (NEs) are biphasic dispersions of two immiscible liquids: either water in oil (W/O) or oil in water (O/W) droplets stabilized by an amphiphilic surfactant, whereby very small emulsion droplets (generally oil droplets in water) with sizes in the order of 100 nm (usually <500 nm) and occurring from a thermodynamic point of view in a non-equilibrium state, are characterized by slow kinetics of destabilization, and therefore, they could be considered as kinetically stable. Small size of NEs results in convenient features such as high surface area per unit volume, robust stability, optically transparent appearance, and tunable rheology (Anton and Vandamme 2011; Gupta et al. 2016; Singh et al. 2017b; Sabry and Ragaei 2018). Very small size of NEs pronouncedly contributes to the prevention of droplet flocculation and coalescence, and thus, the destabilizing process is governed alone by the Ostwald ripening (Anton and Vandamme 2011). As the destabilization of NE occurs due to the change in the droplet size by Ostwald ripening, surface functionalization of NEs could be used to stabilize the O/W interface during emulsification

(Qadir et al. 2016). NEs prepared using biocompatible and biodegradable constituents could be used in medicine and food industry for drug/active ingredient encapsulation, and sustained and controlled release of encapsulated compounds also belong to their advantages.

On the other hand, according to the IUPAC definition, the microemulsion (ME) is "dispersion made of water, oil, and surfactant(s) that is an isotropic and thermodynamically stable system with dispersed domain diameter varying approximately from 1 to 100 nm, usually from 10 to 50 nm, whereby the term "oil" refers to any water-insoluble liquid" (Slomkowski et al. 2011).

In general, NEs could be formulated using so-called "high-energy" methods that utilize specific devices (e.g., ultrasound generators or high-pressure homogenizers) supplying enough energy to increase the W/O interfacial area for generating nanoscale droplets or by "low-energy" methods, in which spontaneous emulsification without requiring any device or energy generate nanoscale droplets (e.g., Jaiswal et al. 2015; Håkansson and Rayner 2018; Hadžiabdić et al. 2017). The high-pressure homogenization, microfluidization, sonication or ultrasonic homogenization, jet disperser high-amplitude ultrasonic method, or membrane emulsification belong to the high-energy methods used for preparation of NEs. The high-energy methods are designed to supply the energy required for emulsification by subjecting it to a disruptive hydrodynamic stress, that is, laminar or turbulent shear or cavitations (e.g., Håkansson and Rayner 2018; Hadžiabdić et al. 2017). For example, a dual-channel microfluidization is a suitable method to fabricate fine NEs with high oil loading levels, which may be advantageous for many commercial applications, while the single-channel method was found to be effective only at producing NEs at relatively low oil concentrations (10%) (Bai and McClements 2016). In β-carotene NEs prepared by microfluidization technique, droplet size decreased from 416.0 to 97.2 nm with increasing microfluidization pressure, number of cycles, and emulsifier concentration, and as the optimum conditions for fabricating such NEs, homogenization pressure of 120 MPa and three cycles were estimated (Jo and Kwon 2014). Abbas et al. (2013) outlined the principles and production technology of high-intensity ultrasound, analyzed the role of acoustic cavitation in the preparation of food-grade O/W NEs, and discussed technical hurdles, issues, and future prospects of this technology.

Low-energy transitional emulsification methods to prepare NEs based on phase transitions of nonionic surfactants (PEGylated surfactants) related to a sudden change in their relative solubility in the oily phase and aqueous phases were overviewed by Anton et al. (2018). Spontaneous emulsification is a low-energy method that simply involves addition of an organic phase (oil + surfactant) into an aqueous phase, and the produced droplet size is affected by surfactant-to-oil ratio, surfactant type, surfactant location, and oil type (Komaiko and McClements 2015). The catastrophic phase inversion method for fabrication of NE is based on gradually diluting, under mild flow conditions, one liquid (such as water) with another immiscible liquid (such as oil) until phase inversion occurs, and a NE is formed (e.g., O/W) that could be used to encapsulate many active compounds (Perazzo and Preziosi 2018).

2.3 Nanoemulsions Applicable in Agriculture

The primary object of preparing chemical pesticide colloids is to significantly increase the apparent solubility of these molecules in water and consequently the delivery of the active compound homogeneously to the pests and/or plants. However, recent research has shown that the production of these colloids can also improve other properties of pesticides, such as bioavailability and increased physico-chemical stability, and promote sustained release (Bhattacharyya et al. 2016).

Pesticides formulated with NEs use less organic solvents when compared with conventional formulations (emulsifiable concentrate) (Chin et al. 2012) and also have lower surfactant concentration (between 3% and 10%) than microemulsions (20% or higher) (McClements 2012). In comparison to the emulsions, the advantages of the NEs due to the small droplet size are the greater spreadability, wettability, and superior mechanical stability (McClements 2012). Consequently, NEs incorporating pesticides settle evenly on the leaves of plants. Some studies have shown that NE improves the solubility of poorly water-soluble pesticides, resulting in the increasing of bioactivity of the pesticides in comparison with conventional pesticides (Leng et al. 2014; Wang et al. 2017).

Using a *Quillaja* saponin as a natural surfactant, Kumari et al. (2018) prepared and characterized thymol NE. The authors tested the antibacterial activity of the NE against *Xanthomonas axonopodis* pv. *glycine*, which is the organism responsible for bacterial pustule disease in soybean. The most stable NE was produced with 50 min of sonication and resulted in a NE with mean diameter 274 ± 2 nm, polydispersity index (PDI) 0.1, and zeta potential −31 mV. Thymol NE showed strong antibacterial activity, and no bacterial colony was observed in the concentration range of 0.02–0.06 (v/v) of NE as compared to control and thymol and saponin. The incidence of the disease was evaluated in vivo, and the results showed that the control plants and plants treated with thymol and saponin alone presented a higher disease incidence (64–78%), while lower disease incidence (3.3–29%) was observed in plants treated with thymol NE. Treatment with NE also resulted in enhancements of plant growth.

In the study by Lim et al. (2013), the authors prepared and characterized a water-soluble herbicide formulation based on NEs of glyphosate isopropylamine. The herbicidal activity was evaluated against three weeds: *Asystasia gangetica*, *Diodia ocymifolia*, and *Paspalum conjugatum*. They found that the glyphosate NEs showed a lower spray deposition in all species when compared to a commercial formulation (Roundup®). However, after 14 days of the treatment, the injury rates of the weeds treated with NE were similar to those observed with plants that were treated with commercial formulation. The similar herbicidal activities of NE and the commercial formulation, even with the lower NE deposition, could be attributed to potentiation of the biological activity when glyphosate was formulated as NE.

In another paper, Du et al. (2016) described the preparation and characterization of O/W NEs using laurate as oily phase and a mixture of two surfactants, alkyl polyglycoside (AL) and polyoxyethylene 3-lauryl ether (PLE). In order to evaluate the potential application of the NEs, a model insoluble molecule, β-cypermethrin, was incorporated into the NEs. Two NEs were analyzed, the first one containing laurate/AL:PLE(6:4)/water in a ratio 10:5:85, and in the second one, this ratio was 20:6:74. NE prepared with the lowest laurate concentration showed lower mean diameter (~110 nm) in comparison with NE prepared with higher concentration of laurate (~200 nm). However, in both cases, there were no significant changes in droplet size between the NEs prepared with and without β-cypermethrin. Even after dilution of the β-cypermethrin, the NE remained as single phase and homogeneous. After dilution, an increase in droplet size was observed. As consequence, diluted NE showed excellent wetting and spreading properties on the hydrophobic surface.

Badawy et al. (2017) prepared O/W NE with an insecticide (diazinon) using Tween® or lecithin as surfactants. In the prepared optimized NE, the dependent variables were droplet size, PDI, dynamic viscosity, and pH. NE prepared using synthetic surfactant (Tween®) showed a size droplet varying from 30 to 138 nm and PDI values from 0.084 to 0.256. The NEs prepared using natural surfactant (lecithin) showed a broader range of droplet size varying from 56.2 to 920.2 nm and also broader range of PDI values (0.122–0.946). Diazinon NE prepared using Tween® showed low viscosity ranging from 2.33 to 21.33 cP, whereas viscosity of lecithin NE incorporating diazinon were higher (55–1039.33 cP). Both NEs prepared with natural or synthetic surfactant showed an acid pH value. In another study aiming to NE applications in agriculture, Zhao et al. (2017) developed and characterized a positive charged O/W NE as a carrier system for lambda-cyhalothrin and evaluated the influence of addition of externally ionic liquids on Z-average, size distribution, zeta potential, viscosity, and stability of the NEs. The measurement of contact angles of NEs on wheat leaves (*Triticum aestivum*) was evaluated. The zeta potential of the NE was affected by the addition of ionic liquids with different alkyl chain length. Increases in alkyl chain length resulted in a change of droplet charge from negative to positive. The droplet size showed no increases in diameter after 90 days of storage, and NE exhibited a non-Newtonian fluid behavior. NE showed a tough adsorption on negative surface decreasing the contact angle on wheat leaves. The addition of ionic liquids provided an increase in the stability due to electrostatic repulsion, and the NE could be considered as a potential carrier system for agrochemicals.

Choupanian et al. (2017) studied the preparation and characterization of neem oil NE and evaluated the increase in the NE stability by addition of two nonionic surfactants (naturally based polysorbate and alkyl polyglucoside). The biological activity of neem oil NE was evaluated against two important pests of stored products: *Sitophilus oryzae* (L.) and *Tribolium castaneum* (Herbst). All NEs showed a droplet size ranging from 200 to 600 nm, and the formulation, which showed higher ratio of nonionic surfactants, had smaller droplet size in comparison to the formulations showing the same composition of surfactants, however with lower concentration. The addition of nonionic surfactants resulted in increases in viscosity of the NE. Both pests treated with 1% of NE showed 100% of mortality after 2 days of treatment, except for the NE with higher droplet size (507 nm) causing

85% and 74% mortality of *S. oryzae* and *T. castaneum*, respectively. However, all NEs showed more toxic effects against both pests compared with free neem oil and Neemix®.

Díaz-Blancas et al. (2016) prepared O/W NEs entrapping tebuconazole using nonionic surfactant (Tween® 80) or AG54, which is a surfactant composed by a mixture of nonionic and anionic amphiphilic components. According to a pseudo-ternary diagram, the phase equilibrium area was achieved in the range of 0.49–0.90, 0.01–0.23, and 0.07–0.49 of organic phase, aqueous phase, and surfactant, respectively. The viscosity of both NEs depended on aqueous phase percentage and showed the same behavior. When the concentration of the aqueous phase increased from 4 to 30 wt%, a 4-fold decrease in the viscosity was observed, whereas increase in aqueous phase concentration from 30 to 50 wt% resulted in viscosity increase; however, NE prepared with Tween® 80 showed higher viscosity than NE fabricated with AG54. The droplet size of the NE containing Tween® 80 was 9 ± 1 nm, and the size remained constant with the increase of concentration in aqueous phase from 1 to 50 wt%. In contrast, NE produced with AG54 showed a strong dependence on the aqueous phase concentration up to 20 wt%. However, further increases in the percentage of the aqueous phase did not influence the size droplet that remained around 250 nm.

Hazrati et al. (2017) studied O/W NE containing *Satureja hortensis* essential oil (EO) and evaluated the herbicidal activity of the NE against worldwide weeds *Amaranthus retroflexus* and *Chenopodium album*. The NE showed a droplet size of 92.7 ± 2.6 nm and PDI 0.29 ± 0.01, and after 30 days of storage, an increase of mean diameter and a decrease in PDI were observed. Under laboratory conditions, the germination percentage showed a dose-dependent response, and the best germination inhibition was observed for the concentration 800 μL/L NE. In this assay, the root elongation was more affected in comparison with shoot elongation. In greenhouse assays, a decrease in growth of the plants in a dose-dependent manner was observed as well. In addition, a decrease in chlorophyll content and increase in relative in electrolyte leakage 5 days after the treatment were estimated, which could be attributed to cell membrane disruption and increased membrane permeability. This NE presented herbicidal activity against weeds due to the presence of carvacrol that is a phenolic monoterpene showing strong phytotoxic effects.

Feng et al. (2016) evaluated the effect of adding the aqueous phase to the organic phase and vice versa, as well as the initial location of the surfactant (organic phase or aqueous phase) on the stability of NE using β-cypermethrin as pesticide model. According to the authors, the NE stability depended strongly on the emulsification process. NEs in which the emulsifier agent was diluted in organic phase showed higher stability than NEs in which it was diluted in aqueous phase. In addition, the most stable NE was prepared by adding the organic phase containing the emulsifier in the aqueous phase. This NE showed a droplet size of 166 nm and PDI of 0.16. Mossa et al. (2017) fabricated NEs of camphor EO and evaluated their insecticide activity against *Sitophilus granaries*. NEs were produced with fixed concentration of camphor oil (5%), different ratios of Tween® 20 (w/w), and different sonication time. Optimized NEs were achieved at a ratio 1:1.5 of Tween® 20-camphor EO and 40 min of sonication, showing a droplet size of 99.0 ± 0.605 nm and remained unchanged during 3 months of storage. Both camphor EO and NE showed insecticidal activity against *S. granaries* in a dose-dependent manner. After 72 h of exposure to the highest concentrations of the NE tested (250 and 300 μg/g), 100% of insect mortality and reduction in the progeny (98% and 100%) was observed, and higher concentration of the free EO was necessary to reach the same effect. The estimated LC_{50} for EO and NE were 282.01 μg/g and 181.49 μg/g, respectively. The insect mortality caused by EO was increased by 36.5% when this oil was incorporated into the NE.

In addition, Fernandes et al. (2014) developed a NE with an extract consisting of the apolar fraction from fruit extract of *Manilkara subsericea* and evaluated its insecticidal activity against *Dysdercus peruvianus*. The best oil phase was found to be octyldodecyl myristate that was able to solubilize equal amount of apolar extract (1:1 w/w). Optimized emulsion was composed of the apolar extract (5%), surfactants (5%), octyldodecyl myristate (5%), and water phase (85%). The NE without extract showed a droplet size of 57 ± 0.3 nm and zeta potential of −59.6 ± 4.1 mV, while the NE containing the extract showed an increase in mean diameter (155.2 ± 3.8 nm) and a decreased zeta potential value (−47.4 ± 3.2 mV). The mortality of the insects treated with NE containing the extract that started in the first day reached 12.23 ± 0.58%, after 30 days of exposure was enhanced to 44.43 ± 6.66%. It could be mentioned that NE containing extract did not induce effect against acetylcholinesterase or mortality in mice.

In order to improve the permeability and efficiency of delivery of antibacterial (ampicillin) into the citrus phloem by foliar spray, Yang et al. (2015) designed a NE using eight adjuvants. Among the adjuvants tested, Brij® 35 showed the highest increases (3.33-fold) in the cuticular permeability when compared with control (water). Two O/W NEs were prepared with different physicochemical properties; however, both formulations showed good thermodynamic stability. One NE showed droplet size of 5.26 ± 0.04 nm and pH value 7.76 ± 0.03 (NE-1), and another one had higher droplet size (94 ± 1.48 nm) and higher pH value (8.31 ± 0.05) (NE-2). After addition of the adjuvant (Brij® 35) to both NEs, a laboratory assay with *Bacillus subtilis* was performed in order to evaluate the antibacterial activity. The NEs showed higher inhibitory zone diameters of 5.75 and 6.6 mm for NE-1 and NE-2, respectively, in comparison with Brij® 35 alone (4.34 mm) and free ampicillin (2.83 mm). In addition, in vivo assays also showed that the NEs were more efficient at suppressing or eliminating Las bacterium when compared with free ampicillin and Brij® 35 alone.

Ali et al. (2017) developed NEs containing neem and citronella oil using a method of low-energy spontaneous emulsification. Antifungal activity of NEs was evaluated against two phytopathogenic fungi *Rhizoctonia solani* and *Scletotium rolfsii*. Optimized primary emulsion for both oils was achieved using the ratio of oil, surfactant, and water of 0.50:1:8.50. This primary emulsion was employed to make neem NE with diverse amounts of citronella oil (0.5–5%) or citronella oil NE with diverse amounts of neem oil (0.5–5%). Neem NEs with different percentage of citronella oil showed a droplet size varying between 11.23 ± 3.86 nm and 17.80 ± 4.52 nm, whereas citronella oil with diverse amounts of neem oil showed a droplet size varying between 8.12 ± 2.80 nm and 12.04 ± 3.74 nm. In vitro antifungal activity of the different NEs against *R. solani* and *S. rolfsii* was screened by poisoned food method. Neem NE with the highest percentage of citronella oil and citronella NE with the highest percentage of neem oil showed the best antifungal activity against the both fungi tested. The estimated LC_{50} related to *R. solani* were 13.67 and 25.64 mg/L, and those related to *S. rolfsii* were 14.71 and 20.88 mg/L, respectively, for neem NE and citronella NE, respectively.

Liu et al. (2011) studied O/W NEs composed of different mixture of nonionic surfactant polyoxyethylene 3-lauryl ether ($C_{12}E_3$) and anionic surfactant dipotassium monododecyl phosphate (MAPK) containing the insecticide bifenthrin. According to the authors, the most stable emulsion was achieved with a mixed ratio of 6:4 (MAPK:$C_{12}E_3$) and 10 wt% surfactant mixture. The NE showed a droplet size of 200.7 nm that increased after 180 days of storage to 218.6 nm but phase separation was not observed. Pant et al. (2014) prepared NEs of eucalyptus oil and evaluated the potentiation of their insecticidal activity by addition of the aqueous filtrate of *Pongamia glabra* and *Jatropha curcas*. Insecticidal activity was estimated against *Tribolium castaneum* that is an important pest of stored grains. Four formulations, all containing the same composition and the same concentration of eucalyptus oil (10% w/w), were prepared, whereby the only component that varied was the aqueous filtrate concentration of karanja and jatropha (0%, 20%, 49%, and 60%). The droplet size of the NEs decreased with increasing concentration of the aqueous filtrate in the formulation. NE produced only with water as continuous phase showed the highest PDI (0.8), while the NE with 20% of aqueous filtrate showed PDI of 0.113, which increased to 0.278 at the highest concentration of aqueous filtrate utilized. The percentage of insecticidal activity of *T. castaneum* did not show significant difference among all formulations tested; however, it was observed a significant decrease in the value of LC_{50} from 5.49 mg/L estimated for the NE without plant aqueous filtrate to 0.1646 mg/L for the NE with 60% of aqueous filtrate. In addition, the volatilization of the eucalyptus oil was stabilized in the NE with aqueous filtrate.

Sharma et al. (2018) evaluated NEs prepared using nonionic surfactants and containing a mixture of two EOs, clove oil and lemongrass oil, for antifungal activity against *Fusarium oxysporum* f.sp. *lycopersici* (FOL). Optimized formulation contained 5% of a blend of clove oil and lemongrass oil (1:1), 10% of surfactant mixture (Tween® 20-CoE-40, 7:3), 5% propylene glycol, and 80% water. This NE showed droplet size 76.73 ± 4.8 nm, PDI 0.207 ± 0.02, and viscosity 26.9 ± 1.9 cP. Such NE was more efficient (48.5% more fungistatic activity) in inhibiting the mycelial growth (MIC 4000 mg/L) in comparison with free oil mixture (MIC 7000 mg/L). In addition, NE showed more pronounced fungicidal activity (5000 mg/L) than free oil mixture (9000 mg/L) against FOL, whereby the nanoemulsion disrupted the membrane integrity of FOL.

Hashem et al. (2018) developed a NE containing *Pimpinella anisum* L. EO and evaluated its insecticidal activities against *Tribolium castaneum* adults and progeny. The NE showed a droplet size of 198.9 nm, PDI 0.303, zeta potential of −25.4 mV, and low viscosity 0.8872 cP. Adult mortality increased proportionally with increasing NE concentrations as well as time of exposure, while the number of progeny and grain weight showed gradual decrease. After 12 h, there was 81.33 ± 0.08% of adult mortality and reduction of 70.85% if the progeny when exposed to 10% NE, and the LC_{50} value after 72 h of exposure was 9.84%. The NE adhered to different parts of the insect body, such as the head, thorax, abdomen, elytra, mouth, and legs. In addition, the NE caused a distinct kind of alteration in the midgut cells of the insects. Abd-Elsalam and Khokhlov (2015) prepared eugenol NE using Tween® 20, a non-ionic surfactant, and water as continuous water phase and evaluated its antifungal activities in vitro and in vivo against *Fusarium oxysporum* f. sp. *vasinfectum* (FOV) isolates. Eugenol NE showed a droplet size of 80 nm, spherical shape, and no increases in droplet size after 30 days of storage at room temperature (25 °C). In vitro antifungal activity (zone inhibition) showed that among the four FOV tested, the NE was more efficient against DQO86833 (5 cm) and AY264267 (4.5 cm) at application of 2% NE, whereas treatment with 5%, NE inhibited the mycelial growth of all FOV isolates. Fungal morphology was also affected by the NE, and reduction in the size and number of conidiospores and reduction in hyphae pigmentation were observed.

As a summary, in literature, there are many examples of systems with potential to be used in agriculture, however, a lot of work are necessary to become all initiatives in products to be used in crop protection.

2.4 Nanoemulsions Applicable in Medicine

NEs as stabilized heterogeneous systems of two immiscible liquids have a great potential in various biomedical applications; especially, they are very attractive for drug delivery. Encapsulation in particular lipophilic drugs leads to the formation of therapeutic nanoformulations providing modification of bioavailability, alternative administration routes, release of drugs, and, thus, a reduction of the required drug amount (e.g., Jampílek et al. 2015; Jampílek and Kráľová 2017b, 2018a, 2019b, c; Pentak et al. 2016; Patra et al. 2018; Prasad et al. 2017b, 2018; Tayeb and Sainsbury 2018).

2.4.1 Transdermal Nanoemulsion Formulations

Drugs with low oral bioavailability due to the first pass metabolism are good candidates for transdermal delivery. Transdermal permeation of majority drugs is hindered by the upper layer of the epidermis (*stratum corneum*); therefore for topical delivery of drugs, chemical permeation enhancers that are able to make the *stratum corneum* more permeable for drugs and reduce the primary skin barrier by different mechanisms are used. In addition, physical methods (e.g., sonophoresis, iontophoresis, electroporation, microneedles, etc.) can be used to increase permeability of drugs (Jampílek and Brychtová 2012). For enhancing epidermal and dermal drug deposition, also nanoscale drug delivery systems, including NEs, are widely applied (Iqbal et al. 2018). Although the skin represents a natural physical barrier against particle penetration, the therapeutic NPs could be delivered especially in diseased skin and to the openings of hair follicles (Prow et al. 2011), and good candidates for transdermal delivery are drugs with low oral bioavailability due to the first pass metabolism. It is advantageous that NEs are vehicles acting also as transdermal permeation enhancers without utilizing additional permeation enhancers (Shakeel et al. 2010).

The progressive advancement in the delivery of drugs via NE with special reference to the dermal and transdermal administration and the most suitable semisolid dosage forms for the particular type of NEs (O/W, W/O, and others), including the effects of particle size and zeta potential on the delivery of drugs through dermal or transdermal route, was overviewed by Rai et al. (2018). Low viscosity of NEs that might be unsuitable for topical application could be overcome by hydrogel-thickened NEs using thickening polymer, e.g., CS (Barradas et al. 2017). Challenges and future prospects of NEs as a drug delivery system were presented by Yukuyama et al. (2017). Applications of NEs in the field of dermatology highlighting the advantages over the other dermatological therapies connected with increased contact surface area by the particle size, which leads to increased drug efficacy, were overviewed by de Souza et al. (2018). For example, Salim et al. (2016) discussed the potential of drug-loaded NEs for the treatment of psoriasis to achieve better efficacy and eliminate side effects and noted that the delivery and penetration of a drug through the psoriasis skin layer could be enhanced by a small droplet size. Nastiti et al. (2017) in their review paper focused their attention on the composition and characterization of MEs and NEs for topical and transdermal delivery and the mechanism of skin delivery across *stratum corneum* and via hair follicles.

Nanoemulgels that are basically O/W NEs gelled with the use of some gelling agent in it, in which the gel phase in the formulation is nongreasy and stabilizes the formulation through reduction in surface as well as interfacial tension, could be considered as a novel transdermal delivery system that is able to overcome poor oral bioavailability of drugs, more specifically target to the site of action, can avoid first-pass metabolism, and relieve the user from gastric/systemic incompatibilities (Choudhury et al. 2017). In nanoemulgels, the NE containing drug is incorporated into a gel base. Lipophilic drugs could be easily incorporated, and the skin permeability of the incorporated drugs can be enhanced in several folds due to the finely distributed droplets of NE phase resulting in notably improved pharmacokinetic and pharmacodynamic profiles of the lipophilic drugs. Consequently, nanoemulgel formulations could be considered as potential and promising candidates for topical delivery of lipophilic drugs in the future (Sengupta and Chatterjee 2017). A review paper of Pawar and Babu (2014) is devoted to various lipid materials (vegetable oils, fatty acids, fatty alcohols, medium chain glycerides, and fatty acid esters) used in the preparation of NEs for topical and transdermal drug delivery.

Integral NEs with particle size of 80 nm can diffuse into but not penetrate the viable epidermis; however, they can efficiently fill the whole hair follicle canals and reach as deep as 588 μm underneath the dermal surfaces and the "cargos" released from the NEs diffuse into the surrounding dermal tissues. On the other hand, NEs with mean particle size of 500 nm cannot penetrate the *stratum corneum* and can only migrate along the hair follicle canals, while NEs with median size, e.g., 200 nm, show moderate transdermal permeation effect (Su et al. 2017). Capsaicin O/W NEs with droplet sizes 20–62 nm were reported to permeate all skin layers from the *stratum corneum* to the dermis (Kim et al. 2014a).

Permeation flux of antifungal drug fluconazole from optimized drug-loaded olive oil NEs through artificial skin was approximately 3-fold higher than the control (Ansaril et al. 2017). A testosterone transdermal delivery system developed using a palm oil base (HAMIN™) with particle sizes 97–774.0 nm was tested using in vivo skin permeability test, and it was found that testosterone was well absorbed with a mean C_{max} and T_{max} of 60.94 ng/mL and 2.29 h after in vivo application on rabbit skin indicating that such nanoformulation could have great potential for topical delivery of testosterone (Haron et al. 2015). The optimized clove oil-based olmesartan NE (droplet size of 53.11 ± 3.13 nm, PDI 0.335 ± 0.008) showing a 1.23-fold increase in the bioavailability compared with oral formulation of drug due to better permeation through rat skin could be used as an antihypertensive dosage form for effective transdermal delivery of olmesartan (Aqil et al. 2016). O/W NEs containing the skin penetration enhancer oleic acid or eucalyptol as oil pronouncedly enhanced the skin penetration of encapsulated caffeine and naproxen compared to their aqueous control solutions. Caffeine maximum flux enhancement was connected with a synergistic increase in both caffeine *stratum corneum* solubility and skin diffusivity, while increased solubility in the *stratum corneum* was the dominant determinant for higher naproxen fluxes (Abd et al. 2016). Transdermal NE formed of 2% *Foeniculum vulgare* Mill. EO, 5.6% oleic acid, 68% S_{mix} (1:1), and distilled water showed a high potential of reducing plasma glucose levels in rats that continued for 7 days after a single topical application of a dose of 120 mg/kg of fennel EO, bringing glucose to normal levels in diabetic rats (Mostafa et al. 2015a). O/W NEs containing minoxidil, an antihypertensive vasodilator, and the skin penetration enhancer oleic acid or eucalyptol as oil phases pronouncedly enhanced drug permeation through skin compared with control solutions. Minoxidil retention in the *stratum corneum* and deeper skin layers was promoted to higher extent with eucalyptol NEs, while oleic acid formulations gave the greatest hair follicle penetration. The increases in both minoxidil *stratum corneum* solubility and skin diffusivity in both nanoemulsion systems were reflected in drug maximum flux enhancement connected with enhanced fluidity and disruption of *stratum corneum* lipids (Abd et al. 2018). NEs containing isoflavone-rich soybean extracts that are considered as promising skin antiaging products due to their antioxidant activity could be applied as suitable topical formulations to protect skin from UVA/UVB oxidative damage (Back et al. 2018). Antioxidant hydrogels containing an *Achyrocline satureioides* extract-loaded NEs aimed at topical application were found to be suitable to protect the porcine ear skin against oxidative stress generated by UVA/UVB light (Balestrin et al. 2016). Brownlow et al. (2015) developed vitamin E-enriched NE vehicles loaded with genistein for chemoprevention against UVB-induced skin damage showing enhanced dermal delivery of the drug. Also Nam et al. (2018) developed W/O NE (diameter of <5 nm) that contained nitric oxide (NO) by mixing surfactant with vitamin E antioxidant body oil (Product No. 04800, Cococare, Dover, NJ) and $NaNO_2$ solutions. After spreading of this NEs with NO on penis skin of the middle-aged dogs, blood NOx concentration in the penis increased, resulting in penile erection without any notable topical and systemic side effects, suggesting that such W/O NEs could be used in noninvasive medication for patients suffering in erectile dysfunction with low response to phosphodiesterase type 5 (PDE5) inhibitors such as Viagra® or Cialis®.

Formulations of hydrogels containing negatively or positively charged Copaiba oil NEs exhibited anti-inflammatory effects, which was reflected in mouse ear edema (69% and 67%) and rat paw edema inhibition (32% and 72%) and decrease of inflammatory factors, such as dermis and epidermis hyperplasia and inflammatory cells infiltration in histological cuts were estimated as well (Lucca et al. 2018). Skin permeation with the positively charged Copaiba oil NEs increased 3-fold the retention of the major component in copaiba oil, β-caryophyllene, in the epidermis, and also in the receptor fluid compared to the negatively charged NEs (Lucca et al. 2017). Coffee oil-algae oil-based NEs with a particle size of 30 nm, zeta potential −72.72 mV, and 100% encapsulation efficiency (EE) of docosahexaenoic acid (an important component of algae oil) applied at dose of 0.1% efficiently mitigated trans-epidermal water loss, skin erythema, melanin formation, and subcutaneous blood flow in animal experiments and were found to inhibit the growth of melanoma cells B16-F10 (IC_{50}: 26.5 μg/mL) and arrest the cell cycle G_2/M phase, whereby the apoptosis pathway of melanoma cells may involve both mitochondria and death receptor (Yang et al. 2017).

Transdermal administration of nanogel based on optimized catechin NEs showed sustained release profile of catechin and enhanced photoprotection potential due to its improved permeability as well as bioavailability compared to the conventional gel, and it could represent an effective strategy for decreasing UV-induced oxidative damage in the skin tissues (Harwansh et al. 2016). Natural pentacyclic triterpenes-loaded NEs showed greater ability to inhibit inflammation than NEs loaded with the synthetic mixture of triterpenes (Alvarado et al. 2015).

NEs of CS oleate encapsulating α-tocopherol with particle size of 220 nm were reported to be suitable for topical application in wound healing (Bonferoni et al. 2018). Astaxanthin (ASX)-loaded carboxymethyl CS functionalized NE formulation with spherical droplets showing mean diameter >100 nm and a small negative surface charge exhibited higher ASX chemical stability and skin permeability than ASX NEs and ASX solution and was characterized with low cytotoxicity (Hong et al. 2017).

Genistein-loaded cationic NEs (mean droplet size of ca. 200–300 nm) prepared by spontaneous emulsification and using hydroxyethyl cellulose as a thickening agent (at 3%) showed considerable increase of drug retention in mucosa compared to the genistein propylene glycol solution and exhibited antiherpetic activity in vitro against herpes simplex virus 1 (HSV-1, strain 29R) (Argenta et al. 2016). Factorial design applied to the optimization of lipid composition of topical antiherpetic NEs containing isoflavone genistein was presented by Argenta et al. (2014). The 10,11-methylenedioxycamptothecin loaded hyaluronic acid (HA) NEs were reported to perform desirable skin permeable capacity across human keloid skin, whereby the drug was transferred directly to keloid lesion area. The growth-inhibitory effect was further clarified upon cell cycle regulation, which arrested cells at G_1/S and prevented them entry into mitosis (Gao et al. 2014).

NEs with co-encapsulated C6 ceramide (0.35%) and paclitaxel (0.50%) containing tributyrin delivered 2- and 2.4-fold more paclitaxel into viable skin layers of porcine skin in vitro at 4 and 8 h post-application than the MEs and 1.9-fold more C6 ceramide at 8 h, whereby the drugs were co-localized mainly in the epidermis. The EC_{50} values related to melanoma cells viability estimated for individually encapsulated paclitaxel and ceramide in NEs were 4- and 13-fold lower than for unencapsulated formulation, and at co-encapsulation of both active ingredients, synergic effect was observed, because further decrease of EC_{50} by 2.5–4.5-fold was estimated, and calculated combination index also indicated a synergistic effect. Topical application of NEs on 3D bioengineered melanoma models for 48 h stimulated marked epidermis destruction, and only few cells remained in this layer (Carvalho et al. 2017).

The olein-based β-D-glucan-loaded NE prepared using ultrasound exhibited higher antioxidant activity as compared to free antioxidant β-D-glucan (Alzorqi et al. 2016). In sulpiride ME formulations containing glyceryl monooleate, Labrafil, and Avocado as oily phases, drug solubility increased to 43.35 mg/mL with drug content >97%, and based on pharmacodynamic performance and antipsychotic activity of sulpiride, this formulations could be used as antipsychotic nasal drug delivery to overcome sulpiride low oral bioavailability (Ayoub et al. 2016). Optimized eugenol-NE prepared using Tween® 80 and Labrasol® as surfactant and cosurfactant, respectively, showing hydrodynamic diameters of 89.98 ± 6.48 nm, PDI of 0.238 ± 0.021, and zeta potential of −10.05 ± 0.11 mV were found to enhance the transdermal delivery of eugenol without causing skin irritation (erythema and edema) in vivo suggesting their potential to be used in wound healing and anti-inflammatory treatments (Ahmad et al. 2018a). The NE containing cumin EO/oleic acid and Tween® 20/ethanol (2:1) showed high phenolic encapsulation efficiency and remarkable cumulative phenols permeation through rat skin as well as high in vitro and in vivo antioxidant efficiency and provided high hepatoprotective potential and reserved rats' body weight after a period of 7 days of a single transdermal application (Mostafa et al. 2015b).

Staphylococcus aureus treated with optimized NE containing eucalyptus oil as organic phase, water as continuous phase, and nonionic surfactant, Tween® 80, as emulsifier prepared by 30-min sonication and showing the mean droplet diameter of 3.8 nm resulted in complete loss of viability within 15 min of interaction, and the membrane of treated bacterial cells was damaged. This NE was not irritant and exhibited higher wound contraction rate in Wistar rats with respect to control and neomycin-treated rats (Sugumar et al. 2014).

The optimized W/O NE of thiocolchicoside (TCC), an effective therapeutic agent against the orthopedic, traumatic, and rheumatologic disorders, containing in ratio 1:1 linseed and sefsol (propylene glycol caprylate) as the oil phase, showed mean globule diameter of 117 nm, PDI of 0.285, and the steady-state flux (J_{ss}) and permeability coefficient (K_p) of 30.63 ± 4.18 μg/cm²/h and 15.21 × 10^{-3} ± 2.81 cm²/h, respectively, in in vitro permeation experiment using porcine skin suggesting that W/O NEs, which are compatible with the lipophilic sebum environment of the hair follicle, facilitate the transport of TCC, which might be predominantly transfollicular in nature (Kumar et al. 2016).

Curcumin (CUR) NE for transdermal application with mean droplet diameter, PDI, and zeta potential of optimized NE 85.0 ± 1.5 nm, 0.18 ± 0.0, and − 5.9 ± 0.3 mV, respectively, notably improved the permeation flux of CUR from the hydrophilic matrix gel Viscolam AT 100P (sodium polyacryloyldimethyl taurate, hydrogenated polydecene, polyoxyethylene), whereby NE formulation not only improved CUR permeability but also protected the drug from chemical degradation (Rachmawati et al. 2015). CUR-loaded NE (droplet size of 41.13 ± 3.34 nm and zeta potential of −33.1 ± 1.45 mV) that was incorporated into gel using Carbopol® 980 (1% w/v) and tested on Freund's complete adjuvant-induced arthritic rat model after topical application of CUR-NE gel in Wistar rats showed substantial reversal of arthritic symptoms suggesting that the nanoformulation could exhibit therapeutic effects locally in inflammatory arthritic disorders with improved topical bioavailability (Naz and Ahmad 2015). Study of CUR distribution in neonate pig skin using CUR-loaded myristic acid microemulsions showed dermal CUR accumulation (326 μg/g skin) and transdermal CUR penetration (87 μg/cm²/d). CUR encapsulated in ME inhibited bacterial growth of *Staphylococcus epidermidis* (EC_{50} of 0.86 μg/mL) and was found to be 12-fold more effective than CUR dissolved in dimethyl sulfoxide suggesting that such MEs could be used as alternative treatment for *S. epidermidis*-associated diseases and acne vulgaris (Liu and Huang 2012).

NEs composed of propylene glycol, Transcutol®, water, Labrasol®, Plurol Oleique®, isostearyl isostearate, oleic acid, and D-limonene with incorporated imipramine or doxepin in the NE system (3% w/w) were tested for an analgesic and anti-allodynic activity at transdermal delivery of drugs, and it was found that in vivo analgesic and anti-allodynic activity in rats was stronger for the doxepin-loaded NE suggesting that such nanoformulation could be an alternative analgesic therapy with a potential clinical application (Sandig et al. 2013).

NEs consisting of monoammonium glycyrrhizinate, Span® 80, Brij® 35, isopropyl alcohol, soybean oil, and distilled water were reported as appropriate vehicles for transdermal delivery of glycyrrhizin through human cadaver skin, and while excipients of NEs acted as permeation enhancers themselves, the use of additional permeation enhancers was not necessary (Harwansh et al. 2011).

2.4.2 Nanoemulsions for Cancer Therapy

Surface modification of nanocarriers could contribute to their enhanced functions in imaging, targeting, and delivery, increase drug bioavailability, reduce undesirable side effects, and minimize non-specific uptake, thus allowing specific cancer targeting to certain target (Yu and Zhang 2009). NEs in the translational research and their role in targeted cancer therapy were summarized by Ganta et al. (2014). The current status of NEs in the cancer therapeutics and commercial field on the basis of morphology, formulation, characteristics, and characterization parameters was overviewed by Sahu et al. (2017). Sasikumar and Kamalasanan (2017) analyzed the possibilities of exploring NE platform for targeted drug delivery to prostate cancer. Recent advances in lipid nanocarriers, including NEs, applicable in the fight against cancer were summarized by Jampílek and Kráľová (2019b).

Zhao et al. (2018b) reported that woody oil-based emulsive nanosystems could efficiently deliver poorly soluble natural alkaloids resulting in increases in the sensitivity of lung cancer cells. O/W cinnamon oil NE (40.52 nm) and vitamin D encapsulated cinnamon oil NE (48.96 nm) showed anti-cancerous activity in human alveolar carcinoma cells; they induced DNA damage along with corresponding increase in micronucleus frequency, arrested the cell cycle progression in G_0/G_1 phase, and showed increased expression of Bax, capase-3, and caspase-9 and decreased expression of BcL2 proteins along with considerable increase in apoptotic cell population and loss of mitochondrial membrane potential (Meghani et al. 2018). The antitumor potential of both O/W NE with encapsulated vitamin K_2 (VK_2) and NE incorporating VK_2 with a ligand conjugate sialic acid-cholesterol (showing enhanced affinity toward the membrane receptors overexpressed in tumors) anchored on the surface was evaluated in S180 murine sarcoma tumor cells. It was found that i.v. or intragastric administration of VK_2 NE to syngeneic mice with subcutaneously established S180 tumors resulted in considerable tumor growth suppression, higher effect being observed with surface-modified NE, and both NEs were nontoxic (Shi et al. 2018).

Ahmad et al. (2018b) reported that the optimized silymarin NE with mean particle size of 21.24 nm reduced the cancer cell viability, increased the intensity of reactive oxygen species (ROS) and chromatin condensation, and could be considered as an efficient carrier for oral delivery of silymarin against human hepatocellular carcinoma without damaging normal cells. Methyl jasmonate loaded NE with mean droplet size of 75.06 nm and PDI 0.017 was found to be more effective in killing cancer cells; it induced a stronger sub-G_1 arrest than methyl jasmonate solution and showed a considerable absence of toxicity in human umbilical vein endothelial cells (Habibi et al. 2017). The EC_{50} values related to cytotoxic activity against A549 tumor cell line (human lung carcinoma) estimated with pure *Casearia sylvestris* Sw. extract and its NE were 4.0 μg/ mL and 1.0 μg/mL, respectively, suggesting 4-fold higher efficiency of the nanoformulation (Pereira et al. 2017).

The cytotoxic effect of exopolysaccharides extracted from brown seaweed (*Sargassum longifolium*) encapsulated in orange oil NE with particle size 178 nm and zeta potential 43.9 mV estimated by MTT method in colon (HCT 116) cell lines was lower (70%) than that of seaweed polysaccharide encapsulated with nanostructured lipid carrier (Shofia et al. 2018). Weekly i.v. administration of O/W NE encapsulating DHA-SBT-1214, a novel ω-3 fatty acid conjugated taxoid prodrug against prostate cancer stem cells, to NOD/SCID mice bearing subcutaneous PPT2 tumor xenografts resulted in strong suppression of tumor growth compared to Abraxane®

and placebo NE, and viable cells that survived from this in vivo treatment regimen were no longer able to induce floating spheroids and holoclones (Ahmad et al. 2017). Migotto et al. (2018) developed a cationic bioadhesive NE surface modified with CS and particle sizes of 46.3 nm for intraductal administration of C6 ceramide showing 4.5-fold lower EC_{50} value of C6 ceramide related to the reduction of MCF-7 cells viability, and this NE prolonged drug localization for more than 120 h in the mammary tissue following intraductal administration compared to its solution. NEs prepared using tanshinone extract of *Salvia miltiorrhiza* with the mean particle size of 14.2 nm inhibited human lung carcinoma cell (A549) proliferation more effectively than the extract alone, and they penetrated into cytoplasm through endocytosis and caused upregulation of p-JNK, p53, and p21 and downregulation of CDK2, cyclin D1, and cyclin E1 expressions in a dose-dependent manner and caused cell cycle arrest at G_0/G_1 phase (Lee et al. 2016). NE formulation of *Nigella sativa* L. EO with droplet diameter 20–50 nm notably reduced the viability of MCF-7 breast cancer cells, whereby the treated cells included cell membrane blebbing, cytoplasmic vacuolation, marginalization of chromatin, and fragmentation of the nucleus suggesting induction of apoptosis in MCF-7 cells (Periasamy et al. 2016).

O/W NE coated with a thiol-modified CS designed for co-delivery of piperine (weight ratio 100: 1) was fabricated, and high degree of CS modification with particle sizes of 110 nm did not show any cytotoxic effect on normal fibroblasts and promoted death in colon cancer cells (Vecchione et al. 2016). Formulation of diallyl disulfide and α-linolenic acid prepared as protein NEs showing antioxidant and radical scavenging property and acting also as optimal H_2S slow-release donors exhibited considerable anti-proliferative effect on MCF-7 breast cancer cell lines and HuT 78 T-cell lymphoma cells, induced apoptosis and cell cycle arrest at the G_0/G_1 phase, and improved the Lin^- $Sca1^+$ human cardiac progenitor cells proliferation suggesting that they could be used in selective cancer therapy and for promoting the muscle tissue repair (Ciocci et al. 2016).

Evaluation of anticancer activities of ginger EO (GEO) and frankincense EO (FEO) NEs prepared by a high-pressure homogenization technique with incorporated antineoplastic agent, mitomycin C (MMC), showed that NE-based EOs ameliorated the apoptotic effects of MMC on the cancer cells (IC_{50} values of GEO-MMC and FEO-MCC NEs estimated for HeLa cells were reduced by 44.12- and 29.42-folds, respectively, while those for MCF-7 cells were decreased by 29.29- and 55.3-folds when compared to MMC solution). FEO-MMC NE caused also the greatest change on the HeLa cellular morphology, while MCF-7 cells were most damaged and their nuclei were segmented when subjected into high GEO-MMC NE. Consequently, mixing of drug with GEO NE and FEO NE considerably improved its cytotoxicity on the MCF-7 and HeLa cells (Al-Otaibi et al. 2018). MMC formulated into NEs based on EOs of chamomile (ChEO) and garlic (GarEO) reduced the cell viabilities of HeLa cervical cancer cells by 42- and 20-fold compared to free MMC, whereby treatment with GarEO NE or GarEO-MMC NEs resulted in stronger alteration of the cell membrane of the HeLa cells than treatment with ChEO or ChEO-MMC NEs. Using staining with 4′,6-diamidino-2-phenylindole, it was found that NEs of GarEO and GarEO-MMC have got attached to the cell membrane causing damage to the cell, while those of ChEO or ChEO-MMC passed the cell membrane and affected the nucleus directly (Alkhatib et al. 2018a).

In vitro cytotoxicity against MCF-7 and HeLa cells of an albumin anchored docetaxel (DTX) lipid NE was higher than that of the plain NE, and in in vivo experiment, it caused 80.01 ± 2.74% inhibition of solid tumors induced in C57BL/6 mice compared to 55.62 ± 5.41% inhibition caused by administration of the plain NE, whereby also its tumor-targeting activity was 3-fold higher than that of plain NE (Afzal et al. 2016b). Transferrin coupled DTX lipid NE (200–393 nm) prepared by homogenization and ultra-sonication process caused 84.66 ± 4.29% tumor inhibition in tumor-induced C57BL/6 mice and was found to have also 3.54-fold higher tumor-targeting activity compared to the plain lipid NE (Afzal et al. 2016c).

Carvacrol NE showing a negative surface charge of −29.89 mV and 99.1 nm mean droplet size powerfully induced apoptosis in doxorubicin (DOX)-resistant A549 lung carcinoma cells (A549DR), displayed cell senescence leading to cell cycle arrest, and inhibited the autophagy suggesting that it could be used as a potential candidate for lung cancer therapy (Khan et al. 2018).

Paclitaxel (PTX) NEs, in which α-tocopherol oil core of Tocosol™ was substituted with γ-tocotrienol, and D-α-tocopheryl polyethylene glycol 1000 succinate (vitamin E TPGS) with PEGylated γ-tocotrienol showing droplet size <300 nm were tested against Bx-PC-3 and PANC-1 pancreatic tumor cells. Fastest release was observed with NEs loaded with free PTX, when γ-tocotrienol was used as the core and anticancer activity was improved also by substituting α-tocopherol with γ-tocotrienol and pronounced increase in activity was estimated when PTX lipid conjugates were used (Abu-Fayyad et al. 2018). Entrapment of gemcitabine-γ-tocotrienol conjugates into NEs pronouncedly improved their anticancer activity against Bx-PC-3 and PNAC-1 pancreatic cancer cells when compared to the free drug. Moreover, it was found that gemcitabine-γ-tocotrienol conjugates were least affected by deamination deactivation reaction in vitro when compared with the free and conjugated gemcitabine in solution (Abu-Fayyad and Nazzal 2017). CUR/δ-T3 tocotrienol NEs with average particle size of 261 nm, PDI 0.27, and zeta potential of 35 mV significantly suppressed constitutive NF-κ B activation, and significantly induced apoptosis in breast and ovarian cancer cells, MCF-7 and OVCAR-8 (Steuber et al. 2016). Hyaluronic acid-complexed PTX NEs containing DL-α-tocopheryl acetate and soybean oil with a diameter of 85.2 ± 7.55 nm and a zeta potential of −35.7 ± 0.25 mV administered at a dose of 25 mg/kg to nude mice transplanted with CD44-overexpressing non-small cell lung carcinoma cells (NCI-H460) xenografts suppressed cancer cell growth more than the Taxol® and strongly inhibited tumor growth, which could be connected with the specific tumor-targeting affinity of HA for CD44-overexpressed cancer cells (Kim and Park 2017a). HA-coated NEs composed of DL-α-tocopheryl acetate, soybean oil, polysorbate 80, and ferric chloride incorporating PTX with particle diameter and zeta potential of 65 ± 15 nm and − 39.5 ± 0.33 mV, respectively, showed superb targeting of ovarian SK-OV-3 tumor cells overexpressing CD44 (Kim and Park 2017b).

The study of the antitumor activity and cardiotoxicity of the garlic oil (GarO) NE and GarO NE with incorporated DTX with particle sizes 63.19 ± 1.85 nm and 110. ± 14.37 nm, respectively, in Ehrlich ascites carcinoma (EAC)-bearing mice showed that the administration of GarO NE enhanced the lactate dehydrogenase activity in the ascetic fluid and ameliorated the heart enzymes, while treatment with DTX-Gar NE improved the mean survival time of the mice (27.7 ± 11.63 days) that at treatment with aqueous DTX solution reached 23.1 ± 1.52 days (Alkhatib et al. 2017a). It is important to note that incorporating the DTX into NEs based on orange oil also improved its antitumor efficacy and reduced its cardiotoxicity in female Swiss Albino mice-bearing Ehrlich tumor in their ascetic fluid (Alkhatib et al. 2017b). Efficient accumulation of NEs incorporating fraxinellone (Frax), an anti-fibrotic drug, prepared by an ultrasonic emulsification method with a particle size approx. 145 nm in the tumor site after systemic administration was observed, and NE was taken up by tumor-associated fibroblasts (TAFs) and tumor cells. Following i.v. administration of Frax NE, a notable decrease in TAFs and stroma deposition was estimated, and an increase of natural-killer cells and cytotoxic T cells as well as a decrease of regulatory B cells and myeloid-derived suppressor cells in the tumor microenvironment suggested that this treatment remolded the tumor immune microenvironment as well. Enhanced tumor-specific T-cell infiltration, activated death receptors on the tumor cell surface, and increased apoptotic tumor cell death were achieved using a combination of a tumor-specific peptide vaccine with Frax NE suggesting that such approach could be an effective and safe strategy to remodel fibrotic tumor microenvironment resulting in enhanced immune response activation and, thus, in a prolonged efficiency for advanced desmoplastic melanoma (Hou et al. 2018). Improved in vivo antitumor activity in female Swiss Albino mice inoculated with Ehrlich ascites and reduced hepatotoxicity was observed also with sorafenib incorporating NE formulated with flaxseed oil with particle diameter and zeta potential of 77.46 ± 8.28 nm and − 3.4 ± 1.2 mV, respectively (Alkhatib et al. 2017c). Similarly, incorporation of sorafenib into carrot seed oil NE improved the antitumor efficacy of drug and reduced its hematoxicity and hepatotoxicity (Alkhatib et al. 2018b).

Acid-sensitive lipidated DOX prodrug (C16-DOX) entrapped in lipid NE improved chemotherapeutic index and tumor-control efficacy in an in vivo murine 4T1 breast cancer model compared to free drug and caused considerable reduction in lung metastasis due to possibility using higher dose of DOX (Camara et al. 2017). Folate-functionalized soy lecithin lipid NEs of DOX and yttrium 90 (^{90}Y) were found to inhibit growth of folate receptor-rich nasopharyngeal carcinoma (NPC) cells CNE1 in vitro and pronouncedly decrease tumor volume in NPC-induced nude mice compared to DOX + ^{90}Y-lipid NE, causing massive (89.9%) necrosis and hemorrhage of CNE1 cells but lower growth inhibition (21%) of folate-deficient nasal epithelial cells (RPMI 2650) than DOX + ^{90}Y-lipid NE (43.65%) (Liu et al. 2017). DTX NE containing as biocomponents soybean oil and lecithin with droplet size, PDI, and zeta potential 233.23 ± 4.3 nm, 0.240 ± 0.010, and − 43.66 ± 1.9 mV, respectively, exhibited strong cytotoxic activity against MCF-7 cancer cells (IC_{50}: 13.55 ± 0.21 μg/mL at 72 h) as well as 2.83-fold higher cell uptake than control, whereby at a dose of 20 mg/kg, no toxicity or necrosis was observed with liver and kidney tissues of mice (Verma et al. 2016). DTX NE surface functionalized with folate with particle sizes <150 nm showed a 270-fold decrease in IC_{50} value in chemoresistant ovarian cancer cells SKOV3TR as compared to DTX alone, while in SKOV3 tumor-bearing mice, these NEs delivered DTX by folate receptor-mediated endocytosis resulting in cytotoxicity capable of overcoming ABC transporter-mediated taxane resistance (Ganta et al. 2016). Folate-PEG-decorated DTX lipid NE was superior in tumor targeting by 4.81- and 2.08-fold over controls and in tumor regression as well providing better results as PEGylation, albumin, and transferrin strategies (Afzal et al. 2016a).

Theranostic NEs incorporating PTX and contrast agents prepared using linear polyglycerol-poly(ε-caprolactone) diblock copolymers and ethiodized poppy-seed oil, lipiodol, as a core oil exhibited superb anticancer activities against HeLa ovarian cancer cells as well as the ability to be used as a contrast agent (Le Kim et al. 2017). NE with co-encapsulated PTX and baicalein (5,6,7-trihydroxyflavone) showed better antitumor efficacy in MCF-7/Tax cells exhibiting progressive resistance to PTX in vitro as well as much higher antitumor efficacy in vivo than other PTX formulations suggesting that this NE formulation could be used to overcome multidrug resistance via oxidative stress augmentation and P-glycoprotein inhibition (Meng et al. 2016).

Multifunctional smart CUR-loaded CS/perfluorohexane nanodroplets of 101.2 nm and 77.8% CUR entrapment designed for contrast-ultrasound imaging and on-demand drug delivery showed notably higher inhibition of 4 T1 human breast cancer cells in vitro by ultrasound exposure, suggesting that they could have a great potential for image-guided cancer therapy (Baghbani et al. 2017). Ultrasound-responsive multifunctional smart alginate/perfluorohexane nanodroplets with the mean particle size of 55.1 nm designed for co-delivery of DOX and CUR and showing 92.3% entrapment efficiency of DOX exhibited enhanced cytotoxicity in adriamycin-resistant A2780 ovarian cancer cells compared to DOX nanodroplets due to synergistic effects of DOX and CUR, and their combined application with ultrasound irradiation leads to strong tumor regression (Baghbani and Mortarzadeh 2017). MEs of CUR with docosahexaenoic acid (DHA)-rich oil showed strong cytotoxicity on human glioblastoma U-87MG cell line (IC_{50}: 3.755 ± 0.24 ng/mL), which could be a attributed to the synergistic effect of CUR and DHA in the ME, and following administration of this ME to Sprague-Dawley rats, the CUR concentration at 24 h achieved 466-fold (for intranasal) and 421-fold (for intravenous) of the IC_{50} value estimated in the U-87MG cell line suggesting that designed ME could be used for therapy of brain cancer by both routes of administration (Shinde and Devarajan 2017).

Photodynamic therapy (PDT) is a clinically approved cancer therapy utilizing photochemical reaction between a light activable molecule or photosensitizer, light, and molecular oxygen resulting in formation of reactive oxygen species that directly damage cells and/or vasculature leading to tumor destruction and induce inflammatory and immune responses (van Straten et al. 2017). For both hydro-alcoholic extract from *Tectona grandis* L.f. leaves (TGE) alone and extract incorporated into

O/W NE with particles sizes approx. 20 nm and tested against melanoma B16 F10 cells, photodynamic effect was estimated due to increasing toxicity under illumination with red light; however, NE formulation was much less toxic toward normal cells in the dark compared to free TGE, which exhibited notable dark toxicity toward both B16 F10 and murine fibroblast NIH3T3 cells (Furtado et al. 2017). Acai oil in NE, a novel photosensitizer for PDT, applied at PDT treatment of NIH/3T3 normal cells and B16F10 melanoma cell lines caused 85% cell death for melanoma cells, while maintaining high viability in normal cells and tumor volume reduction of 82% in tumor-bearing C57BL/6 mice was observed following 5-fold treatment with PDT using acai oil in NE (Monge-Fuentes et al. 2017). Magneto low-density NE (MLDNE) that can carry maghemite NPs and chlorin e6 (phytochlorin) as an active photosensitizer drug was designed as a potential vehicle for combined hyperthermia and PDT to treat cancer, because it is selectively taken up by MCF-7 cancer cell surfaces with receptor recognition based on the overexpression of low-density lipoprotein receptor. MLDNE showing particle size <200 nm and high drug encapsulation efficiency was found to enhance tumor damage after minor heat dissipation and/or minimum visible light photosensitization doses by classical magnetic hyperthermia and PDT resulting in notable synergic action on MCF-7 cells reflected in reduced cell viability (Pellosi et al. 2018). de Matos et al. (2018) studied the effects of PDT on cellular viability using CUR NE as a photosensitizing drug in cervical carcinoma cell lines and found that NEs were internalized inside cells and were observed in the intracellular environment for up to 36 h after incubation with cell lines, whereby after the PDT, high phototoxic effect of CUR NE (<5% of viable cells after irradiation) was estimated suggesting that CUR NEs have potential as an alternative treatment to cervical lesions using an endoscopic diode fiber laser setup for in situ activation or cavity activation applying a diffuse fiber delivery system.

2.4.3 Nanoemulsions for Pulmonary Drug Delivery

Optimized quercetin-loaded O/W NE prepared using palm oil ester/ricinoleic acid as oil phase (droplet size of 131.4 nm, PDI 0.257, zeta potential 51.1 mV) showing good stability against phase separation and storage at 4 °C for 3 months induced cytotoxicity toward A549 lung cancer cells without affecting the normal cells, suggesting that NE could be used as a potential carrier system for pulmonary delivery of molecules with low water solubility (Arbain et al. 2018).

Minz and Pandey (2018) designed recombinant hepatitis B surface antigen loaded solid fat NEs as carrier system and monophosphoryl lipid A as an adjuvant-carrier system and evaluated it as multiadjuvanted vaccine system for deep pulmonary vaccination. The observed humoral (sIgA and IgG) and cellular (IL-2 and IF-γ) immune responses considerably exceeded those estimated with naive antigen (recombinant surface antigen without any excipient) solution.

Nasr et al. (2012) used commercially available lipid NE, the Intralipid® or Clinoleic®, to prepare amphotericin B lipid NE aerosols for targeting peripheral respiratory airways via nebulization.

Tea tree oil NEs prepared using Cremophor EL (average size of 12.5 nm) exhibited excellent antimicrobial activities on *Escherichia coli*, *Acinetobacter baumannii*, *Klebsiella pneumoniae*, *S. aureus*, and *Candida albicans* and after inhalation to the lung showed higher anti-fungal effect than fluconazole on the fungal pneumonia rat models, causing reduced lung injury, highly microbial clearance, blocking of leukocyte recruitment, and decrease of pro-inflammatory mediators. Such inhalable formulation could be used in local therapies of fungal and bacterial pneumonia with no obvious adverse events (Li et al. 2016a).

Shah et al. (2017) tested rifampicin-oleic acid first-generation NE and its respective CS and CS-folate conjugate-decorated second- and third-generation NEs showing mean droplet sizes of 40–60 nm related to aerosolization, pulmonary inhalation, intracellular trafficking potential in macrophages, and pharmacokinetics profiles. Chitosan oligosaccharide lactate was employed as mucoadhesive, bioadhesive, and targeting moiety for macrophages, while folic acid was used as targeting ligand of macrophage. It was found that tested NEs exhibited >95% aerosol output and inhalation efficiency >75%, and the size and surface tension of NEs affected the aerosol output and aerosolized and inhaled fine particle fractions in an inverse relationship. Higher cell internalization potential, reduced plasma drug concentration, and higher lung drug content showed third-generation NEs. NE containing highly refined soybean oil together with cetylpyridinium chloride, Tween® 80, and ethanol in water produced by high-speed emulsification showing droplets with mean diameter of 450 nm and safe mucosal adjuvants, when delivered intranasally along with *Mycobacterium tuberculosis*-specific immunodominant antigens, induced potent mucosal IL-17 T-cell responses and conferred protection upon *M. tuberculosis* challenge in mice. Moreover, when such NE tuberculosis vaccine is delivered along with *Mycobacterium bovis* bacillus Calmette-Guerin licensed vaccine, decreased disease severity could be observed (Ahmed et al. 2017).

2.4.4 *Nanoemulsions for Ocular Drug Delivery*

Nanotechnology-based systems for treating and managing various ocular diseases include more accessible formulations for deeper segments of the eyes, which enable the availability of drugs at required site in a required amount without inversely affecting the eye tissues (Lalu et al. 2017). Findings related to topical application of lipid-based systems (MEs, NEs, liposomes, and solid lipid NPs), polymeric systems (hydrogels, contact lenses, polymeric NPs, and dendrimers), and used physical methods (iontophoresis and sonophoresis) were reviewed by Souza et al. (2014). The use of nanocarriers, including NEs for gene therapy of retinal diseases, was overviewed by Al-Halafi (2014).

Coumarin-rich extract from *Pterocaulon balansae* incorporated into NEs composed of medium-chain triglycerides (MCT), and egg lecithin with droplet sizes >300 nm caused a 95% reduction of trophozoite viability after 24 h of incubation with a NE containing 1.25 mg/mL of coumarins, whereby the effect was comparable with that of chlorhexidine suggesting potential of such NEs to be used for the local treatment of ocular keratitis caused by *Acanthamoeba* (Panatieri et al. 2017).

Castor oil and mineral oil NEs with mean particle size of 234 nm, PDI >0.16, zeta potential of −8.562 ± 3.49 mV, slightly acid pH, and viscosity of about 1.2 cP, which were instilled on the surface of a commercial contact lens, were found to be situated inside and on the surface of the lenses and their transparency remained near 100% suggesting their suitability for ocular application, because the contact lens remained transparent and ion-permeable after association with this formulation (Katzer et al. 2014).

The introduction of a positive charged mucoadhesive polymer, CS, into NEs designed for ocular administration of timolol containing the drug as maleate or as ion-pair with bis(2-ethylhexyl) sulfosuccinate increased drug permeation probably due to the interaction of CS with corneal epithelial cells (Gallarate et al. 2013).

A combination of NE with 2% hydroxyethyl β-cyclodextrin resulted in 9.2-fold higher lutein accumulation (119 ± 6 μg/g/h) compared to lutein suspension alone, and the scleral accumulation of lutein increased by increasing the cyclodextrin content, whereby the nanoformulation showed low cytotoxicity in retinal cells (Liu et al. 2015a). Lutein-loaded NE consisting of isopropyl myristate, triacetin, Tween® 80, and ethanol with particle sizes approx. 10–12 nm showed sustained and pronouncedly increased lutein release. Such NE could be considered as a potential alternative delivery system for lutein, an effective drug used in therapy of macular degeneration (Lim et al. 2016).

Dexamethasone acetate (DEX) and polymyxin B sulfate were formulated as a cationic NE, in which soy phosphatidylcholine represented 30% of the lipid phase containing DEX, while polymyxin B sulfate was dissolved in the water phase, and this NE could be considered as a viable alternative to the commercial ophthalmic suspensions designed for the treatment of ophthalmic infections (Li et al. 2016b). Triamcinolone acetonide-loaded NE consisting of MCT oil, soybean lecithin, and Poloxamer 188 released the drug in a fast rate and could provide an adequate drug release rate on the eye surface to be used in topical ophthalmic administration (Silva and Lemos-Senna 2016).

Alternative controlled-release ocular terbinafine hydrochloride-loaded NE in situ gels prepared by NE dispersion in gellan gum solution (0.2%, w/w) were transparent, pseudoplastic, mucoadhesive, showed more retarded zero-order drug release rates and increased bioavailability (Tayel et al. 2013).

Acetazolamide-loaded NEs fabricated using peanut oil, Tween® 80, and/or cremophor EL as surfactant and Transcutol® P or propylene glycol as cosurfactant were incorporated into ion induced in situ gelling systems composed of gellan gum alone and in combination with xanthan gum, hydroxypropyl methylcellulose (HPMC), or Carbopol®. The gels showed a considerably sustained drug release in comparison to the NEs, whereby gellan/xanthan and gellan/HPMC exhibited improved therapeutic efficacy and more prolonged intraocular pressure lowering effect relative to that of commercial eye drops and oral tablet (Morsi et al. 2017).

Cyclosporine A (Cy-A) NEs containing oleic acid, Tween® 20, and Transcutol® P and coated with CS were evaluated as nonirritant and nontoxic, being well tolerated by rabbit eye much more sensitive than human. In corneal penetration studies for low molecular weight CS coatings, 59.2% and 57.5% of drug penetration were estimated, while for medium high molecular weight CS coatings, it was 55.5% and 52.3%, respectively (Shen et al. 2018). Optimized Cy-A loaded polymeric mucoadhesive NE with higher Cy-A payload, in which the concentration of CS was adjusted according to the blinking force of eyelids, was able to maintain the therapeutic concentrations (≥50–300 ng/g) of drug in the rabbit cornea and conjunctiva over the period of 24 h (Akhter et al. 2016).

Celecoxib NE containing 5% oleic acid-Transcutol® P (10:1) in oil phase with droplet sizes ranging from 6.96 to 26.65 nm and pH 6.5–6.9, respectively, showed 82.6% of the drug released in the 24 h of the experiment. NE containing 85% of surfactant (Tween® 80, Span® 20) and cosurfactant (propylene glycol) in a ratio 2:1 showed that 15.73% drug permeated through rabbit cornea and the flux (J_{ss}) of celecoxib from this NE (0.65 mg cm^{-2}/h) was 22-fold higher than this of control (Moghimipour et al. 2017).

2.5 Nanoemulsions Applicable in Food Industry

There are various nanocomposites suitable for the protection of different foodstuffs, smart active or responsive packaging materials, edible coatings, as well as diverse nanosensors applicable for monitoring of food quality, safety, and integrity, and the application of nanocomposites in food packaging systems could diminish the harshness of food processing and the amount of additives and chemical preservatives in food industries (e.g., Jampílek and Kráľová 2018b). Nanoemulsion technology is particularly appropriate for the preparation of encapsulating systems for functional compounds, because it prevents their degradation and improves their bioavailability. An overview of food-grade NEs, their recent applications in the food systems, and patent review of emulsions was presented by Yalcinoz and Ercelebi (2018). Promising advantages reached with the use of NEs as delivery systems of flavoring and preservative agents in the food industry were summarized by Salvia-Trujillo et al. (2015). Adjonu et al. (2014) overviewed emulsifiers/surfactant (ionic, nonionic, phospholipid, polysaccharide, and protein) used in NEs designed for food applications as well as suitability of proteins and protein hydrolysates as nanoemulsifiers and analyzed the potential of whey protein-derived peptides as both emulsifiers and bioactive compounds in NE delivery systems. The influence of particle characteristics such as size and interfacial properties on potential biological fate (digestion and absorption) of ingested NEs was described by McClements and Xiao (2012). The influence of droplet characteristics on the physicochemical and sensory properties of beverage emulsions, with special focus on their influence on product stability as well as developments in the soft drinks area, including fortification with vitamins, reduced calorie beverages, and "all-natural" products were overviewed by Piorkowski and McClements (2014). Although to the potential advantages of food-grade NEs over conventional emulsions for applications in the food industry belong higher stability to particle aggregation and gravitational separation, higher optical transparency and increased bioavailability of encapsulated components, potential risks associated with consumption of lipid nanoparticles that are able to alter the fate of bioactive components within the gastrointestinal tract, and possible toxicity of surfactants and organic solvents could not be marginalized (McClements 2013).

2.5.1 *Nanoemulsions Formulated with Natural Emulsifiers, Biosurfactants, and Biopolymers*

Mungure et al. (2018) overviewed the potential application of pectin, an anionic polysaccharide, for the stabilization of NE systems. Synthesis and applications of pectin-based nanomaterials that combine the advantages of both pectin and the nanoscale particles was overviewed by Zhao and Zhou (2016). Bai et al. (2016) compared the effectiveness of a number natural emulsifiers (whey protein, gum arabic, *Quillaja* saponin, and soy lecithin) at fabricating corn O/W NEs using dual-channel microfluidization and found that whey protein isolate and *Quillaja* saponin more effectively formed NEs with fine droplets than gum arabic and soy lecithin (lower amount of emulsifier needed, fabrication of smaller droplets), which could be connected with faster emulsifier adsorption and a greater reduction in interfacial tension resulting in more effective droplet disruption within the homogenizer for saponins and whey proteins. McClements and Gumus (2016) analyzed the possibility of the replacement of synthetic surfactants with natural emulsifiers, such as amphiphilic proteins, polysaccharides, biosurfactants, phospholipids, and bioparticles, discussed the physicochemical basis of emulsion formation and stabilization by natural emulsifiers, and compared the benefits and limitations of different natural emulsifiers.

Konjac glucomannan, a renewable natural polysaccharide occurring in the tuber *Amorphophallus konjac* K. Koch, composed of β-1,4-linked D-mannose and D-glucose in either a 1.6:1 or 1.4:1 molar ratio, exhibit good swelling, gelling, or emulsifying properties. Its derivative, konjac glucomannan octenyl succinate (KGOS), was found to be a good potential emulsifier and stabilizer for encapsulation of lipophilic bioactive compounds such as β-carotene. Emulsification yield of the KGOSC nanoemulsion containing 0.03% β-carotene, 0.3% KGOS, and 10% MCT exceeded 95%, and after 30 days of storage, the particle size and PDI of the KGOSC NE did not reach 5 nm and 0.5, respectively, whereby the sensitivity of KGOSC NEs to storage conditions decreased in following order: temperature > oxygen > light (Li et al. 2018).

W/O NEs of Sacha Inchi oil (SIO), containing a very high content of the ω-3 fatty acid, α-linolenic acid (approx. 50%), and tocopherols (176–226 mg/100 g) prepared using olive leaf phenolics (OLP), showed droplet size 2.15 ± 0.13 nm and were stable without phase separation during 30-day storage, whereby the release of OLP controlled the oxidative progress of SIO by prolonging the induction time, preventing the production of primary and secondary oxidative products, and also the deterioration of polyunsaturated fatty acids and tocopherols (Liu et al. 2018). Among O/W NEs prepared with 10% avocado oil using natural (lecithin) and synthetic (Tween® 80) emulsifiers at different concentrations (2.5–10%) with zeta potential ranging from −26 to −59 mV and lipid droplets of 103–249 nm, Tween® 80 was found to be more effective than lecithin (Arancibia et al. 2017). Multilayer emulsions produced by a high-pressure homogenization method using an electrostatic layer-by-layer deposition process of lecithin-CS membranes encapsulating linseed oil and α-lipoic acid simultaneously were found to be an effective delivery system to incorporate them into functional foods and beverages. CS encapsulation inhibited the degradation of α-lipoic acid and improved the oxidation stability of linseed oil in multilayer emulsions that showed good centrifugal, dilution, and storage stabilities (Huang et al. 2018).

Using MCT oil, the rhamnolipids biosurfactants can effectively form small droplets ($d_{32} < 0.15$ μm) at low surfactant-to-oil ratios (<1:10). Rhamnolipid-coated droplets were found to be stable to aggregation at pH values 5–9, salt concentrations <100 mM NaCl, and 20–90 °C. Droplet aggregation at pH 2–4 and high ionic strength 200–500 mM NaCl could be connected with a reduction in electrostatic repulsion at low pH and high salt levels. Rhamnolipids could be considered as effective natural surfactants suitable to replace synthetic surfactants in certain commercial applications (Bai and McClements 2016).

Quillaja saponin (Q-Naturale®), a natural food-grade surfactant isolated from the bark of the *Quillaja saponaria* Molina tree, containing saponin-based amphiphilic molecules, was reported to be able to replace synthetic surfactants in food and beverage products (Yang et al. 2013). A promising NE delivery system for oregano as an essential oil model with long-term stability using a sugar-based biosurfactant (*Quillaja* saponin) produced by microfluidization was found to be suitable to be used in antimicrobial as well as flavoring and potential antioxidant applications in food and beverages or in pharmaceutical and cosmetic products (Doost et al. 2018).

The study of high-pressure homogenized whey protein emulsions prepared by mixing whey protein concentrate (10%), soybean oil (0% or 5%), and soy lecithin (0% or 5%) showed that increasing homogenization pressures significantly decreased particle sizes of all samples but increased electrical conductivity of samples. After 25 MPa treatment emulsions with lecithin had smaller particle sizes than those made without lecithin, but at 50 MPa, further reduction of particle sizes in the presence of lecithin was not observed (Yan et al. 2017). Yerramilli et al. (2017) studied partially replacement of sodium caseinate (SCas) in the formation and long-term stabilization of 5 wt% O/W NEs prepared using a high-pressure homogenization by pea protein isolate (PPI) and found that NEs stabilized by 1:1 mixture of SCas and PPI did not display any creaming or aggregation and remained stable for more than 6 months (droplets <200 nm), which could be connected with the interaction of pea proteins disrupted by high-pressure homogenization with SCas in the continuous phase of the NEs. Consequently, the plant proteins have potential to be applicable in the long-term stabilization of NEs in the food and beverage industry. O/W NEs stabilized by high-pressure homogenized lentil proteins isolate (LPI) showed pronouncedly higher in vitro lipid digestibility than unmodified LPI NEs, which could be connected with the higher interfacial area of smaller droplets and weaker interfacial moduli of modified LPI-stabilized interfaces compared to those with unmodified LPI. It could be noted that high-pressure homogenization notably decreased LPI particle size distribution, surface hydrophobicity, and the interfacial storage moduli relative to the unmodified LPI (Primozic et al. 2018).

As an attractive biosurfactant suitable for preparation of NE formulations applicable in food and beverage products, saponin extracted of Brazilian ginseng roots was reported (Rosa et al. 2016).

2.5.2 Nanoemulsions of Essential Oils and Their Constituents

Excellent antimicrobial efficacy of plant EOs predestines them to be used as an alternative of health hazardous synthetic preservatives in food products, and encapsulation of EOs in NEs, MEs, solid-lipid NPs, and liposomes improves their shelf life already at low doses (Prakash et al. 2018a). Donsi and Ferrari (2016) in a review paper focused their attention on EO NEs as antimicrobial agents in food; critically analyzed the reported antimicrobial activity data, both in vitro and in products; and discussed the regulatory issues associated with their use in food systems. An overview related to the different systems for the encapsulation of bioactive oils and the currently applied elaboration methods was presented by Rodriguez et al. (2016).

Although EOs in the food industry are mainly used as flavoring agents, O/W NE of betel (*Piper betle* L.) leaf EO showed antimicrobial activity against five strains of Gram-positive and Gram-negative bacteria with MIC of 0.5–1.25 μL/mL and MBC of 1–2.5 μL/mL suggesting that it can serve as natural antimicrobial agent for food system (Roy and Guha 2018). Mexican oregano (*Lippia graveolens* Kunth) EO containing thymol and γ-terpinene, which was incorporated into active coatings and spread on fresh pork meat as free, nanoemulsified, and microencapsulated EO at a dose of 2.85 mg EO/cm^2, caused growth inhibition of *Lactobacillus plantarum* (5 log population reduction) and *Pseudomonas fragi* (4 log reduction), while ≤1.5 log population reduction was observed for *Brochothrix thermosphacta* and *Salmonella Infantis*, whereby meat microbiota was most efficiently controlled by microencapsulated EO resulting in delayed lipid and oxymyoglobin oxidation of fresh pork meat (Hernandez-Hernandez et al. 2017). Anise oil NE more effectively reduced the population of *E. coli* O157:H7 and *Listeria monocytogenes* (count by 2.51 and 1.64 log cfu/mL, respectively) than bulk anise oil (by 1.48 and 0.47 log cfu/mL) after 6 h of contact time (Topuz et al. 2016). The EO-based NEs could improve the microbial quality of minimally processed fruits and vegetables (Prakash et al. 2018b). Basil oil (*Ocimum basilicum*) NE fabricated by ultrasonic emulsification with droplet diameter 29.3 nm showed antibacterial activity against *E. coli* causing alteration in permeability and surface features of bacterial cell membrane (Ghosh et al. 2013). Long-term stability over 21 days of storage was estimated with *Zataria multiflora* Boiss EO NE (droplet size of 200 nm and PDI 0.2) fabricated by emulsion phase inversion that showed MIC values of 2500 and 5000 μg/mL against *L. monocytogenes* and *Salmonella typhimurium*, respectively, and this NE was more effective in inhibiting the growth of bacteria in milk than in culture media (Shahabi et al. 2017).

Optimized clove oil NEs prepared using SCas (5%) and pectin (0.1%) as coating material by high-speed homogenization with spherical NPs of 172.1 ± 4.39 nm, zeta potential of −37 ± 1.93 mV, and EE of 88% showing stability at all food processing conditions except pH 3.0–5.0 could be recommended as delivery system for antimicrobial bioactive substances in food preservation (Sharma et al. 2017). Optimized ultrasound-mediated nettle oil (1.25 wt%) NEs stabilized by purified jujube polysaccharide with 86.75 nm droplet size inactivated the Gram-positive bacterium more effectively than the Gram-negative one (Gharibzahedi 2017). Stable NEs prepared by mixing eugenol with SCas using shear homogenization observed up to 38.5 mg/mL eugenol showed droplet diameters <125 nm at pH 5–9 after ambient storage for up to 30 days and more effective inhibition of *E. coli* O157:H7 than free eugenol during incubation at 37 °C for 48 h. A greater reduction of intracellular ATP and a greater increase of extracellular ATP were observed in bacteria treated with encapsulated eugenol (20-min interaction at 21 °C) compared to free eugenol suggesting enhanced permeation of eugenol due to its nanoencapsulation, although the possible membrane adaptation could not be excluded (Zhang et al. 2018). Salvia-Trujillo et al. (2014) reported that lemongrass oil-alginate NEs prepared by microfluidization exhibited enhanced antimicrobial activity against *E. coli*, while ultrasound processing of NEs led to loss of their bactericidal action.

Thyme O/W NEs (pH 3.5) were found to be highly unstable to droplet growth and phase separation, which could be connected with Ostwald ripening due to the relatively high water solubility of thyme oil. Inhibition of Ostwald ripening could be achieved by mixing thyme oil with a water-insoluble ripening inhibitor (≥60 wt% corn oil or ≥50 wt% MCT in the lipid phase) before homogenization, yielding NEs with good physical stability. However, ripening inhibitor type and concentration had a notable impact on the antimicrobial activity of EO and increasing the ripening inhibitor levels in the lipid phase resulted in the reduction of the antimicrobial efficacy of NEs (Chang et al. 2012).

Comparison of the in vivo oral bioavailability of conventional emulsion (droplet diameter 1.285 μm) and NEs of vitamin E prepared using sunflower oil and saponin (droplet diameter 0.277 nm) showed that the in vivo oral bioavailability of vitamin E in male Wistar rats at NE application reached a 3-fold higher AUC compared to the conventional emulsion (Parthasarathi et al. 2016). Bovi et al. (2017) reported that NEs of Buriti (*Mauritia flexuosa* L.) oil, one of the richest vegetal sources of carotenoids, could be successfully incorporated in isotonic sports drink and replace the artificial coloring by natural dyes.

In O/W NEs stabilized by soy lecithin and encapsulating carvacrol, the zwitterionic lecithin molecules adsorbed to the O/W interface for 24 h formed a notably viscoelastic layer; at pH 7 the NEs were highly stable, yielding monodispersed droplet size distributions and high resistance to increases in droplet size over 30 days, although the initial size of oil droplets slightly depended on pH (smaller droplets at pH 7 and larger droplets at pH 3) (Nash and Erk 2017).

Eugenol NEs prepared by using ultrasonication as emulsification techniques, CS NPs as carrier, and Tween® 20 surfactant as emulsifier, with regularly spherical shape and sizes ranging from 80 to 100 nm, were characterized with great storage and thermal stability and exhibited superb antioxidant capacity and antimicrobial activity suggesting their great potential to be used in food formulations for extending the shelf life (Shao et al. 2018).

O/W chia seed oil NE systems prepared by spontaneous emulsification and microfluidization and stabilized with Tween® 80 and Span® 80, as well as SCas- and sucrose monopalmitate-stabilized NEs fabricated by microfluidization, were characterized by storage stability at 4 °C during 2 weeks. The NEs prepared with sucrose monopalmitate showed best transparency with droplet diameter ca. 47 nm (Teng et al. 2018). Blended cloves/cinnamon EO NEs fabricated using Tween® 80 surfactant and ethanol (cosurfactant) with oil at the mixed surfactant/cosurfactant ratio of 1:9 showed higher antimicrobial activity against *E. coli*, *B. subtilis*, *S. typhimurium*, and *S. aureus*, even at far lower concentrations than free EOs, and this NE could be applied as a natural antimicrobial agent in food industry (Zhang et al. 2017).

Digested *Hibiscus cannabinus* L. seed O/W NEs stabilized by SCas, Tween® 20, and β-cyclodextrin complexes produced using high-pressure homogenization showed good lipid digestion (85.25%), good bioaccessibility of antioxidants (tocopherols and total phenolic contents), and lower degradation rate of phytosterols compared to digested bulk oil suggesting potential use of such formulation in food and nutraceutical preparations (Cheong et al. 2016). Addition of β-cyclodextrin to primary emulsion containing SCas and Tween® 20 was found to improve the physical stability of kenaf (*Hibiscus cannabinus* L.) seed O/W NEs (Cheong and Nyam 2016).

Negatively charged NEs of black cumin EO fabricated using different ratios of EO with canola and flax seed oils (ripening inhibitors) that were stabilized with octenyl succinic anhydride (OSA)-modified waxy maize starch (mean droplet diameter >200 nm and zeta potential above −30 mV) showed prolonged bactericidal activities against *Bacillus cereus* and *L. monocytogenes* than the pure black cumin EO, which could be explained with its better stability, controlled release, and self-assembly with the cell membrane of Gram-positive bacteria resulting finally in destruction of cellular constituents (Sharif et al. 2017a).

Alexandre et al. (2016) reported about activated films with improved physical properties prepared by incorporation of montmorillonite and nanoemulsioned ginger EO into gelatin-based films with notably better elongation at break, puncture force, and puncture deformation and showing antioxidant activity, which could be used in food packaging applications.

2.5.3 Vitamin Nanoemulsions

The use of natural surfactants, *Quillaja* saponin and lecithin, on the formation and stabilization of NE-based vitamin E delivery systems was described by Ozturk et al. (2014). At simulated small intestine conditions, the rate of lipid digestion and tocotrienol bioaccessibility in bulk oil and within O/W conventional emulsions (>10 μm) and NEs (<350 nm) fabricated using MCT as an oil phase (5–40% wt) and *Quillaja* saponins as a natural surfactant decreased as follows: NEs > emulsions > bulk oil (Xu et al. 2018). The increase of the bioavailability of α-tocopherol as a food supplement could be secured through edible (coconut) oil NE showing a 9.5 mg/mL of encapsulation capacity and almost 100% release of the loaded active ingredient within 24 h, whereby the contribution of kinetic-controlled release was found approx. 70% and that of diffusion-controlled release was found approx. 30%. Besides good stability of NE, a reasonable cell viability (biocompatibility) with apposite antimicrobial activity was estimated suggesting potential application of such edible oil NE in food, beverages, and healthcare industries (Saxena et al. 2018). The α-tocopherol-based O/W NE fabricated using sodium stearoyl lactate and Tween® 80 surfactants by high-energy ultrasonication method with encapsulated benzyl isothiocyanate acted as better antioxidant compared to pure and CUR encapsulated NE; the prepared emulsions exhibited good stability up to 90 days in salt solution (50–200 mM) and different pH conditions, and the degradation of CUR by UV light was successfully controlled by trapping in NE (Kaur et al. 2017).

The impact of antioxidants on the thermal stability of β-carotene encapsulated in diluted O/W emulsions prepared using gum arabic decreased as follows: α-tocopherol > tertiary butyl hydroquinone > ascorbyl palmitate, the most effective antioxidant being α-tocopherol at the concentration of 0.10 wt% under light exposure (Liu et al. 2015b). Digested *Ulva fasciata* polysaccharide-stabilized emulsions encapsulating β-carotene (mean particle size of 0.82 μm) showed higher lipid digestion rate and increased β-carotene bioaccessibility than gum arabic- or beet pectin-stabilized emulsions suggesting that they could be a promising delivery system for β-carotene in functional food and beverage system. Moreover, the stability of β-carotene in the emulsion could be considerably increased by addition of α-tocopherol (Shao et al. 2017). NEs, in which fat-soluble vitamins β-carotene and α-tocopherol were co-encapsulated using flax seed oil and octenyl succinic anhydride modified starch as emulsifier with incorporated eugenol, showed overall higher retention of β-carotene (ca. 42%) and α-tocopherol (ca. 90%) after 4 weeks of storage at 40 °C as compared to NEs in which Tween® 80 was used as emulsifier, which could be connected with the fact that the modified starch emulsifier can form a thicker protective layer around oil droplets (Sharif et al. 2017b).

Increased bioavailability of vitamin D encapsulated in O/W NE prepared by ultrasonication technique showing a droplet size 300–450 nm and shelf life >90 days in simulated gastrointestinal tract was reported by Walia et al. (2017). The ergocalciferol in vitro bioaccessibilities in O/W NEs prepared using emulsifiers with different stabilizing mechanisms was comparable for decaglycerol monooleate (62%; steric mechanism), modified lecithin (64%; electrostatic mechanism), or their combination (65%), but they were significantly higher than that estimated for SCas emulsifier (12%; electrosteric), and also the free fatty acid release rate in the small intestinal phase was the lowest for SCas (Shu et al. 2018).

Folic acid nanoencapsulated by double emulsions having an internal NE composed of W/O system with folic acid present in the water phase and reemulsified within an aqueous phase of pectin-whey protein concentrate complexes showed EE about 88.3%, whereby EE was predominantly affected by the dispersed phase content of double NE and surfactant had the minimum influence (Assadpour et al. 2016).

2.5.4 Nanoemulsions Encapsulating Fatty Acids

Uluata et al. (2015) studied physical stability, autoxidation, and photosensitized oxidation of ω-3 oils in NEs prepared with natural (lecithin and *Quillaja* saponin) and synthetic (Tween® 80 and sodium dodecyl sulfate) surfactants and found that lipid hydroperoxide decreased in the order Tween® 80 > SDS > lecithin > *Quillaja* saponin, whereby *Quillaja* saponin consistently produced the most oxidatively stable emulsions, which could be due to its high free radical scavenging capacity.

NEs encapsulating fish oil prepared using MCT, lemon oil, and thyme oil as carrier oils containing 75% fish oil and 25% carrier oil were physically stable for 42 days at 20 °C, whereby the rate of lipid oxidation in NEs decreased in the following order: MCT >> lemon oil > thyme oil, which could be connected with the presence of high levels of natural antioxidants (phenolics) within both essential oils (Walker et al. 2017). The oxidative stability of fish oil encapsulated in multiple NE with particle sizes 190–210 nm prepared using whey protein concentrate was enhanced, whereby key factors affecting the droplet size of NE and oxidative stability of fish oil were found to be whey protein concentrate concentration level and used antioxidant type (vitamin C and E) (Hwang et al. 2017).

Esquerdo et al. (2018) designed food-grade NEs containing unsaturated fatty acids (UFA) concentrates from carp oil, using CS and gelatin as wall materials, in which these biopolymers provided high stability to the formulations and also behaved as good wall materials, whereby at 90:10 of CS:gelatin ratio the NE was in the acceptable range of the legislation after 7 days of storage, suggesting the increase of the physical and oxidative stability of UFA and such NEs could facilitate the addition of these lipophilic active ingredients in aqueous-based foods or beverages. Application of NE in combination with vacuum packing was found to maintain the polyunsaturated fatty acids content of sea bass (*Dicentrarchus labrax*) fillets stored at 22 °C, since NEs with hazelnut, canola, and soybean oils can be used as a preservative for fish and such NEs together with vacuum packing could prevent the lipid oxidation (Ozogul et al. 2017).

2.5.5 Nanoemulsions Encapsulating Bioactive Compounds

Gelatin-based films incorporated with rutin-loaded O/W NE displayed higher tensile strength and higher elongation at break than the gelatin control film, showed high antioxidant activities, rutin release being mainly governed by Fickian diffusion with simultaneous interfering swelling and disintegration phenomena, and they could be considered as potential active packaging systems to enhance shelf life of food products (Dammak et al. 2017).

Quercetin-loaded NEs fabricated using high-pressure homogenization method with mean droplet size of 152 ± 6 nm, zeta potential of −50 ± 2 mV, and entrapment efficiency of 93.50 ± 0.35% exhibited comparable antioxidant activity to free quercetin, and also the bioaccessibility of quercetin in simulated small intestinal conditions was found to be improved by nanoencapsulation (Ni et al. 2017). Optimized green tea catechins NEs prepared by high-pressure homogenization with droplet diameter of 280 ± 1 nm and 83.16 ± 1.12% EE subjected to different environmental stress (pH, temperature, and salt concentration) were stable for 8 weeks and showed slow and sustained release of polyphenols from lipid matrix in mimicked gastric conditions (Gadkari et al. 2017).

Milk fortified with CUR NEs prepared using high-pressure homogenization exhibited pronouncedly lower lipid oxidation than unfortified (control) milk and milk containing CUR-free NEs suggesting that such CUR NE could be utilized in beverage industry (Joung et al. 2016). Active films based on gelatin-SCas blend containing active compounds (α-tocopherol, garlic essential oil, and cinnamaldehyde) nanoemulsified in water showed good antioxidant activity suggesting their potential to be used as active packaging for shelf life extension of foodstuffs (Cordoba and Sobral 2017). Physically stable NEs enriched with the carotenoid astaxanthin (droplet diameter of 230 nm; zeta potential of −40 mV) prepared using caseinate as emulsifier were stable in the temperature range 5–70 °C, and a wide range of pH (except at pH 4 and 5) and ionic strength could be used in functional foods and beverages (Liu et al. 2016).

Lecithin-containing emulsions of capsaicin (mean particle size 582.63 nm) showed high antimicrobial activity against *S. aureus* with 4.60 log reduction, while with the capsaicin-encapsulated NE prepared using Tween® 80 as surfactant (mean particle size 68.30 nm), a 3.86 log reduction against *E. coli* was estimated (Akbas et al. 2017).

Quercetin trapped saponin stabilized NEs with mean particle size of 52 ± 10 nm exhibited higher stability on exposure to UV light as compared to water/ethanol system showing the degradation rate 9 ± 1% (at pH 7) and 11 ± 1% (at pH 8.0), respectively as compared to 42 ± 2% in water/ethanol system (Kaur et al. 2016).

The thermal resistance of *L. monocytogenes* was reduced from 2- to 5-fold, when 0.5 mM D-limonene was added directly to the heating medium. However, the presence of the same D-limonene concentration in the heating medium in the form of NE reduced heat resistance of *L. monocytogenes* by one hundred times compared to 2–5 reduction at application of free D-limonene suggesting that the addition of nanoemulsified antimicrobials can pronouncedly reduce the intensity of the thermal treatments currently applied in the food processing industry (Mate et al. 2016).

Evaluation of CUR NEs prepared using MCT, canola oil, or linseed oil as oil phases and stabilized by different emulsifiers (Tween® 80, lecithin, whey protein isolate, and acacia) with high-pressure homogenization showed that the increase in oil phase concentration resulted in increased CUR content, particle size, and viscosity of NE but decreased the stability that was pronouncedly affected if the stabilizing agent was lecithin; the maximum CUR content in NE was obtained using MCT as oil phase (Ma et al. 2017). Citral NE with mean particle size 467.83 nm produced using a mixture of gelatin and Tween® 20 as emulsifiers in a ratio 3:1 (total emulsifier concentration of the emulsion system was 10 g/kg) remained stable during storage for 14 days at 30 °C, and under acidic conditions, it was able to protect citral from degradation and decreased the formation of off-flavor compounds (e.g., *p*-cymene, *p*-cresol, and *p*-methylacetophenone) relative to a single emulsifier (Tian et al. 2017).

NEs containing *trans*-cinnamaldehyde as an active agent and 1,8-cineol as the Ostwald ripening inhibitor containing Tween® 80 as emulsifier with surfactant to oil ratio of 2:1 (w/w) that were prepared using ultrasonic technology exhibited notable stability for 6 months with considerably small particle size of 27.76 ± 0.37 nm and superb antibacterial activity against *E. coli*, *S. aureus*, and *Pseudomonas aeruginosa*, and treatment with optimized NE caused dramatic increase of *E. coli*'s membrane fluidity (Moghimi et al. 2017).

Physical stabilities of droplets (344, 173, and 98 nm in diameter) of fucoxanthin NEs prepared using high-pressure microfluidizer and containing as the oil phase the structured lipid that enriched pinolenic acid at sn-2 position decreased with increases in the initial size and storage temperature, while fucoxanthin chemical stability was improved. The reduction of the digestion stability of fucoxanthin NE with decreasing initial particle diameter could be probably attributed to the increased surface area interacting with pancreatic lipase with decreasing droplet size (Huang et al. 2017).

2.5.6 *Nanoemulsions Used in Edible Coatings*

Fresh apples (Golab Kohanz) coated with 0.5% CS NE (<100 nm) pronouncedly reduced weight loss, respiration rate, ethylene production, and peroxidase activity of the samples compared with the control, greatly affected polyphenol oxidase activity, slowed down softening process, maintained the quality of apples, and also improved the flesh color after the climacteric peak (Gardesh et al. 2016). The combination of an antimicrobial edible coating on green beans consisting of modified CS containing a NE of mandarin EO with high hydrostatic pressure notably reduced *Listeria innocua* inoculated on green bean, while it also had a strong impact on green bean firmness during 14-day refrigerated storage at 4 °C (Donsi et al. 2015). The combined use of nanoemulsified lemon EO with modified CS resulted in the remarkable increase in antimicrobial activity, with respect to other EOs, whereby incorporation of nanoencapsulated lemon EO into the modified CS coating prolonged the shelf life of rucola leaves from 3 to 7 d, being more effective than a coating made of modified CS or EO alone (Sessa et al. 2015). Coating of silvery pomfret with citrus EO NEs based on CS NPs loaded EO ensured better preservation effects than coatings with conventional emulsion due to more efficient prevention against microorganisms and lipid oxidation (Wu et al. 2016).

Oleic acid NE as a part of starch-based edible coating suspensions, incorporated with a mixture of three natural antimicrobials, lactic acid, nisin, and lauric arginate, may be used as coating to extend the shelf life of fresh foods (Sanchez-Ortega et al. 2016).

Grape berry (*Vitis labruscana* Bailey) coatings of lemongrass oil incorporating NE exhibited antimicrobial effects against *S. typhimurium* and *E. coli* 0157:H7 during storage at 4 and 25 °C for 28 days, and they did not considerably alter the flavor of the berries, improved their glossiness, reduced losses of weight, and prolonged their shelf life (Kim et al. 2014b). The lemongrass oil/CS NE coating showed effectiveness in improving microbiological safety and preserving grape berries as well (Oh et al. 2017). A carnauba-shellac wax-based NE containing lemongrass oil prepared using high-pressure homogenization and used for coating of "Fuji" apples significantly improved the quality of apples during storage for 5 months compared to apples without coating (Jo et al. 2014). By mixing a carnauba wax-based solution (18%, w/w) with NE containing lemongrass oil coating for plums was developed, which was able to inhibit *S. typhimurium* and *E. coli* during storage; did not pronouncedly alter the flavor, fracturability, or glossiness of the plums; reduced weight loss and ethylene production; and ensured higher firmness of coated plums compared to uncoated ones (Kim et al. 2013).

At 21 days of storage, polyphenol oxidase activity decreased by 65% in the fresh-cut Red Delicious apples coated with α-tocopherol NE (<200 nm; zeta potential < −40 mV, browning indexes of 43.5) and α-tocopherol NE with nopal mucilage (*Opuntia ficus indica*) (browning indexes of 39.3). As determining parameter in controlling texture and the browning index, the particle size of the NE droplets could be considered and application of nopal mucilage helps control the browning index (Zambrano-Zaragoza et al. 2014).

NE-based edible coatings containing oregano EO and mandarin fiber were found to improve the shelf life of low-fat cut cheese (Artiga-Artigas et al. 2017). Orange peel EO (0.5% and 1.0%) ME and NE used in pectin-based coating to extend the shelf life of fresh-cut orange, maintained at 4 °C for 17 days, notably reduced weight and ascorbic acid losses of coated samples, as well as higher antibacterial and antifungal effects were estimated compared to control without changes in senzory parameters (Radi et al. 2018).

Taghavi et al. (2018) investigated the effects of the microfluidic pressure (600–1200 bar) and cycles (2–4) on the inhibitory activity and physicochemical properties of the NE loaded with a natural antibacterial mixture (i.e., citral, *trans*-2-hexen-1-ol, and linalool, 1:1:1 w/w) and found that in general the physicochemical properties of the antibacterial NE were affected by the cycle to greater extent than by the pressure and the microfluidization condition did not considerably affect the antibacterial activity of the NE. Thus, the O/W emulsions could provide an adequate delivery system for these bioactive compounds.

Acknowledgments J. Jampílek and K. Kráľová thank the National Program of Sustainability I (LO1305) of the Ministry of Education of the Czech Republic and the Slovak Research and Development Agency (projects APVV-17-0373 and APVV-17-0318) for support. E.V.R. Campos and L.F. Fraceto would like to thank São Paulo Research Foundation (FAPESP – grant no. 2017/21004-5).

References

Abbas S, Hayat K, Karangwa E, Bashari M, Zhang XM (2013) An overview of ultrasound-assisted food-grade nanoemulsions. Food Eng Rev 5(3):139–157. https://doi.org/10.1007/s12393-013-9066-3

Abd E, Namjoshi S, Mohammed YH, Roberts MS, Grice JE (2016) Synergistic skin penetration enhancer and nanoemulsion formulations promote the human epidermal permeation of caffeine and naproxen. J Pharm Sci 105(1):212–220. https://doi.org/10.1002/jps.24699

Abd E, Benson HAE, Roberts MS, Grice JE (2018) Minoxidil skin delivery from nanoemulsion formulations containing eucalyptol or oleic acid: enhanced diffusivity and follicular targeting. Pharmaceutics 10(1):19. https://doi.org/10.3390/pharmaceutics10010019

Abd-Elsalam KA, Khokhlov AR (2015) Eugenol oil nanoemulsion: antifungal activity against *Fusarium oxysporum* f. sp. *vasinfectum* and phytotoxicity on cottonseeds. Appl Nanosci 5:255–265. https://doi.org/10.1007/s13204-014-0398-y

Abdou ES, Galhoum GF, Mohamed Entsar N (2018) Curcumin loaded nanoemulsions/pectin coatings for refrigerated chicken fillets. Food Hydrocoll 83:445–453. https://doi.org/10.1016/j.foodhyd.2018.05.026

Abu-Fayyad A, Nazzal S (2017) Gemcitabine-vitamin E conjugates: synthesis, characterization, entrapment into nanoemulsions, and *in-vitro* deamination and antitumor activity. Int J Pharm 528(1–2):463–470. https://doi.org/10.1016/j.ijpharm.2017.06.031

Abu-Fayyad A, Kamal MM, Carroll JL, Dragoi AM, Cody R, Cardelli J, Nazzal S (2018) Development and *in-vitro* characterization of nanoemulsions loaded with paclitaxel/γ-tocotrienol lipid conjugates. Int J Pharm 536(1):146–157. https://doi.org/10.1016/j.ijpharm.2017.11.062

Achari GA, Kowshik M (2018) Recent developments on nanotechnology in agriculture: plant mineral nutrition, health, and interactions with soil microflora. J Agric Food Chem 66(33):8647–8661. https://doi.org/10.1021/acs.jafc.8b00691

Adjonu R, Doran G, Torley P, Agboola S (2014) Whey protein peptides as components of nanoemulsions: a review of emulsifying and biological functionalities. J Food Eng 122:15–27. https://doi.org/10.1016/j.jfoodeng.2013.08.034

Afzal SM, Shareef MZ, Dinesh T, Kishan V (2016a) Folate-PEG-decorated docetaxel lipid nanoemulsion for improved antitumor activity. Nanomedicine 11(16):2171–2184. https://doi.org/10.2217/nnm-2016-0120

Afzal SM, Naidu VD, Harishankar N, Kishan V (2016b) Albumin anchored docetaxel lipid nanoemulsion for improved targeting efficiency – preparation, characterization, cytotoxic, antitumor and *in vivo* imaging studies. Drug Deliv 23(4):1355–1363. https://doi.org/10.3109/10717544.2015.1030715

Afzal SM, Shareef MZ, Kishan V (2016c) Transferrin tagged lipid nanoemulsion of docetaxel for enhanced tumor targeting. J Drug Delivery Sci Technol 36:175–182. https://doi.org/10.1016/j.jddst.2016.10.008

Agarwal V, Bajpai M, Sharma A (2018) Patented and approval scenario of nanopharmaceuticals with relevancy to biomedical application, manufacturing procedure and safety aspects. Recent Pat Drug Deliv Formul 12(1):40–52. https://doi.org/10.2174/1872211312666180105114644

Ahmad G, El Sadda R, Botchkina G, Ojima I, Egan J, Amiji M (2017) Nanoemulsion formulation of a novel taxoid DHA-SBT-1214 inhibits prostate cancer stem cell-induced tumor growth. Cancer Lett 406:71–80. https://doi.org/10.1016/j.canlet.2017.08.004

Ahmad N, Alam MA, Ahmad FJ, Sarafroz M, Ansari K, Sharma S, Amir M (2018a) Ultrasonication techniques used for the preparation of novel Eugenol-nanoemulsion in the treatment of wounds healings and anti-inflammatory. J Drug Deliv Sci Technol 46:461–473. https://doi.org/10.1016/j.jddst.2018.06.003

Ahmad U, Akhtar J, Singh SP, Badruddeen B, Ahmad FJ, Siddiqui S, Wahajuddin W (2018b) Silymarin nanoemulsion against human hepatocellular carcinoma: development and optimization. Artif Cells Nanomed Biotechnol 46(2):231–241. https://doi.org/10.1080/21691401.2017.1324465

Ahmed M, Smith DM, Hamouda T, Rangel-Moreno J, Fattom A, Khader SA (2017) A novel nanoemulsion vaccine induces mucosal Interleukin-17 responses and confers protection upon *Mycobacterium tuberculosis* challenge in mice. Vaccine 35(37):4983–4989. https://doi.org/10.1016/j.vaccine.2017.07.073

Akbas E, Soyler UB, Oztop MH (2017) Capsaicin emulsions: formulation and characterization. J Dispers Sci Technol 38(8):1079–1086. https://doi.org/10.1080/01932691.2016.1224717

Akhter S, Anwar M, Siddiqui MA, Ahmad I, Ahmad J, Ahmad MZ, Bhatnagar A, Ahmad FJ (2016) Improving the topical ocular pharmacokinetics of an immunosuppressant agent with mucoadhesive nanoemulsions: formulation development, *in-vitro* and *in-vivo* studies. Colloids Surf B Biointerfaces 148:19–29. https://doi.org/10.1016/j.colsurfb.2016.08.048

Aktar MW, Sengupta D, Chowdhury A (2009) Impact of pesticides use in agriculture: their benefits and hazards. Interdiscip Toxicol 2:1–12. https://doi.org/10.2478/v10102-009-0001-7

Alarcon-Alarcon C, Inostroza-Riquelme M, Torres-Gallegos C, Araya C, Miranda M, Sanchez-Caamano JC, Moreno-Villoslada I, Oyarzun-Ampuero FA (2018) Protection of astaxanthin from photodegradation by its inclusion in hierarchically assembled nano and microstructures with potential as food. Food Hydrocoll 83:36–44. https://doi.org/10.1016/j.foodhyd.2018.04.033

Alexandre EMC, Lourenco RV, Bittante AMQB, Moraes ICF, Sobral PJD (2016) Gelatin-based films reinforced with montmorillonite and activated with nanoemulsion of ginger essential oil for food packaging applications. Food Packag Shelf Life 10:87–96. https://doi.org/10.1016/j.fpsl.2016.10.004

Al-Halafi AM (2014) Nanocarriers of nanotechnology in retinal diseases. Saudi J Ophthalmol 28(4):304–309. https://doi.org/10.1016/j.sjopt.2014.02.009

Ali EOM, Shakil NA, Rana VS, Sarkar DJ, Majumder S, Kaushik P, Singh BB, Kumar J (2017) Antifungal activity of nano emulsions of neem and citronella oils against phytopathogenic fungi, *Rhizoctonia solani* and *Sclerotium rolfsii*. Ind Crop Prod 108:379–387. https://doi.org/10.1016/j.indcrop.2017.06.061

Alkhatib MH, Binsiddiq BM, Backer WS (2017a) *In vivo* evaluation of the anticancer activity of a water-in-garlic oil nanoemulsion loaded docetaxel. Int J Pharm Sci Rev Res 8(12):5373–5379. https://doi.org/10.13040/IJPSR.0975-8232.8(12).5373-79

Alkhatib MH, Alharbi SA, Mahassni SH (2017b) *In vivo* evaluation of the anticancer activity of the docetaxel incorporated into nanoemulsion based on orange oil. Pharmacophore 8(6):41–47

Alkhatib MH, Nori DA, Al-Ghamdi MA (2017c) Antitumor activity and hepatotoxicity effect of Sorafenib incorporated into nanoemulsion formulated with flaxseed oil. Int J Biol Pharm Allied Sci 6(1):175–188

Alkhatib MH, Al-Otaibi WA, Wali AN (2018a) Antineoplastic activity of mitomycin C formulated in nanoemulsions-based essential oils on HeLa cervical cancer cells. Chem Biol Interact 291:72–80. https://doi.org/10.1016/j.cbi.2018.06.009

Alkhatib MH, Alnahdi NS, Backer WS (2018b) Antitumor activity, hematotoxicity and hepatotoxicity of sorafenib formulated in a nanoemulsions based on the carrot seed oil. Int J Life Sci Biotechnol Pharma Res 8(1):50–57

Al-Otaibi WA, Alkhatib MH, Wali AN (2018) Cytotoxicity and apoptosis enhancement in breast and cervical cancer cells upon coadministration of mitomycin C and essential oils in nanoemulsion formulations. Biomed Pharmacother 106:946–955. https://doi.org/10.1016/j.biopha.2018.07.041

Alvarado HL, Abrego G, Souto EB, Garduno-Ramirez ML, Clares B, Garcia ML, Calpena AC (2015) Nanoemulsions for dermal controlled release of oleanolic and ursolic acids: *in vitro*, *ex vivo* and *in vivo* characterization. Colloids Surf B Biointerfaces 130:40–47. https://doi.org/10.1016/j.colsurfb.2015.03.062

Alzorqi I, Ketabchi MR, Sudheer S, Manickam S (2016) Optimization of ultrasound induced emulsification on the formulation of palm-olein based nanoemulsions for the incorporation of antioxidant β-D-glucan polysaccharides. Ultrason Sonochem 31:71–84. https://doi.org/10.1016/j.ultsonch.2015.12.004

Ansaril MJ, Ahmed MM, Anwer MK, Jamil S, Alalaiwe A, Alshetaili AS, Shdefat R, Ali R, Shakeel F (2017) Formulation and characterization of fluconazole loaded oil nanoemulsions. Indo Am J Pharm Sci 4(4):852–860. https://doi.org/10.5281/zenodo.804032

Anton N, Vandamme TF (2011) Nano-emulsions and micro-emulsions: clarifications of the critical differences. Pharm Res 28:978–985. https://doi.org/10.1007/s11095-010-0309-1

Anton N, Akram S, Vandamme F (2018) Transitional nanoemulsification methods. In: Jafari SM, McClements DK (eds) Nanoemulsions formulation, applications, and characterization. AP Academic Press (Elsevier), London, pp 77–100. https://doi.org/10.1016/B978-0-12-811838-2.00004-7

Aqil M, Kamran M, Ahad A, Imam SS (2016) Development of clove oil based nanoemulsion of olmesartan for transdermal delivery: Box-Behnken design optimization and pharmacokinetic evaluation. J Mol Liq 214:238–248. https://doi.org/10.1016/j.molliq.2015.12.077

Arancibia C, Riquelme N, Zuniga R, Matiacevich S (2017) Comparing the effectiveness of natural and synthetic emulsifiers on oxidative and physical stability of avocado oil-based nanoemulsions. Innov Food Sci Emerg Technol 44:159–166. https://doi.org/10.1016/j.ifset.2017.06.009

Arbain NH, Salim N, Masoumi HRF, Wong TW, Basri M, Abdul Rahman MB (2018) *In vitro* evaluation of the inhalable quercetin loaded nanoemulsion for pulmonary delivery. Drug Deliv Transl Res. https://doi.org/10.1007/s13346-018-0509-5

Argenta DF, de Mattos CB, Misturini FD, Koester LS, Bassani VL, Simoes CMO, Teixeira HF (2014) Factorial design applied to the optimization of lipid composition of topical antiherpetic nanoemulsions containing isoflavone genistein. Int J Nanomedicine 9:4737–4747. https://doi.org/10.2147/IJN.S67732

Argenta DF, Bidone J, Misturini FD, Koester LS, Bassani VL, Simoes CMO, Teixeira HF (2016) *In vitro* evaluation of mucosa permeation/retention and antiherpes activity of genistein from cationic nanoemulsions. J Nanosci Nanotechnol 16(2):1282–1290. https://doi.org/10.1166/jnn.2016.11676

Artiga-Artigas M, Acevedo-Fani A, Martin-Belloso O (2017) Improving the shelf life of low-fat cut cheese using nanoemulsion-based edible coatings containing oregano essential oil and mandarin fiber. Food Control 76:1–12. https://doi.org/10.1016/j.foodcont.2017.01.001

Artiga-Artigas M, Guerra-Rosas MI, Morales-Castro J, Salvia-Trujillo L, Martin-Belloso O (2018) Influence of essential oils and pectin on nanoemulsion formulation: a ternary phase experimental approach. Food Hydrocoll 81:209–219. https://doi.org/10.1016/j.foodhyd.2018.03.001

Assadpour E, Maghsoudlou Y, Jafari SM, Ghorbani M, Aalami M (2016) Evaluation of folic acid nano-encapsulation by double emulsions. Food Bioproc Tech 9(12):2024–2032. https://doi.org/10.1007/s11947-016-1786-y

Attia MF, Dieng SM, Collot M, Klymchenko AS, Bouillot C, Serra CA, Schmutz M, Er-Rafik M, Vandamme TF, Anton N (2017) Functionalizing nanoemulsions with carboxylates: impact on the biodistribution and pharmacokinetics in mice. Macromol Biosci 17(7):1600471. https://doi.org/10.1002/mabi.201600471

Ayoub AM, Ibrahim MM, Abdallah MH, Mandy MA (2016) Sulpiride microemulsions as antipsychotic nasal drug delivery systems: in-vitro and pharmacodynamic study. J Drug Delivery Sci Technol 36:10–22. https://doi.org/10.1016/j.jddst.2016.09.002

Back PI, Furtado KR, Nemitz MC, Balestrin LA, Fachel FNS, Gomes HM, Schuh RS, Moreira JC, von Poser GL, Teixeira HF (2018) Skin permeation and oxidative protection effect of soybean isoflavones from topical nanoemulsions – A comparative study of extracts and pure compounds. AAPS PharmSciTech. https://doi.org/10.1208/s12249-018-1133-x

Badawy ME, Saad ASA, Tayeb EHM, Mohammed SA, Abd-Elnabi AD (2017) Optimization and characterization of the formation of oil-in-water diazinon nanoemulsions: modeling and influence of the oil phase, surfactant and sonication. J Environ Sci Health B 52(12):896–911. https://doi.org/10.1080/03601234.2017.1362941

Baghbani F, Mortarzadeh F (2017) Bypassing multidrug resistant ovarian cancer using ultrasound responsive doxorubicin/curcumin co-deliver alginate nanodroplets. Colloids Surf B Biointerfaces 153:132–140. https://doi.org/10.1016/j.colsurfb.2017.01.051

Baghbani F, Chegeni M, Mortarzadeh F, Hadian-Ghazvini S, Raz M (2017) Novel ultrasound-responsive chitosan/perfluorohexane nanodroplets for image-guided smart delivery of an anticancer agent: curcumin. Mater Sci Eng C Mater Biol Appl 74:186–193. https://doi.org/10.1016/j.msec.2016.11.107

Bai L, McClements DJ (2016) Development of microfluidization methods for efficient production of concentrated nanoemulsions: comparison of single- and dual-channel microfluidizers. J Colloid Interface Sci 466:206–212. https://doi.org/10.1016/j.jcis.2015.12.039

Bai L, Huan SQ, Gu JY, McClements DJ (2016) Fabrication of oil-in-water nanoemulsions by dual-channel microfluidization using natural emulsifiers: saponins, phospholipids, proteins, and polysaccharides. Food Hydrocoll 61:703–711. https://doi.org/10.1016/j.foodhyd.2016.06.035

Balestrin LA, Bidone J, Bortolin RC, Moresco K, Moreira JC, Teixeira HF (2016) Protective effect of a hydrogel containing *Achyrocline satureioides* extract-loaded nanoemulsion against UV-induced skin damage. J Photochem Photobiol B 163:269–276. https://doi.org/10.1016/j.jphotobiol.2016.08.039

Barradas TN, Senna JP, Cardoso SA, Nicoli S, Padula C, Santi P, Rossi F, Silva KGDE, Mansur CRE (2017) Hydrogel-thickened nanoemulsions based on essential oils for topical delivery of psoralen: permeation and stability studies. Eur J Pharm Biopharm 116:38–50. https://doi.org/10.1016/j.ejpb.2016.11.018

Bhattacharyya A, Duraisamy P, Govindarajan M, Buhroo AA, Prasad R (2016) Nano-biofungicides: emerging trend in insect pest control. In: Prasad R (ed) Advances and applications through fungal nanobiotechnology. Springer, Cham, pp 307–319

Bhushan B, Luo D, Schricker SR, Sigmund W, Zauscher S (eds) (2014) Handbook of nanomaterials properties. Springer, Berlin-Heidelberg

Bonferoni MC, Riva F, Invernizzi A, Dellera E, Sandri G, Rossi S, Marrubini G, Bruni G, Vigani B, Caramella C, Ferrari F (2018) Alpha tocopherol loaded chitosan oleate nanoemulsions for wound healing. Evaluation on cell lines and *ex vivo* human biopsies, and stabilization in spray dried Trojan microparticles. Eur J Pharm Biopharm 123:31–41. https://doi.org/10.1016/j.ejpb.2017.11.008

Bovi GG, Petrus RR, Pinho SC (2017) Feasibility of incorporating buriti (*Mauritia flexuosa* L.) oil nanoemulsions in isotonic sports drink. Int J Food Sci Technol 52(10):2201–2209. https://doi.org/10.1111/ijfs.13499

Brownlow B, Nagaraj VJ, Nayel A, Joshi M, Elbayoumi T (2015) Development and *in vitro* evaluation of vitamin E-enriched nanoemulsion vehicles loaded with genistein for chemoprevention against UVB-induced skin damage. J Pharm Sci 104(10):3510–3523. https://doi.org/10.1002/jps.24547

Camara ALD, Nagel G, Tschiche HR, Cardador CM, Muehlmann LA, de Oliveira DM, Alvim PQ, Azevedo RB, Calderon M, Longo JPF (2017) Acid-sensitive lipidated doxorubicin prodrug entrapped in nanoemulsion impairs lung tumor metastasis in a breast cancer model. Nanomedicine 12(15):1751–1765. https://doi.org/10.2217/nnm-2017-0091

Campos EVR, de Oliveira JL, Pascoli M, Lima R, Fraceto LF (2016) Neem oil and crop protection: from now to the future. Front Plant Sci 7:1494. https://doi.org/10.3389/fpls.2016.01494

Campos EVR, Proença PLF, Oliveira JL, Bakshi M, Abhilash PC, Fraceto LF (2018) Use of botanical insecticides for sustainable agriculture: future perspectives. Ecol Indic. https://doi.org/10.1016/j.ecolind.2018.04.038

Carvalho VFM, Migotto A, Giacone DV, de Lemos DP, Zanoni TB, Maria-Engler SS, Costa-Lotufo LV, Lopes LB (2017) Co-encapsulation of paclitaxel and C6 ceramide in tributyrin-containing nanocarriers improve co-localization in the skin and potentiate cytotoxic effects in 2D and 3D models. Eur J Pharm 109:131–143. https://doi.org/10.1016/j.ejps.2017.07.023

Chang YH, McLandsborough L, McClements DJ (2012) Physical properties and antimicrobial efficacy of thyme oil nanoemulsions: influence of ripening inhibitors. J Agric Food Chem 60(48):12056–12063. https://doi.org/10.1021/jf304045a

Chellaram C, Murugaboopathi G, John AA, Sivakumar R, Ganesan S, Krithika S, Priya G (2014) Significance of nanotechnology in food industry. APCBEE Procedia 8:109–113. https://doi.org/10.1016/j.apcbee.2014.03.010

Cheong AM, Nyam KL (2016) Improvement of physical stability of kenaf seed oil-in-water nanoemulsions by addition of β-cyclodextrin to primary emulsion containing sodium caseinate and Tween 20. J Food Eng 183:24–31. https://doi.org/10.1016/j.jfoodeng.2016.03.012

Cheong AM, Tan CP, Nyam KL (2016) *In vitro* evaluation of the structural and bioaccessibility of kenaf seed oil nanoemulsions stabilised by binary emulsifiers and β-cyclodextrin complexes. J Food Eng 189:90–98. https://doi.org/10.1016/j.jfoodeng.2016.06.002

Chin CP, Lan CW, Wu HS (2012) Application of biodiesel as carrier for insecticide emulsifiable concentrate formulation. J Taiwan Inst Chem Eng 43:578–584. https://doi.org/10.1016/j.jtice.2012.02.003

Choudhury H, Gorain B, Pandey M, Chatterjee LA, Sengupta P, Das A, Molugulu N, Kesharwani P (2017) Recent update on nanoemulgel as topical drug delivery system. J Pharm Sci 106(7):1736–1751. https://doi.org/10.1016/j.xphs.2017.03.042

Choupanian M, Omar D, Basri M, Asib N (2017) Preparation and characterization of neem oil nanoemulsion formulations against *Sitophilus oryzae* and *Tribolium castaneum* adults. J Pest Sci 42:158–165. https://doi.org/10.1584/jpestics.D17-032

Ciocci M, Iorio E, Carotenuto F, Khashoggi HA, Nanni F, Melino S (2016) H_2S-releasing nanoemulsions: a new formulation to inhibit tumor cells proliferation and improve tissue repair. Oncotarget 7(51):84338–84358. https://doi.org/10.18632/oncotarget.12609

Cordoba JPL, Sobral PJA (2017) Physical and antioxidant properties of films based on gelatin, gelatin-chitosan or gelatin-sodium caseinate blends loaded with nanoemulsified active compounds. J Food Eng 213:47–53. https://doi.org/10.1016/j.jfoodeng.2017.05.023

Dammak I, de Carvalho RA, Trindade CSF, Lourenco RV, Sobral PJD (2017) Properties of active gelatin films incorporated with rutin-loaded nanoemulsions. Int J Biol Macromol 98:39–49. https://doi.org/10.1016/j.ijbiomac.2017.01.094

de Matos RPA, Calmon MF, Amantino CF, Villa LL, Primo FL, Tedesco AC, Rahal P (2018) Effect of curcumin-nanoemulsion associated with photodynamic therapy in cervical carcinoma cell lines. Biomed Res Int 2018:4057959. https://doi.org/10.1155/2018/4057959

de Souza ML, Oliveira DD, Pereira NP, Soares DM (2018) Nanoemulsions and dermatological diseases: contributions and therapeutic advances. Int J Dermatol 57(8):894–900. https://doi.org/10.1111/ijd.14028

Díaz-Blancas V, Medina D, Padilla-Ortega E, Bortolini-Zavala R, Olvera-Romero M, Luna-Bárcenas G (2016) Nanoemulsion formulations of fungicide Tebuconazole for agricultural applications. Molecules 21:1271. https://doi.org/10.3390/molecules21101271

Dolez PI (2015) Nanoengineering: global approaches to health and safety issues. Elsevier, Amsterdam

Donsi F, Ferrari G (2016) Essential oil nanoemulsions as antimicrobial agents in food. J Biotechnol 233:106–120. https://doi.org/10.1016/j.jbiotec.2016.07.005

Donsi F, Marchese E, Maresca P, Pataro G, Vu DK, Salmieri S, Lacroix M, Ferrari G (2015) Green beans preservation by combination of a modified chitosan based-coating containing nanoemulsion of mandarin essential oil with high pressure or pulsed light processing. Postharvest Biol Technol 106:21–32. https://doi.org/10.1016/j.postharvbio.2015.02.006

Doost AS, Devlieghere F, Dirckx A, Van der Meeren P (2018) Fabrication of *Origanum compactum* essential oil nanoemulsions stabilized using Quillaja Saponin biosurfactant. J Food Process Preserv 42(7):e13668. https://doi.org/10.1111/jfpp.13668

Du Z, Wang C, Tai X, Wang GY, Liu XY (2016) Optimization and characterization of biocompatible oil-in-water nanoemulsion for pesticide delivery. ACS Sustain Chem Eng 4(3):983–991. https://doi.org/10.1021/acssuschemeng.5b01058

Esquerdo VM, Silva PP, Dotto GL, Pinto LAA (2018) Nanoemulsions from unsaturated fatty acids concentrates of carp oil using chitosan, gelatin, and their blends as wall materials. Eur J Lipid Sci Technol 120(2):1700240. https://doi.org/10.1002/ejlt.201700240

Fachel FNS, Medeiros-Neves B, Dal Pra M, Schuh RS, Veras KS, Bassani VL, Koester LS, Henriques AT, Braganhol E, Teixeira HF (2018) Box-Behnken design optimization of mucoadhesive chitosan-coated nanoemulsions for rosmarinic acid nasal delivery-*In vitro* studies. Carbohydr Polym 199:572–582. https://doi.org/10.1016/j.carbpol.2018.07.054

Feng J, Shi Y, Yu Q, Sun CC, Yang GT (2016) Effect of emulsifying process on stability of pesticide nanoemulsions. Colloids Surf A Physicochem Eng Asp 497:286–292. https://doi.org/10.1016/j.colsurfa.2016.03.024

Fernandes CP, de Almeida FB, Silveira AN, Gonzalez MS, Mello CB, Feder D, Apolinário R, Santos MG, Carvalho JC, Tietbohl LA, Rocha L, Falcão DQ (2014) Development of an insecticidal nanoemulsion with *Manilkara subsericea* (Sapotaceae) extract. J Nanobiotechnol 12:22. https://doi.org/10.1186/1477-3155-12-22

Fountain ED, Wratten SD (2013) Conservation biological control and biopesticides in agricultural. In: Reference module in earth systems and environmental sciences. Elsevier, Amsterdam. https://doi.org/10.1016/B978-0-12-409548-9.00539-X

Furtado CDM, de Faria FSEV, Azevedo RB, Py-Daniel K, Camara ALD, da Silva JR, Oliveira ED, Rodriguez AFR, Degterev IA (2017) *Tectona grandis* leaf extract, free and associated with nanoemulsions, as a possible photosensitizer of mouse melanoma B16 cell. J Photochem Photobiol B 167:242–248. https://doi.org/10.1016/j.jphotobiol.2017.01.004

Gadkari PV, Shashidhar MG, Balaraman M (2017) Delivery of green tea catechins through oil-in-water (O/W) nanoemulsion and assessment of storage stability. J Food Eng 99:65–76. https://doi.org/10.1016/j.jfoodeng.2016.12.009

Gallarate M, Chirio D, Bussano R, Peira E, Battaglia L, Baratta F, Trotta M (2013) Development of O/W nanoemulsions for ophthalmic administration of timolol. Int J Pharm 440(2):126–134. https://doi.org/10.1016/j.ijpharm.2012.10.015

Ganta S, Talekar M, Singh A, Coleman TP, Amiji MM (2014) Nanoemulsions in translational research-opportunities and challenges in targeted cancer therapy. AAPS PharmSciTech 15(3):694–708. https://doi.org/10.1208/s12249-014-0088-9

Ganta S, Singh A, Rawal Y, Cacaccio J, Patel NR, Kulkarni P, Ferris CF, Amiji MM, Coleman TP (2016) Formulation development of a novel targeted theranostic nanoemulsion of docetaxel to overcome multidrug resistance in ovarian cancer. Drug Deliv 23(3):968–980. https://doi.org/10.3109/10717544.2014.923068

Gao YY, Cheng XJ, Wang ZG, Wang J, Gao TT, Li P, Kong M, Chen XG (2014) Transdermal delivery of 10,11-methylenedioxycamptothecin by hyaluronic acid based nanoemulsion for inhibition of keloid fibroblast. Carbohydr Polym 112:376–386. https://doi.org/10.1016/j.carbpol.2014.05.026

Gardesh ASK, Badii F, Hashemi M, Ardakani AY, Maftoonazad N, Gorji AM (2016) Effect of nanochitosan based coating on climacteric behavior and postharvest shelf-life extension of apple cv. Golab Kohanz. LWT-Food Sci Technol 70:33–40. https://doi.org/10.1016/j.lwt.2016.02.002

Geng TY, Xu C, Xu GH (2016) Albumin conjugated lipid nanoemulsion for site specific delivery of rapamycin at inflammatory site of spinal cord injury. Int J Clin Exp Med 9(11):21028–21037. https://doi.org/10.1208/s12249-017-0867-1

Gharibzahedi SMT (2017) Ultrasound-mediated nettle oil nanoemulsions stabilized by purified jujube polysaccharide: process optimization, microbial evaluation and physicochemical storage stability. J Mol Liq 234:240–248. https://doi.org/10.1016/j.molliq.2017.03.094

Ghosh V, Mukherjee A, Chandrasekaran N (2013) Ultrasonic emulsification of food-grade nanoemulsion formulation and evaluation of its bactericidal activity. Ultrason Sonochem 20(1):338–344. https://doi.org/10.1016/j.ultsonch.2012.08.010

Gupta A, Eral HB, Hatton TA, Doyle PS (2016) Nanoemulsions: formation, properties and applications. Soft Matter 12(11):2826–2841. https://doi.org/10.1039/c5sm02958a

Habibi K, Sepehri H, Delphi L, Mirjalili MH, Rafati H (2017) Design and production of methyl jasmonate nanoemulsions using experimental design technique and evaluation of its anticancer efficacy. Pharmazie 72(11):652–662. https://doi.org/10.1691/ph.2017.7078

Hadžiabdić J, Orman D, Elezović A, Vranić E, Rahić O (2017) Preparation of nanoemulsions by high-energy and low-energy emulsification methods. In: Badnjevic A (ed) CMBEBIH 2017, IFMBE proceedings 62, Springer Nature Singapore Pte Ltd, pp 317–322. https://doi.org/10.1007/978-981-10-4166-2_48

Håkansson A, Rayner M (2018) General principles of nanoemulsion formation by high-energy mechanical methods. In: Jafari SM, McClements DK (eds) Nanoemulsions formulation, applications, and characterization, vol 103. AP Academic Press (Elsevier), London, p 139. https://doi.org/10.1016/B978-0-12-811838-2.00005-9

Haron DEM, Chik Z, Noordin MI, Mohamed Z (2015) *In vitro* and *in vivo* evaluation of a novel testosterone transdermal delivery system (TTDS) using palm oil base. Iran J Basic Med Sci 18(12):1167–1175

Harwansh RK, Patra KC, Pareta SK, Singh J, Rahman MA (2011) Nanoemulsions as vehicles for transdermal delivery of glycyrrhizin. Braz J Pharm Sci 47(4):769–778. https://doi.org/10.1590/S1984-82502011000400014

Harwansh RK, Mukherjee PK, Kar A, Bahadur S, Al-Dhabi NA, Duraipandiyan V (2016) Enhancement of photoprotection potential of catechin loaded nanoemulsion gel against UVA induced oxidative stress. J Photochem Photobiol B 160:318–329. https://doi.org/10.1016/j.jphotobiol.2016.03.026

Hashem AS, Awadalla SS, Zayed GM, Maggi F, Benelli G (2018) *Pimpinella anisum* essential oil nanoemulsions against *Tribolium castaneum*—insecticidal activity and mode of action. Environ Sci Pollut Res 25(19):18802–18812. https://doi.org/10.1007/s11356-018-2068-1

Hayles J, Johnson L, Worthley C, Losic D (2017) Nanopesticides: a review of current research and perspectives. In: Grumezescu AM (ed) New pesticides and soil sensors in nanotechnology in the agri-food industry, vol 10. Elsevier, London, pp 193–225

Hazra DK, Karmakar R, Poi R, Bhattacharya S, Mondal S (2017) Recent advances in pesticide formulations for eco-friendly and sustainable vegetable pest management: a review. Arch Agric Environ Sci 2(3):232–237. https://doi.org/10.13140/rg.2.2.30036.91527

Hazrati H, Saharkhiz MJ, Niakousari M, Moein M (2017) Natural herbicide activity of *Satureja hortensis* L. essential oil nanoemulsion on the seed germination and morphophysiological features of two important weed species. Ecotox Environ Saf 142:423–430. https://doi.org/10.1016/j.ecoenv.2017.04.041

Hemraj C (2017) Nanopesticide: current status and future possibilities. ARTOAJ 5(1):555651. https://doi.org/10.19080/ARTOAJ.2017.05.555651

Hernandez-Hernandez E, Lira-Moreno CY, Guerrero-Legarreta I, Wild-Padua G, Di Pierro P, Garcia-Almendarez BE, Regalado-Gonzalez C (2017) Effect of nanoemulsified and microencapsulated Mexican oregano (*Lippia graveolens* Kunth) essential oil coatings on quality of fresh pork meat. J Food Sci 82(6):1423–1432. https://doi.org/10.1111/1750-3841.13728

Hong L, Zhou CL, Chen FP, Han D, Wang CY, Li JX, Chi Z, Liu CG (2017) Development of a carboxymethyl chitosan functionalized nanoemulsion formulation for increasing aqueous solubility, stability and skin permeability of astaxanthin using low-energy method. J Microencapsul 34(8):707–721. https://doi.org/10.1080/02652048.2017.1373154

Hou L, Liu Q, Shen LM, Liu Y, Zhang XQ, Chen FQ, Huang L (2018) Nano-delivery of fraxinellone remodels tumor microenvironment and facilitates therapeutic vaccination in desmoplastic melanoma. Theranostics 8(14):3781–3796. https://doi.org/10.7150/thno.24821

Huang ZH, Xu LQ, Zhu XM, Hu JN, Peng HL, Zeng ZL, Xiong H (2017) Stability and bioaccessibility of fucoxanthin in nanoemulsions prepared from pinolenic acid-contained structured lipid. Int J Food Eng 13(1):20160273. https://doi.org/10.1515/ijfe-2016-0273

Huang J, Wang Q, Li T, Xia N, Xia Q (2018) Multilayer emulsions as a strategy for linseed oil and α-lipoic acid micro-encapsulation: study on preparation and *in vitro* characterization. J Sci Food Agric 98(9):3513–3523. https://doi.org/10.1002/jsfa.8870

Hwang JY, Ha HK, Lee MR, Kim JW, Kim HJ, Lee WJ (2017) Physicochemical property and oxidative stability of whey protein concentrate multiple nanoemulsion containing fish oil. J Food Sci 82(2):437–444. https://doi.org/10.1111/1750-3841.13591

Iqbal B, Ali J, Baboota S (2018) Recent advances and development in epidermal and dermal drug deposition enhancement technology. Int J Dermatol 57(6):646–660. https://doi.org/10.1111/ijd.13902

Jaiswal M, Dudhe R, Sharma PK (2015) Nanoemulsion: an advanced mode of drug delivery system. 3 Biotech 5:123–127. https://doi.org/10.1007/s13205-014-0214-0

Jampílek J, Brychtová K (2012) Azone analogues: classification, design, and transdermal penetration principles. Med Res Rev 32(5):907–947. https://doi.org/10.1002/med.20227

Jampílek J, Kráľová K (2015) Application of nanotechnology in agriculture and food industry, its prospects and risks. Ecol Chem Eng S 22(3):321–361. https://doi.org/10.1515/eces-2015-0018

Jampílek J, Kráľová K (2017a) Nanopesticides: preparation, targeting and controlled release. In: Grumezescu AM (ed) New pesticides and soil sensors in nanotechnology in the agri-food industry, vol 10. Elsevier, London, pp 81–127. https://doi.org/10.1016/B978-0-12-804299-1.00004-7

Jampílek J, Kráľová K (2017b) Nano-antimicrobials: activity, benefits and weaknesses. In: Ficai A, Grumezescu AM (eds) Nanostructures in therapeutic medicine, vol. 2 – Nanostructures for antimicrobial therapy. Elsevier, Amsterdam, pp 23–54. https://doi.org/10.1016/B978-0-323-46152-8.00002-0

Jampílek J, Kráľová K (2017c) Nanomaterials for delivery of nutrients and growth-promoting compounds to plants. In: Prasad R, Kumar M, Kumar V (eds) Nanotechnology: an agricultural paradigm, Springer, pp 177–226. https://doi.org/10.1007/978-981-10-4573-8_9

Jampílek J, Kráľová K (2018a) Application of nanobioformulations for controlled release and targeted biodistribution of drugs. In: Sharma AK, Keservani RK, Kesharwani RK (eds) Nanobiomaterials: applications in drug delivery. Apple Academic Press & CRC Press, Warentown, pp 131–208.

Jampílek J, Kráľová K (2018b) Nanomaterials applicable in food protection. In: Rai RV, Bai JA (eds) Nanotechnology applications in food industry. CRC Press/Taylor & Francis Group, Boca Raton, pp 75–96.

Jampílek J, Kráľová K (2018c) Benefits and potential risks of nanotechnology applications in crop protection. In: Abd-Elsalam K, Prasad R (eds) Nanobiotechnology applications in plant protection. Springer, Cham, pp 189–246. https://doi.org/10.1007/978-3-319-91161-8_8

Jampílek J, Kráľová K (2019a) Nano-biopesticides in agriculture: state of art and future opportunities. In: Koul O (ed) Nano-biopesticides today and future perspectives. Academic Press & Elsevier, Amsterdam, pp 397–447.

Jampílek J, Kráľová K (2019b) Recent advances in lipid nanocarriers applicable in the fight against cancer. In: Grumezescu AM (ed) Nanoarchitectonics in biomedicine – recent progress of nanoarchitectonics in biomedical science. Elsevier, Amsterdam, pp 219–294.

Jampílek J, Kráľová K (2019c) Natural biopolymeric nanoformulations for brain drug delivery. In: Keservani RK, Sharma AK, Kesharwani RK (eds) Nanocariers for brain targeting. Principles and applications. Apple Academic Press & CRC Press, Warentown, in press. http://appleacademicpress.com/nanocarriers-for-brain-targeting-principles-and-applications/9781771887304

Jampílek J, Záruba K, Oravec M, Kuneš M, Babula P, Ulbrich P, Breyinová I, Tříska J, Suchý P (2015) Preparation of silica nanoparticles loaded with nootropics and their *in vivo* permeation through blood–brain barrier. Biomed Res Int 2015:812673

Jo YJ, Kwon YJ (2014) Characterization of β-carotene nanoemulsions prepared by microfluidization technique. Food Sci Biotechnol 23(1):107–113. https://doi.org/10.1007/s10068-014-0014-7

Jo WS, Song HY, Song NB, Lee JH, Min SC, Song KB (2014) Quality and microbial safety of 'Fuji' apples coated with carnauba-shellac wax containing lemongrass oil. LWT-Food Sci Technol 55(2):490–497. https://doi.org/10.1016/j.lwt.2013.10.034

Joung HJ, Choi MJ, Kim JT, Park SH, Park HJ, Shin GH (2016) Food beverage system: antioxidant property and *in vitro* digestion. J Food Sci 81(3):N745–N753. https://doi.org/10.1111/1750-3841.13224

Kaci M, Belhaffef A, Meziane S, Dostert G, Menu P, Velot E, Desobry S, Arab-Tehrany E (2018) Nanoemulsions and topical creams for the safe and effective delivery of lipophilic antioxidant coenzyme Q10. Colloids Surf B Biointerfaces 167:165–175. https://doi.org/10.1016/j.colsurfb.2018.04.010

Kah M, Walch H, Hofmann T (2018) Environmental fate of nanopesticides: durability, sorption and photodegradation of nanoformulated clothianidin. Environ Sci Nano 5(4):882–889. https://doi.org/10.1039/C8EN00038G

Katzer T, Chaves P, Bernardi A, Pohlmann AR, Guterres SS, Beck RCR (2014) Castor oil and mineral oil nanoemulsion: development and compatibility with a soft contact lens. Pharm Dev Technol 19(2):232–237. https://doi.org/10.3109/10837450.2013.769569

Kaur K, Kumar R, Mehta SK (2016) Formulation of saponin stabilized nanoemulsion by ultrasonic method and its role to protect the degradation of quercitin from UV light. Ultrason Sonochem 31:29–38. https://doi.org/10.1016/j.ultsonch.2015.11.017

Kaur K, Kaur J, Kumar R, Mehta SK (2017) Formulation and physiochemical study of α-tocopherol based oil in water nanoemulsion stabilized with non toxic, biodegradable surfactant: sodium stearoyl lactate. Ultrason Sonochem 38:570–578. https://doi.org/10.1016/j.ultsonch.2016.08.026

Khan I, Bahuguna A, Bhardwaj M, Pal Khaket T, Kang SC (2018) Carvacrol nanoemulsion evokes cell cycle arrest, apoptosis induction and autophagy inhibition in doxorubicin resistant-A549 cell line. Artif Cells Nanomed Biotechnol. https://doi.org/10.1080/21691401.2018.1434187

Kim JE, Park YJ (2017a) Improved antitumor efficacy of hyaluronic acid-complexed paclitaxel nanoemulsions in treating non-small cell lung cancer. Biomol Ther 25(4):411–416. https://doi.org/10.4062/biomolther.2016.261

Kim JE, Park YJ (2017b) High paclitaxel-loaded and tumor cell-targeting hyaluronan-coated nanoemulsions. Colloids Surf B Biointerfaces 150:362–372. https://doi.org/10.1016/j.colsurfb.2016.10.050

Kim IH, Lee H, Kim JE, Song KB, Lee YS, Chung DS, Min SC (2013) Plum coatings of lemongrass oil-incorporating carnauba wax-based nanoemulsion. J Food Sci 78(10):E1551–E1559. https://doi.org/10.1111/1750-3841.12244

Kim JH, Ko JA, Kim JT, Cha DS, Cho JH, Park HJ, Shin GH (2014a) Preparation of a capsaicin-loaded nanoemulsion for improving skin penetration. J Agric Food Chem 62(3):725–732. https://doi.org/10.1021/jf404220n

Kim IH, Oh YA, Lee H, Bin Song K, Min SC (2014b) Grape berry coatings of lemongrass oil-incorporating nanoemulsion. LWT-Food Sci Technol 58(1):1–10. https://doi.org/10.1016/j.lwt.2014.03.018

Knowles A (2008) Recent developments of safer formulations of agrochemicals. Environmentalist 28:35–44. https://doi.org/10.1007/s10669-007-9045-4

Komaiko J, McClements DJ (2015) Low-energy formation of edible nanoemulsions by spontaneous emulsification: factors influencing particle size. J Food Eng 146:122–128. https://doi.org/10.1016/j.jfoodeng.2014.09.003

Kumar M, Misra A, Pathak K (2009) Formulation and characterization of nanoemulsion of olanzapine for intranasal delivery. PDA J Pharm Sci Technol 63(6):501–511.

Kumar D, Ali J, Baboota S (2016) Omega 3 fatty acid-enriched nanoemulsion of thiocolchicoside for transdermal delivery: formulation, characterization and absorption studies. Drug Deliv 23(2):591–600. https://doi.org/10.3109/10717544.2014.916764

Kumari S, Kumaraswamy RV, Choudhary RC, Sharma SS, Pal A, Raliya R, Biswas P, Saharan V (2018) Thymol nanoemulsion exhibits potential antibacterial activity against bacterial pustule disease and growth promotory effect on soybean. Sci Rep 8:6650. https://doi.org/10.1038/s41598-018-24871-5

Lalu L, Tambe V, Pradhan D, Nayak K, Bagchi S, Maheshwari R, Kalia K, Tekade RK (2017) Novel nanosystems for the treatment of ocular inflammation: current paradigms and future research directions. J Control Release 268:19–39. https://doi.org/10.1016/j.jconrel.2017.07.035

Le Kim TH, Jun H, Kim JH, Park K, Kim JS, Nam YS (2017) Lipiodol nanoemulsions stabilized with polyglycerol-polycaprolactone block copolymers for theranostic applications. Biomater Res 21:21. https://doi.org/10.1186/s40824-017-0108-4

Lee WD, Liang YJ, Chen BH (2016) Effects of tanshinone nanoemulsion and extract on inhibition of lung cancer cells A549. Nanotechnology 27(49):495101. https://doi.org/10.1088/0957-4484/27/49/495101

Leng P, Zhang Z, Li Q, Zhao M, Pan G (2014) Microemulsion formulation of carbendazim and its *in vitro* antifungal activities evaluation. PLoS One 9:e109580. https://doi.org/10.1371/journal.pone.0109580

Li M, Zhu LF, Liu BM, Du LN, Jia XD, Han L, Jin YG (2016a) Tea tree oil nanoemulsions for inhalation therapies of bacterial and fungal pneumonia. Colloids Surf B Biointerfaces 141:408–416. https://doi.org/10.1016/j.colsurfb.2016.02.017

Li X, Mueller RH, Keck CM, Bou-Chacra NA (2016b) Mucoadhesive dexamethasone acetate-polymyxin B sulfate cationic ocular nanoemulsion – novel combinatorial formulation concept. Pharmazie 71(6):327-333. https://doi.org/10.1691/ph.2016.5190

Li YC, Zhong G, Meng FB, Yu H, Liu DY, Peng LX (2018) Konjac glucomannan octenyl succinate (KGOS) as an emulsifier for lipophilic bioactive nutrient encapsulation. J Sci Food Agric. https://doi.org/10.1002/jsfa.9122

Lim CJ, Basri M, Omar D, Abdul Rahman MB, Salleh AB, Raja Abdul Rahman RN (2013) Green nanoemulsion-laden glyphosate isopropylamine formulation in suppressing creeping foxglove (*A. gangetica*), slender button weed (*D. ocimifolia*) and buffalo grass (*P. conjugatum*): glyphosate isopropylamine formulation for suppressing various weed species. Pest Manag Sci 69:104–111. https://doi.org/10.1002/ps.3371

Lim C, Kim DW, Sim T, Hoang NH, Lee JW, Lee ES, Youn YS, Oh KT (2016) Preparation and characterization of a lutein loading nanoemulsion system for ophthalmic eye drops. J Drug Delivery Sci Technol 36:168–174. https://doi.org/10.1016/j.jddst.2016.10.009

Liu CH, Huang HY (2012) Antimicrobial activity of curcumin-loaded myristic acid microemulsions against *Staphylococcus epidermidis*. Chem Pharm Bull 60(9):1118–1124. https://doi.org/10.1248/cpb.c12-00220

Liu Y, Wei F, Wang Y, Zhu G (2011) Studies on the formation of bifenthrin oil-in-water nano-emulsions prepared with mixed surfactants. Colloids Surf A Physicochem Eng Asp 389:90–96. https://doi.org/10.1016/j.colsurfa.2011.08.045

Liu CH, Lai KY, Wu WC, Chen YJ, Lee WS, Hsu CY (2015a) *In vitro* scleral lutein distribution by cyclodextrin containing nanoemulsions. Chem Pharm Bull 63(2):59–67. https://doi.org/10.1248/cpb.c14-00318

Liu YQ, Hou ZQ, Yang J, Gao YX (2015b) Effects of antioxidants on the stability of β-carotene in O/W emulsions stabilized by gum arabic. J Food Sci Tech-Mys 52(6):3300–3311. https://doi.org/10.1007/s13197-014-1380-0

Liu XJ, McClements DJ, Cao Y, Xiao H (2016) Chemical and physical stability of astaxanthin-enriched emulsion-based delivery systems. Food Biophys 11(3):302–310. https://doi.org/10.1007/s11483-016-9443-6

Liu Y, Yu XM, Sun RJ, Pan XL (2017) Folate-functionalized lipid nanoemulsion to deliver chemo-radiotherapeutics together for the effective treatment of nasopharyngeal carcinoma. AAPS PharmSciTech 18(4):1374–1381. https://doi.org/10.1208/s12249-016-0595-y

Liu LY, Jin Z, Wang M, Shen WY, Zhu ZZ, Wang Z, Liu LL (2018) W/O nano-emulsions with olive leaf phenolics improved oxidative stability of Sacha Inchi oil. Eur J Lipid Sci Technol 120(5):1700471. https://doi.org/10.1002/ejlt.201700471

Lucca LG, de Matos SP, de Mattos CB, Teixeira HF, Limberger RP, Veiga VF, de Araujo BV, Koester LS (2017) Nanoemulsification potentiates *in vivo* antiedematogenic effect of copaiba oil. J Biomed Nanotechnol 13(5):583–590. https://doi.org/10.1166/jbn.2017.2366

Lucca LG, de Matos SP, Kreutz T, Teixeira HF, Veiga VF, de Araujo BV, Limberger RP, Koester LS (2018) Anti-inflammatory effect from a hydrogel containing nanoemulsified copaiba oil (*Copaifera multijuga* Hayne). AAPS PharmSciTech 19(2):522–530. https://doi.org/10.1208/s12249-017-0862-6

Ma PH, Zeng QH, Tai KD, He XY, Yao YY, Hong XF, Yuan F (2017) Preparation of curcumin-loaded emulsion using high pressure homogenization: impact of oil phase and concentration on physicochemical stability. LWT-Food Sci Techol 84:34–46. https://doi.org/10.1016/j.lwt.2017.04.074

Malhotra BD, Srivastava S, Ali MA, Singh C (2014) Nanomaterial-based biosensors for food toxin detection. Appl Biochem Biotechnol 174:880–896. https://doi.org/10.1007/s12010-014-0993-0

Marei GIK, Rabea EI, Badawy MEI (2018) Preparation and characterizations of chitosan/citral nanoemulsions and their antimicrobial activity. Appl Food Biotechnol 5(2):69–78. https://doi.org/10.22037/afb.v5i2.19005

Mate J, Periago PM, Palop A (2016) When nanoemulsified, D-limonene reduces *Listeria monocytogenes* heat resistance about one hundred times. Food Control 59:824–828. https://doi.org/10.1016/j.foodcont.2015.07.020

McClements DJ (2012) Nanoemulsions versus microemulsions: terminology, differences, and similarities. Soft Matter 8:1719–1729. https://doi.org/10.1039/C2SM06903B

McClements DJ (2013) Edible lipid nanoparticles: digestion, absorption, and potential toxicity. Prog Lipid Res 52(4):409–423. https://doi.org/10.1016/j.plipres.2013.04.008

McClements DJ, Gumus CE (2016) Natural emulsifiers – Biosurfactants, phospholipids, biopolymers, and colloidal particles: molecular and physicochemical basis of functional performance. Adv Colloid Interf Sci 234:3–26. https://doi.org/10.1016/j.cis.2016.03.002

McClements DJ, Xiao H (2012) Potential biological fate of ingested nanoemulsions: influence of particle characteristics. Food Funct 3(3):202–220. https://doi.org/10.1039/c1fo10193e

Meghani N, Patel P, Kansara K, Ranjan S, Dasgupta N, Ramalingam C, Kumar A (2018) Formulation of vitamin D encapsulated cinnamon oil nanoemulsion: its potential anti-cancerous activity in human alveolar carcinoma cells. Colloids Surf B Biointerfaces 166:349–357. https://doi.org/10.1016/j.colsurfb.2018.03.041

Mendes C, Buttchevitz A, Kruger JH, Caon T, Benedet PO, Lemos-Senna EMT, Silva MAS (2017) Chitosan microencapsulation of the dispersed phase of an O/W nanoemulsion to hydrochlorothiazide delivery. J Microencapsul 34(7):611–622. https://doi.org/10.1080/02652048.2017.1373155

Meng LH, Xia XJ, Yang YF, Ye J, Dong WJ, Ma PP, Jin YQ, Liu YL (2016) Co-encapsulation of paclitaxel and baicalein in nanoemulsions to overcome multidrug resistance via oxidative stress augmentation and P-glycoprotein inhibition. Int J Pharm 513(1–2):8–16. https://doi.org/10.1016/j.ijpharm.2016.09.001

Migotto A, Carvalho VFM, Salata GC, da Silva FWM, Yan CYI, Ishida K, Costa-Lotufo LV, Steiner AA, Lopes LB (2018) Multifunctional nanoemulsions for intraductal delivery as a new platform for local treatment of breast cancer. Drug Deliv 25(1):654–667. https://doi.org/10.1080/10717544.2018.1440665

Mihindukulasuriya SDF, Lim LT (2014) Nanotechnology development in food packaging: a review. Trends Food Sci Tech 40:149–167. https://doi.org/10.1016/j.tifs.2014.09.009

Minz S, Pandey RS (2018) Development of adjuvanted solid fat nanoemulsions for pulmonary hepatitis B vaccination. J Pharm Sci 107(6):1701–1712. https://doi.org/10.1016/j.xphs.2018.02.007

Moghimi R, Aliahmadi A, Rafati H (2017) Ultrasonic nanoemulsification of food grade *trans*-cinnamaldehyde: 1,8-Cineol and investigation of the mechanism of antibacterial activity. Ultrason Sonochem 35(Part A):415–421. https://doi.org/10.1016/j.ultsonch.2016.10.020

Moghimipour E, Salimi A, Yousefvand T (2017) Preparation and evaluation of celecoxib nanoemulsion for ocular drug delivery. Asian J Pharm 11(3):S543–S550

Monge-Fuentes V, Muehlmann LA, Longo JPF, Silva JR, Fascineli ML, de Souza P, Faria F, Degterev IA, Rodriguez A, Carneiro FP, Lucci CM, Escobar P, Amorim RF, Azevedo RB (2017) Photodynamic therapy mediated by acai oil (*Euterpe oleracea* Martius) in nanoemulsion: a potential treatment for melanoma. J Photochem Photobiol B 166:301–310. https://doi.org/10.1016/j.jphotobiol.2016.12.002

Morsi N, Ibrahim M, Refai H, El Sorogy H (2017) Nanoemulsion-based electrolyte triggered in situ gel for ocular delivery of acetazolamide. Eur J Pharm Sci 104:302–314. https://doi.org/10.1016/j.ejps.2017.04.013

Mossa ATH, Abdelfatta NAH, Mohafrash SMM (2017) Nanoemulsion of camphor (*Eucalyptus globulus*) essential oil, formulation, characterization and insecticidal activity against wheat weevil, *Sitophilus granarius*. Asian J Crop Sci 9:50–62. https://doi.org/10.3923/ajcs.2017.50.62

Mostafa DM, Abd El-Alim SH, Asfour MH, Al-Okbi SY, Mohamed DA, Awad G (2015a) Transdermal nanoemulsions of *Foeniculum vulgare* Mill. essential oil: preparation, characterization and evaluation of antidiabetic potential. J Drug Delivery Sci Technol 29:99–106. https://doi.org/10.1016/j.jddst.2015.06.021

Mostafa DM, Kassem AA, Asfour MH, Al Okbi SY, Mohamed DA, Hamed TES (2015b) Transdermal cumin essential oil nanoemulsions with potent antioxidant and hepatoprotective activities: *in-vitro* and *in-vivo* evaluation. J Mol Liq 212:6–15. https://doi.org/10.1016/j.molliq.2015.08.047

Mungure TE, Roohinejad S, Bekhit AE, Greiner R, Mallikarjunan K (2018) Potential application of pectin for the stabilization of nanoemulsions. Curr Opin Food Sci 19:72–76. https://doi.org/10.1016/j.cofs.2018.01.011

Nam E, Yoo S, Kim HY, Kim YR, Heo YJ (2018) Transdermal water-in-oil nanocarriers of nitric oxide for triggering penile erection. Sci Rep 8:7312. https://doi.org/10.1038/s41598-018-25786-x

Nash JJ, Erk KA (2017) Stability and interfacial viscoelasticity of oil-water nanoemulsions stabilized by soy lecithin and Tween 20 for the encapsulation of bioactive carvacrol. Colloids Surf A Physicochem Eng Asp 517:1–11. https://doi.org/10.1016/j.colsurfa.2016.12.056

Nasr M, Nawaz S, Elhissi A (2012) Amphotericin B lipid nanoemulsion aerosols for targeting peripheral respiratory airways via nebulization. Int J Pharm 436(1–2):611–616. https://doi.org/10.1016/j.ijpharm.2012.07.028

Nastiti CMRR, Ponto T, Abd E, Grice JE, Benson HAE, Roberts MS (2017) Topical nano and microemulsions for skin delivery. Pharmaceutics 9(4):UNSP 37. https://doi.org/10.3390/pharmaceutics9040037

National Nanotechnology Initiative (2008) Nanotechnology: big things from a tiny world. National Nanotechnology Coordination Office, Arlington

Naz Z, Ahmad FJ (2015) Curcumin-loaded colloidal carrier system: formulation optimization, mechanistic insight, *ex vivo* and *in vivo* evaluation. Int J Nanomedicine 10:4293–4307. https://doi.org/10.2147/IJN.S82788

Ni SL, Hu CB, Sun R, Zhao GD, Xia Q (2017) Nanoemulsions-based delivery systems for encapsulation of quercetin: preparation, characterization, and cytotoxicity studies. J Food Process Eng 40(2):e12374. https://doi.org/10.1111/jfpe.12374

Oh YA, Oh YJ, Song AY, Won JS, Song KB, Min SC (2017) Comparison of effectiveness of edible coatings using emulsions containing lemongrass oil of different size droplets on grape berry safety and preservation. LWT Food Sci Technol 75:742–750

Ozogul Y, Durmus M, Ucar Y, Kosker AR, Ozogul F (2017) The combined impact of nanoemulsion based on commercial oils and vacuum packing on the fatty acid profiles of sea bass fillets. J Food Process Preserv 41(6):e13222. https://doi.org/10.1111/jfpp.13222

Ozturk B, Argin S, Ozilgen M, McClements DJ (2014) Formation and stabilization of nanoemulsion-based vitamin E delivery systems using natural surfactants: Quillaja saponin and lecithin. J Food Eng 142:57–63. https://doi.org/10.1016/j.jfoodeng.2014.06.015

Ozturk B, Argin S, Ozilgen M, McClements DJ (2015) Nanoemulsion delivery systems for oil-soluble vitamins: influence of carrier oil type on lipid digestion and vitamin D_3 bioaccessibility. Food Chem 187:499–506. https://doi.org/10.1016/j.foodchem.2015.04.065

Panatieri LF, Brazil NT, Faber K, Medeiros-Neves B, von Poser GL, Rott MB, Zorzi GK, Teixeira HF (2017) Nanoemulsions containing a coumarin-rich extract from *Pterocaulon balansae* (Asteraceae) for the treatment of ocular *Acanthamoeba keratitis*. AAPS PharmSciTech 18(3):721–728. https://doi.org/10.1208/s12249-016-0550-y

Pant M, Dubey S, Patanjali PK, Naik SN, Sharma S (2014) Insecticidal activity of eucalyptus oil nanoemulsion with karanja and jatropha aqueous filtrates. Int Biodeterior Biodegrad 91:119–127. https://doi.org/10.1016/j.ibiod.2013.11.019

Parthasarathi S, Muthukumar SP, Anandharamakrishnan C (2016) The influence of droplet size on the stability, *in vivo* digestion, and oral bioavailability of vitamin E emulsions. Food Funct 7(5):2294–2302. https://doi.org/10.1039/c5fo01517k

Patra JK, Das G, Fraceto LF, Campos EVR, Rodriguez-Torres MDP, Acosta-Torres LS, Diaz-Torres LA, Grillo R, Swamy MK, Sharma S, Habtemariam S, Shin HS (2018) Nano based drug delivery systems: recent developments and future prospects. J Nanobiotechnol 16(1):71. https://doi.org/10.1186/s12951-018-0392-8

Pawar KR, Babu RJ (2014) Lipid materials for topical and transdermal delivery of nanoemulsions. Crit Rev Ther Drug Carrier Syst 31(5):429–458. https://doi.org/10.1615/CritRevTherDrugCarrierSyst.2014010663

Pellosi DS, Macaroff PP, Morais PC, Tedesco AC (2018) Magneto low-density nanoemulsion (MLDE): a potential vehicle for combined hyperthermia and photodynamic therapy to treat cancer selectively. Mater Sci Eng C Mater Biol Appl 92:103–111. https://doi.org/10.1016/j.msec.2018.06.033

Pentak D, Kozik V, Bak A, Dybal P, Sochanik A, Jampílek J (2016) Methotrexate and cytarabine – loaded nanocarriers for multidrug cancer therapy. Spectroscopic study. Molecules 21(12):1689. https://doi.org/10.3390/molecules21121689

Perazzo A, Preziosi V (2018) Catastrophic phase inversion techniques for nanoemulsification. In: Jafari SM, McClements DK (eds) Nanoemulsions formulation, applications, and characterization. AP Academic Press (Elsevier), London, pp 53–76. https://doi.org/10.1016/B978-0-12-811838-2.00003-5

Pereira FG, Marquete R, Oliveira-Cruz L, Quintanilha-Falcao D, Mansur E, Moreira DD (2017) Cytotoxic effects of the essential oil from leaves of *Casearia sylvestris* Sw. (Salicaceae) and its nanoemulsion on A549 tumor cell line. Bol Latinoam Caribe Plant Med Aromat 16(5):506–512

Periasamy VS, Athinarayanan J, Alshatwi AA (2016) Anticancer activity of an ultrasonic nanoemulsion formulation of *Nigella sativa* L. essential oil on human breast cancer cells. Ultrason Sonochem 31:449–455. https://doi.org/10.1016/j.ultsonch.2016.01.035

Pinheiro AC, Coimbra MA, Vicente AA (2016) *In vitro* behaviour of curcumin nanoemulsions stabilized by biopolymer emulsifiers – effect of interfacial composition. Food Hydrocoll 52:460–467. https://doi.org/10.1016/j.foodhyd.2015.07.025

Piorkowski DT, McClements DJ (2014) Beverage emulsions: recent developments in formulation, production, and applications. Food Hydrocoll 42(Part:1):5–41. https://doi.org/10.1016/j.foodhyd.2013.07.009

Pisarčík M, Lukáč M, Jampílek J, Bilka F, Bilková A, Pašková Ľ, Devínsky F, Horáková R, Opravil T (2018) Silver nanoparticles stabilised with cationic single-chain surfactants. Structure-physical properties-biological activity relationship study. J Mol Liq 272:60–72. https://doi.org/10.1016/j.molliq.2018.09.042

Prakash B, Kujur A, Yadav A, Kumar A, Singh PP, Dubey NK (2018a) Nanoencapsulation: an efficient technology to boost the antimicrobial potential of plant essential oils in food system. Food Control 89:1–11. https://doi.org/10.1016/j.foodcont.2018.01.018

Prakash A, Baskaran R, Paramasivam N, Vadivel V (2018b) Essential oil based nanoemulsions to improve the microbial quality of minimally processed fruits and vegetables: a review. Food Res Int 111:509–523. https://doi.org/10.1016/j.foodres.2018.05.066

Prasad R, Kumar V, Prasad KS (2014) Nanotechnology in sustainable agriculture: present concerns and future aspects. Afr J Biotechnol 13(6):705–713

Prasad R, Pandey R, Barman I (2016) Engineering tailored nanoparticles with microbes: quo vadis. Wiley Interdiscip Rev Nanomed Nanobiotechnol 8:316–330. https://doi.org/10.1002/wnan.1363

Prasad R, Bhattacharyya A, Nguyen QD (2017a) Nanotechnology in sustainable agriculture: recent developments, challenges, and perspectives. Front Microbiol 8:1014. https://doi.org/10.3389/fmicb.2017.01014

Prasad R, Pandey R, Varma A, Barman I (2017b) Polymer based nanoparticles for drug delivery systems and cancer therapeutics. In: Kharkwal H, Janaswamy S (eds) Natural polymers for drug delivery. CAB International, Wallingford, pp 53–70.

Prasad R, Kumar V and Kumar M (2017c) Nanotechnology: Food and Environmental Paradigm. Springer Nature Singapore Pte Ltd. (ISBN 978-981-10-4678-0)

Prasad M, Lambe UP, Brar B, Shah I, J Manimegalai J, Ranjan K, Rao R, Kumar S, Mahant S, Khurana SK, Iqbal HMN, Dhama K, Misri J, Prasad G (2018) Nanotherapeutics: an insight into healthcare and multi-dimensional applications in medical sector of the modern world. Biomed Pharmacother 97:1521–1537. https://doi.org/10.1016/j.biopha.2017.11.026

Primozic M, Duchek A, Nickerson M, Ghosh S (2018) Formation, stability and *in vitro* digestibility of nanoemulsions stabilized by high-pressure homogenized lentil proteins isolate. Food Hydrocoll 77:126–141. https://doi.org/10.1016/j.foodhyd.2017.09.028

Prow TW, Grice JE, Lin LL, Faye R, Butler M, Becker W, Wurm EMT, Yoong C, Robertson TA, Soyer HP, Roberts MS (2011) Nanoparticles and microparticles for skin drug delivery. Adv Drug Deliv Rev 63(6):470–491. https://doi.org/10.1016/j.addr.2011.01.012

Qadir A, Faiyazuddin MD, Hussain MDT, Alshammari TM, Shakeel F (2016) Critical steps and energetics involved in a successful development of a stable nanoemulsion. J Mol Liq 214:7–18. https://doi.org/10.1016/j.molliq.2015.11.050

Qian C, McClements DJ (2011) Formation of nanoemulsions stabilized by model food-grade emulsifiers using high-pressure homogenization: factors affecting particle size. Food Hydrocoll 25(5):1000–1008. https://doi.org/10.1016/j.foodhyd.2010.09.017

Rachmawati H, Budiputra DK, Mauludin R (2015) Curcumin nanoemulsion for transdermal application: formulation and evaluation. Drug Dev Ind Pharm 41(4):560–566. https://doi.org/10.3109/03639045.2014.884127

Radi M, Akhavan-Darabi S, Akhavan HR, Amiri S (2018) The use of orange peel essential oil microemulsion and nanoemulsion in pectin-based coating to extend the shelf life of fresh-cut orange. J Food Process Preserv 42(2):e13441. https://doi.org/10.1111/jfpp.13441

Rai VK, Mishra N, Yadav KS, Yadav NP (2018) Formulation development, stability issues, basic considerations and applications. J Control Release 270:203–225. https://doi.org/10.1016/j.jconrel.2017.11.049

Raliya R, Saharan V, Dimpka C, Biswas P (2018) Nanofertilizer for precision and sustainable agriculture: current state and future perspectives. J Agric Food Chem 66(26):6487–6503. https://doi.org/10.1021/acs.jafc.7b02178

Raviadaran R, Chandran D, Shin LH, Manickam S (2018) Optimization of palm oil in water nanoemulsion with curcumin using microfluidizer and response surface methodology. LWT-Food Sci Technol 96:58–65. https://doi.org/10.1016/j.lwt.2018.05.022

Rodriguez J, Martin MJ, Ruiz MA, Clares B (2016) Current encapsulation strategies for bioactive oils: from alimentary to pharmaceutical perspectives. Food Res Int 83:41–59. https://doi.org/10.1016/j.foodres.2016.01.032

Rosa MTMG, Silva EK, Santos DT, Petenate AJ, Meireles MAA (2016) Obtaining annatto seed oil miniemulsions by ultrasonication using aqueous extract from Brazilian ginseng roots as a biosurfactant. J Food Eng 168:68–78. https://doi.org/10.1016/j.jfoodeng.2015.07.024

Roy A, Guha P (2018) Formulation and characterization of betel leaf (*Piper betle* L.) essential oil based nanoemulsion and its *in vitro* antibacterial efficacy against selected food pathogens. J Food Process Preserv 42(6):e13617. https://doi.org/10.1111/jfpp.13617

Sabry AH, Ragaei M (2018) Nanotechnology and their applications in insect's pest control. In: Abd-Elsalam KA, Prasad R (eds) Nanobiotechnology applications in plant protection, nanotechnology in the life sciences. Springer International Publishing AG, Part of Springer Nature, Cham, pp 1–28. https://doi.org/10.1007/978-3-319-91161-8_1

Sahu P, Das D, Mishra VK, Kashaw V, Kashaw SK (2017) Nanoemulsion: a novel eon in cancer chemotherapy. Mini Rev Med Chem 17(18):1778–1792. https://doi.org/10.2174/1389557516666160219122755

Salim N, Ahmad N, Musa SH, Hashim R, Tadros TF, Basri M (2016) Nanoemulsion as a topical delivery system of antipsoriatic drugs. RSC Adv 6(8):6234–6250. https://doi.org/10.1039/c5ra14946k

Salvia-Trujillo L, Rojas-Graue MA, Soliva-Fortuny R, Martin-Belloso O (2014) Impact of microfluidization or ultrasound processing on the antimicrobial activity against *Escherichia coli* of lemongrass oil-loaded nanoemulsions. Food Control 37:292–297. https://doi.org/10.1016/j.foodcont.2013.09.015

Salvia-Trujillo L, Rojas-Graue A, Soliva-Fortuny R, Martin-Belloso O (2015) Physicochemical characterization and antimicrobial activity of food-grade emulsions and nanoemulsions incorporating essential oils. Food Hydrocoll 43:547–556. https://doi.org/10.1016/j.foodhyd.2014.07.012

Sanchez-Ortega I, Garcia-Almendarez BE, Santos-Lopez EM, Reyes-Gonzalez LR, Regalado C (2016) Characterization and antimicrobial effect of starch-based edible coating suspensions. Food Hydrocoll 52:906–913. https://doi.org/10.1016/j.foodhyd.2015.09.004

Sandig AG, Campmany ACC, Campos FF, Villena MJM, Naveros BC (2013) Transdermal delivery of imipramine and doxepin from newly oil-in-water nanoemulsions for an analgesic and anti-allodynic activity: development, characterization and *in vivo* evaluation. Colloids Surf B Biointerfaces 103:558–565. https://doi.org/10.1016/j.colsurfb.2012.10.061

Sasikumar A, Kamalasanan K (2017) Nanomedicine for prostate cancer using nanoemulsion: a review. J Control Release 260:111–123. https://doi.org/10.1016/j.jconrel.2017.06.001

Saxena V, Hasan A, Sharma S, Pandey LM (2018) Edible oil nanoemulsion: an organic nanoantibiotic as a potential biomolecule delivery vehicle. Int J Polym Mater Polym Biomater 67(7):410–419. https://doi.org/10.1080/00914037.2017.1332625

Sekhon BS (2014) Nanotechnology in agri-food production: an overview. Nanotechnol Sci Appl 7:31–53. https://doi.org/10.2147/NSA.S39406

Sengupta P, Chatterjee B (2017) Potential and future scope of nanoemulgel formulation for topical delivery of lipophilic drugs. Int J Pharm 526(1–2):353–365. https://doi.org/10.1016/j.ijpharm.2017.04.068

Sessa M, Ferrari G, Donsi F (2015) Novel edible coating containing essential oil nanoemulsions to prolong the shelf life of vegetable products. In: Pierucc S, Klemes JJ (eds) ICHEAP12: 12th international conference on chemical & process, engineering book series: chemical engineering transactions, vol 43, pp 55–60. https://doi.org/10.3303/CET1543010

Severino R, Vu KD, Donsi F, Salmieri S, Ferrari G, Lacroix M (2014) Antibacterial and physical effects of modified chitosan based-coating containing nanoemulsion of mandarin essential oil and three non-thermal treatments against *Listeria innocua* in green beans. Int J Food Microbiol 191:82–88. https://doi.org/10.1016/j.ijfoodmicro.2014.09.007

Severino R, Ferrari G, Vu KD, Donsi F, Salmieri S, Lacroix M (2015) Antimicrobial effects of modified chitosan based coating containing nanoemulsion of essential oils, modified atmosphere packaging and gamma irradiation against *Escherichia coli* O157:H7 and *Salmonella Typhimurium* on green beans. Food Control 50:215–222. https://doi.org/10.1016/j.foodcont.2014.08.029

Shah K, Chan LW, Wong TW (2017) Critical physicochemical and biological attributes of nanoemulsions for pulmonary delivery of rifampicin by nebulization technique in tuberculosis treatment. Drug Deliv 24(1):1631–1647. https://doi.org/10.1080/10717544.2017.1384298

Shahabi N, Tajik H, Moradi M, Forough M, Ezati P (2017) Physical, antimicrobial and antibiofilm properties of *Zataria multiflora* Boiss essential oil nanoemulsion. Int J Food Sci Technol 52(7):1645–1652. https://doi.org/10.1111/ijfs.13438

Shakeel F, Ramadan W, Faisal MS, Rizwan M, Faiyazuddin M, Mustafa G, Shafiq S (2010) Transdermal and topical delivery of anti-inflammatory agents using nanoemulsion/microemulsion: an updated review. Curr Nanosci 6(2):184–198. https://doi.org/10.2174/157341310790945740

Shao P, Qiu Q, Xiao J, Zhu YG, Sun PL (2017) Chemical stability and *in vitro* release properties of β-carotene in emulsions stabilized by *Ulva fasciata* polysaccharide. Int J Biol Macromol 102:225–231. https://doi.org/10.1016/j.ijbiomac.2017.03.186

Shao Y, Wu CH, Wu TT, Li Y, Chen SG, Yuan CH, Hu YQ (2018) Eugenol-chitosan nanoemulsions by ultrasound-mediated emulsification: formulation, characterization and antimicrobial activity. Carbohydr Polym 193:144–152. https://doi.org/10.1016/j.carbpol.2018.03.101

Sharif HR, Abbas S, Majeed H, Safdar W, Shamoon M, Khan MA, Shoaib M, Raza H, Haider J (2017a) Formulation, characterization and antimicrobial properties of black cumin essential oil nanoemulsions stabilized by OSA starch. J Food Sci Technol Mys 54(10):3358–3365. https://doi.org/10.1007/s13197-017-2800-8

Sharif HR, Goff HD, Majeed H, Liu F, Nsor-Atindana J, Haider J, Liang R, Zhong F (2017b) Physicochemical stability of β-carotene and α-tocopherol enriched nanoemulsions: influence of carrier oil, emulsifier and antioxidant. Colloids Surf A Physicochem Eng Asp 529:550–559. https://doi.org/10.1016/j.colsurfa.2017.05.076

Sharma M, Mann B, Sharma R, Bajaj R, Athira S, Sarkar P, Pothuraju R (2017) Sodium caseinate stabilized clove oil nanoemulsion: physicochemical properties. J Food Eng 212:38–46. https://doi.org/10.1016/j.jfoodeng.2017.05.006

Sharma A, Sharma NK, Srivastava A, Kataria A, Dubey S, Sharma S, Kundu B (2018) Clove and lemongrass oil based non-ionic nanoemulsion for suppressing the growth of plant pathogenic *Fusarium oxysporum* f.sp. *lycopersici*. Ind Crop Prod 123:353–362. https://doi.org/10.1016/j.indcrop.2018.06.077

Shen J, Liu XD, Zhou M, Liu HJ (2018) Novel nanoemulsion formulation for ocular drug delivery for cataract therapeutics. J Biomater Tissue Eng 8(5):647–657. https://doi.org/10.1166/jbt.2018.1790

Shi J, Zhou SL, Kang L, Ling H, Chen JP, Duan LL, Song YZ, Deng YH (2018) Evaluation of the antitumor effects of vitamin K_2 (menaquinone-7) nanoemulsions modified with sialic acid-cholesterol conjugate. Drug Deliv Transl Res 8(1):1–11. https://doi.org/10.1007/s13346-017-0424-1

Shinde RL, Devarajan PV (2017) Docosahexaenoic acid-mediated, targeted and sustained brain delivery of curcumin microemulsion. Drug Deliv 24(1):152–161. https://doi.org/10.1080/10717544.2016.1233593

Shofia SI, Jayakumar K, Mukherjee A, Chandrasekaran N (2018) Efficiency of brown seaweed (*Sargassum longifolium*) polysaccharides encapsulated in nanoemulsion and nanostructured lipid carrier against colon cancer cell lines HCT 116. RSC Adv 8(29):15973–15984. https://doi.org/10.1039/c8ra02616e

Shu GF, Khalid N, Tan TB, Zhao YG, Neves MA, Kobayashi I, Nakajima M (2018) *In vitro* bioaccessibility of ergocalciferol in nanoemulsion-based delivery system: the influence of food-grade emulsifiers with different stabilising mechanisms. Int J Food Sci Technol 53(2):430–440. https://doi.org/10.1111/ijfs.13601

Silva LFC, Lemos-Senna E (2016) Development of triamcinolone acetonide-loaded nanoemulsion for ophthalmic drug delivery. Lat Am J Pharm 35(10):2159–2168

Singh OV (2015) Bio-nanoparticles: biosynthesis and sustainable biotechnological implications. Wiley-Blackwell, Hoboken.

Singh T, Shukla S, Kumar P, Wahla V, Bajpai VK, Rather IA (2017a) Application of nanotechnology in food science: perception and overview. Front Microbiol 8:1501. https://doi.org/10.3389/fmicb.2017.01501

Singh Y, Meher JG, Raval K, Khan FA, Chaurasia M, Jain NK, Chourasia MK (2017b) Nanoemulsion: concepts, development and applications in drug delivery. J Control Release 252:28–49. https://doi.org/10.1016/j.jconrel.2017.03.008

Slomkowski S, Alemán JV, Gilbert RG, Hess M, Horie K, Jones RG, Kubisa P, Meisel I, Mormann W, Penczek S, Stepto RFT (2011) Terminology of polymers and polymerization processes in dispersed systems (IUPAC Recommendations 2011). Pure Appl Chem 83:2229–2259. https://doi.org/10.1351/PAC-REC-10-06-03

Souza JG, Dias K, Pereira TA, Bernardi DS, Lopez RFV (2014) Topical delivery of ocular therapeutics: carrier systems and physical methods. J Pharm Pharmacol 66(4):507–530. https://doi.org/10.1111/jphp.12132

Steuber N, Vo K, Wadhwa R, Birch J, Iacoban P, Chavez P, Elbayoumi TA (2016) Tocotrienol nanoemulsion platform of curcumin elicit elevated apoptosis and augmentation of anticancer efficacy against breast and ovarian carcinomas. Int J Mol Sci 17(11):1792. https://doi.org/10.3390/ijms17111792

Su R, Fan WF, Yu Q, Dong XC, Qi JP, Zhu QG, Zhao WL, Wu W, Chen ZJ, Li Y, Lu Y (2017) Size-dependent penetration of nanoemulsions into epidermis and hair follicles: implications for transdermal delivery and immunization. Oncotarget 8(24):38214–38226. https://doi.org/10.18632/oncotarget.17130

Sugumar S, Ghosh V, Nirmala MJ, Mukherjee A, Chandrasekaran N (2014) Ultrasonic emulsification of eucalyptus oil nanoemulsion: antibacterial activity against *Staphylococcus aureus* and wound healing activity in Wistar rats. Ultrason Sonochem 21(3):1044–1049. https://doi.org/10.1016/j.ultsonch.2013.10.021

Taghavi E, Mirhosseini H, Rukayadi Y, Radu S, Biabanikhankahdani R (2018) Effect of microfluidization condition on physicochemical properties and inhibitory activity of nanoemulsion loaded with natural antibacterial mixture. Food Bioprocess Technol 11(3):645–659. https://doi.org/10.1007/s11947-017-2037-6

Tayeb HH, Sainsbury F (2018) Nanoemulsions in drug delivery: formulation to medical application. Nanomedicine 13(19):2507–2525. https://doi.org/10.2217/nnm-2018-0088

Tayel SA, El-Nabarawi MA, Tadros MI, Abd-Elsalam WH (2013) Design, *in vitro* characterization and *in vivo* estimation of the ocular irritation and drug pharmacokinetics in the aqueous humor of rabbits. Int J Pharm 443(1–2):293–305. https://doi.org/10.1016/j.ijpharm.2012.12.049

Teng J, Hu XQ, Wang MF, Tao NP (2018) Fabrication of chia (*Salvia hispanica* L.) seed oil nanoemulsions using different emulsifiers. J Food Process Preserv 42(1):e13416. https://doi.org/10.1111/jfpp.13416

Tian HX, Li DF, Xu T, Hu J, Rong YZ, Zhao B (2017) Citral stabilization and characterization of nanoemulsions stabilized by a mixture of gelatin and Tween 20 in an acidic system. J Sci Food Agric 97(9):2991–2998. https://doi.org/10.1002/jsfa.8139

Topuz OK, Ozvural EB, Zhao Q, Huang QR, Chikindas M, Golukcu M (2016) Physical and antimicrobial properties of anise oil loaded nanoemulsions on the survival of foodborne pathogens. Food Chem 203:117–123. https://doi.org/10.1016/j.foodchem.2016.02.051

Uluata S, McClements DJ, Decker EA (2015) Physical stability, autoxidation, and photosensitized oxidation of ω-3 oils in nanoemulsions prepared with natural and synthetic surfactants. J Agric Food Chem 63(42):9333–9340. https://doi.org/10.1021/acs.jafc.5b03572

van Straten D, Mashayekhi V, de Bruijn HS, Oliveira S, Robinson DJ (2017) Oncologic photodynamic therapy: basic principles, current clinical status and future directions. Cancers 9(2):19. https://doi.org/10.3390/cancers9020019

Vecchione R, Quagliariello V, Calabria D, Calcagno V, De Luca E, Iaffaioli RV, Netti PA (2016) Curcumin bioavailability from oil in water nano-emulsions: *in vitro* and *in vivo* study on the dimensional, compositional and interactional dependence. J Control Release 233:88–100. https://doi.org/10.1016/j.jconrel.2016.05.004

Ventola CL (2017) Progress in nanomedicine: approved and investigational nanodrugs. Pharm Ther 42(12):742–755.

Verma P, Meher JG, Asthana S, Pawar VK, Chaurasia M, Chourasia MK (2016) Perspectives of nanoemulsion assisted oral delivery of docetaxel for improved chemotherapy of cancer. Drug Deliv 23(2):479–488. https://doi.org/10.3109/10717544.2014.920430

Walia N, Dasgupta N, Ranjan S, Chen LY, Ramalingam C (2017) Fish oil based vitamin D nanoencapsulation by ultrasonication and bioaccessibility analysis in simulated gastro-intestinal tract. Ultrason Sonochem 39:623–635. https://doi.org/10.1016/j.ultsonch.2017.05.021

Walker RM, Gumus CE, Decker EA, McClements DJ (2017) Improvements in the formation and stability of fish oil-in-water nanoemulsions using carrier oils: MCT, thyme oil, & lemon oil. J Food Eng 211:60–68. https://doi.org/10.1016/j.jfoodeng.2017.05.004

Wang L, Li X, Zhang G, Dong JF, Eastoe J (2007) Oil-in-water nanoemulsions for pesticide formulations. J Colloid Interface Sci 314:230–235. https://doi.org/10.1016/j.jcis.2007.04.079

Wang C, Tai X, Du Z, Liu X (2017) Formulation and physicochemical properties of promising avermectin microemulsion with biodegradable surfactant and oil. J Dispers Sci Technol 38:409–415. https://doi.org/10.1080/01932691.2016.1172315

Wu CH, Wang LP, Hu YQ, Chen SG, Liu DH, Ye XQ (2016) Edible coating from citrus essential oil-loaded nanoemulsions: physicochemical characterization and preservation performance. RSC Adv 6(25):20892–20900. https://doi.org/10.1039/c6ra00757k

Xu F, Pandya JK, Chung C, McClements DJ, Kinchla A (2018) Emulsions as delivery systems for gamma and delta tocotrienols: formation, properties and simulated gastrointestinal fate. Food Res Int 105:570–579. https://doi.org/10.1016/j.foodres.2017.11.033

Yalcinoz S, Ercelebi E (2018) Potential applications of nano-emulsions in the food systems: an update. Mater Res Express 5(6):062001. https://doi.org/10.1088/2053-1591/aac7ee

Yan B, Park SH, Balasubramaniam VM (2017) Influence of high pressure homogenization with and without lecithin on particle size and physicochemical properties of whey protein-based emulsions. J Food Process Eng 40(6):UNSP e12578. https://doi.org/10.1111/jfpe.12578

Yang Y, Leser ME, Sher AA, McClements DJ (2013) Formation and stability of emulsions using a natural small molecule surfactant: Quillaja saponin (Q-Naturale®). Food Hydrocoll 30(2):589–596. https://doi.org/10.1016/j.foodhyd.2012.08.008

Yang C, Powell CA, Duan Y, Shatters R, Zhang M (2015) Antimicrobial nanoemulsion formulation with improved penetration of foliar spray through citrus leaf cuticles to control citrus huanglongbing. PLoS One 10(7):e0133826. https://doi.org/10.1371/journal.pone.0133826

Yang CC, Hung CF, Chen BH (2017) Preparation of coffee oil-algae oil-based nanoemulsions and the study of their inhibition effect on UVA-induced skin damage in mice and melanoma cell growth. Int J Nanomedicine 12:6559–6580. https://doi.org/10.2147/IJN.S144705

Yerramilli M, Longmore N, Ghosh S (2017) Improved stabilization of nanoemulsions by partial replacement of sodium caseinate with pea protein isolate. Food Hydrocoll 64:99–111. https://doi.org/10.1016/j.foodhyd.2016.10.027

Yu WY, Zhang N (2009) Surface modification of nanocarriers for cancer therapy. Curr Nanosci 5(2):123–134. https://doi.org/10.2174/157341309788185370

Yukuyama MN, Kato ETM, Loebenberg R, Bou-Chacra NA (2017) Challenges and future prospects of nanoemulsion as a drug delivery system. Curr Pharm Des 23(3):495–508. https://doi.org/10.2174/1381612822666161027111957

Zambrano-Zaragoza ML, Gutierrez-Cortez E, Del Real A, Gonzalez-Reza RM, Galindo-Perez MJ, Quintanar-Guerrero D (2014) Fresh-cut Red Delicious apples coating using tocopherol/mucilage nanoemulsion: effect of coating on polyphenol oxidase and pectin methylesterase activities. Food Res Int 62:974–983. https://doi.org/10.1016/j.foodres.2014.05.011

Zhang SJ, Zhang M, Fang ZX, Liu YP (2017) Preparation and characterization of blended cloves/cinnamon essential oil nanoemulsions. LWT-Food Sci Technol 75:316–322. https://doi.org/10.1016/j.lwt.2016.08.046

Zhang Y, Pan K, Zhong QX (2018) Eugenol nanoencapsulated by sodium caseinate: physical, antimicrobial, and biophysical properties. Food Biophys 13(1):37–48. https://doi.org/10.1007/s11483-017-9509-0

Zhao XJ, Zhou ZQ (2016) Synthesis and applications of pectin-based nanomaterials. Curr Nanosci 12(1):103–109. https://doi.org/10.2174/1573413711666150818224020

Zhao X, Zhu Y, Zhang C, Kei JM, Yu M, Du FP (2017) Positive charge pesticide nanoemulsions prepared by the phase inversion composition method with ionic liquids. RSC Adv 7:48586–48596. https://doi.org/10.1039/C7RA08653A

Zhao CH, Shen X, Guo MR (2018a) Stability of lutein encapsulated whey protein nano-emulsion during storage. PLoS One 13(2):e0192511. https://doi.org/10.1371/journal.pone.0192511

Zhao J, Liu S, Hu XY, Zhang YM, Yan SL, Zhao H, Zeng M, Li Y, Yang L, Zhang JQ (2018b) Improved delivery of natural alkaloids into lung cancer through woody oil-based emulsive nanosystems. Drug Deliv 25(1):1426–1437. https://doi.org/10.1080/10717544.2018.1474970

Chapter 3
Application of Gum Arabic in Nanoemulsion for Safe Conveyance of Bioactive Components

Sanju Bala Dhull, M. Anju, Sneh Punia, Ravinder Kaushik, and Prince Chawla

Contents

3.1 Introduction

Due to the remarkable applications and properties, substances of miniscule and subatomic sizes such as nanoemulsions have emerged as a boon for several industries (Aguiar et al. 2016; Shishir et al. 2018). Nanoemulsion is an even mixture with the range of droplet size (10–200 nm) formed from two or more immiscible liquids (mainly oil and water phase) (Ha et al. 2015; Taarji et al. 2018; Gupta et al.

S. B. Dhull · S. Punia
Department of Food Science and Technology, Chaudhary Devi Lal University, Sirsa, Haryana, India

M. Anju
Department of Energy and Environment Sciences, Chaudhary Devi Lal University, Sirsa, Haryana, India

R. Kaushik · P. Chawla (✉)
School of Bioengineering and Food Technology, Shoolini University, Solan, Himachal Pradesh, India

R. Prasad et al. (eds.), *Nanobiotechnology in Bioformulations*, Nanotechnology in the Life Sciences, https://doi.org/10.1007/978-3-030-17061-5_3

2016; Shishir et al. 2018). Unique properties such as surface activity and thermodynamic stability make nanoemulsions isotropic and transparent heterogeneous systems that are used mainly in food and pharmaceutical industries for the delivery of significant hydrophobic and hydrophilic drugs and bioactive constituents (Binsi et al. 2017; Shariffa et al. 2016; Yao et al. 2013). Furthermore, the major components of an emulsion system are oil and emulsifier, and the incorporation of an emulsifier plays a vital role in the fabrication of nanosized droplets by decreasing the interfacial tension. Emulsifiers also play a significant part in alleviation of the nanoemulsions through repulsive electrostatic interactions and steric hindrances (Shariffa et al. 2016; Binsi et al. 2017; Ozturk et al. 2015). Therefore, a suitable emulsifier always fascinates researchers' interest to formulate stable and active nanoemulsion system for the delivery of bioactive ingredients up to tissue level (Su et al. 2018; Ha et al. 2015; Li et al. 2016; Taarji et al. 2018; Gupta et al. 2015). Predominantly, several categories of proteins and polysaccharides are used as active emulsifiers in an emulsion system (Aguiar et al. 2016; Jia et al. 2016; de Souza Simões et al. 2017).

Surface charge of emulsifier is an important tool to maintain the repulsive electrostatic interactions and steric hindrance between oil droplets and water; hence, either polyanionic or polycationic biopolymers are used for the formulation of stable nanoemulsion (Shariffa et al. 2016; Yao et al. 2018; Ozturk and McClements 2016; Zhang et al. 2015; Bai et al. 2017). Among several biopolymers, gum arabic is abundantly used emulsifier for the formulation of nanoemulsion. Polyanionic structure of gum arabic is highly suitable to strengthen the emulsion system for a long time (Gupta et al. 2015; Ha et al. 2015; Li et al. 2016; Taarji et al. 2018; Ozturk et al. 2015; Bai et al. 2017). Several researchers explored the efficacy of gum arabic as a stable emulsifier for safe and effective delivery of bioactive components (Shariffa et al. 2016; Yao et al. 2018; Ozturk and McClements 2016; Zhang et al. 2015; Bai et al. 2017; Yao et al. 2013).

Therefore, this chapter aims to acme the significant evolvement of gum arabic in improving the nanoemulsion.

3.2 Nanoemulsion: An Effective Delivery System

A uniform mixture of two immiscible (oil and water phase) liquid phases attributes to the formation of an emulsion system. Two types of emulsion systems (O/W and W/O) and different methods (high- and low-energy emulsification methods) for the synthesis of emulsion have already been explored by several authors (Ha et al. 2015; Gupta et al. 2016; Li et al. 2016; Taarji et al. 2018; Shariffa et al. 2016; Binsi et al. 2017; Ozturk et al. 2015). High-energy methods always exhibit smooth types of emulsion system, whereas low-energy methods due to the least interference of emulsifier in surface interfacial tension formulate coarse form of the emulsion (Ha et al. 2015; Li et al. 2016). Furthermore, emulsions are also categorized as macroemulsion, microemulsion, and nanoemulsion based on droplet size and charge distribution (Mehrnia et al. 2017; Niu et al. 2016; Shahgholian and Rajabzadeh 2016). Nanoemulsions have distinct properties and applications as effective carriers for several lipophilic and hydrophilic bioactive components and drugs (Gupta et al. 2016; Hosseini et al. 2013; Esfanjani and Jafari 2016; Wang et al. 2011). Over the past years, the principal research focus has been on synthesizing nanoemulsion through both low- and high-energy methods (Shariffa et al. 2016; Yao et al. 2018; Ozturk and McClements 2016). In addition, several destabilization processes such as coalescence, flocculation, gravitational separation, creaming, and Ostwald ripening alter the droplet size, charge potential, and physicochemical properties of nanoemulsion (Gupta et al. 2016; Ha et al. 2015; Li et al. 2016; Taarji et al. 2018; Shariffa et al. 2016; Binsi et al. 2017; Ozturk et al. 2015; Shishir et al. 2018). Consequently, formulation of excellent quality nanoemulsion that could persist kinetic stability for a long period is desirable. Furthermore, both stabilizer and emulsifier are equally required for a persistent emulsion system. According to Dickinson and Stainsby (1988), an emulsifier could be a chemical or mixture of chemicals consisting of efficacy for the formulation of emulsion with short duration stabilization. On the other hand, stabilizers are the components that can attribute long-duration stability of the emulsion system. Hence, steric repulsive and electrostatic forces attributed by a suitable emulsifier direct the stability to the nanoemulsion. Both forces also significantly depend upon the type of emulsifier and ionic and nonionic surfactants (Ha et al. 2015; Li et al. 2016; Taarji et al. 2018; Shariffa et al. 2016; Binsi et al. 2017; Ozturk et al. 2015; Rachmawati et al. 2017). Addition of emulsifier during fabrication of the nanoemulsion maintains the kinetic stability against agglomeration and coalescence that makes nanoemulsion an effective vehicle than that of macro- and microemulsion system (Mehrnia et al. 2017; Niu et al. 2016; Shahgholian and Rajabzadeh 2016). Furthermore, stability of nanoemulsion is always closely related with droplet size of the emulsion system. Appearance of the nanoemulsion significantly depends upon droplet size of the emulsion system. Wavelength of visible light larger than droplet size of the nanoemulsion often attributes transparent appearance, which correlates with the higher stability of the nanoemulsion. Preparation methods always influence the drop size of the emulsion, and appearance of emulsion always ranges from transparent to milky white in contrast to method of synthesis (Abbas et al. 2015; Sari et al. 2015; Devi et al. 2017; Fu et al. 2018; Lv et al. 2014; Shishir et al. 2018).

3.3 Structure and Physicochemical Properties of Gum Arabic

Predominantly, gum arabic is obtained from *Acacia senegal* and *Acacia seyal*, and the physicochemical properties and chemical structure are always influenced by the environmental conditions, tree age, and undeniably the processing conditions (Al Assaf et al. 2005; Flindt et al. 2005; Hassan et al. 2005; Siddig et al. 2005). Over the past years, several researchers explored chemical composition and molecular structure of gum arabic at large extent. Exclusive emulsifying and rheological properties and application of gum arabic in context to stability were also identified (Abbas et al. 2015; Sari et al. 2015; Devi et al. 2017; Fu et al. 2018; Lv et al. 2014; Shishir et al. 2018; Mehrnia et al. 2017; Niu et al. 2016; Shahgholian and Rajabzadeh 2016). In general, gum arabic is a plant-based exudate gum obtained from Acacia trees. Predominantly, gum arabic is composed of two basic entities, i.e., oligosaccharides and arabinogalactan, which contribute to approximately 97% of the total composition of gum arabic. Less than 3% proportion of gum arabic is composed of protein structures (Sari et al. 2015; Gupta et al. 2015; Devi et al. 2017).

The mainstay of gum arabic is recognized to comprise α-(1→3)-linked d-galactopyranosyl units, whereas side chains of gum arabic are composed of 2–5 β-(1→3)-linked D-galactopyranosyl units that are directly attached with the main chain by 1–6-linkages, respectively (Flindt et al. 2005; Hassan et al. 2005; Siddig et al. 2005). Both the main chain and side chain are composed of units of α-larabinofuranosyl and α-l-rhamnopyranosyl (Dauqan and Abdullah 2013). Physical appearance of gum arabic is pale to orange brown in color, and after disruption, it produces translucent material (Shariffa et al. 2016; Binsi et al. 2017; Ozturk et al. 2015). Excellent quality gum attributes vitreous appearance, and after shattering, pieces of gum arabic become shiny pale in color. Furthermore, chemically, gum arabic is slightly acidic in nature due to intrinsic components such as glycoproteins and magnesium, calcium, and potassium salts of polysaccharides. Proximately, gum arabic exhibits moisture content (13–15%), ash content (2–4%), internal energy (30–39%), volatile matter (51–65%), nitrogen content (0.26–0.39%), and mineral composition consisting of iron (730–2490 ppm), copper (52–66 ppm), zinc (45–110 ppm), and manganese (69–117 ppm) (Dauqan and Abdullah 2013). In addition, gum arabic also consists of vital amino acids (nm/mg of gum arabic) such as serine (28.70), threonine (15.90), proline (15.60), hydroxyproline (54.20), leucine (15.10), histidine (10.70), glutamic acid (8.29), aspartic acid (10.60), valine (7.29), phenylalanine (6.33), lysine (5.13), alanine (5.07), isoleucine (2.38), tyrosine (2.30), arginine (2.12), and methionine (0.11) (Dauqan and Abdullah 2013; Montenegro et al. 2012). Apart from physicochemical composition and properties, gum arabic could be used as both emulsifier and stabilizer. Intrinsic composition of gum arabic attributes significant emulsifying properties, and several authors explored gum arabic as effective agent in stabilizing and delivery of vital bioactive components in emulsion system (Ha et al. 2015; Li et al. 2016; Taarji et al. 2018; Shariffa et al. 2016).

3.4 Stabilization of Nanoemulsion Using Gum Arabic

Several factors are controlled in order to formulate smooth and stable emulsion through either high-energy or low-energy methods (Taarji et al. 2018; Shariffa et al. 2016; Binsi et al. 2017; Ozturk et al. 2015). As discussed earlier, effective suitable gum plays a significant role to stabilize the emulsion system; therefore, gum arabic has been excessively used as emulsifier and stabilizer. Composition of emulsion is an important criterion in contrast to formulated stable emulsion. Herein, gum arabic controls the method of shear and slows down the Ostwald ripening as dispersed phase is highly insoluble in continuous phase. Furthermore, in stabilization of nanoemulsion, surfactant led to formation of lyotropic liquid crystals. Significant excess of ionic and nonionic surfactant in continuous phase empowers new surface area for nanodroplets and gum arabic attributes rapid coating by impeding shear-induced coalescence. In addition, surfactants form micelles, which directly adsorb on to oil drops, and stability of nanoemulsion occurs. Whereas, high shear stress in combination with appropriate emulsifier and surfactant generate nanodroplets (Laplace pressure). Furthermore, protein moiety is also exclusively responsible for the stabilization of nanoemulsion. Due to this surface activity, gum arabic adsorbs favorably on to the surface of the oil phase. It is predicted that the lipophilic polypeptide restraints adsorb and anchor the molecules to the surface, whereas carbohydrate moiety prevents destabilization (amalgamation and flocculation) through steric repulsion and electrostatic behavior (Abbas et al. 2015; Sari et al. 2015; Devi et al. 2017; Fu et al. 2018; Lv et al. 2014; Shishir et al. 2018). In this context, Randall et al. (1988) unveiled that gums with highest nitrogen content attribute significantly highest emulsion capacity and stability as well.

3.5 Food Applications of Nanoemulsions

Over the years, proficient technologies have made conceivable the large-scale fabrication of nanoemulsions and their broad applications in food system. In diverse food applications, nanoemulsions proficiently exist in several products such as spreads, butter, salad creams, beverages, mayonnaise, and desserts in both water-in-oil and oil-in-water emulsion forms (McClements 2017). Significant contribution of nanoemulsion systems to improve the organoleptic properties and physicochemical attributes has already been revealed. According to Donsì and Ferrari (2016), to distribute convinced notch of functionality, nanoemulsion system should be fabricated according to the bioactive components that directly correlate with physicochemical attributes and size of the component in the product and body. Therefore, anticipated in-product and in-body behavior directly influence the application of the nanoemulsion in food system. Delivery of micronutrients, bioactive components (flavanols, flavones, flavanones, isoflavones, hydroxybenzoic acids, hydroxycinnamic acids, stilbenes, curcuminoids, carotenes, and xanthophylls), colors, flavors, and antimicrobial constituents is the principal application achieved by the nanoemulsion delivery system (Salvia-Trujillo et al. 2015, 2017; Yucel et al. 2015; Dario et al. 2016) (Fig. 3.1).

In this context, Gadkari et al. (2017) formulated epigallocatechin gallate (EGCG), catechin, and epicatechin (from green tea) containing nanoemulsions and identified their anticancer, free radical scavenging, cholesterol-lowering, and prevention of arterial sclerosis abilities. Enhanced storage stability of highly degradable catechins was also revealed. Dario et al. (2016) revealed improved thermal

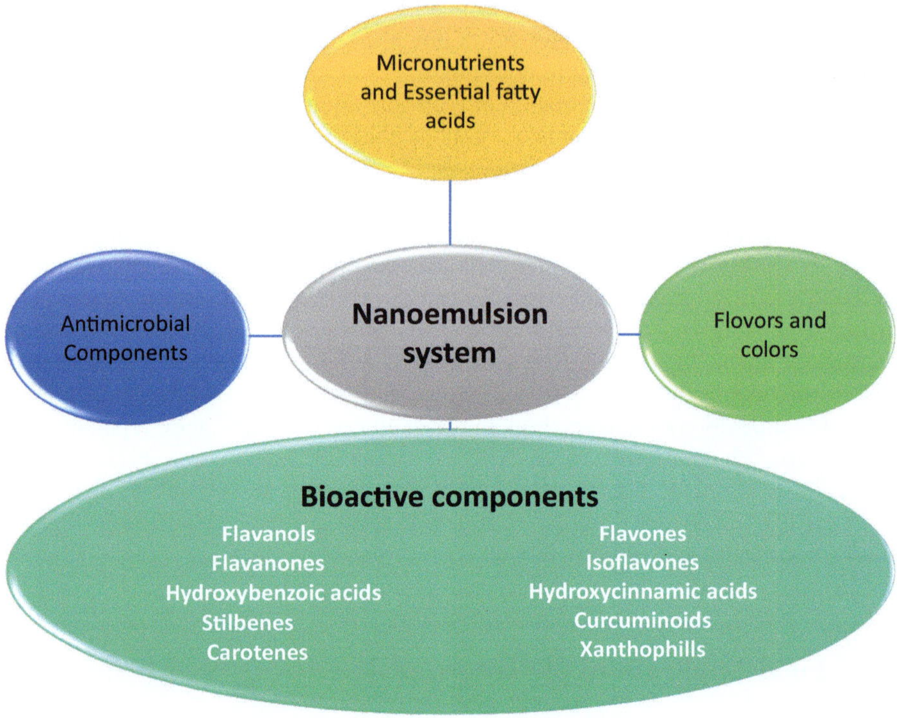

Fig. 3.1 Application of nanoemulsion for the delivery of distinctive components in food system

stability and photostability of quercetin-, kaempferol-, and myricetin-loaded nanoemulsion and evaluated free radical scavenging, anti-aging, and anti-inflammatory abilities of the respective nanoemulsion. Furthermore, formulation of bioactive component-loaded nanoemulsion faces several challenges, and one of the significant problem is formulation of crystals by supersaturated flavones in oil phases. Biopolymers slow down the formation of crystals in oil phase. Wan et al. (2017) fabricated apigenin-, luteolin-, rutin-, and tangeretin-loaded nanoemulsions and observed antimutagenic, antiproliferative, and antiinvasive efficiency of nanoemulsion, respectively. Similarly, Khan et al. (2015) also synthesized naringenin- and hesperidin-loaded nanoemulsion with significantly improved release and bioavailability of these bioactive components, respectively. They successfully evaluated anticarcinogenic, anti-inflammatory, hepatoprotective, and antilipid peroxidation activity of the nanoemulsion. In addition, Avachat and Patel (2015) formulated gallic acid-, ellagic acid-, and p-hydroxybenzoic acid-hindered nanoemulsion and demonstrated their antiproliferative and antioxidant efficacy with significantly increased bioavailability and reduced interaction with the food matrix. Furthermore, cinnamic-, coumaric-, ferulic-, and caffeic acid-encumbered nanoemulsion were formulated by Harwansh et al. (2015), and its antioxidant, prevention of cell damage, anti-aging, and antidiabetic activities were observed.

Stilbenes are a vital class of bioactive components in food system. Sessa et al. (2014) formulated trans-resveratrol-loaded nanoemulsion and found improved UV-light stability and enhanced transcellular absorption and permeation of these bioactive components. In addition, emulsion was evaluated for anti-inflammatory, lowering of low-density lipoproteins, and anti-cancer properties.

Curcuminoids are another important class of bioactive components. Yucel et al. (2015) and Vecchione et al. (2016) formulated curcumin-embedded nanoemulsion. It was found that curcumin exhibits poor water solubility, and in oil phase, curcumin requires amphiphilic components such as lecithin to fabricate stable emulsion system. Some abilities of curcumin-loaded nanoemulsion, viz., boosting brain-derived neurotrophic factor, lowering the risk of heart disease, anticancer, and fighting age-related chronic diseases ability, were also examined. Fan et al. (2017) fabricated nanoemulsions containing lycopene and β-carotene with considerably improved physical stability and control of lipolysis and release of pigments.

Apart from the bioactive components, nanoemulsions have been widely explored to convey vigorous micronutrients and essential fatty acids. Distinctive fat-soluble vitamins (i.e., A, D, and E) and polyunsaturated fatty acids (omega-3, omega-6, and conjugated linoleic acid) (Figs. 3.2 and 3.3) have been widely conveyed using nanoemulsion technique. Owing to the fact, the nanoemulsion system effectively protect these vital constituents against environmental stress and deliver these compounds at specific targeted areas in the body. Several researchers have explored significance of nanoemulsion system for delivery of vital components against several diseases.

In this context, Chen et al. (2016), Walker et al. (2015), and Karthik and Anandharamakrishnan (2016) formulated nanoemulsions loaded with omega-3, omega-6, and omega-9 with significantly enhanced oxidative stability and organoleptic properties. They explored the efficient applications of nanoemulsion against arthritis, diabetes, asthma, osteoporosis, obesity, bad cholesterol, macular degeneration, and contrasting digestive difficulties, respectively. In addition, Fernandez-Avila

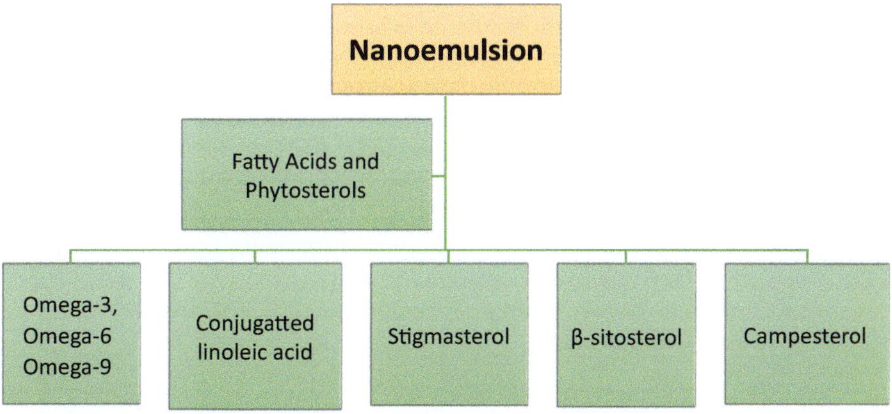

Fig. 3.2 Essential fatty acids and phytosterols delivered through nanoemulsion system

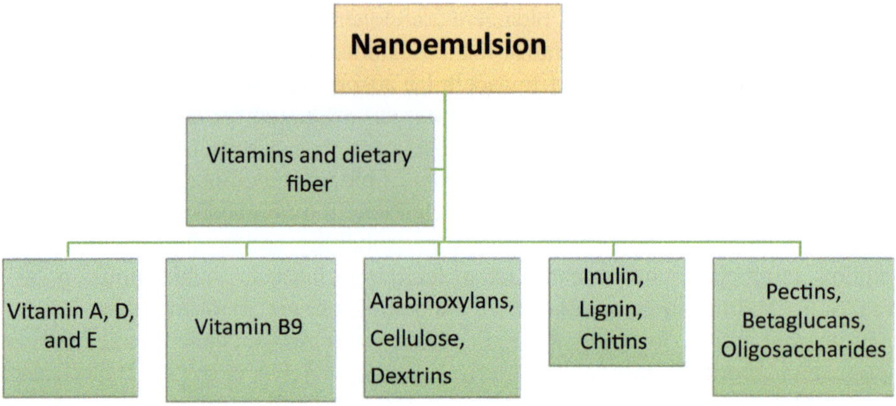

Fig. 3.3 Essential vitamins and dietary fiber delivered through nanoemulsion system

et al. (2017) revealed antidiabetic, antilipogenic, anticancer, and antiatherosclerotic effects of conjugated linoleic acid nanoemulsion potentially affecting immune system and bone modeling. Oxidative stability and influence of food matrix in emulsion system were also explored. Similarly, Lane et al. (2014) also formulated conjugated linoleic acid-loaded nanoemulsion with significantly enhanced in-product bioavailability and observed significant functionality against several diseases. Furthermore, Chen et al. (2016) and Panpipat et al. (2013) fabricated β-sitosterol-loaded nanoemulsion and observed cholesterol-lowering efficacy with significantly improved storage and oxidative stability. Artiga-Artigas et al. (2017) formulated dietary fiber-loaded nanoemulsion and found that the fibers located at the oil-water interface have tendency to form an edible coating with significant health attributes. Several authors also formulated vitamin A-, D-, E-, and B9-loaded nanoemulsion and explored vital functionalities of the nanoemulsion system with significantly enhanced bioavailability and storage stability (Ozturk et al. 2014; Ricaurte et al. 2016; Guttoff et al. 2015; Assadpour et al. 2016).

In contrast, stability of coloring and flavoring compounds gained the attention of food scientists, and over the last few years, effective approaches to overcome this problem have been explored. Essential oils are excessively used as aromatic components in food system, and use of nanoemulsion has been approved as an effective approach to deliver valuable aromatic flavors under drastic conditions (Majeed et al. 2016; Tian et al. 2017). The droplet size is important factor during delivery of coloring and flavoring components in transparent beverages. Therefore, optically transparent nanoemulsions (below 60 nm size) could be used as effective system forthe delivery of coloring compounds.

3.6 Gum Arabic-Stabilized Nanoemulsion as Delivery Vehicle

Over the past years, several researchers formulated nanoemulsion stabilized with gum arabic or combination of gum arabic with other biopolymers. Herein, recent studies for the effective delivery of bioactive components and hydrophobic micronutrients using nanoemulsion are discussed. Yao et al. (2013) fabricated gum arabic-stabilized oil-in-water nanoemulsion of conjugated linoleic acid (CLA). For evaluation of effective delivery of CLA, three types of gums, i.e., conventional gum arabic (GA) and two matured gum arabic samples (EM2 and EM10), were used, and the oil-in-water nanoemulsion was formulated using all the three types of gum arabic samples. In this context, it was observed that higher gum arabic concentration attributes least physical stability that causes more lipid oxidation. Major reason for the least physical and chemical stability of nanoemulsion was coalescence and depletion prompted flocculation of the nanoemulsion droplets. Furthermore, it was found that matured EM10 gum arabic attributed effective protection of CLA. Possible reason was significant interfacial properties of matured EM10 gum arabic in comparison with conventional gum arabic and EM2. As discussed earlier gum arabic provides stability to nanoemulsion by adsorption onto oil droplets, and absorption kinetics of all gum arabic samples at CLA interface was evaluated. Yao et al. (2013) found rapid decrease in interfacial tension during initial adsorption state and also found decrease in onset time with increasing gum arabic concentration. In contrast, Zhang et al. (2015) formulated nanoemulsion of weighted orange oil terpenes using four types of biopolymers, i.e., gum arabic, modified gum arabic, whey protein isolates, and modified starch. Physicochemical and optical properties of nanoemulsions were examined, and mean droplet diameter in volume was the criterion of selection of appropriate nanoemulsion for the delivery of the orange oil terpenes. Ozturk et al. (2015) successfully formulated nanoemulsion for the delivery of vitamin E using gum arabic and whey protein isolates and revealed that gum arabic attributed effective stability to nanoemulsion against environmental stresses. Gum arabic-stabilized nanoemulsion exhibited higher ionic strength against ionic solution and wider range of pH values. Ionic stability of gum arabic-stabilized nanoemulsion was mainly stabilized by electrostatic repulsion and steric repulsion. At higher temperatures (30–90 °C), gum arabic-stabilized nanoemulsion was highly resistant to agglomeration and coalescence of the emulsion system. In addition, after evaluation of the effect of shear rate and temperature on apparent viscosity, gum arabic-stabilized nanoemulsion did not show any inferior changes in terms of apparent viscosity. Gum arabic-stabilized nanoemulsion also attributed higher bioavailability of vitamin E during in vitro experiments. Herein, according to their observation, gum arabic was suitable and appropriate emulsifier and stabilizer for the stabilization of vitamin E-enriched nanoemulsion. Bai et al. (2017) synthesized corn oil, fish oil, and lemon oil nanoemulsion stabilized with gum arabic, beet pectin, and corn fiber gum. Based upon the surface activity and interfacial tension of the emulsifier concentration and homogenization pressure on the emulsion formation, efficiency was evaluated. It was observed that the gum arabic and beet pectin attributed efficient emulsifying properties than corn fiber gum, with a least amount of emulsifier requisite and smaller droplets being produced. Greater reduction in interfacial tension and stronger adsorption led to the formation of more efficient disruption and low recoalescence inside the homogenizer for beet pectin and gum arabic. Study inferred valuable evidence for selecting polysaccharide-based emulsifiers for utilization in food and pharmaceutical application. Correspondingly, Binsi et al. (2017) examined synergistic ability of gum arabic and sage polyphenols in stabilization and protection of fish oil from high heat interruption and oxidative stability during spray-drying process. Higher stability and lower lipid oxidation during storage of the nanoemulsion were reported. During in vitro release action of oil, significantly higher release of oil in simulated gastrointestinal fluids was observed. Furthermore, Silva et al. (2015) fabricated annatto seed oil and gum arabic-stabilized nanoemulsion and found electrostatic and steric stabilization of gum arabic-stabilized nanoemulsion during assessment of ionic strength, creaming stability, and effect of pH and temperature. Similarly, Shariffa et al. (2016) also revealed significance of gum arabic on the physicochemical properties of lycopene nanodispersions. Su et al. (2018) studied the effect of gum arabic on storage stability and antimicrobial efficacy of β-lactoglobulin-stabilized D-limnene nanoemulsion. All mentioned studies revealed steric and electrostatic stability of gum arabic-stabilized nanoemulsion during different ionic, pH, and temperature conditions. Release of bioactive compounds during in vitro studies revealed significant efficiency.

3.7 Conclusion and Future Trends

Nanoemulsion technique is a well-established mode for the delivery of bioactive compounds and fat-soluble micronutrients. This technique exhibits improved functionality than macro- and microemulsion techniques in terms of greater protection, augmented stability, persistent release profile, and enhanced bioavailability of bioactive compounds. The efficacious application of nanoemulsion ultimately hinges on the selection of suitable emulsifier with desired rheological and physicochemical properties in sustenance to realm the target bioactive compounds. Polysaccharides particularly gum arabic have attracted considerable international attention due to its excellent emulsifying and stabilizing functionalities. On the contrary, high-energy and low-energy nanoemulsion formulation technologies are emerging that can provide excellent release profile and enhanced thermal, light, and oxidative stability. The future studies are required to cope with the limitations of nanoemulsion processes and improve the existing methods, as well as meet the commercial demands for their industrial-scale production. Further research should be emphasized on the application of bioactive-loaded nanoemulsion in food and biological systems in order to discover their influence on cell viability as well as adsorption, distribution, and metabolism.

References

Abbas S, Karangwa E, Bashari M, Hayat K, Hong X, Sharif HR, Zhang X (2015) Fabrication of polymeric nanocapsules from curcumin-loaded nanoemulsion templates by self-assembly. Ultrason Sonochem 23:81–92

Aguiar J, Estevinho BN, Santos L (2016) Microencapsulation of natural antioxidants for food application–The specific case of coffee antioxidants-A review. Trends Food Sci Technol 58:21–39

Al Assaf S, Phillips GO, Williams PA (2005) Studies on acacia exudate gums. Part I: the molecular weight of *Acacia senegal* gum exudate. Food Hydrocoll 19(4):647–660

Artiga-Artigas M, Acevedo-Fani A, Martín-Belloso O (2017) Improving the shelf life of low-fat cut cheese using nanoemulsion-based edible coatings containing oregano essential oil and mandarin fiber. Food Control 76:1–2

Assadpour E, Maghsoudlou Y, Jafari SM, Ghorbani M, Aalami M (2016) Evaluation of folic acid nano-encapsulation by double emulsions. Food Bioprocess Technol 9(12):2024–2032

Avachat AM, Patel VG (2015) Self nanoemulsifying drug delivery system of stabilized ellagic acid–phospholipid complex with improved dissolution and permeability. Saudi Pharm J 23(3):276–289

Bai L, Huan S, Li Z, McClements DJ (2017) Comparison of emulsifying properties of food-grade polysaccharides in oil-in-water emulsions: Gum arabic, beet pectin, and corn fiber gum. Food Hydrocoll 66:144–153

Binsi PK, Nayak N, Sarkar PC, Jeyakumari A, Ashraf PM, Ninan G, Ravishankar CN (2017) Structural and oxidative stabilization of spray dried fish oil microencapsulates with gum arabic and sage polyphenols: characterization and release kinetics. Food Chem 219:158–168

Chen XW, Chen YJ, Wang JM, Guo J, Yin SW, Yang XQ (2016) Phytosterol structured algae oil nanoemulsions and powders: improving antioxidant and flavor properties. Food Funct 7(9):3694–3702

Dario MF, Oliveira CA, Cordeiro LR, Rosado C, Inês de Fátima AM, Maçôas E, Santos MS, da Piedade ME, Baby AR, Velasco MV (2016) Stability and safety of quercetin-loaded cationic nanoemulsion: in vitro and in vivo assessments. Colloids Surf A Physicochem Eng Asp 506:591–599

Dauqan E, Abdullah A (2013) Utilization of gum arabic for industries and human health. Am J Appl Sci 10(10):1270

de Souza Simões L, Madalena DA, Pinheiro AC, Teixeira JA, Vicente AA, Ramos OL (2017) Micro-and nano bio-based delivery systems for food applications: in vitro behavior. Adv Colloid Interf Sci 243:23–45

Devi N, Sarmah M, Khatun B, Maji TK (2017) Encapsulation of active ingredients in polysaccharide-protein complex coacervates. Adv Colloid Interf Sci 239:136–145

Dickinson E, Stainsby G (1988) Advances in food emulsions and foams. Elsevier Science Publishers, New York

Donsì F, Ferrari G (2016) Essential oil nanoemulsions as antimicrobial agents in food. J Biotechnol 233:106–120

Esfanjani AF, Jafari SM (2016) Biopolymer nano-particles and natural nano-carriers for nano-encapsulation of phenolic compounds. Colloids Surf B: Biointerfaces 146:532–543

Fan Y, Yi J, Zhang Y, Wen Z, Zhao L (2017) Physicochemical stability and in vitro bioaccessibility of β-carotene nanoemulsions stabilized with whey protein-dextran conjugates. Food Hydrocoll 63:256–264

Fernandez-Avila C, Gutierrez-Merida C, Trujillo AJ (2017) Physicochemical and sensory characteristics of a UHT milk-based product enriched with conjugated linoleic acid emulsified by Ultra-High Pressure Homogenization. Innovative Food Sci Emerg Technol 39:275–283

Flindt C, Al-Assaf S, Phillips GO, Williams PA (2005) Studies on acacia exudate gums. Part V. Structural features of *Acacia seyal*. Food Hydrocoll 19(4):687–701

Fu C, Ding C, Sun X, Fu A (2018) Curcumin nanocapsules stabilized by bovine serum albumin-capped gold nanoclusters (BSA-AuNCs) for drug delivery and theranosis. Mater Sci Eng C 87:149–154

Gadkari PV, Shashidhar MG, Balaraman M (2017) Delivery of green tea catechins through Oil-in-Water (O/W) nanoemulsion and assessment of storage stability. J Food Eng 199:65–76

Gupta C, Chawla P, Arora S, Tomar SK, Singh AK (2015) Iron microencapsulation with blend of gum arabic, maltodextrin and modified starch using modified solvent evaporation method-milk fortification. Food Hydrocoll 43:622–628

Gupta A, Eral HB, Hatton TA, Doyle PS (2016) Nanoemulsions: formation, properties and applications. Soft Matter 12(11):2826–2841

Guttoff M, Saberi AH, McClements DJ (2015) Formation of vitamin D nanoemulsion-based delivery systems by spontaneous emulsification: factors affecting particle size and stability. Food Chem 171:117–122

Ha TVA, Kim S, Choi Y, Kwak HS, Lee SJ, Wen J, Ko S (2015) Antioxidant activity and bioaccessibility of size-different nanoemulsions for lycopene-enriched tomato extract. Food Chem 178:115–121

Harwansh RK, Mukherjee PK, Bahadur S, Biswas R (2015) Enhanced permeability of ferulic acid loaded nanoemulsion based gel through skin against UVA mediated oxidative stress. Life Sci 141:202–111

Hassan EA, Al-Assaf S, Phillips GO, Williams PA (2005) Studies on Acacia gums: Part III molecular weight characteristics of *Acacia seyal* var. seyal and *Acacia seyalvar fistula*. Food Hydrocoll 19(4):669–677

Hosseini SF, Zandi M, Rezaei M, Farahmandghavi F (2013) Two-step method for encapsulation of oregano essential oil in chitosan nanoparticles: preparation, characterization and in vitro release study. Carbohydr Polym 95(1):50–56

Jia Z, Dumont MJ, Orsat V (2016) Encapsulation of phenolic compounds present in plants using protein matrices. Food Biosci 15:87–104

Karthik P, Anandharamakrishnan C (2016) Enhancing omega-3 fatty acids nanoemulsion stability and in-vitro digestibility through emulsifiers. J Food Eng 187:92–105

Khan AW, Kotta S, Ansari SH, Sharma RK, Ali J (2015) Self-nanoemulsifying drug delivery system (SNEDDS) of the poorly water-soluble grapefruit flavonoid Naringenin: design, characterization, in vitro and in vivo evaluation. Drug Deliv 22(4):552–561

Lane KE, Li W, Smith C, Derbyshire E (2014) The bioavailability of an omega 3 rich algal oil is improved by nanoemulsion technology using yogurt as a food vehicle. Int J Food Sci Technol 49(5):1264–1271

Li J, Hwang IC, Chen X, Park HJ (2016) Effects of chitosan coating on curcumin loaded nano-emulsion: study on stability and in vitro digestibility. Food Hydrocoll 60:138–147

Lv Y, Yang F, Li X, Zhang X, Abbas S (2014) Formation of heat-resistant nanocapsules of jasmine essential oil via gelatin/gum arabic based complex coacervation. Food Hydrocoll 35:305–314

Majeed H, Antoniou J, Hategekimana J, Sharif HR, Haider J, Liu F, Ali B, Rong L, Ma J, Zhong F (2016) Influence of carrier oil type, particle size on in vitro lipid digestion and eugenol release in emulsion and nanoemulsions. Food Hydrocoll 52:415–422

McClements DJ (2017) The future of food colloids: next-generation nanoparticle delivery systems. Curr Opin Colloid Interface Sci 28:7–14

Mehrnia MA, Jafari SM, Makhmal-Zadeh BS, Maghsoudlou Y (2017) Rheological and release properties of double nano-emulsions containing crocin prepared with Angum gum, Arabic gum and whey protein. Food Hydrocoll 66:259–267

Montenegro MA, Boiero ML, Valle L, Borsarelli CD (2012) Gum Arabic: more than an edible emulsifier. In: Products and applications of biopolymers. InTech., Dr. Johan Verbeek (Ed.), ISBN: 978-953-51-0226-7, Available from: http://www.intechopen.com/books/products-and-applications-of-biopolymers/gum-arabic-more-than-an-edible-emulsifier

Niu F, Pan W, Su Y, Yang Y (2016) Physical and antimicrobial properties of thyme oil emulsions stabilized by ovalbumin and gum arabic. Food Chem 212:138–145

Ozturk B, McClements DJ (2016) Progress in natural emulsifiers for utilization in food emulsions. Curr Opin Food Sci 7:1–6
Ozturk B, Argin S, Ozilgen M, McClements DJ (2014) Formation and stabilization of nanoemulsion-based vitamin E delivery systems using natural surfactants: quillaja saponin and lecithin. J Food Eng 142:57–63
Ozturk B, Argin S, Ozilgen M, McClements DJ (2015) Formation and stabilization of nanoemulsion-based vitamin E delivery systems using natural biopolymers: whey protein isolate and gum arabic. Food Chem 188:256–263
Panpipat W, Dong M, Xu X, Guo Z (2013) Thermal properties and nanodispersion behavior of synthesized β-sitosteryl acyl esters: a structure–activity relationship study. J Colloid Interface Sci 407:177–186
Rachmawati H, Novel MA, Nisa RM, Berlian G, Tandrasasmita OM, Rahma A, Tjandrawinata RR (2017) Co-delivery of curcumin-loaded nanoemulsion and Phaleriamacrocarpa extract to NIH 3T3 cell for antifibrosis. J Drug Deliv Sci Technol 39:123–130
Randall RC, Phillips GO, Williams PA (1988) The role of the proteinaceous component on the emulsifying properties of gum arabic. Food Hydrocoll 2(2):131–140
Ricaurte L, de Jesús Perea-Flores M, Martinez A, Quintanilla-Carvajal MX (2016) Production of high-oleic palm oil nanoemulsions by high-shear homogenization (microfluidization). Innovative Food Sci Emerg Technol 35:75–85
Salvia-Trujillo L, Rojas-Graü A, Soliva-Fortuny R, Martín-Belloso O (2015) Physicochemical characterization and antimicrobial activity of food-grade emulsions and nanoemulsions incorporating essential oils. Food Hydrocoll 43:547–556
Salvia-Trujillo L, Soliva-Fortuny R, Rojas-Graü MA, McClements DJ, Martín-Belloso O (2017) Edible nanoemulsions as carriers of active ingredients: a review. Annu Rev Food Sci Technol 8:439–466
Sari TP, Mann B, Kumar R, Singh RRB, Sharma R, Bhardwaj M, Athira S (2015) Preparation and characterization of nanoemulsion encapsulating curcumin. Food Hydrocoll 43:540–546
Sessa M, Balestrieri ML, Ferrari G, Servillo L, Castaldo D, D'Onofrio N, Donsì F, Tsao R (2014) Bioavailability of encapsulated resveratrol into nanoemulsion-based delivery systems. Food Chem 147:42–50
Shahgholian N, Rajabzadeh G (2016) Fabrication and characterization of curcumin-loaded albumin/gum arabic coacervate. Food Hydrocoll 59:17–25
Shariffa YN, Tan TB, Abas F, Mirhosseini H, Nehdi IA, Tan CP (2016) Producing a lycopene nanodispersion: the effects of emulsifiers. Food Bioprod Process 98:210–216
Shishir MRI, Xie L, Sun C, Zheng X, Chen W (2018) Advances in micro and nano-encapsulation of bioactive compounds using biopolymer and lipid-based transporters. Trends Food Sci Technol 78:34–60
Siddig NE, Osman ME, Al-Assaf S, Phillips GO, Williams PA (2005) Studies on acacia exudate gums, part IV. Distribution of molecular components in *Acacia seyal* in relation to *Acacia senegal*. Food Hydrocoll 19(4):679–686
Silva EK, Gomes MT, Hubinger MD, Cunha RL, Meireles MA (2015) Ultrasound-assisted formation of annatto seed oil emulsions stabilized by biopolymers. Food Hydrocoll 47:1–3
Su J, Guo Q, Mao L, Gao Y, Yuan F (2018) Effect of gum arabic on the storage stability and antibacterial ability of β-lactoglobulin stabilized d-limonene emulsion. Food Hydrocoll 84:75–83
Taarji N, da Silva CAR, Khalid N, Gadhi C, Hafidi A, Kobayashi I et al (2018) Formulation and stabilization of oil-in-water nanoemulsions using a saponins-rich extract from argan oil press-cake. Food Chem 246:457–463
Tian H, Li D, Xu T, Hu J, Rong Y, Zhao B (2017) Citral stabilization and characterization of nanoemulsions stabilized by a mixture of gelatin and Tween 20 in an acidic system. J Sci Food Agric 7(9):2991–2998
Vecchione R, Quagliariello V, Calabria D, Calcagno V, De Luca E, Iaffaioli RV, Netti PA (2016) Curcumin bioavailability from oil in water nano-emulsions: in vitro and in vivo study on the dimensional, compositional and interactional dependence. J Control Release 233:88–100

Walker RM, Decker EA, McClements DJ (2015) Physical and oxidative stability of fish oil nanoemulsions produced by spontaneous emulsification: effect of surfactant concentration and particle size. J Food Eng 164:10–20
Wan J, Li D, Song R, Shah BR, Li B, Li Y (2017) Enhancement of physical stability and bioaccessibility of tangeretin by soy protein isolate addition. Food Chem 221:760–770
Wang J, Cao Y, Sun B, Wang C (2011) Physicochemical and release characterisation of garlic oil-β-cyclodextrin inclusion complexes. Food Chem 127(4):1680–1685
Yao X, Xu Q, Tian D, Wang N, Fang Y, Deng Z, Lu J (2013) Physical and chemical stability of gum arabic-stabilized conjugated linoleic acid oil-in-water emulsions. J Agric Food Chem 61(19):4639–4645
Yao X, Nie K, Chen Y, Jiang F, Kuang Y, Yan H, Phillips GO (2018) The influence of non-ionic surfactant on lipid digestion of gum Arabic stabilized oil-in-water emulsion. Food Hydrocoll 74:78–86
Yucel C, Quagliariello V, Iaffaioli RV, Ferrari G, Donsì F (2015) Submicron complex lipid carriers for curcumin delivery to intestinal epithelial cells: effect of different emulsifiers on bio accessibility and cell uptake. Int J Pharm 494(1):357–369
Zhang J, Peppard TL, Reineccius GA (2015) Preparation and characterization of nanoemulsions stabilized by food biopolymers using microfluidization. Flavour Fragr J 30(4):288–294

Chapter 4
Nanoemulsions: A Promising Tool for Dairy Sector

Anil Panghal, Navnidhi Chhikara, V. Anshid, Manga Veera Sai Charan, Vinod Surendran, Anju Malik, and Sanju Bala Dhull

Contents

A. Panghal · N. Chhikara · V. Anshid · M. V. Sai Charan · V. Surendran
Department of Food Technology and Nutrition, Lovely Professional University, Jalandhar, Punjab, India

A. Malik
Department of Energy and Environmental Sciences, Chaudhary Devi Lal University, Sirsa, Haryana, India

S. B. Dhull (✉)
Department of Food Science and Technology, Chaudhary Devi Lal University, Sirsa, Haryana, India

R. Prasad et al. (eds.), *Nanobiotechnology in Bioformulations*, Nanotechnology in the Life Sciences, https://doi.org/10.1007/978-3-030-17061-5_4

4.1 Introduction

The definition for emulsion by the International Union of Pure and Applied Chemistry (IUPAC) is: "In the case of emulsion, liquid crystals or liquid droplets are always dispersed in a liquid". Emulsion are formed by mixing two or more immiscible phases or fluids where dispersed phase (in the form of droplets) is suspended in continuous phase (Fig. 4.1). Milk is an oil-in-water emulsion in which fat molecules are mostly situated in spherical globules. A blend of water and oil may result in a rough nonpermanent emulsion, which will isolate into two different systems upon standing because of the mixture of the scattered globules. The characteristics of the emulsion droplets depend upon various factors like pH, temperature, ionic strength, total soluble solids, and type of emulsifiers (mostly proteins, i.e., casein or whey protein) present at the interface. Emulsion can be classified as nanoparticles, microcapsules, and microreservoir as per the particle size. Emulsifying agents, and oil are the most essential components to make a stable emulsion. There are many oils that can be used for the nanoemulsion methods like coconut oil, evening primrose oil, castor oil, corn oil, linseed oil, olive oil, peanut oil, and mineral oils. An emulsifying agent ought to be nontoxic and good in taste, possess chemical stability, and have the capacity to decrease the surface tension beneath 10 dynes/cm. Around the scattered stage, these

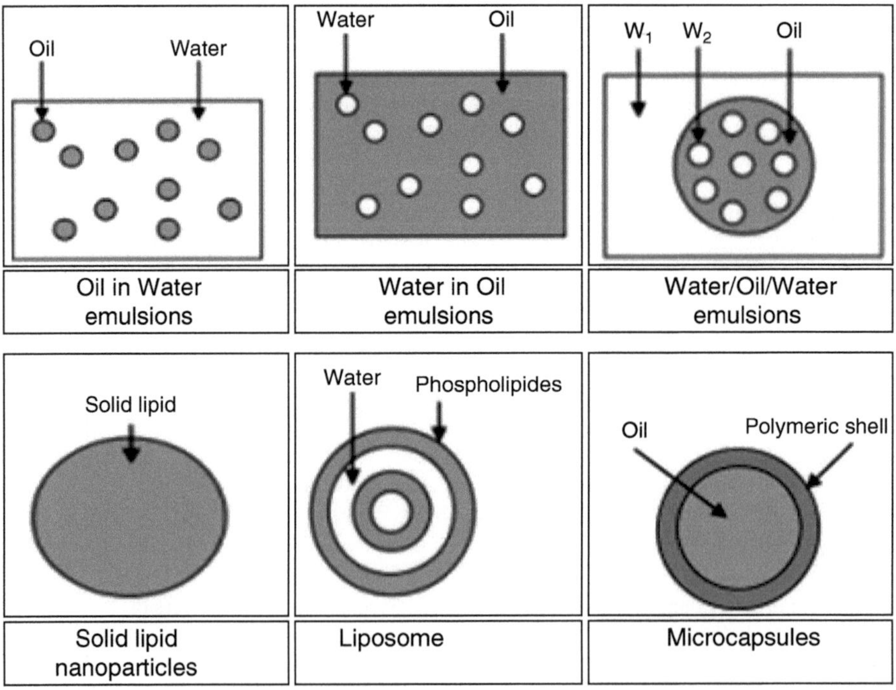

Fig. 4.1 Different nanomaterials prepared by (i) membrane emulsification and (ii) membrane mixing (adapted from Charcosset 2016)

globules ought to be adsorbed quickly and should form a reasonable film on the emulsion mixture. Emulsifies are also called as emulgents which help to stabilize the emulsion by a raise in their kinetic stability. The emulsifying agents should be helpful for maintaining a sufficient viscosity and zeta potential in the system, which ensures optimum stability. Emulsifying agents frames particulate by developing monomolecular as well as multimolecular films around the scattered globules. Whey and casein proteins present in milk act as good emulsifying agents. The adsorption of these proteins takes place at the interface of air-water/oil-water avoiding the coalescence by maintaining the interfacial membranes. Due to the amphiphilic structure, proteins present in milk behave as surface-active components in emulsion, and they stabilize the emulsion droplets by combining steric stabilization and electrostatic mechanism (Corredig 2009).

There are two types of emulsion based on their dispersion phase.

Oil-in-Water Type of Emulsion (O/W)
The emulsion in which water acts as dispersion medium (continuous phase) and oil will be dispersed in it, for example, milk, where fat globules are the dispersed phase and water is the dispersion medium.

Water-in-Oil Type of Emulsion (W/O)
It is a type of emulsion in which oil is the dispersion phase and water is dispersed in it. These types of emulsion are also called as oil emulsion. The best examples for this type of emulsion are butter and cold cream.

Microemulsion (the definition given by IUPAC) is the "dispersion made of water, oil, and surfactant(s) which is isotropic and thermodynamically stable system with diameter varying approximately from 10 to 100 nm and usually 10 to 50 nm." Nanoemulsion are overlapping microemulsions in size but possess better structural components and prolonged thermodynamic stability, or they are thermodynamically stable transparent dispersion of oil and water stabilized by an interfacial film of surfactant and co-surfactant molecules having droplet size of less than 100 nm. Microemulsions are thermodynamically steady, and these are the systems with equilibrium state, whereas nanoemulsion are not in equilibrium but are having a capacity of spontaneous phenomena to divide into the constituent phases, and they can remain stable for several years due to extreme kinetic stability. Nanoemulsion technique is possible to control morphology, surface area, geometry, homogeneity, rheological behavior, and other properties of particle without or with minimum usage of reducing agents or external stabilizing agents. It provides a better absorbability of oil-dissolvable supplements in cell-culture innovation. Other than the stabilization benefits, nanoemulsion might be utilized as substitutes for vesicles and liposomes (Fig. 4.1). These types of emulsion have more surface area and better kinetic stability against coalescence or creaming and can be considered as a good carrier for hydrophobic bioactive compounds. Emulsion in dairy products are less stable toward the variations in ionic strength, pH, and processing situations such as shearing, heating, or cooling. Nanoemulsion technology is offering a number of new advances to milk industry for the improvement of safety, quality, shelf life, and healthiness of products. The lipophilic components can be solubilized using nanoemulsion. The absorption of nanoemulsion will be very high due to their good surface area and minute droplet size. These emulsion can be made into different forms like foams, liquids, sprays, and creams. Nanoemulsion can be utilized for the bioavailability enhancement of valuable nutrients in dairy-related products. Nowadays, the attention has laid on the use of nanoemulsion in food-processing industries to deliver functional properties of ingredients in food matrix mainly in dairy industry.

4.2 Advantages of Nanoemulsion in Dairy Sector

(i) The small-sized droplets account for a vast decline in the gravity compel, and the brownian movement might be adequate for reducing gravity. This implies no creaming or sedimentation happens during the storage period in the milk.
(ii) The flocculation of drops can also be prevented with the help of these small-sized milk droplets which helps in stabilizing the total system.
(iii) It provides an opportunity to improve the texture, mouthfeel richness, and taste of milk by altering the flow ability.
(iv) The major ingredients of milk like proteins, fat, sugars, carbohydrates, vitamins, etc. can be coated using this technique which helps them to cope up with the harsh conditions of pH change in the stomach and also enhance the bioavailability of nutrients.
(v) The encapsulation of the milk components and the process of droplet size reduction will help in increasing the shelf life.
(vi) Such types of emulsion will have a very small size and thus a big surface area which help in the higher rate of penetration through the skin and are also helpful in transferring the necessary components to the body through the skin. Small size also assists in uniform and quick distribution throughout the body. Even their small droplet size can also help in increasing the rate of various processes like wetting, distributing, and penetrating power.
(vii) Nanoemulsion are transparent fluids requiring a small amount of surfactants, whereas microemulsions need high amounts of surfactants.
(viii) They also help in introducing certain good flavoring agents and thus enhance the overall acceptability of products.
(ix) The liposomes and vesicles can be replaced by the use of these types of the emulsions because of higher stability for a longer period. These emulsion can also help in the formation of liquid crystalline films around the drops.

4.3 Methods of Nanoemulsion Formation

Nanoemulsion preparation requires advanced techniques like homogenization and high pressure along with sophisticated instruments. Milk fat due to the effect of buoyancy will float on the surface of milk, to form a layer of cream and produce the phenomenon such as layering, seriously affecting the quality of dairy products. Different methods used in nanoemulsion development are discussed in the following sections.

4.3.1 High-Energy Approaches

High-energy approaches rely on providing intensive energy to disrupt the oil globules and water into very fine droplets with the help of high-pressure homogenizers and microfluidizers. Energy inputs modulate the properties of the nanoemulsion and excellently prevent droplet disruption and coalescence.

4.3.1.1 High-Pressure Homogenizers

Homogenization facilitates the production of dairy products with improved texture, taste, flavor, and shelf life. High-pressure homogenization is the most commonly used method of producing fine emulsion in the food industry. Regular high-weight homogenizers work on weights apparently in the range of 50 and 100 MPa. A coarse emulsion is usually produced using a high-speed blender and is then fed directly into the input of the high-pressure valve homogenizer. The pump of the homogenizer draws coarse emulsion to a chamber on its backstroke and then forces it to pass through a narrow valve at the end of the chamber on its forward stroke. Emulsion experiences a mix of extreme aggravating forces such as cavitation, turbulance and shear which convert bigger droplets into smaller ones. A variety of valves have been designed for different types of applications. Most commercial homogenizers use spring-loaded valves so that the gap through which the emulsion passes can be varied. Decreasing the gap size increases the pressure drop across the valve, which causes a greater degree of droplet disruption and smaller droplets to be produced. On the other hand, narrowing the gap increases the energy input required to form an emulsion, thereby increasing manufacturing costs. The wide range of sorts of spouts will be accessible to upgrade the proficiency of droplet size reduction in homogenizer.

Microfluidizer technology more efficiently converts high fluid pressure into shear forces and helps the industries to meet the performance standards for high-pressure homogenization. During microfluidization, the product enters the system via an inlet reservoir and is delivered by a high-pressure intensifier pump into an interaction chamber at speeds of up to 400 m/s and pressures in the range 20–275 MPa. In the fixed-geometry interaction chamber, the liquid is divided into two or more microstreams, and these high-velocity streams undergo a sudden pressure drop on collision at a perpendicular angle, causing turbulence, cavitation, shear, and ultimately droplet disruption. Through this process, a uniform distribution of small particles is maintained. The product is subjected to instant cooling followed by collection in the output reservoir (Fig. 4.2).

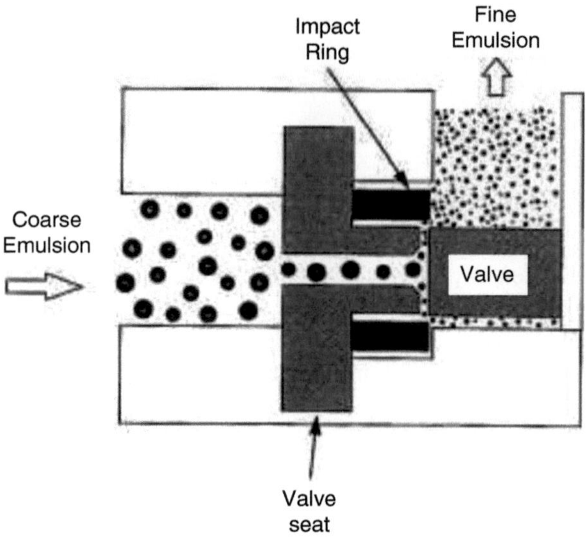

Fig. 4.2 High-pressure valve homogenizers for production of emulsion with fine droplet sizes

4.3.1.2 Microfluidizer

It is a mixing technique, which is somewhat similar in the high-pressure homogenizer. A high-pressure displacement pump is used in the device to force the product through the interaction chamber composing of many small channels called microchannel. Prior to the homogenizer by microfluidizer, the aqueous phase and oil phase were combined together to yield a coarse emulsion, usually obtained by high-speed homogenizer. The coarse emulsion then passed through a microfluidizer to produce stable nanoemulsion (Koroleva and Yurtov 2012; McClements 2012).

4.3.1.3 Ultrasonication

The ultrasound waves when pass through a liquid medium, the interaction between the ultrasonic waves, liquid, and dissolved gas leads to a phenomenon known as acoustic cavitation. Ultrasonic emulsification offers several benefits over conventional emulsification methods used in dairy systems such as mechanical shaking, high- or ultrahigh-pressure homogenization and microfluidization in terms of improved energy efficiency, higher emulsion stability, lowered requirement of surfactants, and controllable size distributions. Ultrasonic cavitation will produce strong mechanical action, under an appropriate ultrasonic frequency, resulting in better emulsification. Ultrasonic homogenizers are especially reasonable for less-consistency liquids and less appropriate for high viscous frameworks. The utilization of ultrasound has numerous points of interest; for example, less power utilization, less utilization of surfactant, droplet size reduction, and more homogeneous nature of emulsion. Omega-3-rich nanoemulsions are arranged by utilizing an ultrasound gadget (Cavazos-Garduño et al. 2015). Strunz et al. (2008) examined

Table 4.1 Cases of nanoemulsion produced by a method of high-energy approaches

Preparation method	Oil phase	Bioactive component	Surfactant	Diameter (nm)	References
High-pressure homogenization	Medium-chain triglycerides	β-Carotene	Tween-20–80	132–184	Yuan et al. (2008)
	Corn oil	–	SDS, Tween-20, sodium caseinate, β -lactoglobulin,	123–245	Qian and McClements (2011)
	Medium-chain triglycerides	–	Tween-80	150–250	Liedtke et al. (2000)
	Sunflower		Tween-80, SDS	90–190	Donsì et al. (2011)
Microfluidizers	Lauroglycol 90	Aspirin	Cremophor EL	150–170	Tang et al. (2013)
	Croton oil	–	Tween-80	42–758	Kuo et al. (2008)
	Soybean oil	Aspirin	Tween-80	70–123	Subramanian et al. (2008)
Ultrasonic homogenizers	D-limonene	–	Polyoxyethylene	20–100	Li and Chiang (2012)
	Basil oil	–	oleyl ether	20–50	Ghosh et al. (2013)
	Medium-chain triglycerides	Polylactic acid, Polyethylene glycol-polylactic acid	Tween-80	About 150	Preetz et al. (2010)
High-speed devices	Liquid lipid	–	Tween-80	210–290	Yilmaz and Borchert (2005)

that the utilization of the Brazil nut (wealthy in monounsaturated fatty acid and polyunsaturated fatty acid) can influence the plasma lipids, apolipoprotein, and other practical characteristics of antiatherogens with high-density lipoproteins, i.e., HDL. The lipid nanoemulsion was set up by delayed ultrasonic illumination in the fluid media and a two-step advanced ultracentrifugation of the unrefined emulsion (Table 4.1, Fig. 4.3).

4.3.2 Low-Energy Methods

Low-energy methods result in the spontaneous formation of fine droplets with controlled variation in developing conditions of incompatible oil/water/emulsifier systems. Some commonly used low-energy methods such as membrane emulsification, spontaneous-emulsification, phase inversion etc., along with their applications are summarized in Table 4.2.

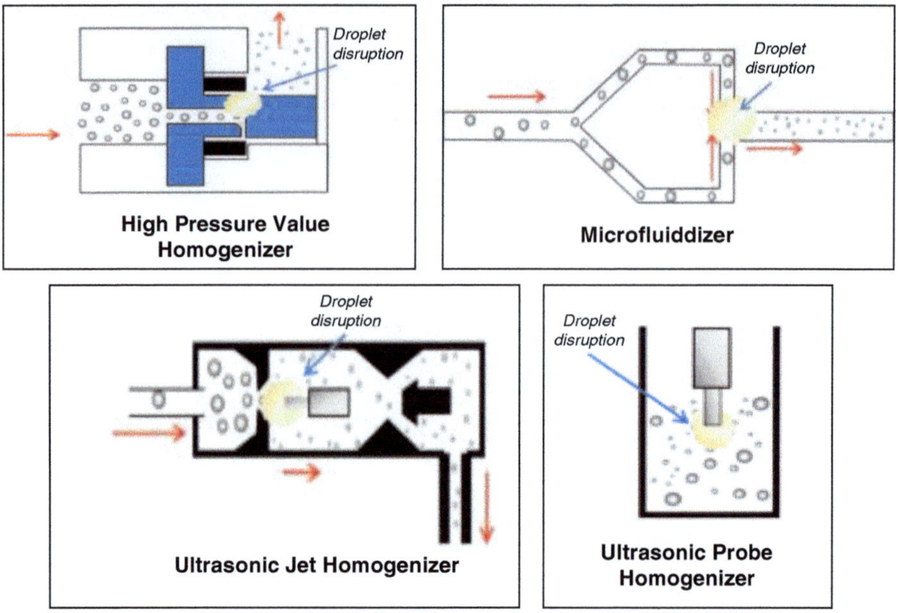

Fig. 4.3 Schematic representation of different mechanical devices that are used for the production of food-grade nanoemulsion using a high-energy approach: high-pressure valve homogenizer; microfluidizer: ultrasonic jet homogenizer; and ultrasonic probe homogenizer

Table 4.2 Cases of nanoemulsion produced by a method of low-energy approaches

Preparation method	Oil phase	Surfactant/ cosurfactant	Bioactive component	Diameter (nm)	References
Membrane emulsification	Methylene chloride	Tween-20/ Tween-80/ polyvinyl alcohol	Flurbiprofen	60–98	Oh et al. (2011)
Membrane emulsification	Medium-chain triglycerides	Tween-80	Vitamin E	76–105	Laouini et al. (2012)
Spontaneous emulsification	Castor oil/ Medium-chain triglycerides	Polysorbate 80	Carbamazepine	148–153	Kelmann et al. (2007)
Spontaneous emulsification	Medium-chain triglycerides	Tween-20, 40, 60, 80	Vitamin E	54–200	Saberi et al. (2013)
Phase inversion composition	Hydrogenated polyisobutane	Polyethylene glycol-400 monoisostearate	–	100	Sonneville-Aubrun et al. (2009)
Phase inversion composition	n-Dodecane	Hexanol or Sodium dodecyl sulfate	–	20–160	Solè et al. (2010)
Phase inversion temperature	Orange oil	Tween-20	β-Carotene	78–100	Qian and McClements (2011)
	Paraffin oil/ Cremophor	Cremophor ® A6		330–500	

4.3.2.1 Membrane Emulsification

This is a quite old method based on low-energy approach developed by Suzuki and others in 1981. The process can produce stable nanoemulsion by controlling the fine droplet size as well as distribution with lower energy input (104–106 J/m^3) and without high mechanical stress (Ribeiro Filho et al. 2005). This approach is considered the best for nanoemulsion formation; however, droplet size is comparatively higher than other methods of nanoemulsion formation.

4.3.2.2 Spontaneous Emulsification

This method involves the mixing of appropriate amounts of oil, water, surfactant, and/or co-surfactant and is the simplest method. Mixing speed, ionic strength, and pH variations are done to form nanoemulsions spontaneously. Fine oil droplets can be formed when an oil/surfactant mixture is added to water. The sequence of addition is not critical because this kind of nanoemulsion is formed spontaneously. But, the system compositions (oil-to-emulsion ratio content, the surfactant-to-emulsion ratio) and preparation conditions (stirring speed) may influence emulsion properties (Kelmann et al. 2007; Saberi et al. 2013).

4.3.2.3 Phase Inversion Composition (PIC)

This method is based on progressive dilution with water or oil, and this composition is optimized at a particular temperature. This phase change is driven by Gibbs free energy of the emulsion resulting in spontaneous inversion of the surfactant's curvature between positive and negative (Sonneville-Aubrun et al. 2009).

4.4 Emulsion Stability of Milk

Fine droplet size of nanoemulsion accounts for its stability against gravitational separation and then alteration in properties. The small droplet size of nanoemulsion confers stability against sedimentation (or creaming) because the Brownian motion and consequently the diffusion rate are higher than the sedimentation (or creaming) rate induced by the gravitational force. Ostwald ripening or molecular diffusion, which arises from emulsion polydispersity and the difference in solubility between small and large droplets, is the main mechanism for nanoemulsion destabilization. The nanoemulsion which are vitamin-rich and stabilized are used to fortify milk, which shows much better health benefits to humans by increasing the bioavailability of that nutrient. The necessary micronutrients like lipophilic vitamins are very important and necessary for human physiology and wellness, so there is a huge growth and demand in the field of food fortifying with vitamins. In emulsion, there should be a

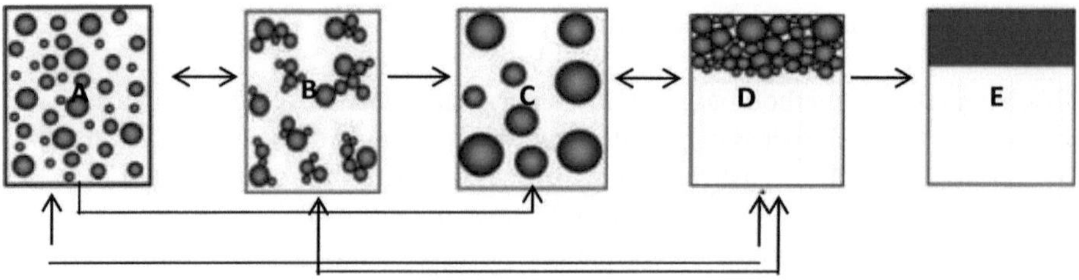

Fig. 4.4 Emulsion stability pathways: This process starts with (**a**) the uniform distribution of drops of oil in water, (**b**) flocculation, (**c**) coalescence, and (**d**) creaming/sedimentation. (**e**) The final step occurs with the separation of the two states, i.e., the entire division of components into two phases. The above-figured arrows represent all the potential pathways and also show the positions where the reversible reactions can be possible by means of simple redistribution

homogeneous distribution of the dispersion-phase medium. However, flocculation, coalescence, and sedimentation result in instability of emulsion (Fig. 4.4).

The flocculation is a process in which formation of network/clump takes place by the accumulation of small drops (droplets). Basically, this formation of network takes place because of the production of uneven forces of repulsion and attraction, which results in the attraction of droplets for accumulation. There may be other factors accounting for the accumulation, but electrostatic force is the major one. Sometimes the electrostatic repulsive forces can also help in the maintenance of stability of emulsion by the process of repulsion of droplets from each other which helps in fine uniform dispersion of drops all over the surface. The next step is coalescence in which two or more small drops merge with each other for the formation of one fine large drop. This process of formation of big-sized particle can occur only by the association of small droplets and can be triggered during creaming and flocculation. Various methods are there for the prevention of coalescence by using proteins and emulsifier. The process of coalescence can be prevented by providing the surface with strong charge which can avoid the association of drops. The mechanism of creaming is caused by the density variation among continuous and dispersed phase which results in the development of buoyancy force. The reduced size of droplets will help in the prevention of creaming. All the abovementioned preventive measures help in the prevention or slowing down of the rate of creaming and hence the extension of shelf life in dairy and dairy-related products.

4.5 Nanoemulsion in Food Industry

In the present aggressive market, new technical innovation is fundamental to keep the administration in food industry as consumer demand is changing rapidly. The future has a place for new items and innovative procedures with the objective of upgrading the execution of processes and enhancing the security and nature of food items by dragging out their time span of usability and freshness. Nanotechnology can possibly change dairy and food sectors which show a bright future in this field. The uses of nanotechnology in dairy and food businesses involve packaging with the help of nanocomposite bottles, nanolaminates, and containers having silver nanoparticles. Nanoparticulate delivery systems involving nanocapsules and nanodispersions have been used as a suitable form to improve the digestibility of food, bioavailability of active components, pharmacological activities of certain compounds, while nanosensors are involved in food safety and biosecurity. Nanoparticles might be utilized for impacting the surface characteristics, nutritious change and also to recognize pathogens in food materials. Food bundling which includes eatables and nano wrapper can broaden timeframe of their realistic usability. "Keen" bundling containing nanosensors which are hostile to the activators of microorganism, counteract gas and humidity and discharge nano against microorganisms can be useful for expanding the time span of food usability.

4.5.1 Food Packaging (Nano Outside)

Nowadays, food business operators and consumers are demanding significantly more from packaging regarding ensuring the performance, freshness, and well-being of foods. Food packaging is the area in which utilization of nanotechnology will be more in future (Bhushan 2017). It has been announced that around 400–500 nanopacking items are being utilized in food business in which nanotechnology is utilized as a part of the production of 25% of all food packing in the coming decade. The main function of nanopacking is to extend the shelf life of food products by enhancing the barrier nature of food packets by reducing moisture and gas exchange and also the exposure to ultraviolet light.

4.5.2 Food "Fortification" Through Nanotechnology

Nanotechnological organizations are attempting to strengthen the handling of dairy and food items using nanoencapsulated supplements, their appearance, and taste which are modified by nanocreated hues, the glucose and fat content expelled/disabled with the help of nano-adjustment, and enhanced mouth feel. The fortification of food can be utilized to build dietary demands, for instance. The consideration of "medicinally gainful" nanocontainers will empower chocolate-based chips or hot chips. Nanotech would likewise empower particular food items like dessert and chocolate to decrease the measure of glucose and fat that a body could ingest. This process can be achieved by utilizing nanosized substances to keep the body from processing/retaining these segments of sustenance. Along these lines, the nano business can display the low-quality food as a well-being food.

4.5.3 *Nanotechnology in Association with Dairy Processing*

According to the consultancy of Helmut Kaiser (2009), "nanofoods" are contributing an expanding development in the improvement of nanoproducts related to dairy and food sector and also for the application of patent. It is possible to connect nanotechnology for the creation of materials like nanoscale food components, contaminant detection, and for making nano gadgets for sub-atomic and cell science from how the development of food is possible to how it gets bundled. The nanotechnology used for sustenance and dairy sectors are secured by two important means, viz., substances which are added in food (nano-inside) and dairy bundling/ packaging (nano-outside). Nanocoating is applicable to dairy products like cheese, and the process of nanotechnology has played an important role in the improvement of nanoscale-palatable film having a width of 5 nm which is undetectable to our normal eye. Nanocoating can also act as a barrier against gas, moisture, and lipids. Other applications of nanocoating are the carriers of antioxidants, antimicrobials, colors, nutrients, and flavors which are the important functional agents and improvement of storage life and textural properties of manufactured foods even in open conditions. By 2003, over 90% of nanopackaging was based on nanocomposites, in which nanomaterials were used to improve the barrier properties of plastic wrapping for foods and dairy products. Nanopackaging can also be designed to release antimicrobials, antioxidants, enzymes, flavors, and neutraceuticals to extend the shelf life (Cha and Chinnan 2004).

4.6 Application of Nanoemulsion in Dairy Products

Consumer focus has changed from energy-providing foods to functional foods associated with required health benefits. This may include fortification with bioactive compounds for nutritional enrichment and antioxidant incorporation to enhance the shelf life, stability, etc.

4.6.1 *Ice cream*

Ice cream is a mix of a compound colloid food that includes the globules of fat, crystals of ice, air bubbles, and a serum phase which is unfrozen. Generally, the size of the ice crystals and air bubbles range from 20 to 50 μm. In general, the fat globules make a coat around the air bubbles, and the protein or the emulsifier forms a coat around fat globules. Different equipment and techniques are involved in the development of nanoemulsion in ice cream. Curcumin is the polyphenoilc component of turmeric with various health benefits like antioxidant, anti-inflammatory, antitumorigenic, anticoagulant, antibacterial, anticarcinogenic, etc. However, its wide usage as functional ingredient is limited due to poor water solubility, bioaccessibility, absorption in GI tract, and being quite prone to degradative changes at an alkaline pH, high temperature, and in presence of light (Aditya et al. 2014). This problem can be resolved by the development of nanoemulsion and incorporation in food system acting as carrier. Ice cream is one such product which is liked by almost all the age groups and in all seasons; so, it can be the best vehicle for the encapsulated curcumin (Fig. 4.5).

4.6.2 Yogurt

Yogurt is a type of milk product prepared by the process of fermentation with bacterial addition and also the incorporation of sweeteners and flavors. Yogurt is commonly utilized in daily diets due to the wide range of health benefits. The bacteria used to make yogurt are known as yogurt cultures. Fermentation of lactose by these bacteria produces lactic acid, which acts on milk protein to give yogurt its texture and characteristic tart flavor. To produce yogurt, milk is first heated, usually to about 85 °C (185 °F), to denature the milk proteins so that they do not form curds. After heating, the milk is allowed to cool to about 45 °C (113 °F). The bacterial culture is mixed in, and the temperature of 45 °C is maintained for 4–12 hours to allow fermentation to occur. Fish oil can be used as supplements to different food items, but because of less solubility in water, it is challenging to add fish oil into fortified and functional foods which may result in poor bioavailability. For a higher bioavailability and better protection against the quality declination, nanoemulsion technique is considered as the most advanced and effectual encapsulation method in which lipids are added to the liquid phase which will enhance the dispersion in the food systems. The water solubility of lipophilic bioactive compounds and oils can be increased with the help of nanoemulsion technology. High-energy methods like sonication and high-pressure homogenization are normally used for the preparation of fish oil nanoemulsion. Nanoemulsion is helpful in protecting the antioxidant potential of γ-oryzanol and aids in the health benefits of yogurt. Yogurt fortified with fish oil/γ-oryzanol nanoemulsion can be prepared by the incorporation of 13 g fish oil or γ-oryzanol nanoemulsion in 100 g sample of yogurt and refrigeration at 4 °C for 21 days in a glass jar which is tightly closed. There will be a gradual decrease in the peroxide value, syneresis, and acidity along with a higher retention of docosahexaenoic acid and eicosapentaenoic acid during the incorporation of γ-oryzanol and fish oil-enriched nanoemulsion in yogurt. The viscosity of yogurt fortified with γ-oryzanol/fish oil nanoemulsion will be lower. Incorporation of nanoemulsion will not alter the crystallization and melting nature of yogurt which are fortified.

4.6.3 Quality Preservation

The main reason for the spoilage in milk is due to the action of microorganisms and can be determined by sensory, sanitary, technological, and nutritional parameters. Addition of *Thymus capitatus* essential oil or their nanoemulsion into a variety of

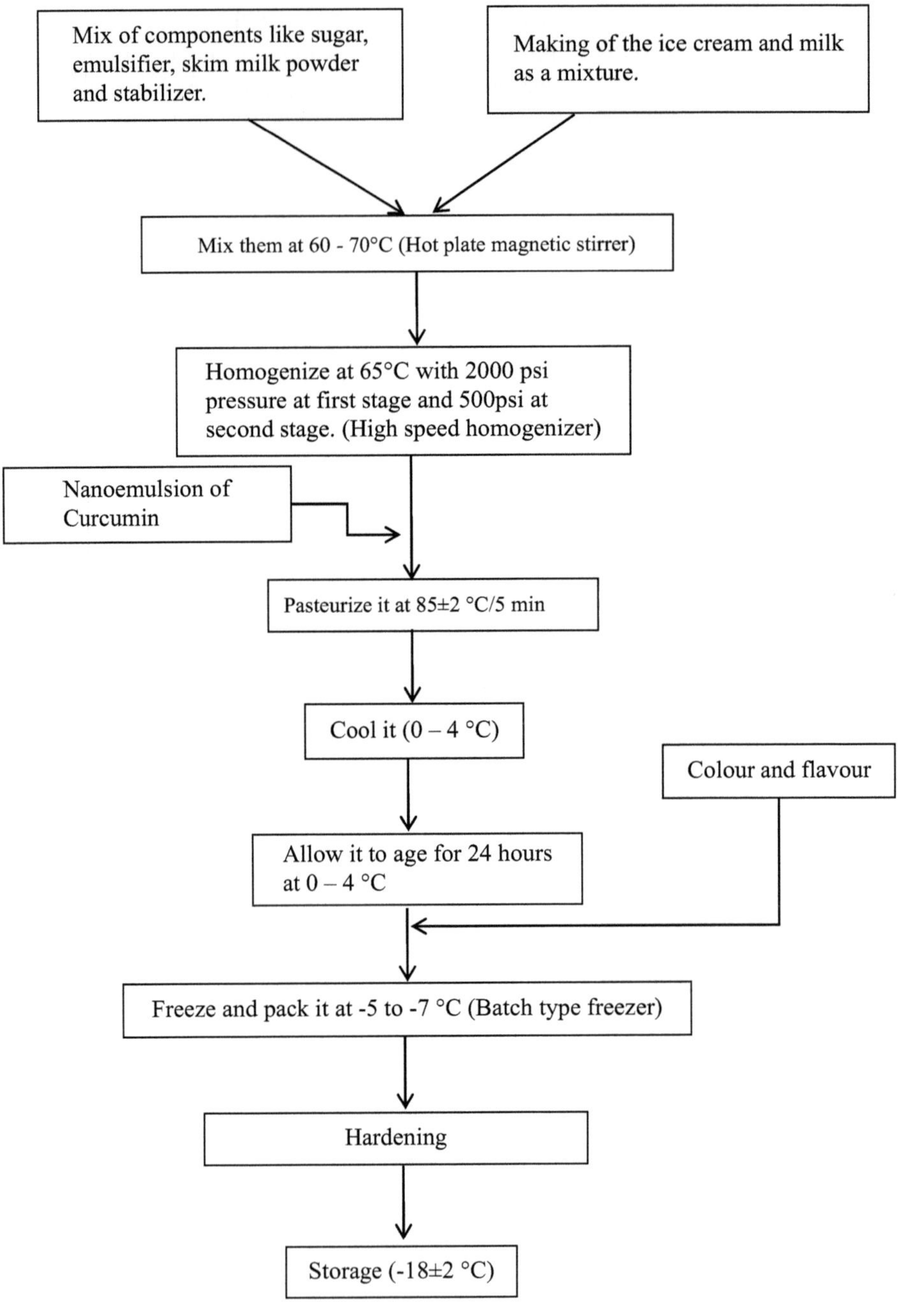

Fig. 4.5 Flowchart showing the preparation of nanoemulsion using curcumin in ice cream

spoiled milk will retard the bacterial growth, thereby improving the microbiological as well as physicochemial qualities, and also their incorporation in semiskimmed ultrahigh-temperature milk will enhance the fermentative and oxidative stability. Degradation of protein by enzymes is responsible for the age gelation of ultrahigh-temperature milk and poor stability of stored milk which is pasteurized. As compared to nanoemulsion, free essential oils are more effective against protein degradation. The quality preservation and shelf-life extension of milk can be easily achieved by nanoencapsulated or bulk *T. capitatus* essential oil.

4.6.4 Cheese

Dairy protein emulsion can act as delivery systems in the fortification of cheese with the help of D_3 vitamin. In the presence of sunlight or after the uptake of supplements and fortified food, essential vitamins like D vitamins will be synthesized in the body. By delivering vitamin D_3 in the oil-in-water emulsion phase, retention of curd can be increased due to the more interaction between the curd and emulsifier. The retention of curd can be improved even more by mixing the emulsion with milk proteins, because of the increased interaction between casein protein of curd and dairy proteins. If vitamin D is used in less amounts for the fortification of cheese and because of the increased vitamin/whey protein ratio, there will be a huge loss of vitamin present in whey. By adding D_3 vitamin as an emulsion and dairy proteins as emulsifying agents, the retention of vitamin in cheese curd can be increased. Nonfat dry milk, whey protein concentrate, and caseinates are the different varieties of protein powder that can be incorporated for the formulation of emulsion which will not alter the D_3 vitamin retention of curd. In an oil carrier, it is better to use milk protein emulsifying agents rather than nanoemulsified D_3 vitamin.

4.6.5 Milk

The fortification of food with necessary micronutrients like vitamin A, D and E have health benefits to humans and improve the quality of life by resisting diseases like osteomalacia, suppressed immune system, osteoporosis, cancer, cardiovascular disease, vision loss etc. The solubility and bioavailability of vitamins can be increased by encapsulating these vitamins using nanoemulsion based on lipids. Oil-in-water-type emulsion are very effective type of emulsion for the encapsulation and stabilization of lipophilic vitamins and thereby increase their bioavailability. Hence, the intake of milk which enriched with vitamin D3 and calcium is very effective in its bioavailability and helps in the reduction of risk factors like aging, bone mass loss, and bone fractures. Gruenwald (2009) stated that skimmed milk is a poor source of vitamins and is fortified with vitamin D_3 which helps to increase calcium absorption in human beings. Basically, vitamin D is a lipid-soluble neutraceutical which helps in the intestinal absorption of calcium, balance of phosphorus, and promotion of bone formation. Because of the water-insoluble nature, vitamin D is not found in food and beverages like low-fat milk and milk products which are the main sources of phosphorus and calcium. Higher efficiency, solubility, and stability can be easily achieved by lipid nanoparticles by acting as a carrier.

4.6.6 Effects of Processing

This portion covers the general effects that occur when emulsifiers related to milk are subjected to common processing conditions of food at the production stage of emulsified foods. In addition to the general processes like shear and heat, processing using high pressure is also noticed. This portion will focus on proteins because they are the major emulsifying agents present in milk. Hydrophobicity, size, solubility, flexibility, and charge are the major physicochemical characteristics that explain the capacity of a protein to stabilize and form emulsion. Any type of process that can influence these parameters can also have impact on the properties of emulsion. Also, there will be a complex intermediate relation between several such attributes. For example, to minimize a direct interaction with fluid solvent, the hydrophobic remains of proteins commonly hide in their own core for maintaining solubility. The exposure of these residues will enhance the surface activity of protein, but through hydrophobic interactions, they will also increase their susceptibility to aggregation. Thus, there will be a reduction in their solubility, and as a result, their diffusing ability and stability toward the interface will gradually fall. The common scheme which explains the processing energy inputs like shear, heat, or high pressure and their effect on solubility, structure of protein, and also functional properties is given in Fig. 4.6.

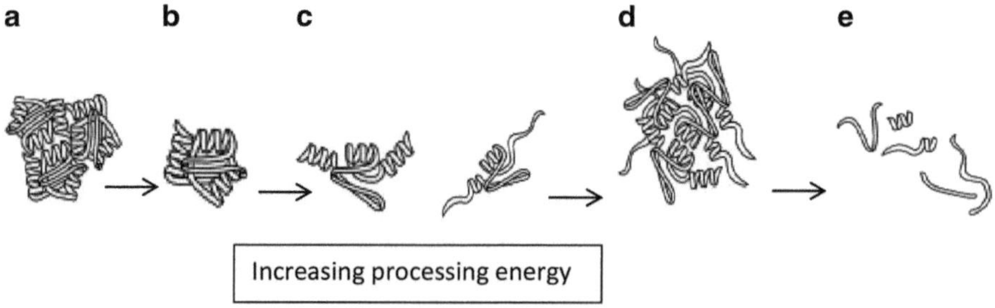

Fig. 4.6 Processing energy effect on the structure of protein and its functionality, a general scheme. (**a**) Quaternary structure of protein. (**b**) Releasing of several surface-active monomers of protein. (**c**) Unfolding of tertiary structure due to further processing results in surface activity and optimal hydrophobicity. (**d**) Occurrence of hydrophobic aggregation due to overprocessing. (**e**) Disruption of the primary structure, which results in the reduction of functionality

Protein ingredient which are produced commercially are normally subjected to treatments like shear and heat, and as a result they form aggregates when rehydrated in the preparation for use. A low surface activity and slower adsorbtion ability are the properties of protein aggregates which may result in a lower effectiveness in the stabilization of emulsion. Thermal processing may cause variations in hydration mainly for higher hydrophobic residuals which decrease the hydrogen bonding, vary the steric parameters and volume, and finally degrade the general structure. A lower temperature can make an impact on the weaker bonds which disrupt the quaternary structure (i.e., separation of oligomers) trailed by the breakdown of the tertiary structure. The emulsification and foaming ability of skimmed milk powder (SMP) will be highly reduced due to thermal treatments. Parris et al. (1991) observed that thermal treatment on SMP stimulates the connection between casein micelles and whey proteins, resulting in higher hydrophilic aggregates having minimal solubility but with an enhanced emulsifying and foaming action. Dairy proteins can stabilize emulsion, and they are commonly oil-in-water emulsion, and water is the main constituent in such systems (Wilde 2009). Milk proteins like caseins are highly sensitive to the strength of ions mainly the range of calcium. So hardness of water has a tremendous effect on the emulsion stability which is stabilized by the action of caseins. Higher level of calcium may cause flocculation of emulsion which are stabilized by casein (Schokker et al. 2000; Dickinson 2003).

4.7 Challenges and Future Trends

Traditionally, milk was widely regarded as a cheap source of functional ingredient, especially with the arrival of price-efficacious extraction and production technologies. The usage of dairy ingredient like caseins, milk fat globules, and whey are gradually increasing in foods due to their beneficial health-supporting features. The present report related to the future trends of food is of more concern regarding the capacity of food to provide a better quality to life and increased health conditions. Phospholipids, proteins, and peptides also have an effect on the health of human beings via the availability of essential type of amino acids, bioactivity, and signaling of homeostatic processes like enhancement in nervous and membrane function, satiety, etc. The proof for ensuring the health-promoting effects of dairy ingredient like peptides derived from casein, whey proteins (Frestedt et al. 2008), and milk phospholipids (Noh and Koo 2004; Dewettinck et al. 2008) shows that their inclusion in foods results in an increased health. Foods which can be emulsified are better dietary fat sources. As compared to conventional emulsifiers, whey proteins can impact on the emulsion rheology, mainly due to charge interactions and their interfacial rheology. Thus, dairy-based products can be used as ingredient which are having different functions that supply physical stability as well as health and nutritional benefits.

References

Aditya NP, Macedo AS, Doktorovova S, Souto EB, Kim S, Chang PS, Ko S (2014) Development and evaluation of lipid nanocarriers for quercetin delivery: a comparative study of solid lipid nanoparticles (SLN), nanostructured lipid carriers (NLC), and lipid nanoemulsions (LNE). LWT Food Sci Technol 59(1):115–121

Bhushan B (ed) (2017) Springer handbook of nanotechnology. Springer, Berlin/Heidelberg

Cavazos-Garduño A, Flores AO, Serrano-Niño JC, Martínez-Sanchez CE, Beristain CI, García HS (2015) Preparation of betulinic acid nanoemulsions stabilized by ω-3 enriched phosphatidylcholine. Ultrason Sonochem 24:204–213

Cha DS, Chinnan MS (2004) Biopolymer-based antimicrobial packaging: a review. Crit Rev Food Sci Nutr 44(4):223–237

Charcosset C (2016) Preparation of nanomaterials for food applications using membrane emulsification and membrane mixing. In: Nanotechnology in the Agri-Food Industry, Emulsions by Grumezescu AM, Academic Press, Vol. 3, pp 37-69. https://doi.org/10.1016/B978-0-12-804306-6.00002-7

Corredig M (2009) Molecular understanding of the interaction of dairy proteins with other food biopolymers. In: Dairy-derived ingredients. Woodhead Publishing, Cambridge, pp 371–393

Dewettinck K, Rombaut R, Thienpont N, Le TT, Messens K, Van Camp J (2008) Nutritional and technological aspects of milk fat globule membrane material. Int Dairy J 18(5):436–457

Dickinson E (2003) Hydrocolloids at interfaces and the influence on the properties of dispersed systems. Food Hydrocoll 17(1):25–39

Donsì F, Annunziata M, Sessa M, Ferrari G (2011) Nanoencapsulation of essential oils to enhance their antimicrobial activity in foods. LWT Food Sci Technol 44(9):1908–1914

Frestedt JL, Zenk JL, Kuskowski MA, Ward LS, Bastian ED (2008) A whey-protein supplement increases fat loss and spares lean muscle in obese subjects: a randomized human clinical study. Nutr Metab 5:8

Ghosh V, Mukherjee A, Chandrasekaran N (2013) Ultrasonic emulsification of food-grade nanoemulsion formulation and evaluation of its bactericidal activity. Ultrason Sonochem 20(1):338–344

Gruenwald J (2009) Novel botanical ingredients for beverages. Clin Dermatol 27(2):210–216

Helmut Kaiser Consultancy (2009) Study: nanobased products inventory 2009 and commercialization. http://www.hkc22.com/nanobasedproducts.html. Accessed 15 June 2018.

Kelmann RG, Kuminek G, Teixeira HF, Koester LS (2007) Carbamazepine parenteral nanoemulsions prepared by spontaneous emulsification process. Int J Pharm 342(1–2):231–239

Koroleva MY, Yurtov EV (2012) Nanoemulsions: the properties, methods of preparation and promising applications. Russ Chem Rev 81(1):21–43

Kuo F, Subramanian B, Kotyla T, Wilson TA, Yoganathan S, Nicolosi RJ (2008) Nanoemulsions of an anti-oxidant synergy formulation containing gamma tocopherol have enhanced bioavailability and anti-inflammatory properties. Int J Pharm 363(1–2):206–213

Laouini A, Fessi H, Charcosset C (2012) Membrane emulsification: a promising alternative for vitamin E encapsulation within nano-emulsion. J Membr Sci 423:85–96

Li PH, Chiang BH (2012) Process optimization and stability of D-limonene-in-water nanoemulsions prepared by ultrasonic emulsification using response surface methodology. Ultrason Sonochem 19(1):192–197

Liedtke S, Wissing S, Müller RH, Mäder K (2000) Influence of high pressure homogenisation equipment on nanodispersions characteristics. Int J Pharm 196(2):183–185

McClements DJ (2012) Nanoemulsions versus microemulsions: terminology, differences, and similarities. Soft Matter 8(6):1719–1729

Noh SK, Koo SI (2004) Milk sphingomyelin is more effective than egg sphingomyelin in inhibiting intestinal absorption of cholesterol and fat in rats. J Nutr 134(10):2611–2616

Oh DH, Balakrishnan P, Oh YK, Kim DD, Yong CS, Choi HG (2011) Effect of process parameters on nanoemulsion droplet size and distribution in SPG membrane emulsification. Int J Pharm 404(1–2):191–197

Parris N, Purcell JM, Ptashkin SM (1991) Thermal denaturation of whey proteins in skim milk. J Agric Food Chem 39(12):2167–2170
Preetz C, Hauser A, Hause G, Kramer A, Mäder K (2010) Application of atomic force microscopy and ultrasonic resonator technology on nanoscale: distinction of nanoemulsions from nanocapsules. Eur J Pharm Sci 39(1–3):141–151
Qian C, McClements DJ (2011) Formation of nanoemulsions stabilized by model food-grade emulsifiers using high-pressure homogenization: factors affecting particle size. Food Hydrocoll 25(5):1000–1008
Ribeiro Filho HM, Delagarde R, Peyraud JL (2005) Herbage intake and milk yield of dairy cows grazing perennial ryegrass swards or white clover/perennial ryegrass swards at low-and medium-herbage allowances. Anim Feed Sci Technol 119(1–2):13–27
Saberi AH, Fang Y, McClements DJ (2013) Fabrication of vitamin E-enriched nanoemulsions by spontaneous emulsification: effect of propylene glycol and ethanol on formation, stability, and properties. Food Res Int 54(1):812–820
Schokker EP, Singh H, Creamer LK (2000) Heat-induced aggregation of β-lactoglobulin A and B with α-lactalbumin. Int Dairy J 10(12):843–853
Solè I, Pey CM, Maestro A, González C, Porras M, Solans C, Gutiérrez JM (2010) Nano-emulsions prepared by the phase inversion composition method: Preparation variables and scale up. J Colloid Interface Sci 344(2):417–423
Sonneville-Aubrun O, Babayan D, Bordeaux D, Lindner P, Rata G, Cabane B (2009) Phase transition pathways for the production of 100 nm oil-in-water emulsions. Phys Chem Chem Phys 11(1):101–110
Strunz CC, Oliveira TV, Vinagre JC, Lima A, Cozzolino S, Maranhão RC (2008) Brazil nut ingestion increased plasma selenium but had minimal effects on lipids, apolipoproteins, and high-density lipoprotein function in human subjects. Nutr Res 28(3):151–155
Subramanian B, Kuo F, Ada E, Kotyla T, Wilson T, Yoganathan S, Nicolosi R (2008) Enhancement of anti-inflammatory property of aspirin in mice by a nano-emulsion preparation. Int Immunopharmacol 8(11):1533–1539
Tang SY, Shridharan P, Sivakumar M (2013) Impact of process parameters in the generation of novel aspirin nanoemulsions–comparative studies between ultrasound cavitation and microfluidizer. Ultrason Sonochem 20(1):485–497
Wilde PJ (2009) Emulsions and nanoemulsions using dairy ingredients. In: Dairy-derived ingredients by Corredig, M, Woodhead Publishing,Cambridge, UK, pp 539–564.
Yilmaz E, Borchert HH (2005) Design of a phytosphingosine-containing, positively-charged nanoemulsion as a colloidal carrier system for dermal application of ceramides. Eur J Pharm Biopharm 60(1):91–98
Yuan Y, Gao Y, Zhao J, Mao L (2008) Characterization and stability evaluation of β-carotene nanoemulsions prepared by high pressure homogenization under various emulsifying conditions. Food Res Int 41(1):61–68

Chapter 5
Nanotechnology: A Successful Approach to Improve Nutraceutical Bioavailability

Sneh Punia, Kawaljit Singh Sandhu, Maninder Kaur, and Anil Kumar Siroha

Contents

5.1 Introduction

Nutraceuticals are isolated or purified from foods and have health benefits besides providing nutrition; they are also used to prevent the occurrence of a disease or are used in its treatment. They are usually sold in medicinal forms not associated with the foods from which they are isolated. The success of nutraceuticals may be

S. Punia (✉) · A. K. Siroha
Department of Food Science and Technology, Chaudhary Devi Lal University, Sirsa, Haryana, India

K. S. Sandhu
Department of Food Science and Technology, Chaudhary Devi Lal University, Sirsa, India

Department of Food Science and Technology, Maharaja Ranjit Singh Punjab Technical University, Bathinda, India

M. Kaur
Department of Food Science and Technology, Guru Nanak Dev University, Amritsar, India

R. Prasad et al. (eds.), *Nanobiotechnology in Bioformulations*, Nanotechnology in the Life Sciences, https://doi.org/10.1007/978-3-030-17061-5_5

attributed to their trait of imparting desirable therapeutic benefits with a reduction in the side effects associated with the pharmaceutical substances used in the prevention and treatment of various ailments (Sahni 2012). Although the list of nutraceuticals used in the treatment of diseases is quite long, over the past few years, phytochemicals with potential health and physiological benefits—such as herbal polyphenols (including curcumin, resveratrol, and rutin) and beta-carotene, present in colored vegetables and fruits—have attracted researchers as well as consumers because they not only can prevent and treat a number of diseases (such as cancers, cardiovascular disorders, and even neurodegenerative disorders) because of their inherent antioxidant property but also can improve immunity. However, the efficacy of many hydrophobic bioactives (pharmaceuticals, supplements, and nutraceuticals) is hindered by limitations in their relative solubility, stability, and bioavailability.

A number of formulation approaches have been utilized for efficient delivery of these nutraceuticals with the aim to enhance their biological activity. Nanoparticle-based delivery systems are one of the promising technologies for the nutraceutical industry.

5.2 System for Delivery of Phenolic Phytochemicals

Nanoparticles interact with phenolic compounds by hydrogen bonds and hydrophobic interactions to encapsulate them in the nanoparticles; consequently, the aqueous solubility of the phytochemicals is enhanced. These particles also prevent oxidation of phenolic compounds encapsulated in the gastrointestinal tract. More importantly, nanoparticles can be taken directly up by epithelial cells in the small intestine, which significantly increases the absorption and bioavailability of phenolic phytochemicals.

5.3 Biocompatible and Biodegradable Nanoparticles

Liposomes, nanoemulsions, micelles, solid lipid nanoparticles (SLNs), nanostructured lipid carriers (NLCs) and poly(lactide-co-glycolic acid) (PLGA) nanoparticles are commonly used biocompatible and biodegradable nanoparticles, and they can be administered via different routes, including oral, intravenous, intraperitoneal, and transdermal administration (Wang et al. 2014; Prasad et al. 2017a). Different types of highly efficient nanomedicines have been used to enhance the physicochemical properties and efficacy of phytochemicals. Nanomedicines can be broadly classified, on the basis of their chief formulation components, into organic and inorganic nanomedicines (Fig. 5.1). Organic nanomedicines can be subclassified into lipid-based nanomedicines (e.g., liposomes, SLNs, NLCs, and lipid micelles) and polymer-based nanomedicines (e.g., polymeric nanoparticles, polymeric micelles, polymer–drug conjugates, and dendrimers) (Table 5.1). Inorganic nanomedicines include magnetic nanoparticles, gold nanoparticles (GNPs), and carbon nanotubes (CNTs) (Rizwanullah et al. 2018).

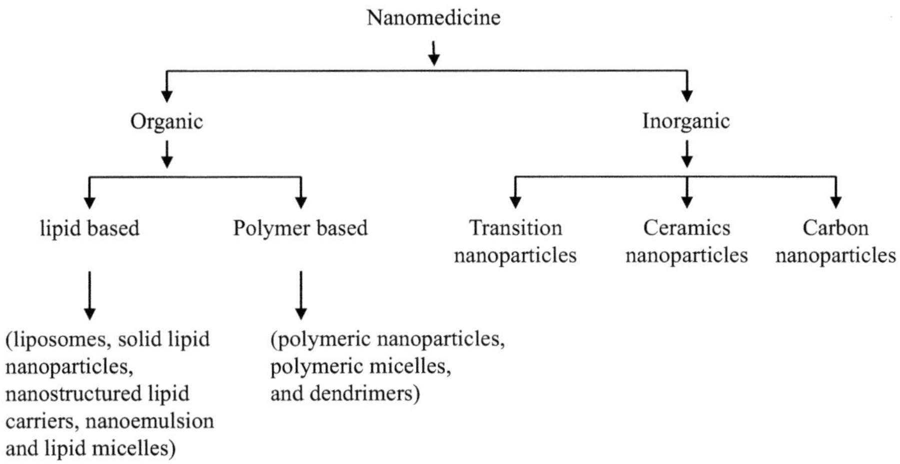

Fig. 5.1 Classification of nanomedicines based on the chief formulation component

Table 5.1 Nanostructured systems for delivery of nutrients

Organic nanoparticles	Characteristics	Potential uses
Lipid-based nanoparticles		
Liposomes	Vesicles formed with a phospholipid bilayer with an aqueous interior, 50 nm–1 μm in diameter (Singh et al. 2012)	Delivery of both hydrophilic and hydrophobic compounds (Singh et al. 2012)
Solid lipid nanoparticles	Prepared from a lipid matrix, particle size 50–1000 nm (Müller et al. 2000)	Delivery and stabilization of hydrophobic materials (Peres et al. 2016)
Nanostructured lipid carriers	Solid matrix at room temperature, variable size from a few nanometers to 3 μm (Müller et al. 2006)	Delivery of both hydrophilic and hydrophobic drugs (Kim et al. 2005)
Microemulsions	Oil-in-water nanoemulsions, droplet radius < 100 nm, diameter < 200 nm (McClements 2013)	Increased bioavailability of encapsulated lipophilic bioactives (McClements 2013)
Lipid-core micelles	Soluble copolymers with lipids, 7–35 nm in size (Torchilin 2005)	Effective solubilization of a broad variety of poorly soluble drugs and diagnostic agents (Torchilin 2005)
Polymer-based nanoparticles		
Polymeric nanoparticles	Fine particles, 100–2500 nm in size; ultrafine particles, 1–100 nm in size (Bennet and Kim 2014)	Stabilization and protection of drug molecules such as proteins, peptides, or DNA molecules from various environmental hazards and degradation (Kawashima 2001; Soppimath et al. 2001; Prasad et al. 2017a)
Polymeric micelles	Spherical, colloidal, supramolecular nanoconstructs, 10–100 nm in size (Yokoyama et al. 1998; Croy and Kwon 2006)	Delivery of drugs and nucleic acids (Jhaveri and Torchilin 2014)
Dendrimers	Three-dimensional, multiply branched, well-organized, nanoscopic macromolecules (Madaan et al. 2014)	Active and passive delivery system for phytochemicals (Madaan et al. 2014)

5.4 Organic Nanoparticles

5.4.1 Lipid-Based Nanoparticles

5.4.1.1 Liposomes

Liposomes are often employed as delivery systems for transporting substances into the body for absorption in the mouth (Silva et al. 2012; Rogers and Anderson 1998) or preventing breakdown by gastric acid (Patel and Ryman 1976). Basically, liposomes are phospholipids, which are amphipathic in nature. When phospholipids (e.g., phosphatidylcholine and dipalmitoyl) are dispersed in water, the molecules aggregate and form a bilayer to minimize contact between the hydrophobic fatty acid chains and the hydrophilic aqueous environment. These structures are called liposomes. They have the same aqueous phase on both sides of the phospholipid bilayer, and their diameter may vary from 50 nm to 1 μm (Singh et al. 2012). They are often employed as delivery systems for bioactive compounds, as during the formation of the liposome, hydrophobic material may be incorporated into the lipid membrane, while hydrophilic molecules present in the aqueous phase may become trapped inside the liposome (Singh et al. 2012). The entrapment of flavors is a major area of research on liposome applications in food systems. Liposomes improve the in vivo biocompatibility and pharmacokinetics of drugs, and their properties may be optimized by simply changing the composition of the bilayer components (Hofheinz et al. 2005; Tattersall and Clarke 2003).

In recent years, several nutraceutical products (vitamins, enzymes, and herbal extracts) have been formulated using liposome technology (Keller 2001). These products have been designed to improve nutrient/bioactive solubilization and enhance absorption. Polyphenols entrapped in liposomes have been shown to be effectively taken up by cancer cells and inhibit their proliferation (Fang et al. 2005; Narayanan et al. 2009). Vitamin A (retinol) entrapped in liposomes may be protected from heat- or light-induced degradation (Lee et al. 2005). Concurrent encapsulation of vitamin E with vitamin C has also been reported (Marsanasco et al. 2011). Further, to offer a new approach to entrap an aqueous soluble drug and an insoluble drug together, liposomes loaded with tea polyphenol (which is water soluble) and vitamin E have been investigated (Fang et al. 2005; Ma et al. 2009) for local delivery, including skin and tumor deposition. Encapsulation efficiencies of 94% and 50% were reported for the hydrophobic and hydrophilic compounds, respectively (Ma et al. 2009). Later, by use of a dispersion method, nanotea liposomes were prepared using lecithin and cholesterol, and these liposomes were found to be stable and suitable for widespread applications (Lu et al. 2011).

5.4.1.2 Solid Lipid Nanoparticles

SLNs were first introduced in the early 1990s and are prepared from a lipid matrix, with final particle sizes ranging between 50 and 1000 nm (Müller et al. 2000). A wide variety of SLNs have been developed, and it has been determined that in their formulation, a lipid, an emulsifier, and water are needed as essential excipients ((Nunes et al. 2017). The lipids normally have a melting point above room and body temperature, and, by definition, they can be triglycerides; mono-, di-, and triglyceride mixtures; waxes; or hard fats. The emulsifiers used to stabilize the lipid dispersion are commonly types of poloxamer and polysorbate, but also may be lecithin, tyloxapol, sodium cholate and sodium glycocholate, taurodeoxycholic acid sodium, butanol and butyric acid, cetylpyridinium chloride, sodium dodecyl sulfate, sodium oleate, polyvinyl alcohol, and cremophor EL (Müller et al. 2000; Severino et al. 2011). Several methods are available to prepare SLNs, such as high-pressure homogenization (hot and cold), microemulsion, solvent emulsion by evaporation or by diffusion, melt dispersion, double emulsion, high-speed stirring, and ultrasonication techniques. Several novel techniques are also used, such as a membrane contactor, solvent injection, multiple emulsion, and supercritical fluid technology (Parhi and Suresh 2010).

The reduction of the particle size and the use of nontoxic materials make SLNs important carriers, combining advantages such as the possibility of controlled drug release and drug targeting, excellent tolerability, capacity for incorporating hydrophilic and lipophilic drugs, increased physical drug stability, and enhancement of bioavailability of the entrapped bioactive, which make them one of the most widely used systems (Weiss et al. 2006; Parhi and Suresh 2010; Severino et al. 2011). In the field of food nanotechnology, and considering the health-promoting properties of phenolic compounds, many studies associated with the development of systems for delivery of phenolic compounds have been published in recent years (Munin and Edwards-Levy 2011), especially to overcome the problem of oral absorption and bioavailability, allowing a greater nutraceutical effect to be associated with these compounds. Along with enzymes (porcine pancreatic lipase and colipase), fatty acids (omega-3), vitamins (E and D_2), and antibiotics (clotrimazole, cyclosporine, and rifampicin), SLNs are also used to load phytochemicals (Nunes et al. 2017). The solid lipid matrix has been shown to protect chemically labile compounds, such as phenolic compounds, from degradation (Trombino et al. 2009). Emulsification–evaporation for benzoic acid (Wang et al. 2012a), hot homogenization for ferulic acid and resveratrol (Trombino et al. 2013; Neves et al. 2013), solvent injection for curcumin (Wang et al. 2012b), microemulsion for sesamol (Geetha et al. 2015), hot homogenization and ultrasonication for quercetin (Han et al. 2014), and microemulsion for β-carotene and α-tocopherol (Trombino et al. 2009) have been adopted.

5.4.1.3 Nanostructured Lipid Carriers

SLNs and lipid emulsions (LEs) are colloidal systems with solid and liquid lipid cores, respectively. These systems exhibit several advantages for delivery of lipophilic drugs, such as the use of biocompatible lipids, large-scale production, protection of drugs from degradation, improved bioavailability, and controlled-release characteristics (Müller et al. 2000). However, these formulation systems have certain drawbacks such as restricted drug-loading capacities, expulsion of the drug from the formulation, and high surfactant concentrations (Khan et al. 2015).

Recently, NLCs have attracted huge attention for oral delivery of lipophilic drugs. Their exclusivity lies in their unique matrix composition, which contains a mixture of incompatible liquid lipids and solid lipids in appropriate and permissible proportions. The presence of solid-cum-liquid lipids in NLCs leads to greater drug encapsulation and loading, and long-term colloidal stability (Muchow et al. 2008; Uner and Yener 2007; Almeida and Souto 2007).

5.4.1.4 Nanoemulsions

Nanoemulsions have advantages over other types of delivery systems for certain applications because of their high physical stability, good optical clarity, rapid digestibility, and effectiveness at enhancing bioavailability (McClements 2011). Nanoemulsions are oil-in-water (O/W) or water-in-oil (W/O) dispersions of two immiscible liquids stabilized using an appropriate surfactant (Mason et al. 2006). This results in an ultrafine dispersion whose differential drug-loading, viscoelastic, and visual properties can cater to a wide range of functionalities, including drug delivery (Singh et al. 2017). Nanoemulsions consisting of small lipid droplets (radius <100 nm) dispersed in water are designed to improve bioavailability (Aboalnaja et al. 2016), and to encapsulate, protect, and deliver various kinds of hydrophobic bioactive agents to improve their bioaccessibility (Ahmed et al. 2012; Pool et al. 2013; Salvia-Trujillo et al. 2013; Yang and McClements 2013; Zheng et al. 2014; McClements 2010; Pathak and Raghuvanshi 2015; Zhang and Wu 2015). O/W nanoemulsions consisting of small oil droplets (radius < 100 nm) dispersed within an aqueous solution (McClements 2011) encapsulate (Porter et al. 2007; McClements 2012), protect, and deliver lipophilic bioactive components, such as drugs and nutraceuticals (food components with specific health benefits) (McClements 2013). W/O nanoemulsions (Porras et al. 2004; Melo et al. 2001) form a class of emulsions that have nanosized water droplets dispersed in organic media through the action of surfactants. Because nanoemulsions have small droplet sizes and narrow distributions, these emulsions may appear transparent or translucent, like microemulsions (Wang et al. 2007). A common surfactant employed in nanoemulsions is lecithin (phosphatidylcholine) derived from egg yolk or soybean (Klang and Valenta 2011). Surfactants such as sodium deoxycholate (bile salt) (Nasr et al. 2012; Vyas et al. 2008) and cremophor EL (polyoxyl-35 castor oil) (Vyas et al. 2008; Zhang et al. 2016) have been used in marketed parenteral products. Tween 20, 40, 60, and 80 (polyoxyethylene sorbitan monolaurate) (Wang et al. 2008; Azeem et al. 2009; Jo and Kwon 2014); Span 20, 40, 60, and 80 (sorbitan monolaurate) (Polychniatou and Tzia 2014; Jadhav et al. 2015); and Solutol HS-15 (polyoxyethylene-660-hydroxystearate) are also regularly used (Scheller et al. 2014).

Nanoemulsions containing droplets of different sizes, charges, compositions, and physical states can be produced by appropriate selection of ingredients and homogenization methods (McClements and Rao 2011; Pouton and Porter 2008; McClements 2011). Nanoscale emulsion systems enhance the oral bioavailability of lipophilic compounds by improving their aqueous solubility, increasing their passive diffusion rate, and facilitating their direct uptake by the intestinal lymphatic system. Bioavailability studies of lipophilic nutraceuticals—such as puerarin (found in the roots of *Pueraria lobata* (Yu et al. 2011)), curcumin (from *Curcuma longa* L. (Yu and Huang 2012)), and resveratrol (extracted from grape skin (Sessa et al. 2014))—in nanoemulsion systems have shown significant improvements in bioavailability in comparison with nonemulsion-based oral formulations. Iron-based nanoparticles have been synthesized using oolong tea extracts for environmental remediation, removing 75.5% of malachite green (50 mg/L) (Huang et al. 2014).

5.4.1.5 Micelles

Micelles are lipid molecules that arrange themselves in a spherical form in aqueous solutions. Lipidic micelles are gaining increasing attention for being effective in improving the solubility and bioavailability of lipophilic drugs. The nanometric size of micelles promotes oral absorption in the intestine (Lukyanov and Torchilin 2004). Micelles formed by conjugates of soluble copolymers with lipids (such as poly(ethylene glycol)–phosphatidyl ethanolamine (PEG-PE) conjugate), with a size of 7–35 nm, are of special interest (Torchilin 2005). These lipid-core micelles have the potential to solubilize poorly soluble drugs and diagnostic agents, and are capable of delivering incorporated drugs directly into the cell cytoplasm (Torchilin 2005). Micelles can increase the aqueous solubility of hydrophobic phytochemicals, enhance their bioavailability, reduce their adverse effects (such as toxicity), enhance their permeation across physiological barriers, and change their biodistribution in the body (Torchilin 2007).

5.4.2 Polymer-Based Nanoparticles

5.4.2.1 Polymer Nanoparticles

These types of nanoparticles are easily made—mostly from biodegradable polymers—and can increase the stability and time of circulation. Synthetic polymers have many advantages over natural polymers, including high purity and reproducibility (Astete and Sabliov 2006). They also have the properties of increasing drug efficacy and sustained release (Kim and Martin 2006), and, in addition to being nontoxic, they are biocompatible and suitable for scale-up methods (Prasad et al. 2017a). Nanoencapsulation with biodegradable polymers is a promising prospect in the food industry for controlled delivery of antimicrobials to help prevent the onset of foodborne illnesses (Weiss et al. 2006). The most widely used polymers are PLGA and poly(lactic acid) (PLA) (Abdollahi and Lotfipour 2012). PLGA is approved by the US Food and Drug Administration (FDA) for use in human therapy (Mahapatro and Singh 2011) and is widely used in pharmaceutical applications in areas such as gene therapy, targeted drug delivery, and delivery of active agents such as proteins, vitamins, and pharmaceutical drugs (Des Rieux et al. 2006; Prasad et al. 2017a). PLGA nanoparticles have also been used as carriers to deliver many phytochemicals such as curcumin, resveratrol, and quercetin (Khalil et al. 2013; Sanna et al. 2015; Pool et al. 2012). PLGA is biocompatible and biodegradable because it yields lactic acid and glycolic acid after it undergoes hydrolysis in the body.

5.4.2.2 Polymer Micelles

Polymeric micelles are spherical, colloidal, supramolecular nanoconstructs, 10–100 nm in size (Yokoyama et al. 1998; Croy and Kwon 2006). They demonstrate various attractive properties as drug carriers (Kwon and Kataoka 1995; Jones and Leroux 1999; Torchilin 2001). They are usually formed by self-assembly of amphiphilic block copolymers (hydrophilic and hydrophobic units) in an aqueous environment (Croy and Kwon 2006). The hydrophobic portion of the block copolymer forms the core of the micelle, while the hydrophilic portion forms the shell (Jhaveri and Torchilin 2014). Poly(ethylene glycol) (PEG) is used for the core (Trubetskoy and Torchilin 1995), and the corona-forming polymers are (N-vinyl pyrrolidone) (PVP) (Bailly et al. 2012) and poly(N-isopropylacrylamide) (PNIPAAm) (Kim et al. 2013). The most promising advantage of polymeric micelles is solubilization of poorly water-soluble or hydrophobic drugs within their core, thus enhancing their bioavailability. Micelles have the ability to slow down opsonization, offering the possibility of longer circulation times and reducing the side effects of the encapsulated drug (Jhaveri and Torchilin 2014).

5.4.2.3 Dendrimers

Dendrimers are three-dimensional, nanosized, radially symmetrical molecules with a well-defined and homogeneous structure consisting of various branches (Srinivasa-Gopalan and Yarema 2007). They are formed by coupling of monomeric molecules that have reactive and protective groups with a multifunctional core moiety, leading to addition of generations around the core followed by removal of protecting groups (Madaan et al. 2014). Of all of the dendrimers, the polyamidoamine (PAMAM) family is the most widely exploited type of dendrimers and was the first to be commercialized (Klajnert and Bryszewska 2001). Others are poly(propylene imine) (PPI), poly(ester amine) (PEA) (Dufès et al. 2005; Wolinsky and Grinstaff 2008), and poly-L-lysine (PLL) (Palmerston Mendes et al. 2017), which have been explored as drug delivery vehicles. They have low polydispersity (Gillies and Frechet 2005) and the ability to easily cross biological barriers (Kievit and Zhang 2011). Their structural and functional properties allow dendrimers to act as carriers/delivery systems for phytochemicals (Patri et al. 2005), making them attractive carriers for anticancer therapeutics (Kesharwani et al. 2014).

5.5 Inorganic Nanoparticles

Inorganic nanoparticles are nontoxic, with good biocompatibility, strong affinity between carriers and biomolecules, and high stability in comparison with organic nanoparticles (Xu et al. 2006). Inorganic nanoparticles—such as GNPs, silver nanoparticles, ceramic nanoparticles, carbon nanoparticles, and CNTs—have been the focus of very extensive research in many fields. Inorganic nanoparticles can be classified into three main categories—(1) transition metal nanoparticles, (2) ceramics nanoparticles, and (3) carbon nanoparticles—in addition to other types. Transition metal nanoparticles are gaining increasing interest in the medical field (Bhattacharyya et al. 2011). GNPs are used in different domains, one of the most important being the biomedical field. They have favorable properties for controlled drug delivery and cancer treatment because of their excellent compatibility with the human organism, low toxicity, high stability, small dimensions, and ability to interact with a variety of substances (Rizwanullah et al. 2018). Silver nanoparticles have been employed in sensor technology (Gomez-Romero 2001), biological leveling (Qiu et al. 2004), and many other biomedical applications (Asha Rani et al. 2009; Li et al. 2011a, b; Pollini et al. 2011; Patil et al. 2012; Prasad 2014). Platinum nanoparticles are widely used as catalysts (Narayanan and El-Sayed 2004) and in many biomedical applications, in combination with other nanoparticles in alloy, core–shell, and bimetallic nanostructures (Hrapovic et al. 2004).

Ceramic nanoparticles are mostly composed from oxides, nitrides, and carbides, with silica (SiO_2) being most frequently used. Mainly, they are used as hollow shells or cores coated with biodegradable and biocompatible polymers. Such surface modifications improve the properties of these nanoparticles, especially for targeted delivery (Odeh et al. 2014).

CNTs have gained increasing attention in the biomedical field in the last two decades because of their unique structure and physicochemical properties, including large aspect ratios, large surface areas, easy surface engineering, and size stability on the nanoscale. Particularly, they are attractive as nanocarriers and mediators for cancer treatment. Through appropriate functionalization, CNTs have been used as nanocarriers for various hydrophilic and lipophilic anticancer drugs (Son et al. 2016).

5.6 Toxicological Aspects

Nanomaterials are rapidly becoming a part of our daily life in the form of cosmetics, food packaging, drug delivery systems, therapeutics, biosensors, etc. (Prasad et al. 2014, 2017b, c) Besides the benefits of nanoparticles, there are open questions about how these particles, used in day-to-day life, may affect humans and impact the environment. Nanostructures in the food sector may not have direct effects on human health; however, they may cause some unavoidable side effects. Nanotoxicity (toxicity produced by nanomaterials) due to overproduction of reactive oxygen species

(ROS) induces oxidative stress, resulting in cells failing to maintain normal physiological redox-regulated functions. Thus, nanotoxicity may lead to DNA damage, unregulated cell signaling, changes in cell motility, cytotoxicity, apoptosis, and cancer initiation (Aziz et al. 2019). Another concern is that (silver and zinc oxide) nanoparticles may migrate into food when they are used in packaging materials. Leaching of silver and zinc ions into food substances may accelerate degradation of nutrients.

5.7 Future Perspectives

As reported in many in vitro and in vivo studies, nanoparticles may enhance the anticancer activities of nutraceuticals, but there are still concerns regarding their cost, safety, side effects, and long-term toxicity. Therefore, a comprehensive assessment of potential risks to human health is essential before nanofood products are made commercially available. Regulations concerning the impacts of these nanomaterials on human and environmental health need to be established to ensure food safety. However, no universal guideline has been specifically developed for safety assessment of nanomaterials in food. Therefore, more studies need to be conducted to evaluate the potential toxicity of nanomaterials or nanostructures that are to be used in food science and related industries. Further research is also needed before antimicrobial nanocomposite packaging materials—which could result in migration of nanoparticles into food—are commercialized.

5.8 Conclusions

Nanotechnology is truly a multidisciplinary science, and its application in drug delivery has already had a significant impact in many areas of medicine. Nanoparticles have been investigated as promising delivery systems for controlled release and improved bioavailability and bioefficacy of nutraceuticals. Nanotechnology can also improve the stability of encapsulated bioactive nutrients and non-nutrients against environmental changes and control their release. As nutraceutical carriers, nanoparticles have promising potential for maintaining and promoting health, as well as for preventing and potentially treating disease.

References

Abdollahi S, Lotfipour F (2012) PLGA- and PLA-based polymeric nanoparticles for antimicrobial drug delivery. Biomed Int 3:1–11

Aboalnaja KO, Yaghmoor S, Kumosani TA, McClements DJ (2016) Utilization of nanoemulsions to enhance bioactivity of pharmaceuticals, supplements, and nutraceuticals: nanoemulsion delivery systems and nanoemulsion excipient systems. Expert Opin Drug Deliv 13:1327–1336

Ahmed K, Li Y, McClements DJ, Xiao H (2012) Nanoemulsion- and emulsion-based delivery systems for curcumin: encapsulation and release properties. Food Chem 132:799–807
Almeida AJ, Souto E (2007) Solid lipid nanoparticles as a drug delivery system for peptides and proteins. Adv Drug Deliv Rev 59:478–490
Asha Rani PV, Mun GLK, Hande MP, Valiyaveettil S (2009) Cytotoxicity and genotoxicity of silver nanoparticles in human cells. ACS Nano 3:279–290
Astete CE, Sabliov CM (2006) Synthesis and characterization of PLGA nanoparticles. J Biomater Sci Polym Ed 17:247–289
Azeem A, Rizwan M, Ahmad FJ, Iqbal Z, Khar RK, Aqil M, Talegaonkar S (2009) Nanoemulsion components screening and selection: a technical note. AAPS PharmSciTech 10:69–76
Aziz N, Faraz M, Sherwani MA, Fatma T, Prasad R (2019) Illuminating the anticancerous efficacy of a new fungal chassis for silver nanoparticle synthesis. Front Chem 7:65. https://doi.org/10.3389/fchem.2019.00065
Bailly N, Thomas M, Klumperman B (2012) Poly(N-vinylpyrrolidone)-block-poly(vinyl acetate) as a drug delivery vehicle for hydrophobic drugs. Biomacromolecules 13:4109–4117
Bennet D, Kim S (2014) Polymer nanoparticles for smart drug delivery. In: Sezer AD (ed) Application of nanotechnology in drug delivery. InTechOpen. https://doi.org/10.5772/58422
Bhattacharyya S, Kudgus R, Bhattacharya R, Mukherjee P (2011) Inorganic nanoparticlcs in cancer therapy. Pharm Res 28:237–259
Croy SR, Kwon GS (2006) Polymeric micelles for drug delivery. Curr Pharm Des 12:4669–4684
Des Rieux A, Fievez V, Garinot M, Schneider YJ, Preat V (2006) Nanoparticles as potential oral delivery systems of proteins and vaccines: a mechanistic approach. J Control Release 116:1–27
Dufès C, Uchegbu IF, Schätzlein AG (2005) Dendrimers in gene delivery. Adv Drug Deliv Rev 57:2177–2202
Fang JY, Hung CF, Hwang TL, Huang YL (2005) Physicochemical characteristics and in vivo deposition of liposome-encapsulated tea catechins by topical and intratumor administrations. J Drug Target 13:19–27
Geetha T, Kapila M, Prakash O, Deol PK, Kakkar V, Kaur IP (2015) Sesamol-loaded solid lipid nanoparticles for treatment of skin cancer. J Drug Target 23:159–169
Gillies ER, Frechet JM (2005) Dendrimers and dendritic polymers in drug delivery. Drug Discov Today 10:35–43
Gomez-Romero P (2001) Hybrid organic–inorganic materials: in search of synergic activity. Adv Mater 13:163–174
Han SB, Kwon SS, Jeong YM, Yu ER, Park SN (2014) Physical characterization and in vitro skin permeation of solid lipid nanoparticles for transdermal delivery of quercetin. Int J Cosmet Sci 36:588–597
Hofheinz RD, Gnad-Vogt SU, Beyer U, Hochhaus A (2005) Liposomal encapsulated anti-cancer drugs. Anti-Cancer Drugs 16:691–707
Hrapovic S, Liu Y, Male KB, Luong JHT (2004) Electrochemical biosensing platform using platinum nanoparticles and carbon nanotubes. Anal Chem 76:1083–1088
Huang L, Weng X, Chen Z, Megharaj M, Naidu R (2014) Synthesis of iron-based nanoparticles using oolong tea extract for the degradation of malachite green. Spectrochim Acta A Mol Biomol Spectrosc 117:801–804
Jadhav C, Kate V, Payghan SA (2015) Investigation of effect of non-ionic surfactant on preparation of griseofulvin non-aqueous nanoemulsion. J Nanostruct Chem 5:107–113
Jhaveri AM, Torchilin VP (2014) Multifunctional polymeric micelles for delivery of drugs and siRNA. Front Pharmacol. https://doi.org/10.3389/fphar.2014.00077
Jo YJ, Kwon YJ (2014) Characterization of β-carotene nanoemulsions prepared by microfluidization technique. Food Sci Biotechnol 23:107–113
Jones M, Leroux J (1999) Polymeric micelles—a new generation of colloidal drug carriers. Eur J Pharmacol Biopharmacol 48:101–111
Kawashima Y (2001) Nanoparticulate systems for improved drug delivery. Adv Drug Deliv Rev 47:1–2

Keller B (2001) Liposomes in nutrition. Trends Food Sci Technol 12:25–31
Kesharwani P, Jain K, Jain NK (2014) Dendrimer as nanocarrier for drug delivery. Prog Polym Sci 39:268–307
Khalil NM, Do Nascimento TCF, Casa DM (2013) Pharmacokinetics of curcumin-loaded PLGA and PLGA–PEG blend nanoparticles after oral administration in rats. Colloids Surf B Biointerfaces 101:353–360
Khan S, Baboota S, Ali J, Khan S, Narang RS, Narang JK (2015) Nanostructured lipid carriers: an emerging platform for improving oral bioavailability of lipophilic drugs. Int J Pharm Investig 5:182–191
Kievit FM, Zhang M (2011) Cancer nanotheranostics: improving imaging and therapy by targeted delivery across biological barriers. Adv Mater 23:H217–H247
Kim DH, Martin DC (2006) Sustained release of dexamethasone from hydrophilic matrices using PLGA nanoparticles for neural drug delivery. Biomaterials 27:3031–3037
Kim BD, Na K, Choi HK (2005) Preparation and characterization of solid lipid nanoparticles (SLN) made of cacao butter and curdlan. Eur J Pharm Sci 24:199–205
Kim DH, Vitol EA, Liu J, Balasubramanian S, Gosztola DJ, Cohen EE, Novosad V, Rozhkova EA (2013) Stimuli-responsive magnetic nanomicelles as multifunctional heat and cargo delivery vehicles. Langmuir 29:7425–7432
Klajnert B, Bryszewska M (2001) Dendrimers: properties and applications. Acta Biochim Pol 48:199–208
Klang V, Valenta C (2011) Lecithin-based nanoemulsions. J Drug Deliv Sci Technol 21:55–76
Kwon GS, Kataoka K (1995) Block-copolymer micelles as long-circulating drug vehicles. Adv Drug Deliv Rev 16:295–309
Lee SC, Lee KE, Kim JJ, Lim SH (2005) The effect of cholesterol in the liposome bilayer on the stabilization of incorporated retinol. J Liposome Res 15:157–166
Li WR, Xie XB, Shi QS, Duan SS, Ouyang YS, Chen YB (2011a) Antibacterial effect of silver nanoparticles on *Staphylococcus aureus*. Biometals 24:135–141
Li X, Xu H, Chen ZS, Chen G (2011b) Biosynthesis of nanoparticles by microorganisms and their applications. J Nanomater. https://doi.org/10.1155/2011/270974
Lu Q, Li DC, Jiang GH (2011) Preparation of a tea polyphenol nanoliposome system and its physicochemical properties. J Agric Food Chem 59:13004–13011
Lukyanov AN, Torchilin VP (2004) Micelles from lipid derivatives of water-soluble polymers as delivery systems for poorly soluble drugs. Adv Drug Deliv Rev 56:1273–1289
Ma QH, Kuang YZ, Hao XZ, Gu N (2009) Preparation and characterization of tea polyphenols and vitamin E loaded nanoscale complex liposome. J Nanosci Nanotechnol 9:1379–1383
Madaan K, Kumar S, Poonia N, Lather V, Pandita D (2014) Dendrimers in drug delivery and targeting: drug–dendrimer interactions and toxicity issues. J Pharm Bioallied Sci 6:139–150
Mahapatro A, Singh DK (2011) Biodegradable nanoparticles are excellent vehicle for site directed in-vivo delivery of drugs and vaccines. J Nanobiotechnol 9:1–11
Marsanasco M, Márquez AL, Wagner JR, Alonso SDV, Chiaramoni NS (2011) Liposomes as vehicles for vitamins E and C: an alternative to fortify orange juice and offer vitamin C protection after heat treatment. Food Res Int 44:3039–3046
Mason TG, Wilking JN, Meleson K, Chang CB, Graves SM (2006) Nanoemulsions: formation, structure, and physical properties. J Phys Condens Matter 18:R635
McClements DJ (2010) Emulsion design to improve the delivery of functional lipophilic components. Annu Rev Food Sci Technol 1(1):241–269
McClements DJ (2011) Edible nanoemulsions: fabrication, properties, and functional performance. Soft Matter 7:2297–2316
McClements DJ (2012) Nanoemulsions versus microemulsions: terminology, differences, and similarities. Soft Matter 8(6):1719–1729
McClements DJ (2013) Nanoemulsion-based oral delivery systems for lipophilic bioactive components: nutraceuticals and pharmaceuticals. Ther Deliv 4:841–857
McClements DJ, Rao J (2011) Food-grade nanoemulsions: formulation, fabrication, properties, performance, biological fate, and potential toxicity. Crit Rev Food Sci Nutr 51:285–330

Melo EP, Aires-Barros MR, Cabral JMS (2001) Reverse micelles and protein biotechnology. Biotechnol Annu Rev 7:87–129
Muchow M, Maincent P, Muller RH (2008) Lipid nanoparticles with a solid matrix (SLN, NLC, LDC) for oral drug delivery. Drug Dev Ind Pharm 34:1394–1405
Müller RH, Mäder K, Gohla S (2000) Solid lipid nanoparticles (SLN) for controlled drug delivery—a review of the state of the art. Eur J Pharm Biopharm 50:161–177
Müller RH, Runge S, Ravelli V, Mehnert W, Thünemann AF, Souto EB (2006) Oral bioavailability of cyclosporine: solid lipid nanoparticles (SLN) versus drug nanocrystals. Int J Pharm 317:82–89
Munin A, Edwards-Levy F (2011) Encapsulation of natural polyphenolic compounds; a review. Pharmaceutics 3:793–829
Narayanan R, El-Sayed MA (2004) Shape dependent catalytic activity of platinum nanoparticles in colloidal solution. Nano Lett 4:1343–1348
Narayanan NK, Nargi D, Randolph C, Narayanan BA (2009) Liposome encapsulation of curcumin and resveratrol in combination reduces prostate cancer incidence in PTEN knockout mice. Int J Cancer 125:1–8
Nasr M, Nawaz S, Elhissi A (2012) Amphotericin B lipid nanoemulsion aerosols for targeting peripheral respiratory airways via nebulization. Int J Pharm 436:611–616
Neves AR, Lúcio M, Martins S, Lima JL, Reis S (2013) Novel resveratrol nanodelivery systems based on lipid nanoparticles to enhance its oral bioavailability. Int J Nanomedicine 8:177–187
Nunes S, Madureira AR, Campos D, Sarmento B, Gomes AM, Pintado M, Reis F (2017) Solid lipid nanoparticles as oral delivery systems of phenolic compounds: overcoming pharmacokinetic limitations for nutraceutical applications. Crit Rev Food Sci Nutr 57:1863–1873
Odeh F, Al-Jaber H, Khater D (2014) Nanoflora—how nanotechnology enhanced the use of active phytochemicals. In: Sezer AD (ed) Application of nanotechnology in drug delivery. InTechOpen. https://doi.org/10.5772/58704
Palmerston Mendes L, Pan J, Torchilin VP (2017) Dendrimers as nanocarriers for nucleic acid and drug delivery in cancer therapy. Molecules. https://doi.org/10.3390/molecules22091401
Parhi R, Suresh P (2010) Production of solid lipid nanoparticles—drug loading and release mechanism. J Chem Pharm Res 2:211–227
Patel HM, Ryman BE (1976) Oral administration of insulin by encapsulation within liposomes. FEBS Lett 62:60–63
Pathak K, Raghuvanshi S (2015) Oral bioavailability: issues and solutions via nanoformulations. Clin Pharmacokinet 54:325–357
Patil RS, Kokate MR, Kolekar SS (2012) Bioinspired synthesis of highly stabilized silver nanoparticles using *Ocimum tenuiflorum* leaf extract and their antibacterial activity. Spectrochem Acta A 91:234–238
Patri AK, Kukowska-Latallo JF, Baker JR (2005) Targeted drug delivery with dendrimers: comparison of the release kinetics of covalently conjugated drug and non-covalent drug inclusion complex. Adv Drug Deliv Rev 57:2203–2214
Peres LB, Peres LB, de Araújo PHH, Sayer C (2016) Solid lipid nanoparticles for encapsulation of hydrophilic drugs by an organic solvent free double emulsion technique. Colloids Surf B Biointerfaces 140:317–323
Pollini M, Paladini F, Catalno M, Taurino A, Licciulli A, Maffezzoli A, Sannio A (2011) Antibacterial coatings on haemodialysis catheters by photochemical deposition of silver nanoparticles. J Mater Sci Mater Med 22:2005–2012
Polychniatou V, Tzia C (2014) Study of formulation and stability of co-surfactant free water–in–olive oil nano- and submicron emulsions with food grade non-ionic surfactants. J Am Oil Chem Soc 91:79–88
Pool H, Quintanar D, De Dios Figueroa J et al (2012) Antioxidant effects of quercetin and catechin encapsulated into PLGA nanoparticle. J Nanomater 2012:86
Pool H, Mendoza S, Xiao H, McClements DJ (2013) Encapsulation and release of hydrophobic bioactive components in nanoemulsion-based delivery systems: impact of physical form on quercetin bioaccessibility. Food Funct 4:162–174

Porras M, Solans C, Gonzalez C, Martínez A, Guinart A, Gutiérrez JM (2004) Studies of formation of W/O nano-emulsions. Colloids Surf A249:115–118
Porter CJH, Trevaskis NL, Charman WN (2007) Lipids and lipid-based formulations: optimizing the oral delivery of lipophilic drugs. Nat Rev Drug Discov 6:231–248
Pouton CW, Porter CJH (2008) Formulation of lipid-based delivery systems for oral administration: materials, methods and strategies. Adv Drug Deliv Rev 60:625–637
Prasad R (2014) Synthesis of silver nanoparticles in photosynthetic plants. Journal of Nanoparticles, Article ID 963961, https://doi.org/10.1155/2014/963961
Prasad R, Kumar V, Prasad KS (2014) Nanotechnology in sustainable agriculture: present concerns and future aspects. Afr J Biotechnol 13(6):705–713
Prasad R, Pandey R, Varma A, Barman I (2017a) Polymer based nanoparticles for drug delivery systems and cancer therapeutics. In: Kharkwal H, Janaswamy S (eds) Natural polymers for drug delivery. Wallingford: CAB, pp 53–70
Prasad R, Bhattacharyya A, Nguyen QD (2017b) Nanotechnology in sustainable agriculture: recent developments, challenges, and perspectives. Front Microbiol 8:1014. https://doi.org/10.3389/fmicb.2017.01014
Prasad R, Kumar V and Kumar M (2017c) Nanotechnology: Food and Environmental Paradigm. Springer Nature Singapore Pte Ltd. (ISBN 978-981-10-4678-0)
Qiu H, Rieger B, Gilbert R, Jerome R, Jerome C (2004) PLA-coated gold nanoparticles for leveling of PLA biocarriers. Chem Mater 16:850–856
Rizwanullah M, Amin S, Mir SR, Fakhri KU, Rizvi MMA (2018) Phytochemical based nanomedicines against cancer: current status and future prospects. J Drug Target 26(9):731–752. https://doi.org/10.1080/1061186X.2017.1408115
Rogers JA, Anderson KE (1998) The potential of liposomes in oral drug delivery. Crit Rev Ther Drug Carrier Syst 15:421–480
Sahni JK (2012) Exploring delivery of nutraceuticals using nanotechnology. Int J Pharm Investig 2:53
Salvia-Trujillo L, Qian C, Martin-Belloso O, McClements DJ (2013) Influence of particle size on lipid digestion and beta-carotene bioaccessibility in emulsions and nanoemulsions. Food Chem 141:1472–1480
Sanna V, Roggio AM, Pala N et al (2015) Effect of chitosan concentration on PLGA microcapsules for controlled release and stability of resveratrol. Int J Biol Macromol 72:531–536
Scheller KJ, Williams SJ, Lawrence AJ, Jarrott B, Djouma E (2014) An improved method to prepare an injectable microemulsion of the galanin-receptor 3 selective antagonist, SNAP 37889, using Kolliphor® HS 15. Methods X 1:212–216
Sessa M, Balestrieri ML, Ferrari G, Servillo L, Castaldo D, D'Onofrio N et al (2014) Bioavailability of encapsulated resveratrol into nanoemulsion-based delivery systems. Food Chem 147:42–50
Severino P, Andreani T, Macedo AS, Fangueiro JF, Santana MH, Silva AM, Souto EB (2011) Current state-of-art and new trends on lipid nanoparticles (SLN and NLC) for oral drug delivery. J Drug Deliv 2012:1–10
Silva AC, Santos D, Ferreira DM, Lopes C (2012) Lipid-based nanocarriers as an alternative for oral delivery of poorly water-soluble drugs: peroral and mucosal routes. Curr Med Chem 19:4495–4510
Singh H, Thompson A, Liu W, Corredig M (2012) Liposomes as food ingredients and nutraceutical delivery systems. In: Garti N, DJ MC (eds) Encapsulation technologies and delivery systems for food ingredients and nutraceuticals. Woodhead, Cambridge, pp 287–318. https://doi.org/10.1533/9780857095909.3.287
Singh Y, Meher JG, Raval K, Khan FA, Chaurasia M, Jain NK, Chourasia MK (2017) Nanoemulsion: concepts, development and applications in drug delivery. J Control Release 252:28–49
Son KH, Hong JH, Lee JW (2016) Carbon nanotubes as cancer therapeutic carriers and mediators. Int J Nanomedicine 11:5163–5185
Soppimath KS, Aminabhavi TM, Kulkarni AR, Rudzinski WE (2001) Biodegradable polymeric nanoparticles as drug delivery devices. J Control Release 70:1–20

Srinivasa-Gopalan S, Yarema KJ (2007) Nanotechnologies for the life sciences: dendrimers in cancer treatment and diagnosis, vol 7. Wiley, New York
Tattersall M, Clarke S (2003) Developments in drug delivery: implications for cancer care. Curr Opin Oncol 15:293–299
Torchilin VP (2001) Structure and design of polymeric surfactant-based drug delivery systems. J Control Release 73:137–172
Torchilin VP (2005) Lipid-core micelles for targeted drug delivery. Curr Drug Deliv 2:319–327
Torchilin VP (2007) Micellar nanocarriers: pharmaceutical perspectives. Pharm Res 24:1–16
Torchilin VP, Trubetskoy VS (1995) Which polymers can make nanoparticulate drug carriers long-circulating. Adv Drug Deliv Rev 16:141–155
Trombino S, Cassano R, Muzzalupo R, Pingitore A, Cione E, Picci N (2009) Stearyl ferulate-based solid lipid nanoparticles for the encapsulation and stabilization of β-carotene and α-tocopherol. Colloids Surf B72:181–187
Trombino S, Cassano R, Ferrarelli T, Barone E, Picci E, Mancuso C (2013) Trans-ferulic acid–based solid lipid nanoparticles and their antioxidant effect in rat brain microsomes. Colloids Surf B Biointerfaces 109:273–279
Uner M, Yener G (2007) Importance of solid lipid nanoparticles (SLN) in various administration routes and future perspectives. Int J Nanomedicine 2:289–300
Vyas TK, Shahiwala A, Amiji MM (2008) Improved oral bioavailability and brain transport of saquinavir upon administration in novel nanoemulsion formulations. Int J Pharm 347:93–101
Wang L, Li X, Zhang G, Dong J, Eastoe J (2007) Design and optimization of a new self-nanoemulsifying drug delivery system. J Colloid Interface Sci 314:230
Wang X, Jiang Y, Wang YW, Huang MT, Ho CT, Huang Q (2008) Enhancing anti-inflammation activity of curcumin through O/W nanoemulsions. Food Chem 108:419–424
Wang M, Qin L, Li K, Zhu R, Wang W, Wang S (2012a) The improvement of the anticancer effect of a novel compound benzoic acid, 2-hydroxy-, 2-D-ribofuranosylhydrazide (BHR) loaded in solid lipid nanoparticles. AAPS Pharm Sci Tech 13:1348–1354
Wang W, Zhu R, Xie Q, Li A, Xiao Y, Li K et al (2012b) Enhanced bioavailability and efficiency of curcumin for the treatment of asthma by its formulation in solid lipid nanoparticles. Int J Nanomedicine 7:3667–3677
Wang S, Su R, Nie S et al (2014) Application of nanotechnology in improving bioavailability and bioactivity of diet-derived phytochemicals. J Nutr Biochem 25:363–376
Weiss J, Takhistov P, McClements DJ (2006) Functional materials in food nanotechnology. J Food Sci 71:R107–R116
Wolinsky JB, Grinstaff MW (2008) Therapeutic and diagnostic applications of dendrimers for cancer treatment. Adv Drug Deliv Rev 60:1037–1055
Xu ZP, Zeng QH, Lu GQ, Yu AB (2006) Inorganic nanoparticles as carriers for efficient cellular delivery. Chem Eng Sci 61:1027–1040
Yang Y, McClements DJ (2013) Vitamin E bioaccessibility: influence of carrier oil type on digestion and release of emulsified alpha-tocopherol acetate. Food Chem 141:473–481
Yokoyama M, Satoh A, Sakurai Y, Okano T, Matsumura Y, Kakizoe T, Kataoka K (1998) Incorporation of water-insoluble anticancer drug into polymeric micelles and control of their particle size. J Control Release 55:219–229
Yu H, Huang Q (2012) Improving the oral bioavailability of curcumin using novel organogel-based nanoemulsions. J Agric Food Chem 60:5373–5379
Yu A, Wang H, Wang J, Cao F, Gao Y, Cui J, Zhai G (2011) Formulation optimization and bioavailability after oral and nasal administration in rabbits of puerarin-loaded microemulsion. J Pharm Sci 100:933–941
Zhang XW, Wu BJ (2015) Submicron lipid emulsions: a versatile platform for drug delivery. Curr Drug Metab 16:211–220
Zhang T, Dong D, Lu D, Wang S, Wu B (2016) Cremophor EL–based nanoemulsion enhances transcellular permeation of emodin through glucuronidation reduction in UGT1A1-overexpressing MDCKII cells. Int J Pharm 501:190–198
Zheng J, Li Y, Song M, Fang X, Cao Y, McClements DJ, Xiao H (2014) Improving intracellular uptake of 5-demethyltangeretin by food grade nanoemulsions. Food Res Int 62:98–103

Chapter 6
Microencapsulation for Delivery of Probiotic Bacteria

Anil Panghal, Sundeep Jaglan, Neelesh Sindhu, V. Anshid, Manga Veera Sai Charan, Vinod Surendran, and Navnidhi Chhikara

Contents

A. Panghal
Department of Food Technology and Nutrition, Lovely Professional University, Phagwara, Punjab, India

AICRP-PHET, Department of Processing and Food Engineering, Chaudhary Charan Singh Haryana Agricultural University, Haryana, India

V. Anshid · M. V. Sai Charan · V. Surendran · N. Chhikara (✉)
Department of Food Technology and Nutrition, Lovely Professional University, Phagwara, Punjab, India

S. Jaglan
Division of Microbial Biotechnology, CSIR-Indian Institute of Integrative Medicine, Jammu, India

N. Sindhu
Department of Veterinary Clinical Complex, Lala Lajpat Rai University of Veterinary and Animal Science, Hisar, India

R. Prasad et al. (eds.), *Nanobiotechnology in Bioformulations*, Nanotechnology in the Life Sciences, https://doi.org/10.1007/978-3-030-17061-5_6

6.1 Introduction

The current food market is experiencing various challenges due to increased consumer awareness, competition in the market, rapidly changing food choices, and the demand for functional foods. Continuous efforts are being made by researchers and industries to incorporate bioactive compounds in food to enhance their appeal and health benefits for consumers. However, thermal processing, which is a widely used processing technique in the food industry, leads to destruction/reduction of these valuable nutrients. Thus, there is a tremendous need for some sort of protective coating material or shield for efficient delivery of nutrients in the human system (Canizales et al. 2018), and this technique is known as microencapsulation. Microencapsulation is the process of entrapping tiny amounts of a core phase material of interest in an exterior/shell material with advanced technology. In the food and beverage industries, the microencapsulation market has been projected to grow at a high compound annual growth rate (CAGR) of 13.1% from 2017 to 2022. Research related to microencapsulation in food-related components was initiated in the USA in the twentieth century, to prevent loss of volatiles and oxidation, and to control flavor release in different foods (Sobel et al. 2014). Microencapsulation is increasingly attracting the attention of the food industry as a means to enhance product value and attain product differentiation.

Microencapsulation is the process of entrapping a component of interest in another material in a miniature form. The encapsulated target component is known as the core material, internal phase, active agent, or payload phase; the encapsulating material is called a coating, shell, external phase, carrier material, or matrix. Different types of encapsulates (viz., reservoir and matrix types) are formed, depending on the preparation method. In the reservoir type, a shell surrounds the core material and is also called a capsule, whereas in the matrix type, the active agent is dispersed over the carrier material and can be found on the surface. Sometimes the active agent is strengthened by a coating, and this is called a capsule in the matrix type (Zuidam and Shimoni 2010).

Microcapsules should be food grade, protective barriers to the inner phase, with the ability to release the inner phase at a controlled rate. This technology can be suitable for delivering bioactive compounds, providing natural flavor and color in food products, controlling oxidative losses, and extending the shelf life of health foods (Mutukumira et al. 2015). Microencapsulation can provide food with promising characteristics to attract consumer interest and promote the relevant industry (Prakash et al. 2016). The microencapsulation technique involves a series of different complex processes through which incorporation of various biologically active materials into different foods is done, providing functional and selected quality attributes to the target food sample (Martín et al. 2015). This novel technique can be used to convey probiotics through severe acidic conditions and ensure their efficient delivery to the gastrointestinal (GI) tract by maintaining cell viability. This microencapsulation can improve viability, control release of encapsulated microorganisms in the gastrointestinal tract, and provide favorable conditions for efficient colonization. The materials used for encapsulation of microorganisms in the microencapsulating agent are soluble only in alkaline conditions, which exist in the intestine. Recent research has focused on microencapsulation of lactic acid bacteria (LAB) to retain their bacteriocinogenic and probiotic characteristics in preparation of diverse food products such as yogurt, cheese, and milk. The various microencapsulation methods differ from one another in different ways, viz., the type of encapsulating agent, both the physical nature and the chemical nature of the food, and the material used. Physical methods include spray drying, spray cooling, and extrusion, resulting in drying of microdroplets, their solidification by cooling, and formation of microspheres by mechanical means. In chemical microencapsulation, molecular inclusion and interfacial polymerization methods are used.

6.1.1 Probiotics

The dependence of the intestinal microflora on food makes it possible to take initiatives to alter the microflora in the human body by replacing harmful microbes with useful microbes. Living microorganisms that are consumed in specific amounts to improve the intestinal microflora are known as probiotics. Probiotic foods provide various health benefits by balancing the microbial count in the intestines; enhancing immunity; promoting synthesis of vitamins such as nicotinic acid, folic acid, and vitamin B; stimulating calcium absorption and reduction of serum cholesterol levels; enhancing nutrient digestibility; restricting the effects of foodborne pathogens; and ameliorating cardiovascular disease, pancreatitis, and hypercholesterolemia (Homayouni et al. 2012; Sanders et al. 2005; Suvarna and Boby 2005; Parvez et al. 2006; Oxman et al. 2001). Probiotic bacteria can be utilized in a wide range of foods, including dairy products (such as cheese, dairy desserts, ice cream, and yogurt) and nondairy products (such as cereals, chocolate, and juices). With the growth in the geriatric population, consumers' consciousness of their well-being, health care costs, and changes in lifestyle are expected to drive growth of the probiotic foods market. The microorganisms contained in probiotic foods should be capable of colonizing the intestine, benign to epithelial cells, bile and acid tolerant,

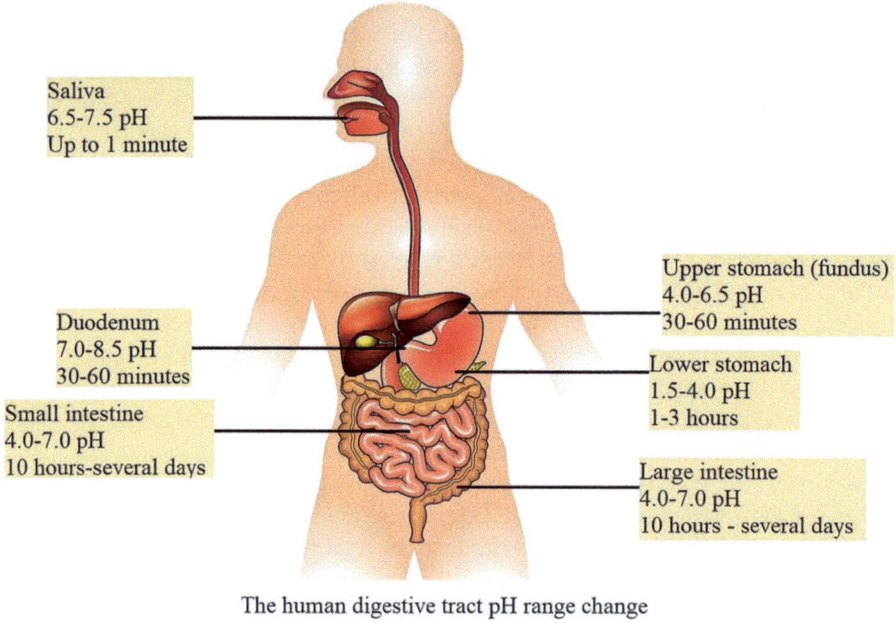

Fig. 6.1 Residence times of food and pH ranges in different parts of the digestive system

nonpathogenic, and categorized as GRAS (Generally Recognized as Safe) (Panghal et al. 2018). The most important criterion for beneficial effects of probiotic cells on the host's health is the viability of the microorganisms. Probiotic microorganisms must survive to reach their target site of action. Many researchers have reported poor survival of probiotic bacteria in products containing free probiotic cells (De Castro-Cislaghi et al. 2012). For effective delivery and stability of microorganisms, encapsulation is required.

For effective probiotic microencapsulation, there is a need to understand the delivery system for controlled release of the microorganisms and the physiology of the gastrointestinal tract, which is composed of the alimentary canal (which runs from the mouth to the anus) and associated glands, chemicals, hormones, and enzymes that assist in digestion (Reed and Wickham. 2009). After ingestion, loss of bacterial viability is observed because of the highly acidic conditions. There are huge variations in the transit time (from 5 min to 2 h) and pH (1–2.5), depending on the age of person, the amount of food intake, and the frequency of intake. Food travels via the stomach before reaching the small intestine; the transit time varies depending on the properties of the food (Fig. 6.1). There is a pH difference between the proximal region (pH 6.15–7.35) and the distal region (pH 6.80–7.88) of the small intestine. Food reaches the large intestine after it passes through the small intestine, and the pH then decreases to pH 5.20–7.02 in the descending colon and pH 5.26–6.72 in the ascending colon. The large intestine contains a large number of microorganisms that are beneficial for human health.

6.2 Probiotic Encapsulation and Applications of Encapsulated Probiotics

The probiotic encapsulation process is done with the involvement of cell immobilization through physical entrapment in a polymeric network. After immobilization, LAB can provide numerous advantages (in terms of metabolic and biomass production) in comparison with a free cell system. The benefits of immobilization of LAB include protection of the cells from chemicals, reusability of biocatalysts, improvement of the plasmid balance and hindrance of loss during the washing process, high cell concentrations, protection of cells from physical hazards, and greater hindrance of bacteriophage invasion through different contamination events (Rubio et al. 2014). The efficiency of microencapsulation is very high in processing of a highly viable bacterial starter culture. Encapsulation also extends the shelf life—as was observed in encapsulated *Lactobacillus rhamnosus* VTT E-97800 at room temperature for 6 months—and the shelf life can be further extended to 18 months in cryogenic storage. Encapsulation with a mixture of alginate and glycerol increases the survivability of a starter culture after deep freezing (Kavitake et al. 2017).

Encapsulation also improves probiotic viability significantly in harsh conditions. For example, cellulose acetate phthalate (CAP)–based encapsulation of *Bifidobacterium pseudolongum* resulted in better viability through the gastrointestinal tract in simulated conditions (Iravani et al. 2015). The ability of *Bifidobacterium longum* to be viable in the gastrointestinal tract in simulated conditions (pH 1.5) was improved by encapsulation with calcium alginate. Coating of the alginate capsules containing *Lactobacillus acidophilus* with calcium chloride improved the ability of *L. acidophilus* to tolerate the strong acid (pH 2) and bile (1%). The nature of the capsule material, the capsule diameter, and the coating thickness also affect the viability of cells. A thinner coating may lead to loss of the protective effects of the encapsulation (Hamaguchi et al. 2018). In acid–bile conditions the viability of probiotics encapsulated in alginate capsules shows a huge difference between a capsule diameter of 20 μm and a diameter of 70 μm (Amine et al. 2014).

Encapsulation of probiotics not only has a protective action against unsupportive acidic pH levels (high acidity, low pH) but also provides protection against drying and against molecular oxygen from obligatory anaerobic microbes, digestive enzymes, H_2O_2, and bacteriophages. During storage in refrigeration, greater acidity and a lower pH decrease the viability of probiotics. A significant increase in the viability of bifidobacteria in frozen ice milk has been observed in alginate-encapsulated probiotics (Kataria et al. 2018).

6.2.1 *Biomass Production for Industrial-Scale Application in Fermenters*

Microencapsulation improves the ability of probiotic microorganisms to tolerate different stress factors (such as bacteriophage infection and chemical poisoning agents), provides protection against genetic mutations, and enhances productivity during severe agitation, with higher biomass production (Alting and Zhaoping 2015).

6.2.2 New Methods in Food Manufacture

New products may be developed by adopting probiotic microencapsulation, which also maintains the cellular metabolic activity of probiotics at a desirable rate. Yogurt manufacturing by encapsulation of traditional yogurt probiotic microbes (such as *Streptococcus delbrueckii* ssp. *bulgaricus* and *Streptococcus salivarius* ssp. *thermophiles*) provides better quality than traditional yogurt manufacturing. The ratio between *Lactobacillus delbrueckii* ssp. *bulgaricus* and *Streptococcus thermophiles* is maintained from the start to the end of the fermentation process, and, as a result, control of the flavor is also possible (De Prisco et al. 2017).

6.2.3 Improved Sensory Properties of Probiotics

This technique also improves and maintains the sensory profile of formulated food. Sharpness of flavor and overacidification of a product restricts the shelf life of the product and its sensory score. Encapsulated probiotic food is comparatively less sour because of lower acid production and decrease in pH even during storage. This flavor fixation of fermented products by encapsulation of probiotics is possible because the encapsulated cells are relatively or totally inactive metabolically; thus, there is no alteration of flavor. In fermented probiotic products the acetic acid produced by *Bifidobacterium* spp. may develop an "off" flavor because of vinegar taint developing during the fermentation within the storage period. Microencapsulation of *Bifidobacterium* spp. is a very effective solution to solve this problem because it controls production of acetic acid by the encapsulated cells (Holkem et al. 2017).

6.3 Microencapsulation Methods

The process used to produce microencapsulated capsules involves different types of methods, including fluidized bed layering, spray congealing, extrusion, and spray drying.

Microcapsules should be prepared with controlled agitation, a moderate pH, a low temperature, and reduced oxygen contact, and they should be functional at room temperature as well as at 4 °C. The size of the particles can be up to 80 μm, so that there is a smooth mouthfeel instead of a sandy mouthfeel. Thus, selection of a suitable method is critical, as it affects the functionality as well as the cost of the product (Table 6.1). Selection of the technique is done on the basis of following aspects:

(i) What conditions can affect the viability of the probiotics?
(ii) What are the techniques used during food production and preservation?
(iii) What storage conditions need to be maintained before consumer use of this encapsulated food product?

Table 6.1 Techniques used for the process of microencapsulation and related characteristics

	Techniques							
Characteristics	Spray drying			Freeze drying	Fluidized bed	Extrusion	Impinging	External gelation
Foods used	Yogurt	Nondairy probiotic food	Dairy dessert	Red beet and grape juices, low-fat pasteurized milk	Formulas for infants	Yogurt	Pear- and peach-based snacks, orange juice	Yogurt
Stability (chemical and physical)	No role for milk fat in increasing the viability of the probiotic	In raspberry juice, during preheating, microbes were killed	In bile salt, microparticles did not improve probiotic survivability	Trehalose in simulated gastric juice	During storage, no improvement in probiotic viability	In yogurt, not much improvement in probiotic survivability with use of chitosan	No improvement in survivability of probiotics with encapsulation	Less protection for bacteria in low-pH environments
Materials used for encapsulation	Whey protein	Raspberry juice pulp	Whey protein	Trehalose, lactose, sucrose, or cellobiose	Whey protein	Alginate	Alginate	Hi-Maize® starch plus alginate
Special treatments	Utilization of milk fat in processing	Bacterial preheating	Gelatin added, biologically active compounds encapsulated	–	Utilization of Cellets® and palm oil	Utilization of chitosan	–	Utilization of glycerol and Hi-Maize® starch
Particle sizes	3–75 μm	–	11.23 μm	–	15–40 μm	1.89*mm*	10–40 μm	0.5–1 *mm*
References	Picot and Lacroix (2004)	Anekella and Orsat (2013)	De Castro-Cislaghi et al. (2012)	Basholli-Salihu et al. (2014)	Weinbreck et al. (2010)	Krasaekoopt et al. (2006)	Sohail et al. (2012)	Sultana et al. (2000)

(iv) What density and particle size are desirable to optimize the appeal of this product?
(v) What are the mechanisms and initiators of release of this product at the target site?
(vi) What are the cost restrictions?

6.3.1 Different Techniques Used for Microencapsulation

6.3.1.1 Spray Drying

Spray drying has primarily been used for flavor encapsulation since the 1900s and is one of the most commonly used techniques because of its highly adaptive, consistent, economical, and quality results. Spray-dried powder particles are relatively small, uniform in size and shape, can be easily transported without any special requirements, and can be stored for prolonged periods. The procedure includes scattering of the core material and preparation of an emulsion, followed by homogenization of the fluid and subsequent atomization of the blend in the drying chamber for evaporation. Matrix-type microcapsules are formed by this process (Table 6.1). The process parameters, such as the feed rate, particle size, air current (counter- and co-current), and temperature difference need to be controlled (Cook et al. 2012). If the feed temperature is very high, volatile and sensitive compounds might be lost before they get microencapsulated. Cracks on the microspheres and premature release and destruction of the ingredients have been observed when high inlet temperatures were employed. High temperatures during the process of splash drying cause the cell pores to release intracellular substances. High temperatures used in drying also affect the encapsulated probiotic bacterial cultures, as a temperature beyond 85 °C is lethal to probiotic organisms (Ray et al. 2016). Appropriate changes in and control of the handling conditions—for example, the inlet temperature (T_i) and the outlet temperature (T_0) can result in better encapsulated cultures with coveted particles. Different components that influence dried probiotic viability are the kind of strain and its resilience in pressurized conditions, the drying temperature, the time of introduction to heating (before a splash drying process), and the water action and capacity conditions (after a shower drying process).

6.3.1.2 Freeze Drying

For production of a powdered probiotic product, freeze drying is used. The process is based on sublimation, which takes place in three different phases of liquid, starting with freezing, followed by primary and secondary drying. This process is better than spray drying because it results in higher probiotic survival rates (Cook et al. 2012). Crystals formed during freezing might damage the cell membrane and also create stress due to high osmolarity. Thus, different compounds such as trehalose, whey protein, maltodextrin, skim milk powder, and glucose are added to the media to protect the cell membrane and thus improve the viability of the probiotic microorganisms. Sometimes a cryoprotectant can also be added to the media before fermentation to reduce the osmotic differences between the external and internal environments.

6.3.1.3 Fluid Bed

The main advantages of this process are based on its level of heat exposure and comparatively low cost. The probiotic culture is encapsulated in a supporting material (such as skim milk, alginate, and fat) before drying. Sometimes shellac—a purified product from the insect *Kerria lacca* (Coccoidea)—is also used as a coating material for the capsules because it provides effective resistance against gastric fluid and is of natural origin (Nedovic et al. 2011). However, because it has low solubility in intestinal fluid, it is less often used as an enteric coating, especially for coatings on hydrophobic substances. Polyvinylpyrrolidone, hydroxypropyl methylcellulose, and sodium alginate have been used by researchers for enteric coating with addition of the water-soluble polymer glyceryl triacetate and glycerol as a plasticizer (Guignon et al. 2002). Probiocap and Duaolac have been used by some companies to develop products and are used in multilayer coatings (Solanki et al. 2013).

6.3.1.4 Extrusion

The extrusion method uses hydrocolloids to encapsulate probiotic living cells. The extrusion method is one of the most commonly used methods because of its low cost, simplicity, gentle formulation conditions (which can include aerobic and anaerobic conditions) and high viability. The extrusion method uses hydrocolloids (alginate and Cu-alginate) to encapsulate probiotic living cells. This process involves the preparation of a hydrocolloid solution, addition of microorganisms, and extrusion of the cell suspension through a syringe needle. The droplets are allowed to drip via a vibrating nozzle into a hardening solution, and formation of the droplets occurs in a controlled manner. This technique is known as prilling (Bidoret et al. 2017). An electrostatic field or coaxial flow is also used to produce small droplets. Application of an electrical field gives rise to electrostatic forces, which disrupt the liquid surface at the needle tip, creating a charged small droplet. For mass production of beads, a multinozzle system or rotating disk system is used. Centrifugal extrusion involves a nozzle with concentric orifices situated on the external boundary of a rotating cylinder. The core material contents and the liquid shell materials are pumped through the inner orifice and the outer orifice, respectively (Burgain et al. 2015). With rotation of the system, capsules are formed by an extrusion rod. However, this process cannot be applied for large-scale production, because microbead formation is slow. This technology produces particles of a large size (500–1000 nm), thus affecting the smooth mouthfeel of the product.

6.3.1.5 Two-Step Drying

Generally, a higher temperature will be used at the inlet and outlets of a spray dryer. This is done to produce a required dry powdered component with the required moisture content—i.e., below 4%, which is helpful for further storage. Advancements in drying conditions with an aim to optimize the survival of probiotics during storage is required. To this end, scientists have developed a framework for utilization of spray drying, with an inlet temperature of 80 °C, an outlet temperature of 48 °C, and vacuum drying at a moderate temperature of 45 °C. This process results in greater viability of probiotics at a lower cost (Ong et al. 2018).

Spray Freeze Drying

This type of drying method includes both freezing and spray drying techniques. The probiotic cells are atomized into a vapor state in a liquid nitrogen environment, producing scattered frozen beads, which are then dried in a dryer. However, this procedure has a few limitations, including high energy requirements, a long handling time, and a high cost (30–50 times the cost of spray drying) (Dolly et al. 2011).

Spray Chilling

Spray cooling is also known as spray chilling and spray congealing. This process is very similar to spray drying, as it also forms small droplets. However, this method is completely dependent on injection of cold air, which makes the substances solid. Molten materials with bioactive components are atomized and become solid when they meet the cold air. In the spray chilling technique, fat matrices are used as a carrier, but this is detrimental to the encapsulation ability and expulsion of the innermost material during extended storage, the polymeric arrangement, and the crystalline structure properties of the lipid components during solidification. Spray chilling is the least expensive encapsulation technology and is also used for generation of smaller beads (de Lara Pedroso et al. 2012).

6.3.1.6 Ultrasonic Vacuum Spray Drying

This technology has been developed to avoid oxidative and thermal stress. The system works under a low temperature, and an ultrasonic nozzle and vacuum atmosphere are also used in the chamber during drying (Semyonov et al. 2011).

6.3.1.7 Hybridization System

A hybridization system is used for a dry encapsulation technique. It contains a high-speed rotor with a stator, a powder recirculation circuit, and six blades. Collisions occur between a high-speed airstream generated by the rotating blades and a powder mixture (containing host and guest particles). The particles form an ordered mixture by embedding of the guest particles on the surfaces of the host particles. This results in a high yield of microcapsules, and heat-induced bacterial damage can be minimized by use of a cooling system in which the temperature is maintained below 30 °C. Prebiotic substances—such as sorbitol, mannitol, lactulose, xylitol, inulin, fructo-oligosaccharide (FOS), and raffinose—have been tested by use of this technique (Fayed et al. 2018). The results showed that double microencapsulation by hybridization has beneficial effects on the probiotic host.

6.3.1.8 Electrospinning

Electrospinning is a combination of electrospray and spinning. An electrical field applied to the fluid melts the solution coming from a dye tip, which acts as an electrode. Droplets are deformed and ejection of the charged jet from the tip leads to formation of continuous fibers. The benefit of the electrospinning technique is that it produces thin fibers or capsules a few nanometers in size with a large surface area. This technique is attractive for various different applications because it allows large-scale production. Electrospinning for encapsulation of probiotics has been done using a protein-based matrix (whey protein concentrate) or a carbohydrate-based matrix (pullulan). The cell viability in the whey protein concentrate microcapsules was greater than that in the pullulan structures (Zaeim et al. 2018).

6.3.1.9 Impinging Aerosol Technology

Two separate aerosols are used in impinging aerosol technology, which includes a microbial suspension in an alginate solution and calcium chloride. A mixture of the microbial suspension and alginate is injected from the top of a cylinder, with calcium chloride being injected from the base. The diameter of the alginate particles (and hence also that of the microcapsules produced by this technique) is <40 μm (Sohail et al. 2011).

6.4 Supporting/Wall Materials

In microencapsulation, the microsphere stability and the encapsulation efficiency are of great importance and are dependent on the material used for encapsulation, known as the wall or supporting material. The supporting/wall material should be soluble in water, since most spray drying suspensions are water based and easy to dry. The supporting material should be compatible with the core materials and of low viscosity, and it should possess good mechanical strength, emulsification properties, and film-forming properties. Biopolymers, low molecular weight carbohydrates, natural gums (acacia, k-carrageenan, and alginates) and proteins (whey protein and gelatin) are generally considered good wall materials. The use of carriers such as FOS, maltodextrin, arabic gum, inulin, polydextrose, skim milk powder, and soy milk protein in the suspension may have significant effects on the viability (Table 6.2). Since these wall materials

Table 6.2 Products, cultures, and techniques involved in the process of microencapsulation

Categories	Probiotic strains	Supporting materials	Techniques	References
Protein	–	Gluten, casein, gelatin, albumin	Emulsion, spray drying	Ono and Oka (1980)
Fresh cheese	*Lactobacillus bulgaricus*, *Streptococcus thermophiles*	Calcium alginate	Extrusion	Prevost and Divies (1987)
Fresh cheddar cheese	*Bifidobacterium bifidum*, *Lactococcus lactis* spp. *lactis*	k-Carrageenan	Emulsification	Dinakar and Mistry (1994), Sodini et al. (1997)
Crescenza cheese	*Bifidobacterium bifidum*, *Bifidobacterium infantis*, *Bifidobacterium longum*	Calcium alginate	Freeze drying	Gobetti and Türp (1998)
Cheddar cheese	*Lactobacillus paracasei*	Skim milk	Spray drying	Gardiner et al. (2002)
Cheddar cheese	*Lactobacillus acidophilus*, *Bifidobacterium infantis*	Alginate/starch	Emulsification	Godward and Kailasapathy (2003)
Feta cheese	*Lactobacillus acidophilus*, *Bifidobacterium lactis*	Alginate	–	Kailasapathy and Masandole (2005)
Kasseri cheese	*Lactobacillus acidophilus*, *Bifidobacterium bifidum*	Alginate	Extrusion, emulsification	Ozer et al. (2008)
White brined cheese	*Lactobacillus acidophilus*, *Bifidobacterium bifidum*	Alginate	Extrusion, emulsification	Ozer et al. (2009)
Mayonnaise	*Bifidobacterium infantis*, *Bifidobacterium bifidum*	Calcium alginate	–	Kailasapathy (2002)
Yogurt	*Lactobacillus casei*	k-Carrageenan, locust bean gum	–	Brinques and Ayub (2011)
Cheese	*Bifidobacterium infantis*, *Bifidobacterium longum*, *Bifidobacterium lactis*, *Bifidobacterium bifidum*, *Lactobacillus paracasei*	–	–	Boylston et al. (2004)

Sausage	*Lactobacillus reuteri, Escherichia coli*	–	–	Muthukumarasamy and Holley (2006)
Yogurt, fermented milk	*Bifidobacterium longum, Bifidobacterium infantis, Bifidobacterium* spp., *Lactobacillus casei, Bifidobacterium breve, Bifidobacterium lactis, Lactobacillus acidophilus, Lactobacillus rhamnosus, Lactococcus lactis*	–	–	Casarotti et al. (2014), Fazilah et al. (2018)
Mayonnaise	*Bifidobacterium bifidum, Bifidobacterium infantis*	–	–	Kailasapathy (2002)
Ice cream, ice milk, frozen desserts	*Bifidobacterium lactis, Bifidobacterium* spp., *Lactobacillus casei, Lactobacillus acidophilus*	–	–	Iravani et al. (2015)
Juice	*Lactobacillus rhamnosus*	–	–	Speranza et al. (2017)
Milk	*Bifidobacterium lactis, Lactobacillus casei, Lactobacillus acidophilus, Bifidobacterium bifidum, Bifidobacterium longum, Lactobacillus bulgaricus, Bifidobacterium breve*	–	–	Iravani et al. (2015)
Cereal-based foods	*Bifidobacterium longum, Bifidobacterium* spp., *Bifidobacterium lactis, Lactobacillus rhamnosus*	–	–	Bampi et al. (2016)

contain prebiotic sources, when they are mixed with probiotics, the resulting powders can be considered synbiotics.

6.4.1 *Alginate*

Alginate, a linear heteropolysaccharide, is composed of D-mannuronic and L-guluronic acids. The composition and the sequence of L-guluronic acid and D-mannuronic vary widely, depending on the source, and thus control the functionality of the alginate. A mixture is made by mixing a cell suspension with a solution made of sodium alginate. This mixture is dripped into a solution containing cations (usually $CaCl_2$ as a form of Ca^{2+}) to form beads. A gel sphere is formed instantly from the droplets, and the gel sphere entraps cells in a three-dimensional (3D) structure. Sodium ions from the guluronic acids are exchanged with divalent cations (Ca^{2+}, Ba^{2+}, or Ca^{2+}) to form the polymer cross-linking (Ahmadi et al. 2018). Microencapsulation of probiotic microbes is also done by using calcium alginate, in a concentration range of 0.5–5%. Microcapsules can also be formed by internal gelation with alginate, using a solution containing calcium carbonate. An organic acid like acetic acid is added to the solution, followed by formation of a water-in-oil emulsion. The organic acid penetrates the water phase and reacts with the calcium carbonate, freeing up carbonic acid and calcium carbonate. The reaction between calcium ions and the alginate forms an egg-box structure (Chávarri et al. 2010). Blending with cornstarch or creation of a semipermeable membrane layer of chitosan surrounding the capsules improves the efficiency and effectiveness of the encapsulation process.

6.4.2 *Carrageenan*

The neutral polysaccharide k-carrageenan is used as a food additive and as a microencapsulation support system. At a 2–5% concentration, carrageenan needs a temperature between 60 °C and 90 °C for proper dissolution. Gelation of the mix is done by adding probiotics to a polymer solution kept at 40–45 °C or by keeping it at room temperature. Beads are formed and, to prevent swelling of the gel, K^+ (in the form of KCl) is added to stabilize the gel, otherwise gelation will be induced. However, KCl has an inhibitory action against some LAB, so other ions such as Cs^+, NH^+, and CS^+ are recommended (Shi et al. 2013).

6.4.3 Whey Protein

Whey protein possesses high nutritional values and excellent functional properties, making it suitable for microencapsulation. Capability of this protein to bind with a wide variety of active molecules makes it an effective transporter for the safe delivery of compounds to the host (Ahmadi et al. 2018).

6.4.4 Pectin

Pectin is a heteropolysaccharide, which can be extracted from almost all fruits. It is used as a dietary fiber source and as a gelling agent in the food industry (Zaeim et al. 2018). Pectin remains intact in the stomach and small intestine; thus, it can be used as a coating material for microencapsulation. Gebara et al. (2013) reported that a microencapsulation system using pectin microparticles with whey protein resulted in far greater protection of *L. acidophilus* in comparison with free cells.

6.4.5 Milk

Pure milk, along with other supporting materials, can be used as an encapsulation carrier. Use of locust bean and carrageenan to coat microparticles of milk has been found to provide increased protection for *L. bulgaricus,* but the microspheres were irregular in shape and possessed poor mechanical strength. This problem can be resolved by using alginates with milk, and these microspheres are effective for probiotic protection in an extreme simulated gastrointestinal environment (Shi et al. 2013).

6.4.6 Sodium Carboxymethyl Cellulose

Sodium carboxymethyl cellulose (NaCMC) is a water-soluble cellulose derivative containing linked glucopyranose with different levels of carboxymethyl substitution. In drug and probiotic delivery, the importance of NaCMC is due to its gastric acid resistance and solubility within the intestine. A type of microcapsule has been developed in which NaCMC and a *Lactobacillus reuteri* cell suspension were used with and without rice bran emulsified by palm oil and cross-linked with aluminum ions. The viability of *Lactobacillus reuteri* was improved by the microencapsulation process utilizing rice bran and NaCMC, even after heat exposure. Hence, these particles can be used for probiotic product development involving thermal treatment at any stage of processing (Martín et al. 2015).

6.4.7 Cellulose Acetate Phthalate

Being safe in nature, CAP is used to control the release of drugs within the intestine. CAP is soluble in a medium with a pH $\geq$6, as it contains a phthalate group, but it is insoluble in a medium with a pH $\leq$ 5. This clearly suggests that microencapsulation using CAP will increase the efficiency of delivery of probiotics in the gastrointestinal tract. Addition of CAP to an emulsion prepared using oil and starch will increase the viability of probiotic microbes in simulated gastrointestinal conditions.

6.4.8 Chitosan

Chitin is a linear polysaccharide obtained from crustacean shells by a process of deacetylation, and has a positive charge. Chitosan has the ability to form a gel similar to alginate, and it is water soluble at a pH < 6 (Chávarri et al. 2010).

6.4.9 Gelatin

Partial hydrolysis of collagen produces gelatin, which has a special structure and versatile functional characteristics. Gelatin forms a solution of high viscosity when mixed with water, which can change its form from a viscous solution to a jelly form on cooling. Synergetic effects are produced when gelatin, which is amphoteric, is used together with gellan gum (Riaz and Masud 2013). At a pH > 6 the aforementioned polymers are miscible; they repel each other as they carry a net negative charge. At a pH < 6, the net charge of gelatin is positive and thus attracts the negatively charged gellan gum. A mixture of gelatin and toluene di-isocyanate used for encapsulation of *Lactobacillus* ssp. *cremoris* improved the crack tolerance of the capsules even at high concentrations (Gbassi and Vandamme 2012).

6.4.10 Chickpea Protein

Because of its remarkable functional properties and nutritional importance, chickpea is utilized as an encapsulation material. Its use also addresses some concerns regarding allergens. Chickpea protein–alginate microcapsules were first created by researchers using emulsion technology. These microcapsules were found to provide protection for *Bifidobacterium adolescentis* in synthetic gastric juice. The size of the beads produced using this design was <100 μm. These microcapsules have no adverse effects on food ingredients for consumers. Research has shown that chickpea protein–alginate capsules are a suitable probiotic carrier and can be implemented in food applications. The combination of chickpea protein and alginate provides protection for *B. adolescentis* against intestinal fluids and synthetic gastric juice (Wang et al. 2014).

6.5 Efficiency Evaluation of Microencapsulated Probiotics

Probiotic efficiency in microencapsulation is estimated through a series of qualitative parameters.

6.5.1 Probiotic Cell Viability

The effectiveness of microencapsulation in association with maintenance of the viability of probiotics against adverse stress factors such as acidity can be easily calculated by considering the cell loss kinetics inside the products and/or simulated body conditions over time. Because of the metabolic activity of the starter culture, fermented products such as yogurt will not be stable throughout the storage period, and a linear relation is not followed, because of logarithmic loss of probiotic cells, which differs from thermal bacteriology rules. For description of the D value in thermal bacteriology (meaning that the microbial population reduces by one logarithm in an equal period of time, under constant conditions of acidity or pH, for example), the beads encounter the artificial and static conditions of the product (lactic acid of a suitable pH) and/or the gastrointestinal tract (chloridric acid with a pH of 1.5–2 for about 2 h, and then phosphate buffer with a neutral pH, bile salts, and digestive enzymes from the pancreas) and the kinetics of the cell loss are evaluated. A temperature of 37 °C should be maintained for all of the stages to be judged. With the help of lactic acid and a temperature of 4 °C, fermented products such as yogurt can be easily simulated. To reduce the duration of the experiment, the lactic acid concentration can be increased by several times (e.g., to 20–25%) (Mortazavian et al. 2007).

6.5.2 Cell Release Ability or Dissolution Ability

Capsules have the ability to release their entrapped cells at the target site, mainly in the intestine. The intestine can also supply suitable conditions for survival and multiplication of probiotics, as well as activity and settlement of probiotics, achieved by release of cells at the target site. The volume of cell release can be calculated by subjecting microbeads to simulated gastrointestinal conditions. Before the encapsulation process, released cells are counted for comparison with the initial cell count (Mortazavian et al. 2007). With the help of an appropriate mesh, capsules that are digested can be easily filtered by a method of membrane processing. The percentage of beads that are not digested can be represented by the weight ratio of the filtrate to the retentate particles. By measuring the ratio of the mean bead diameter before the digestive test to the diameter after the test, the digestibility of the beads can be easily calculated. Bead digestibility can also be assessed by using a light-scattering or scanning electron microscopy (SEM) technique. The digestibility of the beads with exposure to pancreatic enzymes can be assessed with use of suitable enzymes and an incubation temperature of 37 °C.

6.5.3 *Microgeometric Properties of Beads*

The microgeometric characteristics of the beads include their size/diameter, shape, uniformity, and integrity. With the help of light-scattering technology or application of a laser diffractometer, the size of the beads can be calculated. SEM or light microscopy can also be used as an alternative method for direct observation. In addition to this, methods such as membrane filtration or sieving of the encapsulated mix are used (Mortazavian et al. 2007). Usually, the beads are uniformly of an elliptical or spherical shape, and this is mainly observed by use of the aforementioned microscopic techniques. The uniformity and integrity of the beads have a direct impact on their efficiency. From this point of view, assessment of cracks, pores, and voids on the surface of the beads is also necessary. When considering the microgeometric properties of the beads, particles that are self-aggregated and do not carry any cells should not be mistaken for real beads. This evidence can be observed in small bare granules and starch particles.

6.5.4 *Capsule Density*

The compression of the capsules in the beads can be assessed by SEM. A light-diffusing strategy has additionally been recommended, as thick materials have greater light absorption. The surface thickness (mass/surface) of the capsules can be evaluated by breaking/dissolving them, drying them, and finally weighing them, dividing the outcome by the total surface of the beads (Mortazavian et al. 2007). The surface of the beads can be surveyed to assess the average bead diameter and its proportion to the surface.

6.6 Factors Affecting the Efficiency of Microencapsulated Probiotic Microbes in Food

Diverse factors can be considered for evaluation of the adequacy of the probiotic encapsulation process—for example, viability management after exposure to unfavorable environmental conditions, cell release/recovery, the capacity, and the solidification period (the time required for capsule development). The factors that influence the aforementioned parameters are discussed in the following sections.

6.6.1 Characteristics of Capsules Related to Different Environmental Factors

An appropriate choice of capsule material with regard to its intended environment is imperative. For example, spillage of calcium ions from alginate capsules causes them to decompose. Alginate capsules are unsuitable for an environment with high acidity and the presence of chelating agents, but in milk-based media (for example, yogurt, cream, and liquid milk) the abundance of calcium ions may mean that calcium ion leaching from gel bead structures is markedly inhibited; thus, the gel beads may maintain their structure. Use of resistant starch (RS) as the material for the capsules will mean that they resist enzymatic digestion. Since delivery of probiotic cells to the small intestine is the aim, the choice of the capsule material ought to be such that it does not degrade until it is subjected to the pH of the small intestine or to pancreatic enzymes. If the beads are resistant to intestinal digestion, this will bring about their immediate discharge from the body without settling of the probiotic cells in the digestive tract. If the beads must be settled in the large intestine, it is best for them to be tolerant of pancreatic and small-intestinal conditions, but not large-intestinal conditions. However, this generally is not achieved effectively because of limitations in the qualities of synthetic encapsulation materials. When the beads open in the small intestine, the cells they release will normally reach the large intestine. In some cases, pancreatic shock following release of the cells in the small intestine may diminish their viability, but we are not aware of any in-depth research on this topic. In general, every capsule needs to be impervious to acidic gastric juices (Mortazavian et al. 2007). In some cases, it is important to utilize particular kinds of hydrophobic components for encapsulation to make the beads tolerant of high-moisture conditions.

6.6.2 Capsule Coating

The best way to make the capsules effective is by coating them. Alginate capsules can be coated with a shell coating, which helps them to resist the chelating actions of calcium ions and improves their mechanical strength. The strength of the capsules is increased with an increase in the concentration of the coating material.

6.6.3 Capsule-Forming Solution Concentration and Bead Diameter

The most important part of the encapsulation method is the concentration of the solution and the resultant beads. The larger the diameter of the beads, the greater protection they provide against hostile environmental influences. This has been demonstrated in both body and product conditions. Alginate capsules with a diameter of 0.5–1.0 mm have been shown to significantly improve the survival of bifidobacteria in yogurt with a normal pH in refrigerated storage, but not at a pH simulating the conditions in the stomach. Expanding the bead diameter more than is suitable (with respect to the kind of capsule and the product it contains) is not an option because it results in an unattractive mouthfeel and flavor. Besides, an expansion in the capsule size diminishes the ability of pancreatic enzymes to digest it. Any increase in the bead diameter—particularly when it is made of RS—should only be done with due consideration, since this component is difficult for pancreatic enzymes to digest (De Araújo Etchepare et al. 2016). Research regarding the concentrations of solutions used to make capsules showed that an increase in the alginate solution concentration from 0.75% to 1.8% influenced the viability of *L. acidophilus* in simulated gastric conditions, but when a 2% concentration was used, creation of round and homogeneous capsules was not possible because the solution was more viscous and had lower mass diffusivity. It was also shown that an increase in the concentration of a solution of calcium alginate and high-amylose cornstarch (HACS) (beyond 2% and even up to 4%) did not markedly affect the protective properties of the beads when they were exposed to strong environmental influences (Mortazavian et al. 2007). More research is required before a comprehensive conclusion can be drawn.

6.6.4 Environmental Conditions

The kind of adverse environmental factors that will be encountered, and their severity, are probably the most critical parameters that may diminish the viability of encapsulation. For example, capsules endure low-acid conditions (yogurt) significantly more than high-acid conditions (gastric juice). Alginate containers with a mean diameter of 100 μm are adequate for most kinds of fermented products but not for gastric acid (Maleki et al. 2015). This warrants particular consideration because the encapsulation needs to be suitable not only for the conditions required by the product itself but also for the effects that passage through the gastrointestinal tract will have on the capsules and the product they contain (Mortazavian et al. 2007). In view of the fact that alginate is used so frequently for encapsulation of probiotics, precautionary measures are needed to overcome its vulnerability in acidic conditions.

6.6.5 Capsule Material Modification

A standard practice for enhancing the viability of encapsulation is to adjust the chemistry of the capsule materials. By making small changes in the auxillary coordinates, adjustments can be made in the capsules, and additives can be included, such as glycerol for cryoprotection of the cells against cryogenic freezing. By use of this technique, the viability of bifidobacteria and *L. acidophilus* have reportedly been increased to 88.5% and 90%, respectively (Chávarri et al. 2010). A cryoprotectant lessens ice crystallization by attachment to free water particles. The volume of glycerol-containing beads has been shown to be 43% smaller than that of normal beads during freezing, showing the cryoprotective impact of the glycerol. Slowly digestible starch (SDS) and Tween 80 can be added, as they increase the cells' resistance to solidification during drying and change the diameter of the beads. The more SDS and Tween 80 are added to the encapsulating solution, the smaller the resulting beads will be. Likewise, it has been asserted that use of the aforementioned materials gives alginate cases particular attributes, including the capacity for disintegration in the buffer media and greater resistance to lactic acid.

6.6.6 Initial Microbial Cell Concentration

The quantitative effectiveness of the encapsulation and the total number of entrapped cells in each bead in the encapsulation will both increase with an increase in the microbial cell concentration in the encapsulating solution. If the load of the cells exceeds a prespecified limit, softening of the capsule structure may occur. Thus, the process must adhere to specific conditions, and better examination during the process helps to achieve the best resultant formation. Control of process parameters such as the freezing process, drying, micronization, and storage will also reduce damage to the capsules and maintain consistency in their size and shape (Brinques and Ayub 2011).

6.7 Conclusion and Future Prospects

The innovation of smaller-scale encapsulation has evolved from straightforward immobilization or capture to development of complex and precise containers on a miniaturized scale. The advances in this field have been huge with regard to the development of nutraceuticals, functional foods, and encapsulation of live probiotic bacterial cells on a miniature scale. Probiotic treatment (or microbial intervention) depends on the idea of robust gut microflora. Smaller-scale encapsulation processes will target on delivering abundant amount of probiotic microbes to the consumers. This technology will be utilized as an instrument to package both prebiotic fiber and probiotic microscopic organisms inside similar capsules to improve these microorganisms' viability and colonization of the gastrointestinal tract, and to reap the health benefits of their use. In the future, different forms of delivery might be created—for example, combining prebiotics, probiotics, and also nutraceuticals. Along these lines, development of even more advanced dietary frameworks should be researched. In the food preparation industry, conservation, increased capacity, and small-scale formulation will progressively play greater roles to ensure the suitability and improve the survival of microbes in unfavorable natural conditions (Mishra et al. 2018). New food regulations may determine the strains and the quantities of practical probiotic microscopic organisms with stricter expiry dates for realistic usability of foods or supplements that are guaranteed to be probiotic. Studies (clinical information) should be conducted on the impacts of encapsulation on the safety of probiotic microscopic organisms. Fermented dairy, grains, meat, soups, and ready to eat food products can be considered as food vehicles to utilise microencapsulation innovations for better delivery of probiotic microbes as well as nutrients to consumer. In the food industry, containers, tablets, suspensions, creams, and powders will be progressively utilized for miniaturized encapsulation and innovations in coordinated utilization and external use of probiotics. They will progressively be utilized to treat patients with health problems. Smaller-scale capsules should be created with more precise hardware and containers, and better conveyance frameworks. Sooner, rather than later, nanocapsules may be used to create designed probiotic bacterial combinations that can be conveyed to specific parts of the gastrointestinal tract, where they can bind with particular receptors. These designed nanoencapsulated probiotic bacterial combinations may act as antibodies, with the ability to balance insusceptible reactions (Eratte et al. 2017). Enhanced procedures should be introduced to track these smaller-scale or nanoencapsulated probiotic bacterial cells to ensure their delivery, stability, controlled release, and beneficial effects in the gastrointestinal framework. More in vivo studies ought to be conducted in human subjects to establish the viability of small-scale or nanoencapsulated delivery of probiotic microbes and their controlled release in the gastrointestinal system.

References

Ahmadi H, Wang Q, Lim LT, Balamurugan S (2018) Encapsulation of *Listeria* phage A511 by alginate to improve its thermal stability. In: MRJ C, Kropinski AM, Lavigne R (eds) Bacteriophages: methods and protocols, vol 3. Humana Press, New York, pp 89–95

Alting S, Zhaoping ZH (2015) Optimization of bioethanol production by *Saccharomyces cerevisiae* microencapsulated on alginate-delignified cellulose material. Int J Pharma Bio Sci 6(2):1259–1270

Amine KM, Champagne CP, Salmieri S, Britten M, St-Gelais D, Fustier P, Lacroix M (2014) Effect of palmitoylated alginate microencapsulation on viability of *Bifidobacterium longum* during freeze-drying. LWT-Food Sci Technol 56(1):111–117

Anekella K, Orsat V (2013) Optimization of microencapsulation of probiotics in raspberry juice by spray drying. Lebenson Wiss Technol 50(1):17–24

Bampi GB, Backes GT, Cansian RL, de Matos FE, Ansolin IM, Poleto BC, Corezzolla LR, Favaro-Trindade CS (2016) Spray chilling microencapsulation of *Lactobacillus acidophilus* and *Bifidobacterium animalis* subsp. *lactis* and its use in the preparation of savory probiotic cereal bars. Food Bioprocess Technol 9(8):1422–1428

Basholli-Salihu M, Mueller M, Salar-Behzadi S, Unger FM, Viernstein H (2014) Effect of lyoprotectants on β-glucosidase activity and viability of *Bifidobacterium infantis* after freeze-drying and storage in milk and low pH juices. Lebenson Wiss Technol 57(1):276–282

Bidoret A, Guihard L, Cauret L, Poncelet D (2017) Production of κ-carrageenan beads by prilling process. Can J Chem Eng 95(4):799–805

Boylston TD, Vinderola CG, Ghoddusi HB, Reinheimer JA (2004) Incorporation of bifidobacteria into cheeses: challenges and rewards. Int Dairy J 14(5):375–387

Brinques GB, Ayub MA (2011) Effect of microencapsulation on survival of *Lactobacillus plantarum* in simulated gastrointestinal conditions, refrigeration, and yogurt. J Food Eng 103(2):123–128

Burgain J, Corgneau M, Scher J, Gaiani C (2015) Encapsulation of probiotics in milk protein microcapsules. In: LMC S (ed) Microencapsulation and microspheres for food applications. Academic, London, pp 391–406

Canizales JR, Rodríguez GR, Avila JA, Saldaña AM, Parrilla EA, Ochoa MA, Aguilar GA (2018) Encapsulation to protect different bioactives to be used as nutraceuticals and food ingredients. In: Mérillon J-M, Ramawat KG (eds) Bioactive molecules in food. Springer, Cham, pp 1–20

Casarotti SN, Monteiro DA, Moretti MM, Penna AL (2014) Influence of the combination of probiotic cultures during fermentation and storage of fermented milk. Food Res Int 59:67–75

Chávarri M, Marañón I, Ares R, Ibáñez FC, Marzo F, Del C, Villarán M (2010) Microencapsulation of a probiotic and prebiotic in alginate–chitosan capsules improves survival in simulated gastro-intestinal conditions. Int J Food Microbiol 142(1–2):185–189

Cook MT, Tzortzis G, Charalampopoulos D, Khutoryanskiy VV (2012) Microencapsulation of probiotics for gastrointestinal delivery. J Control Release 162(1):56–67

De Araújo Etchepare M, Raddatz GC, Cichoski AJ, Flores ÉM, Barin JS, Zepka LQ, Jacob-Lopes E, Grosso CR, de Menezes CR (2016) Effect of resistant starch (Hi-Maize) on the survival of *Lactobacillus acidophilus* microencapsulated with sodium alginate. J Funct Foods 21:321–329

De Castro-Cislaghi FP, Carina Dos Reis ES, Fritzen-Freire CB, Lorenz JG, Sant'Anna ES (2012) *Bifidobacterium* Bb-12 microencapsulated by spray drying with whey: survival under simulated gastrointestinal conditions, tolerance to NaCl, and viability during storage. J Food Eng 113(2):186–193

De Lara Pedroso D, Thomazini M, Heinemann RJ, Favaro-Trindade CS (2012) Protection of *Bifidobacterium lactis* and *Lactobacillus acidophilus* by microencapsulation using spray-chilling. Int Dairy J 26(2):127–132

De Prisco A, Van Valenberg HJ, Fogliano V, Mauriello G (2017) Microencapsulated starter culture during yoghurt manufacturing, effect on technological features. Food Bioprocess Tech 10(10):1767–1777

Dinakar P, Mistry VV (1994) Growth and viability of *Bifidobacterium bifidum* in cheddar cheese. J Dairy Sci 77(10):2854–2864

Dolly P, Anishaparvin A, Joseph GS, Anandharamakrishnan C (2011) Microencapsulation of *Lactobacillus plantarum* (MTCC 5422) by spray-freeze-drying method and evaluation of survival in simulated gastrointestinal conditions. J Microencapsul 28(6):568–574

Eratte D, Dowling K, Barrow CJ, Adhikari B (2017) Recent advances in the microencapsulation of omega-3 oil and probiotic bacteria through complex coacervation: a review. Trends Food Sci Technol 71:121–131

Fayed B, Abood A, El-Sayed HS, Hashem AM, Mehanna NS (2018) A synbiotic multiparticulate microcapsule for enhancing inulin intestinal release and *Bifidobacterium* gastro-intestinal survivability. Carbohydr Polym 193:137–143

Fazilah NF, Ariff AB, Khayat ME, Rios-Solis L, Halim M (2018) Influence of probiotics, prebiotics, synbiotics and bioactive phytochemicals on the formulation of functional yogurt. J Funct Foods 48:387–399

Gardiner GE, Bouchier P, O'Sullivan E, Kelly J, Collins JK, Fitzgerald G, Ross RP, Stanton C (2002) A spray-dried culture for probiotic cheddar cheese manufacture. Int Dairy J 12(9):749–756
Gbassi GK, Vandamme T (2012) Probiotic encapsulation technology: from microencapsulation to release into the gut. Pharmaceutics 4(1):149–163
Gebara C, Chaves KS, Ribeiro MCE, Souza FN, Grosso CR, Gigante ML (2013) Viability of *Lactobacillus acidophilus* La5 in pectin–whey protein microparticles during exposure to simulated gastrointestinal conditions. Food Res Int 51(2):872–878
Gobetti JP, Türp JC (1998) Fibrosarcoma misdiagnosed as a temporomandibular disorder: a cautionary tale. Oral Surg Oral Med Oral Pathol Oral Radiol 85(4):404–409
Godward G, Kailasapathy K (2003) Viability and survival of free, encapsulated and co-encapsulated probiotic bacteria in ice cream. Milchwissenschaft 58(3–4):161–164
Guignon B, Duquenoy A, Dumoulin ED (2002) Fluid bed encapsulation of particles: principles and practice. Dry Technol 20(2):419–447
Hamaguchi S, Zafar MA, Cammer M, Weiser JN (2018) Capsule prolongs survival of *Streptococcus pneumoniae* during starvation. Infect Immun. https://doi.org/10.1128/IAI.00802-17
Holkem AT, Raddatz GC, Barin JS, Flores ÉM, Muller EI, Codevilla CF, Jacob-Lopes E, Grosso CR, De Menezes CR (2017) Production of microcapsules containing *Bifidobacterium* BB-12 by emulsification/internal gelation. LWT-Food Sci Technol 76:216–221
Homayouni A, Payahoo L, Azizi A (2012) Effects of probiotics on lipid profile: a review. Am J Food Technol 7(5):251–265
Iravani S, Korbekandi H, Mirmohammadi SV (2015) Technology and potential applications of probiotic encapsulation in fermented milk products. J Food Sci Technol 52(8):4679–4696
Kailasapathy K (2002) Microencapsulation of probiotic bacteria: technology and potential applications. Curr Issues Intest Microbiol 3(2):39–48
Kailasapathy K, Masondole L (2005) Survival of free and microencapsulated *Lactobacillus acidophilus* and *Bifidobacterium lactis* and their effect on texture of feta cheese. Aust J Dairy Tech 60(3):252
Kataria A, Achi SC, Halami PM (2018) Effect of encapsulation on viability of *Bifidobacterium longum* CFR815j and physiochemical properties of ice cream. Indian J Microbiol 58(2):248–251
Kavitake D, Kandasamy S, Devi PB, Shetty PH (2017) Recent developments on encapsulation of lactic acid bacteria as potential starter culture in fermented foods—a review. Food Biosci 21:34–44
Krasaekoopt W, Bhandari B, Deeth HC (2006) Survival of probiotics encapsulated in chitosan-coated alginate beads in yoghurt from UHT-and conventionally treated milk during storage. Lebenson Wiss Technol 39(2):177–183
Maleki D, Azizi A, Vaghef E, Balkani S, Homayouni A (2015) Methods of increasing probiotic survival in food and gastrointestinal conditions. La Prensa Medica 101:4. https://doi.org/10.4172/lpma.1000154
Martín MJ, Lara-Villoslada F, Ruiz MA, Morales ME (2015) Microencapsulation of bacteria: a review of different technologies and their impact on the probiotic effects. Innov Food Sci Emerg Technol 27:15–25
Mishra SS, Behera PK, Kar B, Ray RC (2018) Advances in probiotics, prebiotics and nutraceuticals. In: Panda SK, PKS H (eds) Innovations in technologies for fermented food and beverage industries. Springer, Cham, pp 121–141
Mortazavian A, Razavi SH, Ehsani MR, Sohrabvandi S (2007) Principles and methods of microencapsulation of probiotic microorganisms. Iran J Biotechnol 5(1):1–18
Muthukumarasamy P, Holley RA (2006) Microbiological and sensory quality of dry fermented sausages containing alginate-microencapsulated *Lactobacillus reuteri*. Int J Food Microbiol 111(2):164–169
Mutukumira AN, Ang J, Lee SJ (2015) Microencapsulation of probiotic bacteria. In: Liong MT (ed) Beneficial microorganisms in food and nutraceuticals. Springer, Cham, pp 63–80
Nedovic V, Kalusevic A, Manojlovic V, Levic S, Bugarski B (2011) An overview of encapsulation technologies for food applications. Procedia Food Sci 1:1806–1815

Ong YX, Lee LY, Davoodi P, Wang CH (2018) Production of drug-releasing biodegradable microporous scaffold using a two-step micro-encapsulation/supercritical foaming process. J Supercrit Fluids 133:263–269

Ono M, Oka T (1980) The differential actions of cortisol on the accumulation of α-lactalbumin and casein in midpregnant mouse mammary gland in culture. Cell 19(2):473–480

Oxman T, Shapira M, Klein R, Avazov N, Rabinowitz B (2001) Oral administration of *Lactobacillus* induces cardioprotection. J Altern Complement Med 7(4):345–354

Ozer B, Uzun YS, Kirmaci HA (2008) Effect of microencapsulation on viability of *Lactobacillus acidophilus* LA-5 and *Bifidobacterium bifidum* BB-12 during Kasar cheese ripening. Int J Dairy Technol 61(3):237–244

Ozer B, Kirmaci HA, Şenel E, Atamer M, Hayaloğlu A (2009) Improving the viability of *Bifidobacterium bifidum* BB-12 and *Lactobacillus acidophilus* LA-5 in white-brined cheese by microencapsulation. Int Dairy J 19(1):22–29

Panghal A, Janghu S, Virkar K, Gat Y, Kumar V, Chhikara N (2018) Potential non-dairy probiotic products—a healthy approach. Food Biosci 21:80–89

Parvez S, Malik KA, Ah Kang S, Kim HY (2006) Probiotics and their fermented food products are beneficial for health. J Applmicrobiol 100(6):1171–1185

Picot A, Lacroix C (2004) Encapsulation of bifidobacteria in whey protein–based microcapsules and survival in simulated gastrointestinal conditions and in yoghurt. Int Dairy J 14(6):505–515

Prakash KS, Chavan R, Mishra V (2016) Microencapsulation of probiotics and its applications. In Frontier Discoveries and Innovations in Interdisciplinary Microbiology, 33–44, Springer, New Delhi.

Prevost H, Divies C (1987) Fresh fermented cheese production with continuous pre fermented milk by a mixed culture of mesophilic lactic streptococci entrapped in calcium alginate. Biotechnol Lett 9(11):789–794

Ray S, Raychaudhuri U, Chakraborty R (2016) An overview of encapsulation of active compounds used in food products by drying technology. Food Biosci 13:76–83

Reed KK, Wickham R (2009) Review of the gastrointestinal tract: from macro to micro. Semin Oncol Nurs 25(1):3–14

Riaz QU, Masud T (2013) Recent trends and applications of encapsulating materials for probiotic stability. Crit Rev Food Sci Nutr 53(3):231–244

Rubio R, Jofré A, Martín B, Aymerich T, Garriga M (2014) Characterization of lactic acid bacteria isolated from infant faeces as potential probiotic starter cultures for fermented sausages. Food Microbiol 38:303–311

Sanders ME, Tompkins T, Heimbach JT, Kolida S (2005) Weight of evidence needed to substantiate a health effect for probiotics and prebiotics. Eur J Nutr 44(5):303–310

Semyonov D, Ramon O, Shimoni E (2011) Using ultrasonic vacuum spray dryer to produce highly viable dry probiotics. LWT-Food Sci Technol 44(9):1844–1852

Shi LE, Li ZH, Zhang ZL, Zhang TT, Yu WM, Zhou ML, Tang ZX (2013) Encapsulation of *Lactobacillus bulgaricus* in carrageenan-locust bean gum coated milk microspheres with double layer structure. LWT-Food Sci Technol 54(1):147–151

Sobel R, Versic R, Gaonkar AG (2014) Introduction to microencapsulation and controlled delivery in foods. In: Gaonkar AG, Vasisht N, Khare AR, Sobel R (eds) Microencapsulation in the food industry: a practical implementation guide. Academic, Amsterdam, pp 3–12

Sodini I, Boquien CY, Corrieu G, Lacroix C (1997) Use of an immobilized cell bioreactor for the continuous inoculation of milk in fresh cheese manufacturing. J Ind Microbiol Biotechnol 18(1):56–61

Sohail A, Turner MS, Coombes A, Bostrom T, Bhandari B (2011) Survivability of probiotics encapsulated in alginate gel microbeads using a novel impinging aerosols method. Int J Food Microbiol 145(1):162–168

Sohail A, Turner MS, Prabawati EK, Coombes AG, Bhandari B (2012) Evaluation of *Lactobacillus rhamnosus* GG and *Lactobacillus acidophilus* NCFM encapsulated using a novel impinging aerosol method in fruit food products. Int J Food Microbiol 157(2):162–166

Solanki HK, Pawar DD, Shah DA, Prajapati VD, Jani GK, Mulla AM, Thakar PM (2013) Development of microencapsulation delivery system for long-term preservation of probiotics as biotherapeutics agent. Biomed Res Int 2013:1–21

Speranza B, Petruzzi L, Bevilacqua A, Gallo M, Campaniello D, Sinigaglia M, Corbo MR (2017) Encapsulation of active compounds in fruit and vegetable juice processing: current state and perspectives. J Food Sci 6:1291–1301

Sultana K, Godward G, Reynolds N, Arumugaswamy R, Peiris P, Kailasapathy K (2000) Encapsulation of probiotic bacteria with alginate–starch and evaluation of survival in simulated gastrointestinal conditions and in yoghurt. Int J Food Microbiol 62(1–2):47–55

Suvarna VC, Boby VU (2005) Probiotics in human health: a current assessment. Curr Sci 11:1744–1748

Wang J, Korber DR, Low NH, Nickerson MT (2014) Entrapment, survival and release of *Bifidobacterium adolescentis* within chickpea protein–based microcapsules. Food Res Int 55:20–27

Weinbreck F, Bodnár I, Marco ML (2010) Can encapsulation lengthen the shelf-life of probiotic bacteria in dry products? Int J Food Microbiol 136(3):364–367

Zaeim D, Sarabi-Jamab M, Ghorani B, Kadkhodaee R, Tromp RH (2018) Electrospray-assisted drying of live probiotics in acacia gum microparticles matrix. Carbohydr Polym 183:183–191

Zuidam NJ, Shimoni E (2010) Overview of microencapsulates for use in food products or processes and methods to make them. In: Zuidam NJ, Nedovic V (eds) Encapsulation technologies for active food ingredients and food processing. Springer, New York, pp 3–29

Chapter 7
Herbonanoceuticals: A Novel Beginning in Drug Discovery and Therapeutics

Nidhi Saini, Abhilasha Thakur, Pawan Kaur, and Suresh Kumar Gahlawat

Contents

7.1 Introduction: Drug Discovery and Drug Design

A typical drug discovery cycle (from lead identification through clinical trials) takes around 14 years, with a cost of US$800 million, which makes it costly and tedious (Lavecchia and Di Giovanni 2013). Because of progress in frontline advancement, busy lifestyle, and less nutritious and unbalanced eating regimens nowadays, new generation is going up against various critical restorative issues for the duration of life. The pharmaceutical industries are fabricating high-volume and low-value/low-volume and high-value manufactured drugs with a specific end goal to have positive or safe reaction to the body. In any case, as of late, these medications are regarded in a specific way for their indications or warnings, higher costs, nonspecificity, and other real issues like antagonism. Drug hinderance or blocking is an important issue, which may lead to death in a person. In this way, people are choosing traditional methods with a particular ultimate objective to beat these issues. Throughout Indian history, individuals have had confidence in Ayurveda, and at present Ayurveda is

N. Saini · A. Thakur · P. Kaur · S. K. Gahlawat (✉)
Department of Biotechnology, Chaudhary Devi Lal University, Sirsa, Haryana, India

R. Prasad et al. (eds.), *Nanobiotechnology in Bioformulations*, Nanotechnology in the Life Sciences, https://doi.org/10.1007/978-3-030-17061-5_7

drawing world's attention towards itself because it offers important restorative properties and fewer symptoms. India is a biodiverse nation, rich in restorative plants (totaling some 45,000 species—almost 7% of the world's flowering plants) (Ali and Choudhary 2011). All parts of a plant contain bioactive compounds, which have positive physiological activity in the human body. Numerous phytochemicals are frequently gathered in the external layer of different plant tissues. These bioactive compounds are otherwise called phytoconstituents/phytochemicals/secondary metabolites, which are not supplements but assume a huge part in the plant's protective components (Doughari 2012). They are considered to have increased supplement value to be used more gainfully by the body. A celebrated expression—"Prevention is better than cure"—is conceivable because of the vicinity of bioactive compounds in plants. These give shade, enhance the smell of the plants, and have advantageous effects in the body, such as elimination of free radicals (Zhang et al. 2015). Some beneficial roles of phytochemicals (which make them unique) are low toxicity, greater affordability, easy availability, biological and therapeutic properties, and raw material enabling extraction of semisynthetic chemical compounds for use in cosmetics, perfumes, etc. (Dillard and German 2000). The process of computer-aided drug design is shown in Fig. 7.1.

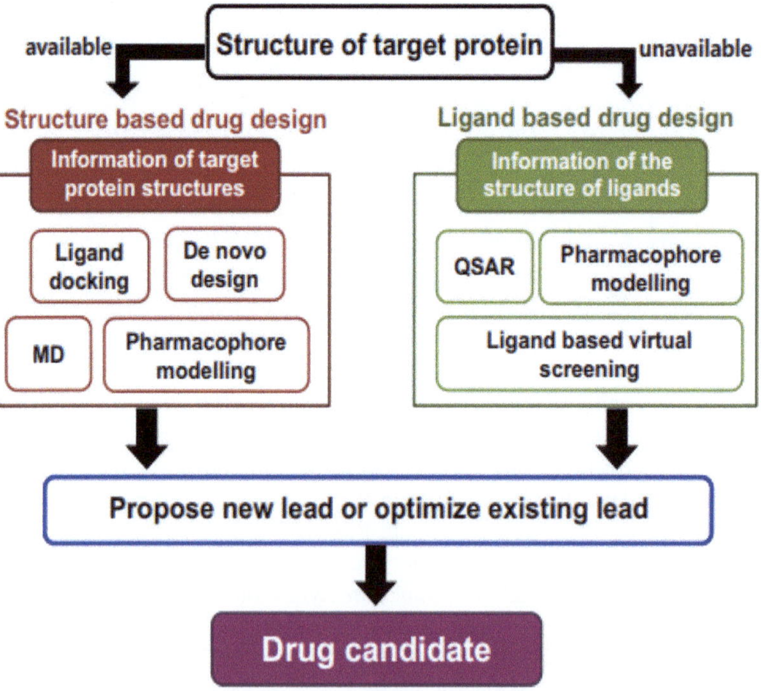

Fig. 7.1 Computer-aided drug design process (Hassan et al. 2016), showed the *MD* molecular docking, *QSAR* quantitative structure–activity relationship

7.1.1 Herbal Phytoconstituents of Drugs

According to the World Health Organization (WHO), herbal medicines are finished, labeled, therapeutic products that contain bioactive ingredients, aerial or underground parts of a plant or other plant material, or combinations of these (Hariharan and Subburaju 2012). According to WHO reports, 80% of people in developing countries use natural medicines for their primary health care needs (Yadav et al. 2014). Bioactive herbal compounds are used in the manufacturing of plant-based drugs, which have unique properties with long-term effects and good compatibility in the body. Alkaloids, flavonoids, phenols, tannins, terpenes, and sterols are the major phytochemicals that can be used as herbal medicines if taken in appropriate amounts. They are present in fruits and vegetables (Saxena et al. 2013). Some plant-based drugs that have been introduced in the USA over time are vincristine, vinblastine, reserpine, and deserpidine (Fabricant and Farnsworth 2001).

7.1.2 Activeness

The phytochemical and phytopharmacological sciences have officially established the composition and biological activities of various medicinal plant products. The vast majority of the naturally active constituents of concentrates—for example, flavonoids, tannins, and terpenoids—are very water soluble yet exhibit low assimilation, since they cannot cross lipid films, have high molecular weights, and show poor retention, causing losses of bioavailability and viability. A few investigations have demonstrated strong activity of natural medicines in vitro that is not reproducible in vivo. Besides, some fundamental substances are only occasionally utilized, in light of the fact that they are not compatible with other parts of the formulation or have bothersome properties (Bonifacio et al. 2014)

7.1.3 Methods for Discovering New Herbal Medicines

There are a great many phytoconstituents in nature, but they generally take many years to be identified and isolated as active compounds. Huge numbers of those that have been discovered have failed the optimization process because of their absorption, distribution, metabolism, excretion, and toxicity (ADME/Tox) inadequacies. Some may be dismissed because they fail other toxicity tests technological advances in combinational science and high-throughput screening have enabled synthesis and screening of huge libraries of compounds in a short time, reinvigorating the drug discovery process (Lavecchia and Di Giovanni 2013). To overcome these problems, researchers are working hard on various strategies for identification of promising compounds, such as virtual screening, which was first described in a peer-reviewed publication in 1997 (Hughes et al. 2011).

Virtual screening is an exact, information-driven, compound database searching methodology, which has become increasingly prominent and has undergone rapid development in pharmaceutical research. It endeavors to discover novel compounds and chemotypes that have the necessary therapeutic and biological activities to be used as substitutes for existing ligands or, in some cases, may lead to the discovery of new ligands, structures, and homology modeling. Virtual screening is a well-ordered strategy with a course of successive channels that can limit and pick an arrangement of lead-like hits with potential biological activity against proposed drug targets (Lavecchia and Di Giovanni 2013).

One of the elementary and most generally utilized methods of virtual screening is similarity searching, in which a known bioactive reference structure is sought against a database to recognize the nearest neighbor particles, since these are the most likely to display the bioactivity of interest (Holliday et al. 2011). Virtual screening can be utilized to choose compounds for screening from in-house databases and to choose which compounds to manufacture next (Kaserer et al. 2015).

Virtual screening can be divided into two general types: ligand-based virtual screening (LBVS) and structure-based virtual screening (SBVS) (Lavecchia and Di Giovanni 2013).

7.1.3.1 Ligand-Based Virtual Screening

Ligand-based virtual screening utilizes structure–activity information from an arrangement of known actives with a specific end goal to choose competitor mixes for experimental interpretation. LBVS techniques involve pharmacophore and three-dimensional (3D) shape matching, quantitative structure–activity relationships (QSARs), and likeness and substructure searches. A standout among the most well known ligand-based drug design approaches is QSAR. The objective of QSAR is to determine the connection between the basic/physicochemical properties of active compounds and their biological activity. The use of virtual screening for hit and lead identification involves a group of procedures based upon the data that have already been collected regarding the targeted and/or existing ligand earlier in the screening process. Machine learning is rapidly becoming popular in LBVS as novel calculations are proposed to manufacture precise and strong QSARs. Diverse techniques are proposed, and every technique has its own points of interest and detriments. Among these strategies, regression and classification methods—for example, multiple linear regression (MLR), nearest neighbor, naïve Bayesian classification, support vector machines, neural networks, and decision trees—are effectively connected. Discovery of novel lead compounds through computational use of experimentally determined protein structures, obtained from screening of databases through focused design exercises, is well established (Jain 2004).

Prescient models of ligand-target binding active sites can be valuable in the following circumstances:

1. Where numerous current ligands are known, but they share natural properties that reduce their organic utility

2. Where a few ligands have been found for a target (e.g., by high-throughput screening), but they have not been widely examined, and expansion of the set is an essential objective of a medical chemistry effort (Jain 2004)

The execution of this strategy relies upon different factors—for example, preparing the planned diversity, their capacity to manage imbalanced data sets (inactive compounds typically outnumber active compounds), and parameter runs in covering active and inactive chemical space.

7.1.3.2 Structure-Based Virtual Screening

Structure-based virtual screening uses a 3D structure of the natural target (either determined tentatively by x-ray crystallography or nuclear magnetic resonance (NMR), or determined computationally through homology modeling) to dock the candidate molecules and rank them on the basis of their predicted binding affinity or complementarity to the binding site (Anderson 2003).The protein structure of interest is accessible in different databases; for example, the Research Collaboratory for Structural Bioinformatics (RCSB) Protein Data Bank (PDB) and a compound library of small molecules (available commercially or by synthesis) is investigated by docking at the active site of the biochemical target, using computer algorithms and scoring functions. Nowadays, various docking programs are commercially (or freely) accessible, with various conformational sampling algorithms and an assortment of scoring functions. Scoring of ligands is a pivotal advance in the achievement of SBVS. In spite of the fact that prediction of the ligand binding site is generally conceivable with the accessible strategies, scoring is still extremely difficult, and it is consequently hard to distinguish the right binding site or to rank compounds.

The commonly utilized scoring functions can be divided into three general classifications: force field based, knowledge based, and empirical. These capacities have ended up being fruitful for some protein–ligand associations. The basic adaptability of the objective is another essential thing to be considered as top priority because of the computational cost and multifaceted approach required to demonstrate it well (Lionta et al. 2014)

The ongoing advances and applications in SBVS—from an issue-driven point of view, with attention to docking-based virtual screening (DBVS)—are enhancing the screening library before docking; considering target adaptability, metal particles, water molecules, and other key ligand target interactions and environmental factors involved in docking; and enhancing posture/compound determination after docking (Cheng et al. 2012).

Preferably, the target site is a pocket or projection with an assortment of likely hydrogen bond donors and acceptors, hydrophobic qualities, and molecular adherence surfaces. Determining the structure of an objective protein by NMR, x-ray crystallography, or homology modeling is a major step in SBVS (Reddy et al. 2007). There are major differences between LBVS and SBVS. These differences are briefly summarized in Table 7.1.

Table 7.1 Major differences between ligand-based virtual screening and structure-based virtual screening

Aspect	Ligand-based virtual screening	Structure-based virtual screening	References
Three-dimensional protein structure	Unknown	Known	Lee et al. (2011); Hassan et al. (2016)
Drug design information and starting point	Drug design can be based on processes using the known ligands of a target protein as the starting point	Reliable information on the three-dimensional structure and active sites of the target protein can be obtained from x-ray crystallography, nuclear magnetic resonance, or three-dimensional structure databases, and incorporated into a computer model; compounds binding to the target can be designed	Lyne (2002); Ferreira et al. (2015)
Frequently used techniques	Molecular similarity approaches, quantitative structure–activity relationships, pharmacophore models	Docking, molecular dynamics simulation	Lee et al. (2011); Ekins et al. (2007)
Screening	Databases can be screened to find molecules with similar fingerprints by using the molecular fingerprints of known ligands	Molecule database screening can be done by various types of docking software	Lee et al. (2011); Xie (2010);
Process involved	Selection of the data set and extraction of structural/empirical descriptors, variable selection, model construction, validation/ evaluation	An assortment of successive computational stages is involved, including target and database preparation, docking and postdocking examination, and prioritization of compounds for biological testing	Lee et al. (2011); Cheng et al. (2012)
Software programs used to perform virtual screening	uNITY, MACCS-3D, Catalyst, Phase, ROCS	Glide, FlexX, Gold, AutoDock Vina, AutoDock 4.0	Hassan et al. (2016)

7.1.3.3 Virtual Screening Methods: Quantitative Structure–Activity Relationships

Virtual screening of combinatorial libraries has attracted scrutiny lately, as its methods integrate high-throughput screening and combinatorial chemistry. A QSAR-based approach is an established chemoinformatic technique for predicting and classifying biological activities of untested chemicals (Dudek et al. 2006). It is based on the general principles of medicinal science that the biological activity of a ligand or compound is identified with its molecular structure or properties, and that structurally similar molecules may have similar biological activities (Tong et al. 1998). QSAR models can be utilized as a part of outlining new chemical entities (NCEs) and are presently viewed as basic tools in pharmaceutical enterprises to recognize promising hits and create good-quality leads in the initial phase of drug discovery (Myint and Xie 2010).

QSARs depend on the presumption that the structure of a particle (i.e., its geometric, steric, and electronic properties) must contain the aspects that govern its physical, chemical, and organic properties, and on the capacity to represent the chemical by at least one numerical descriptor (Gramatica 2008). QSARs occur within a congeneric arrangement of compounds, affinities of ligands for their binding sites, rate constants, inhibition constants, and other biological activities either with certain structural features (Free Wilson analysis) or with atomic, group, or molecular properties—for example, lipophilicity, polarizability, electronic properties, and steric properties (Hansch analysis) (Kubinyi 1997; Mannhold et al. 2008). Together, structure–activity relationships and QSARs are referred to as (Q)SARs and fall within a scope of strategies known as in silico approaches. A (Q)SAR includes three sections: the (activity) information to be demonstrated and subsequently anticipated, information with which to show it, and a technique to figure out the model. The reasons for in silico studies are as follows (Sethi 2012):

1. To predict biological action and physicochemical properties by rational means
2. To understand and justify the mechanism of action in a series of chemicals

QSAR equations have been utilized to depict a huge number of biological activities in various series of drugs and drug ligands. Predominantly, information on enzyme inhibition is connected with the physicochemical properties of the ligands. QSAR studies can reduce the likelihood of costly failures of drug ligand research in preliminary clinical studies by separating the combinatorial libraries (i.e. seperating the collection of chemical compounds, small molecules or macromolecules like proteins which are synthesized by combinational chemistry) (Mannhold et al. 2008).

Most molecular discoveries today are the result of an iterative, three-stage process of outline, synthesis, and investigation (testing). A typical investigational process is the development of a type of model that describes the observed activity or properties to be identified with the molecular structure. Such models are typically referred to as QSARs. A QSAR includes a methodical procedure with various stages, including data set planning, molecular descriptor selection and generation, numerical or statistical model derivation, preparation and approval of a demonstration utilizing a preparation data set, and testing on a testing data set (Anderson 2003).

The general numerical formula for a QSAR is represented by the following equation

The data driving the drug design effort are progressively quantitative, expanding on new advancements in molecular structure depiction, combinatorial arithmetic, statistics, and computer simulations. Collectively these areas have led to a new QSAR paradigm in drug design (Kubinyi 1997).

The molecular descriptors utilized as a part of QSAR can be characterized as a numerical portrayal of chemical data encoded in a molecular structure through the scientific method. This scientific portrayal must be invariant to the size of the molecule anBiological Activity = f (Physicochemical Property).

The data substance of structural descriptors relies upon two central aspects: (1) the molecular representation of the compound, and (2) the algorithm utilized for calculation of the descriptor.

The three noteworthy kinds of parameters first proposed are (1) hydrophobic, (2) electronic, and (3) steric (Hansch et al. 2001; Mekenyan and Veith 1994; Simon et al. 1984)

QSARs endeavor to relate physical and chemical properties of particles to their biological activities by basically utilizing easily measurable descriptors and basic statistical techniques, such as MLR, to construct a model that both depicts the activity of the data index and can anticipate activities for additional sets of untested compounds (Novotarskyi 2013). The impact of physicochemical properties on absorption (retention), distribution (dispersion), metabolism (digestion), and excretion (discharge) (ADME) is generally assessed by use of QSAR procedures (Malik et al. 2013).There are two fundamental targets for improvement of QSARs:

1. Development of predictive and robust QSARs, with a predetermined man made area, for prediction of the activity of untested molecules.
2. Creation of an informative tool by separating many examples of descriptors identified with the relevant biological activity, allowing comprehension of the components of the given biological activity. This may be helpful for identifying novel molecules with enhanced activity profiles (Nantasenamat et al. 2009).

7.1.3.4 Applications of Quantitative Structure–Activity Relationships

QSARs provide significant contributions to our understanding of the following aspects:

1. Biological activity and physicochemical parameters
2. Mechanisms of activity
3. Recognition of drug-like versus non-drug-like particles
4. Drug resistance
5. Likelihood of toxicity
6. Prediction of physicochemical properties (e.g., water solubility, lipophilicity)
7. Prediction of ADME properties (e.g., gastrointestinal retention, blood–brain barrier permeation, drug metabolism)
8. The actions of peptides

7.1.4 Molecular Docking

In structure-based drug design, molecular docking is a critical system, which can give an evaluation of the binding mode and, in a few applications, the binding affinity of a ligand with its receptor complex. In early docking strategies, only translational and orientational degrees of freedom of the ligand and the receptor are considered. With a rapid increase in the number of known protein crystal structures, the enthusiasm for molecular docking has increased, and many advances have been made in recent years (Baxter et al. 2000). Molecular docking can be achieved through two interrelated advances:

1. Testing of adaptations of the ligand at the active site of the protein
2. Ranking of these conformations by means of a scoring function (Meng et al. 2011).

The data procured against the docking framework can be used to determine the energies by active site stability, Gibb's free energy and binding energy (the energy required during the ligand-drug stabilization).

Molecular docking is helpful to clarify biomolecular interactions for rational drug design (RDD) and discovery, together with mechanical examination by setting a molecule (ligand) into the correct binding site in the target region of the DNA/protein (receptor), usually in a noncovalent form, to outline a complex of potential practicality and more prominent specificity (Dar and Khan 2016). The principal goal of molecular docking is to accomplish a ligand–receptor complex with an enhanced conformation and the least free binding energy. Appropriate use of molecular docking requires an information bank for pursuit of the main target with an appropriate PDB format and a system to plan the ligand as a PDB document (Dar and Khan 2016).

In the molecular docking strategy for investigating interactions between a ligand and a protein, the ligand is cocrystallized with the protein, and then acquisition of an x-ray structure of the complex is attempted. Despite the fact that both synthesis and crystallography can sometimes be very erratic and tedious, the technique may be feasible for small accumulations of ligands. In the event that production and crystallization attempts fail, numerous potential ligands for the protein will likely be screened. Computational molecular docking is the primary strategy chosen, and it is now well known in both the research world and industry.

In assessment of the strategy for binding a ligand protein, or demonstrating parts of the ligand, practical or beneficial theory is essential when its structure is not effectively determined by crystallography. More than 60 docking programs have been created, of which about 10 are widely utilized. The principal key element of molecular docking is a 3D structure of the protein as information. Various programs produce 3D structures of ligands and ascertain their binding parameters with the protein by figuring out the binding affinity (scoring) between the two (Andersson et al. 2010).

There are different databases available that provide data on small ligand particles—for example, the Cambridge Structural Database (CSD), Available Chemical Directory (ACD), MDL [Molecular Design Limited] Drug Data Report (MDDR), and National Cancer Institute (NCI) Database. At the time of performing docking, different interacting conformers are created and correlated with each other. In the dismissal condition, new conformers are obtained and again the search protocol proceeds until it reaches its end point after establishment of one conformation. As indicated by docking and experimental binding, affinity and free binding energy appear to be more problematic than the binding orientation. To overcome this issue, distinctive scoring functions are utilized—for example, a consensus scoring tool with a number of score capacities for the same docking posture with a specific end goal to eliminate the false positives. A huge number of attempts have been made in the advancement of successful docking conventions, and noteworthy advances have been made in computational forecasting of docking modes (Dar and Khan 2016).

In the present docking programs, a ligand is regarded as an adaptable structure, yet the protein conformation is generally considered rigid, and water particles are not recognized in any way. The inclusion of basic water has been carefully considered, and recent investigations have demonstrated that the precision of docking is enhancing by this. The precision of docking program bundles is regularly assessed by supposed redocking tests. Redocking results are computed and assessed using the root mean square deviations (RMSDs) of the local ligand adaptation (as seen in the x-ray structure) and the ligand conformation proposed by the docking programming (docking posture) (Andersson et al. 2010). The most essential part of molecular docking is in the drug disclosure/exposure by the target, and lead enhancement. Depending on the nature and type of the disease of interest, the targets can be either protein or DNA, while drugs are generally organic small molecules.

7.1.4.1 Types of Docking

There are different sorts of molecular docking systems, including:

1. Flexible or rigid ligands/targets, depending on the goal of the docking simulations, such as adaptable ligand docking (where the target is a rigid molecule)
2. Rigid body docking, with both the target and the ligand as rigid molecules
3. Flexible docking, with both associated particles being adaptable (Dar and Khan 2016)

Table 7.2 Commonly used docking programs for virtual screening

Docking method	Principles	Scoring functions	Speed	References
DOCK	Incremental construction method	Force field, chemical complementarity score, contact score	Fast	Meng et al. (2011); Ferreira et al. (2015)
FlexX	Incremental construction method	Empirical score	Fast	Meng et al. (2011); Ferreira et al. (2015)
eHiTS	Systematic search	Empirical score	Fast	Zsoldos et al. (2007); Ferreira et al. (2015)
FRED	Exhaustive search	Gaussian score	Fast	McGann (2011)
ICM-Dock	Stochastic global optimization	Empirical score, force field based	Fast	Meng et al. (2011); Ferreira et al. (2015)
Affinity	Monte Carlo	Force field	Slow	Meng et al. (2011); Ferreira et al. (2015)
Glide	Systematic search	Empirical score	Fast	Ferreira et al. (2015)
AutoDock	Monte Carlo, genetic algorithms	Empirical score, force field based	Slow	Meng et al. (2011); Ferreira et al. (2015)
Gold	Genetic algorithms	Empirical score	Fast	Meng et al. (2011); Ferreira et al. (2015)
QXP	Monte Carlo	Force field	Slow	Meng et al. (2011); Ferreira et al. (2015)

Various docking methods have recently been in use for virtual screening all over the world. These methods work on particular algorithms (scoring functions), which distinguish them from each other. Because of their characteristics (such as accuracy and speed), these are powerful tools for screening of various lead molecules/bioactive molecules. Some of the common methods, along with their various properties, are listed in Table 7.2.

7.2 Drug Delivery

Drug delivery is a technique for administrating a medication/pharmaceutical compound/drug to accomplish a remedial action in a living being. For a medication to have the capacity to achieve its planned clinical impact, it must first have the capacity to reach its target site of activity in the body, at an effective concentration. In the event that the medication is be applied to an outer surface (e.g., skin, ears, or eyes), it may be applied directly to the relevant surface. However, if the medication is intended to create an impact inside the body—regardless of whether it is intended to be systemic (e.g., basic antimicrobials) or aimed at particular tissues (e.g., thyroid medication)—the medication must be formulated such that it can reach the circulation and be transported to the intended site(s) of its action. Medications can be delivered in various different ways. These can be mainly divided into enteral routes (e.g., sublingual, oral, and rectal); parenteral routes (e.g., subcutaneous, intramuscular, and intravenous); inhalations; and topical, transdermal, and intranasal applications. Certain unusual routes may be utilized to yield more effective results, such as intra-arterial chemotherapy administration for malignancies and intrathecal administration for focal sensory system contamination or for spinal anesthesia. These routes are significantly more perilous and clearly require more skill and care for administration of medication. There is an emerging demand for controlled release of drugs to enhance their efficacy. In this regard, the main focus of nanotechnology is on targeted and controlled-release systems for drug delivery, as summarized in Table 7.3.

7.2.1 Nanobased Drug Delivery Systems

The ideal drug delivery system will ensure that the conjugated or bound drug–carrier complex reaches and acts specifically at the chosen target. The focus of the drug–nanocarrier complex can be active, meaning that it targets a ligand specific to the receptor or an epitope of the target tissue. Key properties of any nanomaterial utilized as a part of drug delivery are its biocompatibility and biodegradability, with an aim that the emptied transporter degrades or is transformed into nontoxic substances and eliminated from the body. The objectives of a nanoscale tranquilizer delivery system are as follows:

Table 7.3 Targeted delivery and controlled-release drug delivery system definitions, according to experts in drug delivery

Definitions		
Targeted drug delivery	Controlled drug delivery	References
A targeted drug delivery system is intended to specifically transport medications to target tissues, organs, and cells through use of an assortment of drug transporters	Controlled delivery of medications improved the drug delivery approaches	Tiwari et al. (2012)
It is generally intended to enhance the pharmacological and remedial properties of traditional medications and to overcome problems such as limited solubility, drug aggregation, poor biodispersion, and absence of selectivity, controlling the drug delivery carrier and lessening associated tissue injury	Controlled delivery controls the site and the rate of medication release	Tiwari et al. (2012)
It can be intended to limit drug degradation, increment bioavailability, permit targeting of particular cells, and diminish the total amount of medication required; it can also control the delivery of medication to minimize lethality and unsafe reactions	Conventional dosing regimens cause fluctuations in plasma drug concentrations; these changes depend upon what happens to the medication inside the body, such as its retention, dispersion, digestion, and discharge; controlled release eliminates such fluctuations in plasma drug concentrations	Tiwari (2016); Allen and Cullis (2004)
Targeted medications initially are mostly utilized as a part of disease treatment; optimization of the adequacy of a medication is of prime significance in the choice of a delivery system	A controlled drug delivery system enhances patient compliance, particularly in long-term treatment for persistent infections	Zhan et al. (2017); Tiwari (2016)
Targeted drug delivery is a strategy for administering medicine to a patient in a way that restricts the concentration of the medication to particular tissues	These systems have advantages over conventional drug delivery systems, including limiting the time of medication release, safeguarding fragile medications, and increasing patient comfort and compliance	Manish and Vimukta (2011); Zadbuke et al. (2013)
This improves the efficacy of the medication while reducing its side effects	This reduces drug doses and dosing frequencies	Tiwari (2016)
Drug targeting is the delivery of drugs to receptors or organs or any other specific part of the body to which one wishes to deliver the drug exclusively	Controlled drug delivery occurs when a polymer is combined with a drug or active agent such that the release from the bulk material is predesigned	Manish and Vimukta (2011); Anupama et al. (2016)
Targeted delivery of the drugs enhance the efficacy of the drug to the target site and prevents adverse effects, and increases the bioavailability of the medication.	Doses and dosing frequencies can be reduced, and the required plasma drug concentrations can be maintained, eliminating the risk of treatment failure and enhancing the proficiency of treatment	Farokhzad and Langer (2009); Tiwari (2016)

(continued)

Table 7.3 (continued)

Definitions		
Targeted drug delivery	Controlled drug delivery	References
The focus of a drug delivery system can be either passive or active; a case of passive focusing is the aggregation of chemotherapeutic agents in tumors because of differences in the vascularization of the tumor tissue versus healthy tissue; active focusing includes delivery of the drug on the surface of a drug transporter, enabling the drug to attach specifically to diseased cells	Controlled discharge can be maintained or pulsatile; maintained (or persistent) delivery of a medication involves polymers that release the medication at a controlled rate by dissemination from the polymer or by degradation of the polymer over time; pulsatile release is frequently favored, as it purposely mirrors the path by which the body normally creates hormones (e.g., insulin); it is accomplished by utilizing drug-releasing polymers that react to particular stimuli (e.g., exposure to light, changes in pH or temperature)	Farokhzad and Langer (2009)

(i) Targeting, to ensure that drug concentrations are highest at the intended site of action, with lower drug concentrations and fewer adverse effects in healthy tissues
(ii) Improved solubility, to optimize parenteral drug efficacy
(iii) Controlled release of drug to the target site enhace the efficacy of drug with prolonged curative activity
(iv) Sustained release of drug also extended the half life of the drug
(v) Drug movement through the blood–brain barrier and blood–cochlear barrier (Malam et al. 2009)
(vi) A delivery system that enhances patient compliance, particularly with long-term medication
(vii) An appropriate delivery system for drugs that have a short organic half-life

7.2.2 Nanoparticles and Their Roles in Drug Delivery

Nanoparticles are very small materials with a size range from 1 to 100 nm. Their classification depends upon their properties, shapes, or sizes. They are mainly categorized into four groups: fullerenes, metal nanoparticles, ceramic nanoparticles, and polymeric nanoparticles. Nanoparticles are composed of three different layers: (1) a surface layer modified by various small molecules, metal ions, surfactants, and polymers; (2) a shell layer, chemically different from the core material; (3) the core of the nanoparticle (Khan et al. 2017). Nanoparticles possess remarkable physical and chemical properties, which differ from those of their bulk-sized forms, as their small size, large surface area, and consequently large surface-to-volume ratio improve their contact with their surroundings (Khan et al. 2011; Tiwari et al. 2012).

Nanoparticles are polymeric particles made of chemical polymers or biopolymers ranging in size from about 10 to 1000 nm (1 mm). Drug can be encapsulated or cross linked or linked via covelant bond with polymers for sustained delivery of drug with the help of polymers. Nanoparticles and nanoformulations have already been applied as drug delivery systems with notable results and still have even greater potential for various applications, including antitumor treatment; gene therapy; acquired immune deficiency syndrome (AIDS) therapy; radiotherapy; delivery of proteins, anti-infection agents, virostatics, and vaccines; and vesicles to pass through the blood–brain barrier (Tiwari et al. 2012; Prasad et al. 2017). Herbal products have many unique properties with fewer side effects; hence, herbal nanoparticles are frequently used in drug delivery. Table 7.4 summarizes the applications of herbal nanoparticles in drug delivery.

Synthesis of nanoparticles with properties such as high monodispersity and biocompatibility is needed for a wide range of applications, from environmental to biomedical applications (Harshiny et al. 2015). Nanoparticles are the nanosized components most frequently used in nanotechnology (with a size range of 1–100 nm) and in various applications such as pharmaceuticals, electronics, cosmetics, and photonics (Prasad et al. 2016; Pandey et al. 2017).

Prominent classes of nanoparticles are carbon-based nanoparticles, ceramic nanoparticles, polymeric nanoparticles, lipid-based nanoparticles, and metal-based nanoparticles (Khan et al. 2017). For synthesis of nanoparticles, noble metals such as gold (Au), silver (Ag), palladium (Pd), zinc (Zn), copper (Cu), and iron (Fe) are used because of their unique d–d transition characteristics and high localized surface plasmon resonance (LSPR) phenomena (Khan et al. 2017, Mohan et al. 2014; Dreaden et al. 2012). Gold, platinum, silver, and palladium nanoparticles 20 nm in size have characteristic colors of wine red, yellow, grey-black and dark black, respectively (Khan et al. 2017). Copper nanoparticles have special properties, which have made them important for various applications such as antimicrobials, catalysts, biosensors, nanoscale drugs, drug delivery, and targeted drug delivery systems (Cerda et al. 2017; Rajesh et al. 2018). Gold, silver, copper, etc., can be excellent metals for use in various biomedical applications, such as cellular imaging, photothermal therapy, and biocides (Singh et al. 2016). Nanoparticles can be synthesized to include various types of chemical reducing agents, plant extracts, or enzymes (Cerda et al. 2017; Prasad 2014). These are effective agents with low toxicity, which is especially important in the biomedical field (Rajesh et al. 2018). For synthesis of nanoparticles, a number of reducing agents are used, such as ascorbic acid, lithium aluminum hydride, and sodium borohydride (Cerda et al. 2017; Wu and Chen 2004; Tang et al. 2010). Reducing agents reduce metal ions, followed by agglomeration into clusters. These clusters eventually form metallic colloidal particles (Hussein 2016). Some nanoparticles have excellent physical and chemical properties but do not possess suitable surface properties for specific applications. Apart from the cost effectiveness of copper nanoparticles, the stability of these nanoparticles is matter of concern; therefore, it may be necessary to modify the surface of the metals (Ruckenstein and Li 2005). A common method used for surface modification of these nanoparticles is addition of suitable organic groups to the positively charged surface atoms of the metal (Kickelbick and Schubert 2003).

Table 7.4 Applications of herbal nanoparticles in drug delivery

Application	Purposes	Materials	References
Cancer therapy	Target design, increased solubility and bioavailability, reduced toxicity, greater physiological compatibility, cost effectiveness	Natural products: β-carotene, curcumin, epigallocatechin gallate, genistein, resveratrol, gingerol, capsaicin, paclitaxel, camptothecin	Tiwari et al. (2012); Gomes et al. (2014)
Intracellular targeting	Overall drug avidity, easier internalization into mammalian cells, no major cytotoxicity, specific alterations of cellular signaling and gene expression	DL-lactide-co-glycolide, SIINFEKL, propiolic acid N-conjugated targeting, lipids, polymers, silicon, magnetic materials	Sneh-Edri et al. (2011); Nikalje (2015); Calderon-Colon et al. (2015)
Vaccine adjuvants	Delivery nanodevices are faster and more sensitive with better and safe way; such as antigen entrapping for delivery to specific cells and sustained silver release according to their biodegradation rate; gold nanoparticle adjuvant recognition, absorption of specific biomolecules, improvement of interaction with cells, and enhancement of cellular uptake	Gold nanoshells; nanoparticles silicon-based nanoparticles; polyethylene glycol, polyester biobeads, natural polymers based on polysaccharides such as alginate, inulin, or chitosan; virus-like particles; liposomes; immune-stimulating complexes; polymeric nanoparticles; nondegradable nanoparticles	Gregory et al. (2013)
DNA delivery	Low toxicity, little immunogenicity, and high adaptability; efficient, biocompatible, and modular	Liposomes, dendrimers, gold nanoparticles	Abu-Salah et al. (2010)
Ocular delivery	Improved efficiency with minimal damage to surrounding tissues, improved safety profile, better spreadability on the ocular surface, enhanced bioavailability due to reduced drug binding to pigments, sustained drug release	Liposomes, niosomes, nanomicelles, polymeric nanoparticles, solid lipid nanoparticles, dendrimers, conjugates, calcium phosphate nanoparticles	Agarwal et al. (2018)

In this context, surface modification can stabilize nanoparticle agglomeration, and it also modifies nanoparticles to make them compatible with another phase. For example, metal particles can be made water soluble when suitable groups are attached. A third reason for modifying nanoparticles is to enable their self-organization (Neouze and Schubert 2008; Kickelbick and Schubert 2003; Doty et al. 2005). Metal nanoparticles can be synthesized by various methods such as metal vapor synthesis, photolytic reduction, radiolytic reduction, the sonochemical method, solvent extraction reduction, the microemulsion technique, the polyol process, and alcohol reduction (Cerda et al. 2017). Physical and chemical methods are most

Table 7.5 Different nanosystems, with their various characteristics and their applications

Type of nanosystem	Size	Characteristics	Applications	References
Carbon nanotubes	Diameter 0.5–3 nm, length 20–1000 nm	Third allotropic crystalline form of carbon sheets, in either a single layer (single-walled nanotubes) or multiple layers (multiwalled nanotubes), with remarkable strength and unique electrical properties (conducting, semiconducting, insulating)	Transporter for drug distribution, cell cytoplasm seepage, nucleus seepage, peptide distribution; functionalization boosts solubility and antibody response	Rathod et al. (2011); Liu et al. (2008); Pantarotto et al. (2004); Bianco et al. (2005)
Dendrimers	<10 nm	Highly branched	Controlled bioactive distribution, targeted bioactive distribution to macrophages, liver targeting	Voit and Lederer (2009); Silva (2004); Sahoo et al. (2017)
Liposomes	50–100 nm	Phospholipid vesicles, biocompatible with high efficiency	Active and passive distribution of genes, proteins, peptides	Kano and Fendler (1978); Mallick and Choi (2014); Monnard et al. (1997); Danhier et al. (2010)
Metallic nanoparticles	<100 nm	Small size resulting in large surface area available for functionalization, gold and silver colloids, stable	Gene delivery, conjugation with DNA for delivery, highly stable nanoparticles used as alternative for, radiotherapy	Rai et al. (2009); Sun and Li (2004); Lu et al. (2007); Narayanan and El-Sayed (2003); McBain et al. (2008); Shanmukh et al. (2006)
Nanocrystal quantum dots	2–9.5 nm	Semiconducting material combined with II–IV and III–V section components; bright florescence, limited outflow, expansive ultraviolet excitation, high photostability	In vivo imaging of liver cells, DNA hybridization, immunoassays, receptor-mediated endocytosis, labeling of breast malignancies	Bawendi et al. (2012); Tomasulo and Ramakrishna (1996); Larson et al. (2003); Chen et al. (2013)
Polymeric micelles	10–100 nm	High drug loading and binding capacity biostability	Targeting of specific active and passive drug distribution, diagnostic value	Jones and Leroux (1999); Kataoka et al. (2001); Safari and Zarnegar (2014); Deepagan et al. (2016)
Polymeric nanoparticles	10–1000 nm	Ecological, bioconsonant, offering complete drug protection	Superb transporter for controlled and maintained drug delivery; stealth and surface-altered nanoparticles can be utilized for active and passive dispersion of bioactives	Du et al. (2009); Anderson and Shive (1997); Tang et al. (2012)

widely used for synthesis of different nanoparticles. However, the physical methods are very expensive, while the chemical methods are detrimental to the environment, as well as to living organisms (Shamaila et al. 2016). Moreover, these methods require expensive instruments, high energy, release of harmful chemicals, cell culture, and wasteful purification. The chemical reduction method has some advantages; for example, it is simple and inexpensive, and it is easy to control the size and shape of the nanoparticles. However, it has some disadvantages that can harm the environment, such as the need for disposal of noxious solvents and dangerous chemical reducing agents (Cerda et al. 2017).

Various nanosystems used in drug delivery, with their unique characteristics and their applications, are listed in Table 7.5.

7.3 Herbonanoceuticals

The strategy of applying nanotechnology to plant extracts has been widely cited in the literature, because nanostructured systems may potentiate the actions of plant extracts, promote sustained release of active constituents, reduce the required dose, decrease side effects, and improve activity (Bonifacio et al. 2014).Therefore, green nanosynthesis of nanoparticles has been considered rapid, nontoxic, cost effective, and environmental friendly in comparison with other chemical techniques (Harshiny et al. 2015). In green nanosynthesis of metal nanoparticles, use of (1) nontoxic chemicals used as reducing agents, (2) environmentally friendly solvents, and (3) renewable materials as stabilizing or passivating agents is very important to consider (Mohan et al. 2014; Prasad et al. 2018). Synthesis of nanoparticles using plant and plant-based biopolymers can be considered green chemistry. Plant extracts are used for metal ion reduction to form nanoparticles. Plant metabolites—such as sugars, starches, proteins, polyphenols, alkaloids, terpenoids, phenolic acids, gums, and β-cyclodextrin—play important roles in metal ion reduction into nanoparticles and their stability (Parveen and Rao 2015). In recent decades, some well known examples of carbohydrate polymers, such as chitosan, have been employed for synthesis of metal nanoparticles because of their interactions with metal ions and metal nanoparticles. With the increasing demand for adoption of sustainable and eco-friendly protocols, the use of biodegradable and nonbiodegradable precursors to prepare nanomaterials has been extensively explored. For instance, within the polysaccharide group, cyclodextrins have been applied as stabilizing agents to produce copper nanoparticles (Manivannan and Ramaraj 2012). β-Cyclodextrin is a nonreducing cyclic oligosaccharide with seven units of α-D-glucose. The glucose rings are connected through α-(1–4) linkage with a hydrophilic outer surface and a hydrophobic interior cavity (Navgire et al. 2016). β-Cyclodextrin can form inclusion complexes with various organic compounds (Gogoi and Sarma Chandra 2017) and act as stabilizing agent because particles synthesized with it have the smallest size distribution. This is due to the stiff structure and low water solubility of β-cyclodextrin, causing complex precipitation easily and thereby regulating the growth of the nanoparticles (Cerda et al. 2017).

The drawbacks of nanoparticle application in biological reactions are their poor stability and water solubility. Indeed, to prevent particle aggregation, a capping agent or stabilizing agents are used (Petralia et al. 2012). In one study, synthesis of copper nanoparticles, using β-cyclodextrin as a stabilizing agent and L-ascorbic acid as reducing agent, was done. The morphology of the synthesized Cu nanoparticles was characterized by advanced techniques such as transmission electron microscopy (TEM), powder X-ray diffraction (XRD), NanoDrop analysis, Fourier transform infrared spectroscopy (FTIR), dynamic light scattering (DLS), and ultraviolet–visible spectra. The antimicrobial and antibacterial properties of copper nanoparticles are checked by the use of different strains (Yadav et al. 2017).

Nanoforms of seven metals are generally utilized in Ayurveda: gold (Au), silver (Ag), copper (Cu), iron (Fe), lead (Pb), tin (Sn), and zinc (Zn). Applications of this old system are being rediscovered and could be of immense potential in yielding innovative metal–herb formulations that could hold promise for use in biomedicine (Galib et al. 2011). For developing such preparations, the biocompatibility and therapeutic activity of phytoconstituents present in plants play a determining role in the choice of various herbs. Such herbomineral preparations could have better curative potential because of their physiological compatibility, minimal toxicity, natural origin, easy accessibility, and lower costs. The promising applications of such nanomaterial-based herbal preparations in the field of biomedicine have led to the innovative concept of herbonanoceuticals (Gomes et al. 2014). Many studies have shown that during the processing of Ayurvedic metal-based formulations, the size of the metal particles goes down to the nanometric range (Pavani et al. 2013). Gold is used as *swarna bhasma* (gold ash) in different conventional Indian Ayurvedic preparations and has been characterized as having a particle size of 56–57 nm and a spherical shape. Mercury-based Ayurvedic preparations include crystalline mercuric sulfide in a 25- to 50-nm size range (Pal et al. 2014). Pavani et al. (2013) reported a modified method of *bhasmikaran* (a method for preparing *bhasma*) for preparation of iron oxide nanoparticles, which was based on green synthesis. XRD and TEM analyses have demonstrated that during use of the Ayurvedic *bhasmikaran* method, metal nanoparticles are formed. The metal nanoparticles, along with various phytoconstituents, can lead to better absorption and even tissue-specific targeting of therapeutic ingredients in the body. Because of their small size, these preparations are said to be more effective, as naturally synthesized metal nanoparticles do not cause toxicity in the body (Paul and Chugh 2011).

Green synthesis of metal nanoparticles could be carried out by utilizing many natural compounds, viz., vitamins, carbohydrates (sugars), phytochemicals present in plant extracts, microbes, biodegradable polymers, etc. Plant extracts are used for commercial preparation of metal nanoparticles because of the presence of phytoconstituents, which are potent reducing agents (Iravani 2011). Polyphenols are the main phytoconstituents in plant extracts that possess the ability to act as reducing agents, mainly because of their hydroxyl side chains. These can act as capping and stabilizing agents for metal nanoparticle formation. Gold nanoparticles are among the most widely utilized metal nanoparticles in the biomedical field because of their biocompatibility (Bhattacharya and Mukherjee 2008). Geetha et al. (2013) reported a rapid, economical, and single-step method for formation of gold nanoparticles, utilizing a flower extract from *Couroupita guianensis*. Green synthesis of gold nanoparticles was reported using glucoxylans of *Mimosa pudica* seeds to which no extra chemical stabilizing agent was added (Iram et al. 2014). Rao et al. (2016) discussed different medicinal plants and their active compounds, as well as green-synthesized metallic nanoparticles from medicinal plants, in relation to their anticancer activities. Metal nanoparticles formed using plant extracts showed enhanced tumor specificity, promising activity, and reduced toxic effects on healthy cells. The cytotoxic effects of nanoparticles are predominantly due to their large surface area, which enables efficient drug delivery, and some nanoparticles exhibit anticancer activity. However, further in vivo studies should be conducted to confirm the actual efficacy of herbal nanoparticle–based drugs (Rao et al. 2016).

Metal nanoparticles—including quantum dots, metal oxide nanoparticles, and pure metal nanoparticles—are considered useful in various biomedical applications (Bonifacio et al. 2014; Sharma and Singh 2014; Ambwani et al. 2015). Many researchers have put forward various methods for green synthesis of metal nanoparticles, employing microorganisms and plant extracts, that have been found to be eco-friendly, economical, and nontoxic (Makarov et al. 2014; Prasad et al. 2018). However, there are certain problems with metal nanoparticles, which need to be addressed prior to their commercial usage. The metal nanoparticles used for various purposes are reported to have diverse characteristics and diverse derivations, and are being utilized in various systems. However, their potential side effects in patients and the environment are yet to be thoroughly explored (Krug and Wick 2011). Contradictory biosafety considerations pertaining to metal nanoparticles have been reported in different studies (Tsoi et al. 2012; Edmundson et al. 2014; Ambwani et al. 2015).

Nanoherbal formulations can be employed for site-specific targeting of herbal medicines to enhance their selectivity, solubility, delivery, safety, and effectiveness. Nanosized drugs have an increased surface area, thereby allowing faster distribution in the blood and reduced toxicity while maintaining their therapeutic effects. The enhanced permeation and retention of nanoparticles can also help drugs to cross the blood–brain barrier (Kumar et al. 2015; Ganesan et al. 2017).

There are various therapeutic purposes of nanoparticles, which have been studied by various researchers throughout the world. Some of the applications of both herbal and synthetic nanoparticles are summarized in Table 7.6.

7.4 Conclusion

Since nanotechnology was first introduced and implemented in medical applications, nanomedicine has created expectations in medicinal and health applications to provide superior treatment with greater viability and specificity for certain diseases and conditions, including growth. Over the years, studies have provided details regarding the capability of nanocarriers, utilizing diverse plans to deliver anticancer therapeutic agents that are created artificially or originate from natural sources. This chapter has focused on the collaboration of nanotechnology in progressing the bioavailability of promising

Table 7.6 Applications of various nanoparticles (herbal and synthetic) for diverse therapeutic purposes

Nanoparticles used	Characteristics	Therapeutic purpose	References
Gold nanoshells, which stimulate the appropriate pro- or anti-immunity pathways; silica-based nanoparticles, polyethylene glycol or polyester biobeads; natural polymers based on polysaccharides such as alginate, inulin or chitosan	Faster and more sensitive delivery nanodevices; better, more secure and biocompatible; entrapping of antigens for delivery to specific cells and controlled silver release as indicated by the biodegradation rate; gold nanoparticles as adjuvant, assimilation of particular biomolecules, improved interaction with cells and cell take-up	Vaccine adjuvant	Gregory et al. (2013)
Herbal nanoparticles containing natural compounds such as β-carotene, curcumin, epigallocatechin gallate; gold nanoparticles; nanocapsules; fluorescent nanoparticles; iron oxide nanoparticles	Target definition, increased solubility and bioavailability, toxicity abatement, wider physiological adaptability, cost effectiveness	Cancer therapy	Tiwari et al. (2012); Gomes et al. (2014)
Liposomes, dendrimers, gold nanoparticles	Low toxicity, little immunogenicity, competent, highly bioadaptable	DNA delivery	Abu-Salah et al. (2010)
Liposomes, niosomes, nanomicelles, polymeric nanoparticles, solid lipid nanoparticles, dendrimers, conjugates, calcium phosphate nanoparticles	Enhanced effectiveness with negligible harm to surrounding tissues, enhanced safety profile, better spreadability on the ocular surface, improved bioavailability because of lessened medication binding to pigments, managed drug release	Ocular delivery	Agarwal et al. (2018); Brigger et al. (2012)

phytochemicals, which may encounter delivery issues in their free-drug form. It has also shown that different natural products may yield compounds that are ideal for use in different nanoformulation systems. In any case, progress in nanomedicine—regardless of whether it is phytochemical based or synthetic based—fits to the clinical setting is advancing more slowly than expected, fundamentally because of safety issues and, in particular, the cost effectiveness of nanomedicines when the costs of manufacturing and therapeutic application are taken into consideration. In the event that these issues can be settled, phytochemical-based nanomedicines (herbonanoceuticals) may end up being favored choices for various therapeutic treatments in the future.

References

Abu Salah KM, Ansari AA, Alrokayan SA (2010) DNA-based applications in nanobiotechnology. Bio Med Res Int 715295:1–15

Agarwal P, Huang D, Thakur SS, Rupenthal ID (2018) Nanotechnology for ocular drug delivery. In: Grumezescu AM (ed) Design of nanostructures for versatile therapeutic applications. William Andrew, Oxford, pp 137–188. https://doi.org/10.1016/B978-0-12-813667-6.00004-8
Ali MA, Choudhary RK (2011) India needs more plant taxonomists. Nature 471(7336):37
Allen TM, Cullis PR (2004) Drug delivery systems: entering the mainstream. Science 303(5665):1818–1822
Ambwani S, Tandon R, Gupta A, Ambwani TK, Chauhan RS (2015) Nanoparticles: utility, immuno-toxicology and ethical issues. J Immunol Immunopathol 17:68–78
Anderson AC (2003) The process of structure-based drug design. Chem Biol 10(9):787–797
Anderson JM, Shive MS (1997) Biodegradation and biocompatibility of PLA and PLGA microspheres. Adv Drug Deliv Rev 28(1):5–24
Andersson CD, Chen BY, Linusson A (2010) Mapping of ligand-binding cavities in proteins. Proteins: Struct Funct Bioinf 78(6):1408–1422
Anupama S, Somashekar CN, Tamizhmani T (2016) Recent approach for drug release from matrix tablet: a review. Asian J Res Biol Pharm Sci 4(3):122–132
Bawendi MG, Kim SW, Zimmer JP (2012) Nanocrystals including III–V semiconductors. US Patent No. 8,134,175. US Patent and Trademark Office, Washington, DC
Baxter CA, Murray CW, Waszkowycz BLJ, Sykes RA, Bone RG, Perkins TDJ, Wylie W (2000) New approach to molecular docking and its application to virtual screening of chemical databases. J Chem Inf Comput Sci 40(2):254–262
Bhattacharya R, Mukherjee P (2008) Biological properties of "naked" metal nanoparticles. Adv Drug Deliv Rev 60:1289–1306
Bianco A, Kostarelos K, Partidos CD, Prato M (2005) Biomedical applications of functionalised carbon nanotubes. Chem Commun 5:571–577
Bonifacio BV, Silva PB, Ramos MA, Negri KM, Bauab TM, Chorilli M (2014) Nanotechnology-based drug delivery systems and herbal medicines: a review. Int J Nanomedicine 9:1–15
Brigger I, Dubernet C, Couvreur P (2012) Nanoparticles in cancer therapy and diagnosis. Adv Drug Deliv Rev 64:24–36
Calderon-Colon X, Raimondi G, Benkoski JJ, Patrone JB (2015) Solid lipid nanoparticles (SLNs) for intracellular targeting applications. J Vis Exp 105. https://doi.org/10.3791/53102
Cerda SJ, Gomez EH, Nunez AG, Rivero IA, Ponce GY, Lopez FLZ (2017) A green synthesis of copper nanoparticles using native cyclodextrins as stabilizing agents. J Saudi Chem Soc 21(3):341–348
Chen O, Zhao J, Chauhan VP, Cui J, Wong C, Harris DK, Wei H, Han HS, Fukumura D, Jain RK, Bawendi MG (2013) Compact high-quality CdSe–CdS core–shell nanocrystals with narrow emission linewidths and suppressed blinking. Nat Mater 12(5):445
Cheng T, Li Q, Zhou Z, Wang Y, Bryant SH (2012) Structure-based virtual screening for drug discovery: a problem-centric review. AAPS J 14(1):133–141
Danhier F, Feron O, Preat V (2010) To exploit the tumor microenvironment: passive and active tumor targeting of nanocarriers for anti-cancer drug delivery. J Control Release 148(2):135–146
Dar AM, Khan S (2016) Spectroscopic, viscositic, DNA binding and cytotoxic studies of newly synthesized steroidal imidazolidines. J Fluoresc 26(2):639–649
Deepagan VG, Kwon S, You DG, Um W, Ko H, Lee H, Jo DG, Kang YM, Park JH (2016) In situ diselenide-crosslinked polymeric micelles for ROS-mediated anticancer drug delivery. Biomaterials 103:56–66
Dillard CJ, German JB (2000) Phytochemicals: nutraceuticals and human health. J Sci Food Agric 80(12):1744–1756
Doty RC, Tshikhudo TR, Brust M, Fernig DG (2005) Extremely stable water-soluble Ag nanoparticles. Chem Mater 17(18):4630–4635
Doughari JH (2012) Phytochemicals: extraction methods, basic structures and mode of action as potential chemotherapeutic agents. In: Rao V (ed) Phytochemicals—a global perspective of their role in nutrition and health . InTechOpen. https://doi.org/10.5772/26052
Dreaden EC, Alkilany AM, Huang X, Murphy CJ, El-Sayed MA (2012) The golden age: gold nanoparticles for biomedicine. Chem Soc Rev 41(7):2740–2779

Du WL, Niu SS, Xu YL, Xu ZR, Fan CL (2009) Antibacterial activity of chitosan tripolyphosphate nanoparticles loaded with various metal ions. Carbohydr Polym 75(3):385–389
Dudek AZ, Arodz T, Galvez J (2006) Computational methods in developing quantitative structure–activity relationships (QSAR): a review. Comb Chem High Throughput Screen 9(3):213–228
Edmundson MC, Capeness M, Horsfall L (2014) Exploring the potential of metallic nanoparticles within synthetic biology. New Biotechnol 31(6):572–578
Ekins S, Mestres J, Testa B (2007) In silico pharmacology for drug discovery: applications to targets and beyond. Br J Pharmacol 152(1):21–37
Fabricant DS, Farnsworth NR (2001) The value of plants used in traditional medicine for drug discovery. Environ Health Perspect 109(1):69
Farokhzad OC, Langer R (2009) Impact of nanotechnology on drug delivery. ACS Nano 3(1):16–20
Ferreira LG, dos Santos RN, Oliva G, Andricopulo AD (2015) Molecular docking and structure-based drug design strategies. Molecules 20(7):13384–13421
Galib BM, Mashru M, Jagtap C, Patgiri BJ, Prajapati PK (2011) Therapeutic potentials of metals in ancient India: a review through Charaka Samhita. J Ayurveda Integr Med 2:55–63
Ganesan P, Arulselvan P, Choi DK (2017) Phytobioactive compound–based nanodelivery systems for the treatment of type 2 diabetes mellitus—current status. Int J Nanomedicine 12:1097–1111
Geetha R, Ashokkumar T, Tamilselvan S, Govindaraju K, Sadiq M (2013) Green synthesis of gold nanoparticles and their anticancer activity. Cancer Nanotechnol 4:91–98
Gogoi A, Sarma KC (2017) Synthesis of the novel β-cyclodextrin supported CeO_2 nanoparticles for the catalytic degradation of methylene blue in aqueous suspension. Mater Chem Phys 194:327–336
Gomes A, Ghosh S, Sengupta J, Datta P, Gomes A (2014) Herbonanoceuticals: a new step towards herbal therapeutics. Med Aromat Plants 3:162
Gramatica P (2008) A short history of QSAR evolution. http://qsarworld.com/TempFileupload/Shorthistoryofqsar.pdf. Accessed 25th May, 2018
Gregory AE, Williamson D, Titball R (2013) Vaccine delivery using nanoparticles. Front Cell Infect Microbiol 3:13
Hansch C, Kurup A, Garg R, Gao H (2001) Chem-bioinformatics and QSAR: a review of QSAR lacking positive hydrophobic terms. Chem Rev 101(3):619–672
Hariharan P, Subburaju T (2012) Medicinal plants and its standardization—a global and industrial overview. Global J Med Plant Res 1(1):10–13
Harshiny M, Iswarya CN, Matheswaran M (2015) Biogenic synthesis of iron nanoparticles using *Amaranthus dubius* leaf extract as a reducing agent. Powder Technol 286:744–749
Hassan Baig M, Ahmad K, Roy S, Mohammad Ashraf J, Adil M, Haris Siddiqui M, Khan S, Khan MA, Provaznik I, Choi I (2016) Computer aided drug design: success and limitations. Curr Pharm Des 22(5):572–581
Holliday JD, Kanoulas E, Malim N, Willett P (2011) Multiple search methods for similarity-based virtual screening: analysis of search overlap and precision. J Cheminf 3(1):29
Hughes JP, Rees S, Kalindjian SB, Philpott KL (2011) Principles of early drug discovery. Br J Pharmacol 162(6):1239–1249
Hussein AK (2016) Applications of nanotechnology to improve the performance of solar collectors—recent advances and overview. Renew Sust Energ Rev 62:767–792
Iram F, Iqbal MS, Athar MM, Saeed MZ, Yasmeen A (2014) Glucoxylan mediated green synthesis of gold and silver nanoparticles and their phytotoxicity study. Carbohydr Polym 104:29–33
Iravani S (2011) Green synthesis of metal nanoparticles using plants. Green Chem 13:2638–2650
Jain AN (2004) Ligand-based structural hypotheses for virtual screening. J Med Chem 47(4):947–961
Jones MC, Leroux JC (1999) Polymeric micelles—a new generation of colloidal drug carriers. Eur J Pharm Biopharm 48(2):101–111
Kano K, Fendler JH (1978) Pyranine as a sensitive pH probe for liposome interiors and surfaces: pH gradients across phospholipid vesicles. Biochim Biophys Acta Biomembr 509(2):289–299

Kaserer T, Temml V, Kutil Z, Vanek T, Landa P, Schuster D (2015) Prospective performance evaluation of selected common virtual screening tools. Case study: cyclooxygenase (COX) 1 and 2. Eur J Med Chem 96:445–457

Kataoka K, Harada A, Nagasaki Y (2001) Block copolymer micelles for drug delivery: design, characterization and biological significance. Adv Drug Deliv Rev 47(1):113–131

Khan SB, Faisal M, Rahman MM, Jamal A (2011) Exploration of CeO_2 nanoparticles as a chemi-sensor and photo-catalyst for environmental applications. Sci Total Environ 409(15):2987–2992

Khan I, Saeed K, Khan I (2017) Nanoparticles: properties, applications and toxicities. Arab J Chem. https://doi.org/10.1016/j.arabjc.2017.05.011

Kickelbick G, Schubert US (2003) Advances in nanophase materials and nanotechnology. Functionalization and Surface Treatment of Nanoparticles, MI Baraton, Ed, 91–102

Krug HF, Wick P (2011) Nanotoxicology: an interdisciplinary challenge. Angew Chem Int Ed 50:1260–1278

Kubinyi H (1997) QSAR and 3D QSAR in drug design part 1: methodology. Drug Discov Today 2(11):457–467

Kumar P, Kulkarni PK, Srivastava AA (2015) Pharmaceutical application of nanoparticles in drug delivery system. J Chem Pharm Res 7:703–712

Larson DR, Zipfel WR, Williams RM, Clark SW, Bruchez MP, Wise FW, Webb WW (2003) Water-soluble quantum dots for multiphoton fluorescence imaging in vivo. Science 300(5624):1434–1436

Lavecchia A, Di Giovanni C (2013) Virtual screening strategies in drug discovery: a critical review. Curr Med Chem 20(23):2839–2860

Lee CH, Huang HC, Juan HF (2011) Reviewing ligand-based rational drug design: the search for an ATP synthase inhibitor. Int J Mol Sci 12(8):5304–5318

Lionta E, Spyrou GK, Vassilatis D, Cournia Z (2014) Structure-based virtual screening for drug discovery: principles, applications and recent advances. Curr Top Med Chem 14(16):1923–1938

Liu Z, Chen K, Davis C, Sherlock S, Cao Q, Chen X, Dai H (2008) Drug delivery with carbon nanotubes for in vivo cancer treatment. Cancer Res 68(16):6652–6660

Lu AH, Salabas EE, Schuth F (2007) Magnetic nanoparticles: synthesis, protection, functionalization, and application. Angew Chem Int Ed 46(8):1222–1244

Lyne PD (2002) Structure-based virtual screening: an overview. Drug Discov Today 7(20): 1047–1055

Makarov VV, Love AJ, Sinitsyna OV, Makarova SS, Yaminsky IV, Taliansky ME, Kalinina NO (2014) Green nanotechnologies: synthesis of metal nanoparticles using plants. Acta Nat 6:35–44

Malam Y, Loizidou M, Seifalian AM (2009) Liposomes and nanoparticles: nanosized vehicles for drug delivery in cancer. Trends Pharmacol Sci 30(11):592–599

Malik JK, Soni H, Singhai AK (2013) QSAR—application in drug design. Int J Pharm Res Allied Sci 2(1):1–13

Mallick S, Choi JS (2014) Liposomes: versatile and biocompatible nanovesicles for efficient biomolecules delivery. J Nanosci Nanotechnol 14(1):755–765

Manish G, Vimukta S (2011) Targeted drug delivery system: a review. Res J Chem Sci 1(2):135–138

Manivannan S, Ramaraj R (2012) Synthesis of cyclodextrin-silicate sol–gel composite embedded gold nanoparticles and its electrocatalytic application. Chem Eng J 210:195–202

Mannhold R, Krogsgaard-Larsen P, Timmerman H (2008) QSAR: Hansch analysis and related approaches, vol 1. VCH, Weinheim

McBain SC, Yiu HH, Dobson J (2008) Magnetic nanoparticles for gene and drug delivery. Int J Nanomedicine 3(2):169

McGann M (2011) FRED pose prediction and virtual screening accuracy. J Chem Inf Model 51(3):578–596

Mekenyan OG, Veith GD (1994) The electronic factor in QSAR: MO-parameters, competing interactions, reactivity and toxicity. SAR and QSAR in Environ Res 2(1–2):129–143

Meng XY, Zhang HX, Mezei M, Cui M (2011) Molecular docking: a powerful approach for structure-based drug discovery. Curr Comput Aided Drug Des 7(2):146–157

Mohan S, Oluwafemi OS, George SC, Jayachandran VP, Lewu FB, Songca SP, Kalarikkal N, Thomas S (2014) Completely green synthesis of dextrose reduced silver nanoparticles, its antimicrobial and sensing properties. Carbohydr Polym 106:469–474

Monnard PA, Oberholzer T, Luisi P (1997) Entrapment of nucleic acids in liposomes. Biochim Biophys Acta Biomembr 1329(1):39–50

Myint KZ, Xie XQ (2010) Recent advances in fragment-based QSAR and multi-dimensional QSAR methods. Int J Mol Sci 11(10):3846–3866

Nantasenamat C, Isarankura-Na-Ayudhya C, Naenna T, Prachayasittikul V (2009) A practical overview of quantitative structure–activity relationship. EXCLI J 8:74–88

Narayanan R, El-Sayed MA (2003) Effect of catalysis on the stability of metallic nanoparticles: Suzuki reaction catalyzed by PVP–palladium nanoparticles. J Am Chem Soc 125(27):8340–8347

Navgire ME, Gogoi P, Mallesham B, Rangaswamy A, Reddy BM, Lande MK (2016) β-Cyclodextrin supported MoO_3–CeO_2 nanocomposite material as an efficient heterogeneous catalyst for degradation of phenol. RSC Adv 6(34):28679–28687

Neouze MA, Schubert U (2008) Surface modification and functionalization of metal and metal oxide nanoparticles by organic ligands. Monatshefte fur Chemie-Chem Mon 139(3):183–195

Nikalje AP (2015) Nanotechnology and its applications in medicine. Med Chem 5(2):081–089

Novotarskyi S (2013) QSAR approaches to predict human cytochrome P450 inhibition. Doctoral dissertation, Technische Universität München

Pal D, Sahu CK, Haldar A (2014) *Bhasma*: the ancient Indian nanomedicine. J Adv Pharm Technol Res 5:4–12

Pandey RK, Kawabata Y, Teraji S, Norisuye T, Tran-Cong-Miyata Q, Soh S, Nakanishi H (2017) Metal nanowire–based hybrid electrodes exhibiting high charge/discharge rates and long-lived electrocatalysis. ACS Appl Mater Interfaces 9(41):36350–36357

Pantarotto D, Briand JP, Prato M, Bianco A (2004) Translocation of bioactive peptides across cell membranes by carbon nanotubes. Chem Commun 1:16–17

Parveen A, Rao S (2015) Cytotoxicity and genotoxicity of biosynthesized gold and silver nanoparticles on human cancer cell lines. J Clust Sci 26(3):775–788

Paul S, Chugh A (2011) Assessing the role of Ayurvedic *bhasmas* as ethno-nanomedicine in the metal based nanomedicine patent regime. J Intellect Disabil Res 16:509–515

Pavani T, Chakra CS, Rao KV (2013) A green approach for the synthesis of nano-sized iron oxide, by Indian Ayurvedic modified *bhasmikaran* method. Am J Pharm Sci 1:1–7

Petralia S, Barbuzzi T, Ventimiglia G (2012) Polymerase chain reaction efficiency improved by water soluble β-cyclodextrins capped platinum nanoparticles. Mater Sci Eng C 32(4): 848–850

Prasad R (2014) Synthesis of silver nanoparticles in photosynthetic plants. J Nanopart:963961. https://doi.org/10.1155/2014/963961

Prasad R, Pandey R, Barman I (2016) Engineering tailored nanoparticles with microbes: quo vadis. WIREs Nanomed Nanobiotechnol 8:316–330. https://doi.org/10.1002/wnan.1363

Prasad R, Pandey R, Varma A, Barman I (2017) Polymer based nanoparticles for drug delivery systems and cancer therapeutics. In: Kharkwal H, Janaswamy S (eds) Natural polymers for drug delivery. CAB International, UK, pp 53–70

Prasad R, Jha A, Prasad K (2018) Exploring the realms of nature for nanosynthesis. Springer International Publishing. ISBN:978-3-319-99570-0. https://www.springer.com/978-3-319-99570-0

Rai M, Yadav A, Gade A (2009) Silver nanoparticles as a new generation of antimicrobials. Biotechnol Adv 27(1):76–83

Rajesh KM, Ajitha B, Reddy YAK, Suneetha Y, Reddy PS (2018) Assisted green synthesis of copper nanoparticles using Syzygium aromaticum bud extract: Physical, optical and antimicrobial properties. Optik 154:593–600

Rao PV, Nallappan D, Madhavi K, Rahman S, Wei LJ, Gan S (2016) Phytochemicals and biogenic metallic nanoparticles as anticancer agents. Oxid Med Cell Longevity 2016:3685671, 15. https://doi.org/10.1155/2016/3685671

Rathod KB, Patel MB, Parmar PK, Kharadi SR, Patel PV, Patel KS (2011) Glimpses of current advances of nanotechnology in therapeutics. Int J Pharm Pharm Sci 3:8–12
Reddy AS, Pati SP, Kumar PP, Pradeep HN, Sastry GN (2007) Virtual screening in drug discovery—a computational perspective. Curr Protein Pept Sci 8(4):329–351
Ruckenstein E, Li ZF (2005) Surface modification and functionalization through the self-assembled monolayer and graft polymerization. Adv Colloid Interf Sci 113(1):43–63
Safari J, Zarnegar Z (2014) Advanced drug delivery systems: nanotechnology of health design: a review. J Saudi Chem Soc 18(2):85–99
Sahoo SK, Misra R, Parveen S (2017) Nanoparticles: a boon to drug delivery, therapeutics, diagnostics and imaging. In: Balogh LP (ed) Nanomedicine in cancer. Pan Stanford, Singapore, pp 73–124
Saxena M, Saxena J, Nema R, Singh D, Gupta A (2013) Phytochemistry of medicinal plants. J Pharmacogn Phytochem 1(6):168–182
Sethi NS (2012) A review on computational methods in developing quantitative structure–activity relationship (QSAR). Int J Drug Res Technol 2(2):189–197
Shamaila S, Zafar N, Riaz S, Sharif R, Nazir J, Naseem S (2016) Gold nanoparticles: an efficient antimicrobial agent against enteric bacterial human pathogen. Nano 6(4):71
Shanmukh S, Jones L, Driskell J, Zhao Y, Dluhy R, Tripp RA (2006) Rapid and sensitive detection of respiratory virus molecular signatures using a silver nanorod array SERS substrate. Nano Lett 6(11):2630–2636
Sharma C, Singh C (2014) Nano carriers of novel drug delivery system for Ayurveda herbal remedies need of hour: a bird's eye view. Am J Pharm Tech Res 4:60–69
Silva GA (2004) Introduction to nanotechnology and its applications to medicine. Surg Neurol 61(3):216–220
Simon Z, Chiriac A, Holban S, Ciubotariu D, Mihalas G (1984) Minimum steric difference: the MTD method for QSAR studies. Chemometrics research studies series. Research Studies Press, Letchworth
Singh P, Kim YJ, Zhang D, Yang DC (2016) Biological synthesis of nanoparticles from plants and microorganisms. Trends Biotechnol 34(7):588–599
Sneh-Edri H, Likhtenshtein D, Stepensky D (2011) Intracellular targeting of PLGA nanoparticles encapsulating antigenic peptide to the endoplasmic reticulum of dendritic cells and its effect on antigen cross-presentation in vitro. Mol Pharm 8(4):1266–1275
Sun X, Li Y (2004) Colloidal carbon spheres and their core/shell structures with noble-metal nanoparticles. Angew Chem 116(5):607–611
Tang H, Lin Y, Andrews C, Sodano HA (2010) Nanocomposites with increased energy density through high aspect ratio PZT nanowires. Nanotechnology 22(1):015702
Tang F, Li L, Chen D (2012) Mesoporous silica nanoparticles: synthesis, biocompatibility and drug delivery. Adv Mater 24(12):1504–1534
Tiwari R (2016) Controlled release drug formulation in pharmaceuticals: a study on their application and properties. World J Pharm Res 5:1740–1720
Tiwari G, Tiwari R, Sriwastawa B, Bhati L, Pandey S, Pandey P, Bannerjee SK (2012) Drug delivery systems: an updated review. Int J Pharm Invest 2(1):2
Tomasulo A, Ramakrishna MV (1996) Quantum confinement effects in semiconductor clusters. II. J Chem Phys 105(9):3612–3626
Tong W, Lowis DR, Perkins R, Chen Y, Welsh WJ, Goddette DW, Heritage TW, Sheehan DM (1998) Evaluation of quantitative structure activity relationship methods for large-scale prediction of chemicals binding to the estrogen receptor. J Chem Inf Comput Sci 38(4):669–677
Tsoi KM, Dai Q, Alman BA, Chan WCW (2012) Are quantum dots toxic? Exploring the discrepancy between cell culture and animal studies. Acc Chem Res 46:662–671
Voit BI, Lederer A (2009) Hyperbranched and highly branched polymer architectures synthetic strategies and major characterization aspects. Chem Rev 109(11):5924–5973
Wu SH, Chen DH (2004) Synthesis of high-concentration Cu nanoparticles in aqueous CTAB solutions. J Colloid Interface Sci 273(1):165–169

Xie XQS (2010) Exploiting PubChem for virtual screening. Expert Opin Drug Discovery 5(12): 1205–1220

Yadav M, Bhatia VJ, Doshi G, Shastri K (2014) Novel techniques in herbal drug delivery systems. Int J Pharm Sci Rev Res 28:83–89

Yadav L, Tripathi RM, Prasad R, Pudake RN, Mittal J (2017) Antibacterial activity of cu nanoparticles against *E. coli*, *Staphylococcus aureus* and *Pseudomonas aeruginosa*. Nano Biomed Eng 9(1):9–14. https://doi.org/10.5101/nbe.v9i1.p9-14

Zadbuke N, Shahi S, Gulecha B, Padalkar A, Thube M (2013) Recent trends and future of pharmaceutical packaging technology. J Pharm Bioallied Sci 5(2):98

Zhan J, Ting XL, Zhu J (2017) The research progress of targeted drug delivery systems. IOP Conf Ser Mater Sci Eng 207(1):012017

Zhang YJ, Gan RY, Li S, Zhou Y, Li AN, Xu DP, Li HB (2015) Antioxidant phytochemicals for the prevention and treatment of chronic diseases. Molecules 20(12):21138–21156

Zsoldos Z, Reid D, Simon A, Sadjad SB, Johnson AP (2007) eHiTS: a new fast, exhaustive flexible ligand docking system. J Mol Graph Model 26(1):198–212

Chapter 8
Nanotechnology Advances for the Development of Various Drug Carriers

U. T. Uthappa, Mahaveer D. Kurkuri, and Madhuprasad Kigga

Contents

8.1 Introduction

Nanotechnology is one of the important fields where micromaterials are engineered by changing their physical and chemical properties to produce nanosized materials (Mukherjee et al. 2014). The use of nanomaterials offers immense possibilities to modify the fundamental properties in drug delivery applications such as immunogenicity, blood circulation, half-life, solubility, diffusivity, and drug release characteristics. A number of nanoparticle-based drug carriers have been reported in the past decades for the treatment of various diseases such as cancer, diabetes, asthma, allergy, infections, etc. The nanomaterial-based therapeutic drug delivery systems have been extensively used for both targeted and controlled drug release and proved with promising results in various areas such as cardiology, oncology, neurology, immunology, ophthalmology, pulmonology, and orthopedics. Thus, drug delivery systems using various nanomaterial-based carriers offer great interest in the current pharmaceutical industry in order to deliver the therapeutic agents selectively to specific areas with improved therapeutic effects (Zhang et al. 2007; Prasad et al. 2017).

U. T. Uthappa · M. D. Kurkuri (✉) · M. Kigga (✉)
Centre for Nano and Material Sciences, JAIN (Deemed-to-be University)
Jain Global Campus, Bengaluru, Karnataka, India
e-mail: mahaveer.kurkuri@jainuniversity.ac.in; madhuprasad@jainuniversity.ac.in

R. Prasad et al. (eds.), *Nanobiotechnology in Bioformulations*, Nanotechnology in the Life Sciences, https://doi.org/10.1007/978-3-030-17061-5_8

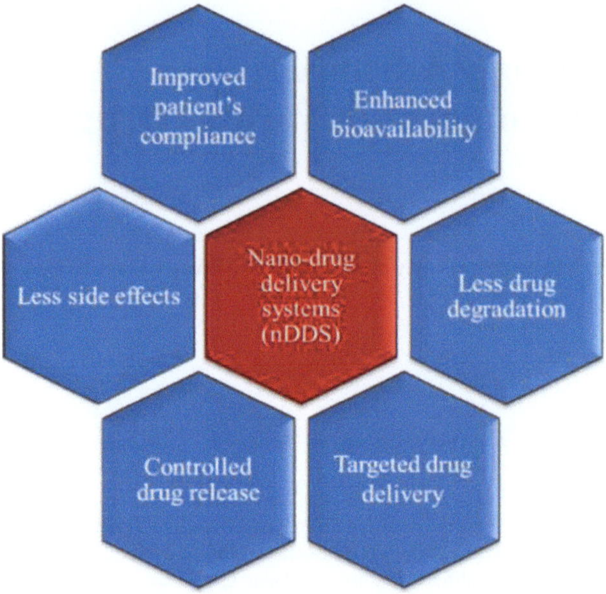

Fig. 8.1 Advantages of nanotechnology- embedded drug delivery systems

The conventional drug delivery systems have limitations due to high drug degradation, less target orientation, low bioavailability, less stability, etc. On the other hand, drug carriers could be designed either by encapsulating therapeutic drugs into the nanoparticles or by adsorbing on the surface followed by tagging them with the targeting ligands to effectively overcome such problems (Jianping Wu and Peng 2018). The nanodrug delivery systems (nDDS) can be easily modified using conventional chemical techniques in order to tune and alter the pharmacokinetic or pharmacodynamic properties (Wen Jiang et al. 2007). The advantages of drug delivery systems using nanotechnology-embedded approach have been illustrated in Fig. 8.1 (Singh et al. 2015).

This chapter focuses recent developments on the synthesis of various drug carriers using different materials such as graphene oxide, metal organic frameworks, and naturally available diatoms. The use of these drug carriers and their surface functionalization to enhance the pharmacokinetic or pharmacodynamic properties were included. In addition to this, important factors like enhanced drug loading capacity, release profiles, biocompatibility, cytotoxicity, and cellular uptake are discussed.

8.2 Graphene Oxide and Surface-Modified Graphene Oxide as a Drug Carrier

Graphene is a 2D material which has obtained substantial attention in the various fields of chemistry and condensed matter physics, due to their unique physical and chemical properties since their first appearance in 2004 (Zhao et al. 2014). Graphene oxide (GO) is an oxidized form of graphite with a layered assembly composed of sp^2 carbon atoms arranged in the form of a honeycomb structure. Each layer comprises of random six-membered aliphatic and aromatic rings. Therefore, in order to synthesize GO, the natural graphite powder is oxidized in acidic media, and later GO was prepared by exfoliation of graphite oxide layers. After oxidation of graphite, different oxygenated functional groups such as carboxyl, hydroxyl, epoxy, etc. can be attached to the graphene layers (Mahkam et al. 2014). GO has emerged as one of the important biomaterials with significant usage in DNA sequencing, probes for cells, biological imaging, etc. (Zhao et al. 2014). These materials being inert to the biological systems with varying sizes from nano- to micrometers and due to their hydrophobic properties result in adsorption of various drug molecules. Due to the graphene-graphene interactions, it leads to poor water solubility and hinders in drug delivery applications. However, in order to make them as efficient candidates, surface functionalization plays a critical role (Maity et al. 2014).

Functionalization of GO through electrostatic-driven self-assembly approach by using two natural polymers, chitosan (CS) and sodium alginate (SA), has been carried out (Lei et al. 2016). The anticancer drug doxorubicin hydrochloride (DOX) was successfully loaded on the functionalized material and studied for pH-dependent drug release actions. The surface modification of GO with these two polymers enhanced the solubility of both GO and anticancer drug used in this study. Further, these DOX-loaded surface functionalized materials GO-CS-SA displayed significant cytotoxicity to MCF-7 cells.

Barahuie et al. designed a novel nanocomposite consisting of GO as a nanocarrier for loading the active anticancer agent called chlorogenic acid (CAGO) which forms nanocomposites, as shown in Fig. 8.2 (Barahuie et al. 2017). The drug loading interaction of chlorogenic acid with GO nanocarrier was predicted as hydrogen-bonding and π-π interaction which was confirmed through Raman spectroscopy, XRD, and FTIR analysis. By using UV-vis spectroscopy, around 13% of drug loading

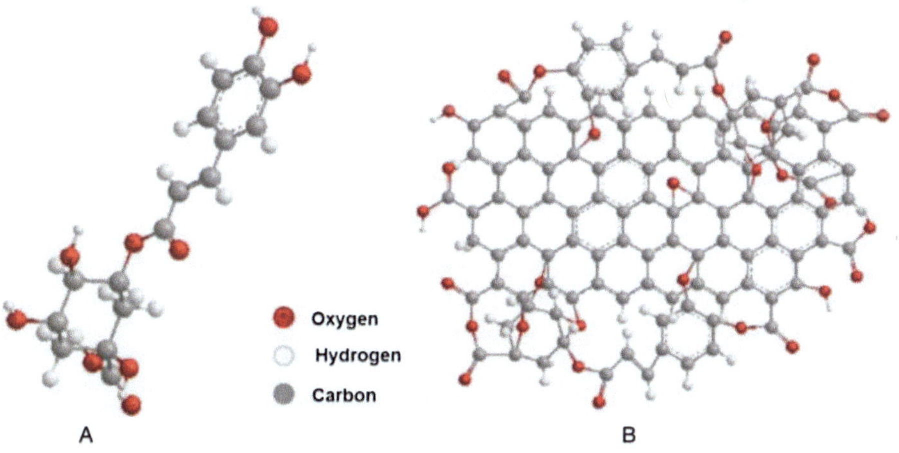

Fig. 8.2 Structure of chlorogenic acid (**a**) and chlorogenic acid-graphene oxide nanocomposite (**b**). (Reproduced with permission Barahuie et al. 2017)

was confirmed. Further, the release profile was depicted as sustained and followed by pH-dependent release of the drug in phosphate conditions at particular pH of 4.8 and 7.4. In order to compare the cytotoxicity, they have tested for all the three different materials such as graphene oxide as nanocarrier, graphene oxide-chlorogenic acid as nanocomposites, and neat chlorogenic acid. Thus, cytotoxicity studies showed negligible toxicity for drug and nanocarrier. However, the enhanced anticancer property was observed for the developed nanocomposite.

A novel drug carrier was designed by functionalizing hydroxypropyl-β-cyclodextrin on carboxylated GO (GO-COO-HP- β-CD) to load the anticancer drug paclitaxel as shown in Fig. 8.3 (Tan et al. 2016).

The results indicated that the stability and solubility of paclitaxel-loaded hydroxypropyl-β-cyclodextrin-GO was prominently enhanced in aqueous media in comparison with the untreated paclitaxel drug. The drug loading capacity was 29.93 % and displayed drug release profile at different pH such as 7.4, 6.5 and 5.0 at 37 °C that can be divided into two stages i.e., for the first 6 h, is due to the surface adhered drug molecules on hydroxypropyl-β-cyclodextrin–graphene oxide and sustained release which continued up to 150 h which is due to hallow inner space created on the hydroxypropyl-β-cyclodextrin-graphene oxide. In addition to this, hemolysis studies revealed that the paclitaxel loaded into the functionalized drug carrier improved the blood compatibility and proliferation of HeLa cells that results in intravenous medicine application.

A targeted drug delivery approach by functionalizing GO with carboxymethyl chitosan (CMC), fluorescein isothiocyanate, and lactobionic acid (LA) was used in the studies as shown in Fig. 8.4 (Pan et al. 2016). The comparative studies were carried out with the similar systems without LA was prepared as controls. The anticancer drug doxorubicin (DOX) was loaded through the adsorption process.

It was noted that the DOX and GO interaction was through π-π stacking, and the drug loading content and efficiency were >96%. The resultant formulation showed pH-sensitive drug release. Though DOX-loaded LA-conjugated GO was able to induce death to cancerous cells, it was nontoxic to noncancerous cells. The reason

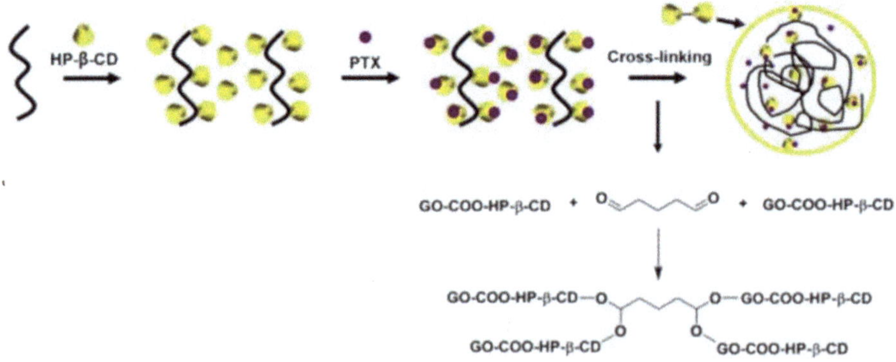

Fig. 8.3 Schematic representation of drug-loaded GO-COO-HP-β-CD nanosphere formation. (Reproduced with permission from Tan et al. 2016)

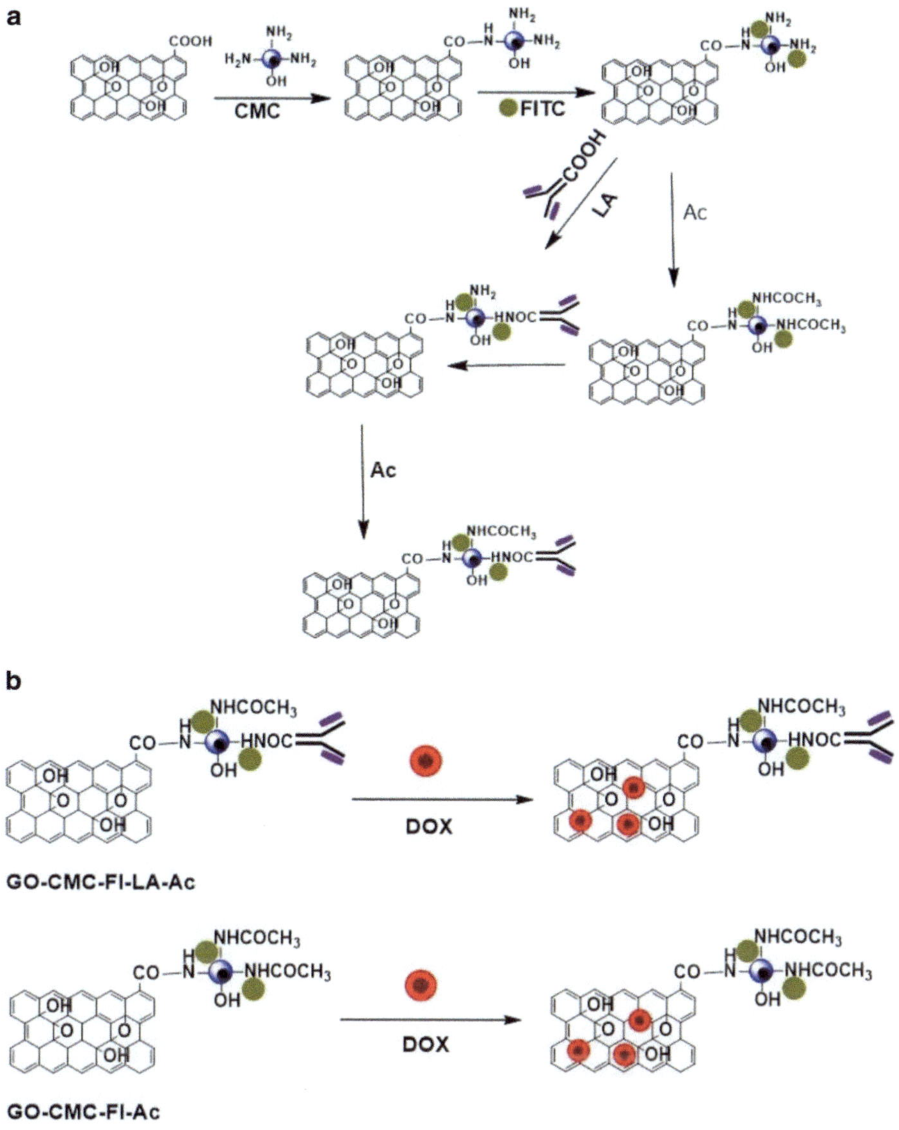

Fig. 8.4 (**a**) Synthesis of functionalized GO materials and (**b**) DOX loading. (Reproduced with permission from Pan et al. 2016)

is due to the selective recognition of LA by the asialoglycoprotein receptors are overexpressed on cancerous hepatic cells.

Huang et al. reported magnetic-guided GO nanocarrier for targeted drug delivery as well as pH-responsive sustained release approach for two different cancer drugs, irinotecan (CPT-11) and doxorubicin (DOX) (Huang et al. 2017). Magnetic nanoparticles on the surface of GO (m-GO) were introduced via chemical co-precipitation process of

Fe_3O_4. Further, m-GO was covalently bound with chitosan (m-GOC) in order to increase the contact of nanocarrier with the cell membrane. To avoid the endocytosis by the reticuloendothelial system (RES) and to extend the blood circulation of the nanocarrier, m-GOC was grafted with polyethylene glycol (PEG) to form mGOC-PEG. The drug loading efficiency and loading content for CPT-11 was 54% and 2.7% and for DOX 65% and 393%, respectively. It was noticed at pH 5.4, 60% and at pH 7.4, 10% of DOX drug was released. In comparison, at pH 5.4, 90% and at pH 7.4, 70% of CPT-11 drug release was observed. In addition, it was possible to kill U87 human glioblastoma cell line by magnetic targeting.

The development of drug carrier depends on tuning the thickness and diameter of the material. Layer-by-layer (LbL) self-assembly supported noncovalent surface functionalization of GO with chitosan (CS) and dextran (Dex), as depicted in Fig. 8.5, and the functionalized material revealed a diameter of 300 nm and a thickness of 60 nm (Xie et al. 2016).

The DOX drug was loaded through π-π stacking and electrostatic attachment and showed pH-sensitive release behaviors in acidic environmental conditions. Therefore, the use of CS and Dex not only improved the dispersibility of both GO and the drug but also decreased nonspecific protein adsorption on GO nanosheets. The cytotoxicity studies on MCF-7 showed strong cytotoxicity for drug-loaded GO-CS-Dex which made a better candidate in terms of anticancer drug delivery.

Wang et al. prepared hollow Fe_3O_4-GO-based nanodrug carrier for the targeted delivery approach. In this method, the surface of Fe_3O_4 was coated with GO via electrostatic interactions (Wang et al. 2017). The drug 5-Fluorouracil (5-FU) was loaded to this nanocarrier which exhibited high drug loading compared to bare

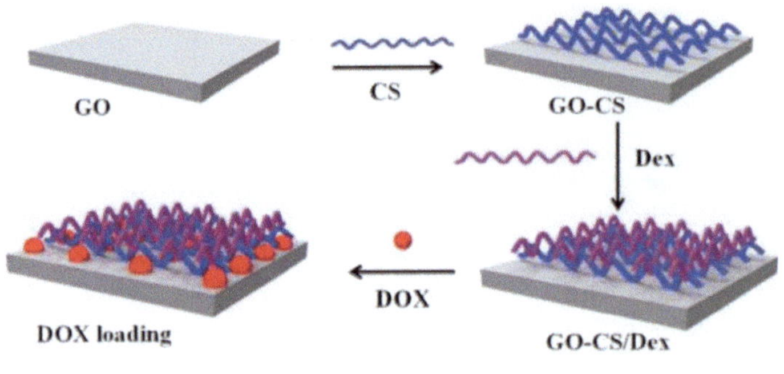

Fig. 8.5 Synthesis of GO-CS-Dex and DOX drug loading. (Reproduced with permission from Xie et al. 2016)

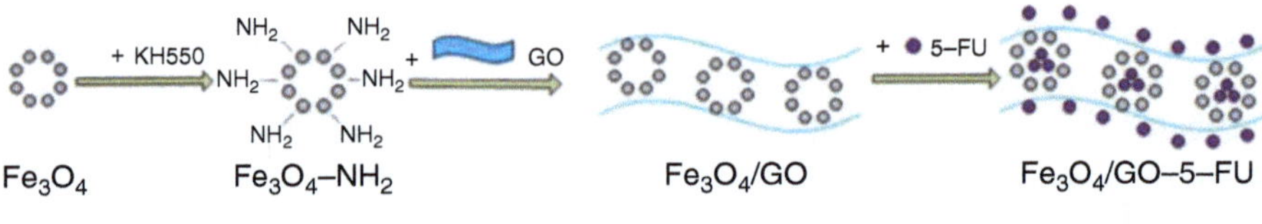

Fig. 8.6 Preparation process of Fe_3O_4/GO composites and drug loading of 5-FU. (Reproduced with permission from Wang et al. 2017)

Fe_3O_4. In addition, this nanocarrier showed strong magnetic properties and therefore could be controlled through the external magnetic field. The toxicity studies proved with good biocompatibility. The preparation process of the drug carrier and loading of the drug are shown in Fig. 8.6.

The conjugation of sodium alginate on the surface of GO as a drug carrier has been reported for the loading and release of DOX. In this technique, GO was modified with adipic acid dihydrazide, and sodium alginate (SA) was covalently conjugated to introduce the amine functional groups. The amount of drug loading capacity was found to be 1.843 mg/mg, and the drug release in the tumor environment at pH 5.0 was comparatively higher in contrast under physiological conditions at pH 6.5 and 7.4. Therefore, in vitro toxicity for GO-SA showed no toxicity for NIH-3T3 cells and for DOX loaded GO-SA high cytotoxicity to HeLa cells (Fan et al. 2016).

Through the process of free-radical polymerization, Hemmati et al. prepared molecularly imprinted polymer and GO conjugates (MMIP-GO) using acrylamido-2-methyl-1-propane sulfonic acid (AMPS) as the functional monomer (Hemmati et al. 2016). The ethylene glycol diacrylate (EGDA) was used as crosslinker in aid of acrylate functionalized Fe_3O_4 nanoparticles and GO and rivastigmine (RIV) used as a template. The rapid burst release was observed for the first 24 h and controlled release of around 58–76% for 7 days.

Synthesis of stable nanohybrid drug carriers is one of the challenges in current drug delivery systems. In this work, using Zinc-clinoptilolite/GO (Zn-Clin/GO) by two different techniques, microwave assisted hydrothermal and reflux method was carried out (Khatamian et al. 2016). The nanocomposites prepared using reflux method were stable, homogeneous, and thus used as a drug carrier. Further, the drug loading capacity for Zn-Clin was 70% in 120 min, GO showed 80% in 30 min, and Zn-Clin/GO showed 90% in 30 min. The MTT assays studies revealed that DOX-loaded Zn-Clin/GO showed more cytotoxic behavior than free DOX and that these drug carriers have potential values in cancer therapy.

Lv et al. prepared multifunctional GO with polyethyleneimine (PEI) in sequence using image dye fluorescein isothiocyanate (FITC) and polyethylene glycol (PEG) linked with target ligand lactobionic acid (LA) as represented in Fig. 8.7 (Lv et al. 2016). This surface-functionalized material was loaded with an anticancer drug, DOX, and 85% loading capacity was observed. Further in vitro studies show that the release profile of DOX was higher at pH 5.8 when compared to physiological conditions. Therefore, these drug-loaded multifunctional materials prominently target SMMC-7221 cells by overexpressing ASGPR receptors.

The conjugation of a novel transferrin (Tf)-poly(allylamine hydrochloride) (PAH) on GO to form a stable and targeted nanocarrier has been reported. In this method, amine group of PAH was conjugated with Tf through EDC/NHS chemistry to form an amide bond between the amine group of PAH and carboxyl group of Tf, and the resultant Tf-PAH were further coated on the surface of GO (Nasrollahi et al. 2016). The anticancer drug docetaxel (DTX) was loaded on the surface of GO through hydrogen bonding and π-π stacking. A high drug loading capacity of 37% was achieved. It was noticed that the drug release at neutral pH is less than in acidic pH, where the dissociation of hydrogen bonds occur. The studies showed that

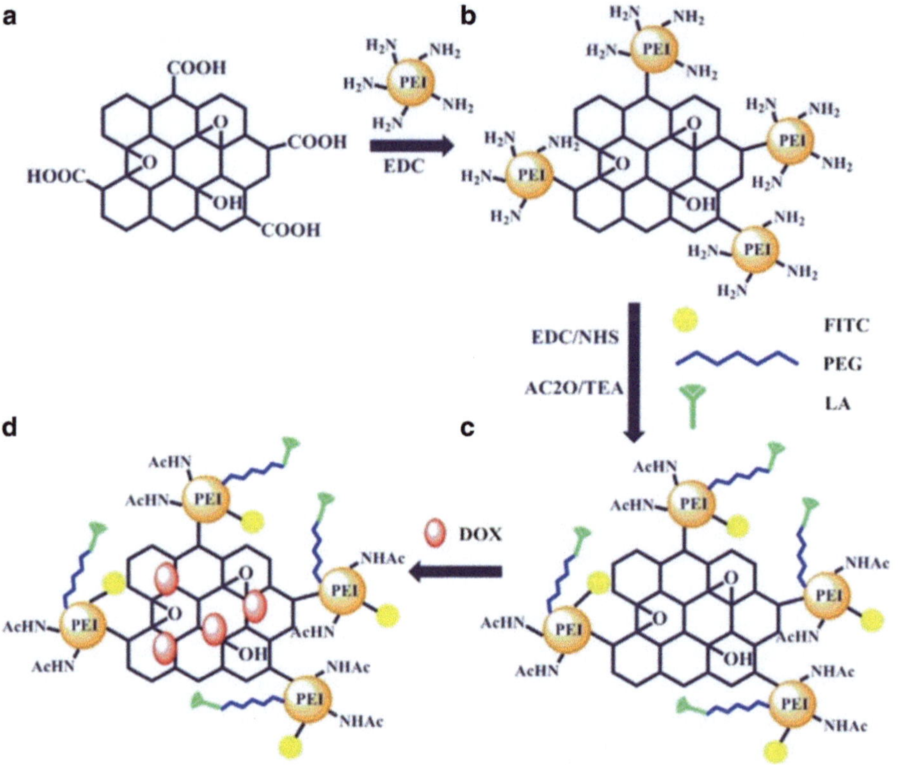

Fig. 8.7 Synthesis of (**a**) GO, (**b**) GO-polyethyleneimine, (**c**) GO-polyethyleneimine-FITC-PEG-lactobionic acid, (**d**) DOX-loaded material. (Reproduced with permission from Lv et al. 2016)

nanocarrier without drug has no cytotoxicity effect on MCF-7 cells. However, drug-loaded nanocarrier killed the cancer cells effectively and could be exhibited as one of the promising nanodrug carriers for the hydrophobic drugs. The PAH conjugation with Tf and DTX loading is shown in Fig. 8.8a, b.

The mucoadhesive properties have been improvized by functionalizing thiol ligands in order to deliver the hydrophobic drug valsartan which showed extended mucus residence time (Pereira de Sousa et al. 2016). Initially, GO was obtained by oxidation of graphite and thiolated in two different methods. In the first method, the hydroxyl group of GO was conjugated with thiourea through the formation of brominated intermediate. The other method includes the carboxylic group conjugated with cysteamine through carbodiimide chemistry, as shown in Fig. 8.9. For this material, arround 31 ± 0.3% of drug loading capacity was observed, and the drug release rate was in a sustained manner. Also, the enhanced mucoadhesive properties were confirmed through rheological measurements and residence time assay.

Dopamine was conjugated with nanographene oxide (DA-Ngo) via amide bonding to synthesize a nanodrug carrier for the anticancer drug methotrexate (Masoudipour et al. 2017). Further, the kinetic studies for drug release from GO in PBS media were investigated. In addition to this, the antineoplastic action of

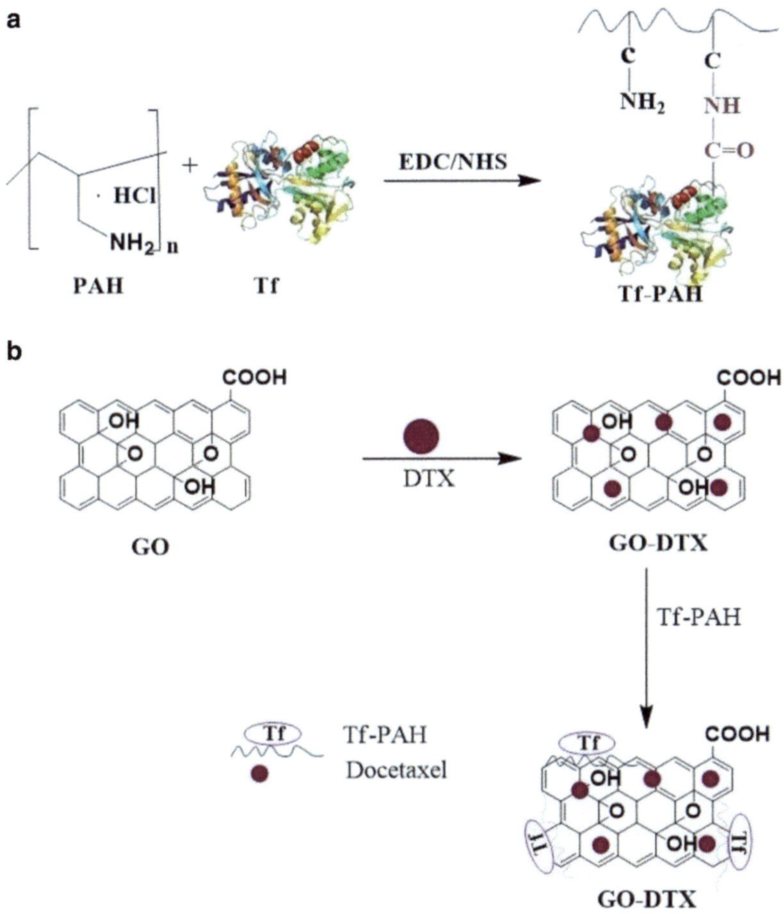

Fig. 8.8 Various steps for preparation of (**a**) Tf-PAH and (**b**) Tf-PAH (GO-DTX). (Reproduced with permission from Nasrollahi et al. 2016)

drug-loaded nanocarrier against the dopamine receptor for both positive and negative cell lines was investigated. The cytotoxicity studies validated that the presence of dopamine on the surface of nanocarrier helps to target the dopamine receptor positive cells and thus, can enhance admit the drug molecules into the cells.

The development of carboxymethyl cellulose/GO (CMC-GO)-based hydrogel beads by physically cross-linking with ferric chloride hexahydrate for controlled release of DOX is shown in Fig. 8.10 (Rasoulzadeh and Namazi 2017). The drug loading was enhanced for CMC-GO and one of the common π-π interaction has occurred between GO and DOX in comparison with hydrogel. However, the controlled drug release was achieved due to strong interactions among the amine group of DOX and carboxylic groups present on GO.

In another study for controlled and targeted applications, initially, GO was functionalized with adipic acid dihydrazide (ADH) to introduce an amine group on the

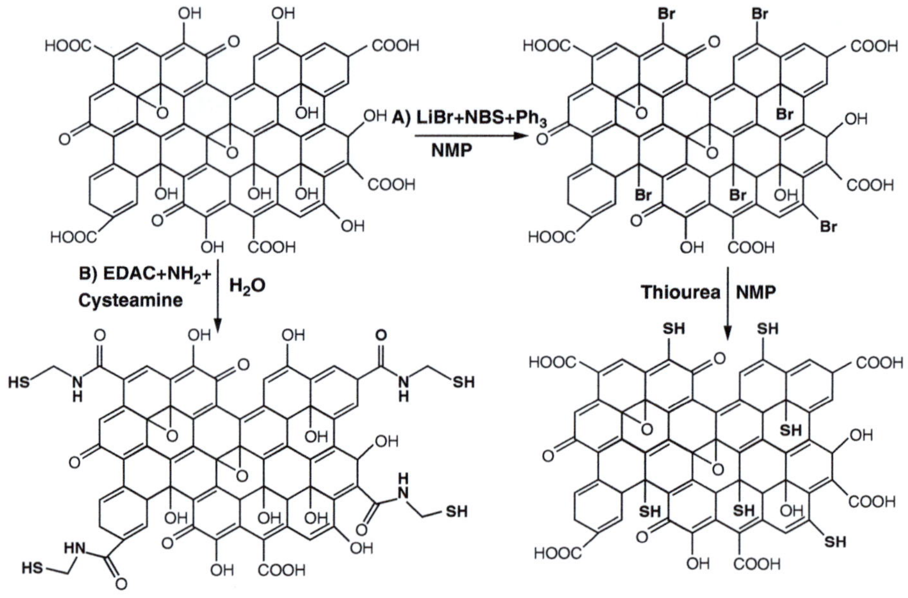

Fig. 8.9 Synthetic scheme of graphene thiolation exploiting the hydroxyl groups (**a**) and the carboxylic acid groups (**b**). (Reproduced with permission from Pereira de Sousa et al. 2016)

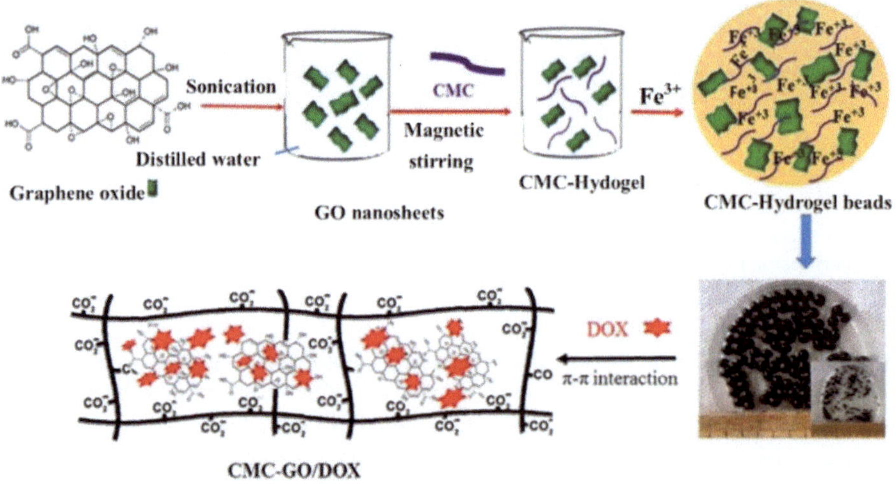

Fig. 8.10 Schematic representation of the whole process. (Reproduced with permission from Rasoulzadeh and Namazi 2017)

GO surface (Rao et al. 2018). Further conjugation with carboxymethyl cellulose (CMC) results in the formation of (GO-ADH-CMC) composites. The anticancer drug DOX was loaded via π-π stacking and hydrogen-bonding interactions to achieve 77.9% of drug loading. The DOX-loaded GO-ADH-CMC system was further used to study the cytotoxicity effects on HeLa and NIH-3T3 cells. It was noticed that

GO-CMC-DOX showed pronounced antitumor activity and it is safer compared to DOX administration.

A multilayer film through layer-by-layer (LbL) assembled nanofilms composed of block copolymer micelles (BCM), and GO was synthesized and used as a drug carrier (Jung et al. 2017). Coumarin-6 was the drug used for the studies and showed rapid release within 24 h. In order to control the drug release, chemical modifications to the inner structure and external structure were applied. For the chemical modification, two different kinds of composition were used for chitosan/GO and GO/GO. In the first method, the inner structure of films modified by crosslinking results in interactions between GO and BCM. In the second method, the outer structure of the film was modified by fabricating the outer layers of the film. However, both the methods and compositions showed different drug release rates, and it was concluded by using simple surface modification techniques the drug release rates can be controlled. The surface modification of the 14 films of GO/BCM by cross-linking Fig. 8.11a and by fabricating a cover layer film by LbL assembly by spin coating Fig. 8.11b.

Aptamer-decorated dextran-coated nanographene, GO-DEX-Apt, for targeted drug delivery has been reported. In this study, dextran (DEX) was covalently conjugated on the surface of nanographene in order to make the stable biocompatible drug carrier, GO-DEX (Alibolandi et al. 2017). The conjugated drug carrier (GO-DEX) was nontoxic to 4T1 mammary carcinoma cell line at a concentration up to 300 μg/mL. The decorated AS1411 aptamer (Apt), a ssDNA, has the ability to improve the intracellular uptake by nucleolin recognition and also can be introduced to the hydroxyl group of DEX in GO-DEX to produce GO-DEX-Apt. *Curcuma longa* with antineoplastic properties was loaded on both the carriers, GO-DEX and GO-DEX-Apt, through π-π stacking interactions with high drug loading capacity of around 29 wt. %. Interestingly, the GO-DEX-Apt could enter efficiently into 4T1 and MCF-7 nucleolin; thus, overexpressed cancer cells showed substantial higher cytotoxicity.

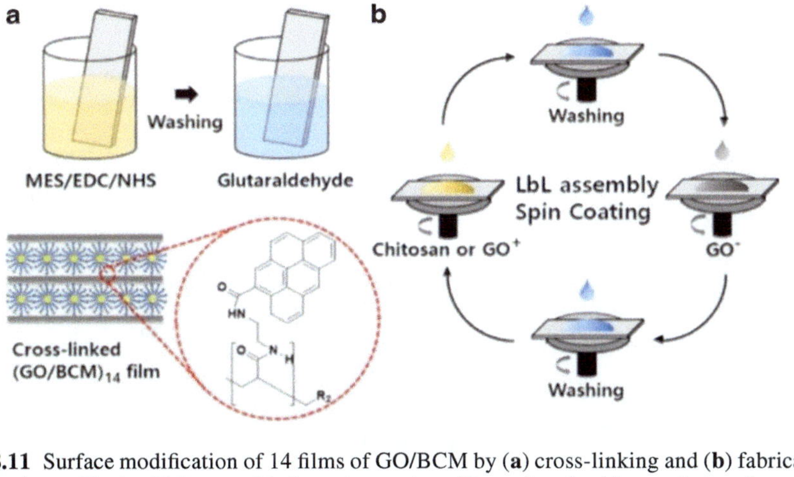

Fig. 8.11 Surface modification of 14 films of GO/BCM by (**a**) cross-linking and (**b**) fabricating a cover layer film by LbL assembly by spin coating. (Reproduced with permission from Jung et al. 2017)

The composites of GO/TiO_2/DOX was prepared and loaded into the chitosan/poly-lactic acid (PLA) nanofiber solutions to form electrospun chitosan/PLA/GO/TiO_2/DOX for the controlled release of DOX drug for up to 2 weeks (Samadi et al. 2018). Therefore, the mean diameter of PLA/chitosan and chitosan/PLA/GO/TiO_2/DOX was 170 nm and 140 nm, respectively. However, the drug loading efficiency for nanofibers containing 10, 25, and 50 mg of carrier was 97, 94, and 92%, respectively. Therefore, DOX drug release from the nanofibers with 10 µm thickness showed non-Fickian diffusion and for 30 and 50 µm exhibited Fickian diffusion. The cell viability showed proliferation inhibition effect on target cancer cells improved by increasing DOX content, and more killing of A549 lung cancer cells was achieved by using chitosan/PLA/GO/TiO_2/DOX due to the existence of an external magnetic field.

Designing targeted drug delivery system to selectively target tumor cells, improve the therapeutic activities, and minimize the side effects is significant. In this method, the carboxylate group of folic acid (FA) was activated to get active ester intermediated (FA-NHS). This FA-NHS was then covalently tethered on the amino group of bovine serum albumin (BSA) in alkaline conditions to form FA-BSA conjugates (Ma et al. 2017). Further, these FA-BSA conjugates were dispersed in a phosphate buffer solution at pH 8.0 and added dropwise to the GO solutions followed by stirring at room temperature. Later, this mixture was sonicated for 30 min and stirred for 24 h at room temperature to form FA-BSA-GO solutions. The DOX drug was loaded to FA-BSA-GO by simple mixing and stirring overnight at room temperature. The in vitro cellular uptake and cytotoxicity studies revealed that the FA-tagged nanohybrid could selectively deliver DOX to the folate receptor-rich MCF-7 cells, indicating the targeted drug delivery approach. The synthesis of FA-BSA is shown in Fig. 8.12.

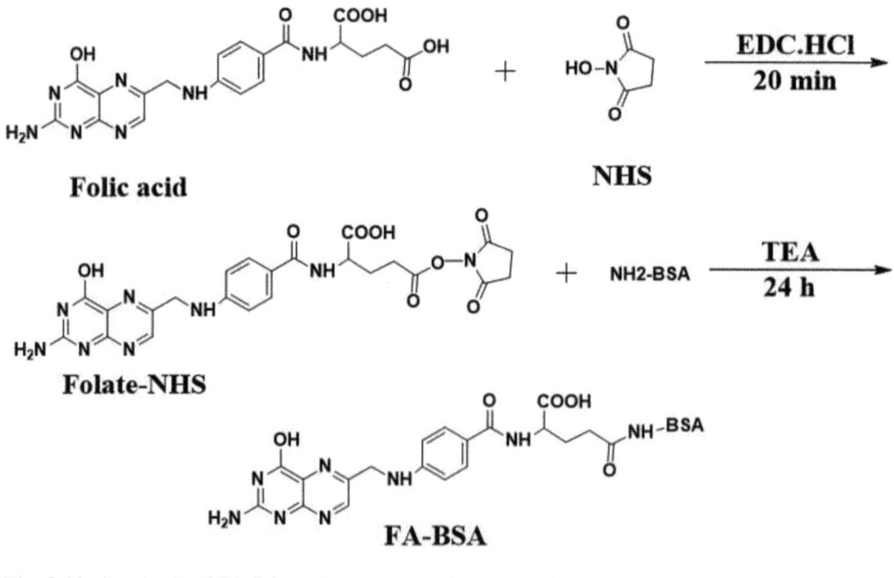

Fig. 8.12 Synthesis of FA-BSA. (Reproduced with permission from Ma et al. 2017)

Yao et al. used a facile hydrothermal method to prepare GO/HA (hydroxyapatite) hybrid designed through creatine phosphate disodium (CPDS)/GO composites template (Yao et al. 2017). In this approach, CPDS acts as a soft template and functions as a phosphate source to induce mineralization of calcium phosphate hydroxyapatite nanoparticles. In addition to this, GO serves as a hard template. The anti-inflammatory drug ibuprofen (IBU) was used for loading and the release studies in simulated body fluid conditions. The results indicated that the sustained release achieved better for the GO-HA system in comparison with HA microspheres. The designed system could be used as sustained release drug carriers for the treatment of postoperative infections of bone replacement.

For antitumor therapy, heparin (Hep) was modified with GO using adipic dihydrazide as a linker to control the release of DOX drug. In this method, the OH group of GO was converted to COOH by using sodium hydroxide and sodium chloroacetate to synthesize GO-COOH (Zhang et al. 2017). To this GO-COOH system, the prepared Hep-ADH was added and agitated for around 24 h to obtain GO-ADH-Hep. The hemolytic and in vitro cytotoxicity studies of GO-ADH-Hep showed better stability, biocompatibility, and blood compatibility. However, the DOX release from GO-ADH-Hep nanodrug carrier exhibited less drug release when compared to free DOX. In addition, the cytotoxicity study indicated GO-ADH-Hep-DOX showed effective cytotoxicities to MCF-7 and HepG2 cells.

GO was extensively used as a drug carrier for drug delivery applications due to its remarkable drug loading ability to the aromatic drugs. However, few of the important factors such as biocompatibility, stability, and blood biocompatibility are the important challenges which need to be addressed for their further applications. In order to solve these issues, Mu et al. designed hyperbranched polyglycerol-modified GO-HPG-GO (Mu et al. 2017). This was prepared through anionic ring opening followed by branching by polymerization using potassium hydride and 18-Crown-6 for deprotonation, as shown in Fig. 8.13. To this, DOX drug was loaded, and the drug loading capacity was found to be 328 mg/g which showed significant cytotoxic property toward tumor cells. However, the prepared HPG-GO showed good stability in aqueous solutions, low toxicity, good blood biocompatibility with an imperceptible effect on hemolysis, and blood coagulation.

Song et al. designed multifunctional GO-based targeted drug delivery system by combining both magnetic and targeting functions such as superparamagnetic iron oxide nanoparticles and lactoferrin (Lf) (Song et al. 2017). The magnetic nanoparticles were loaded to GO through a chemical precipitation method to form GO-Fe_3O_4. Through EDC/NHS chemistry, the target ligand Lf was conjugated; the ligand, an iron-transporting serum glycoprotein, binds to the receptors overexpressed at the surface of glioma cells and vascular endothelial cells of the blood-brain barrier (BBB). The designed nanodrug carrier had high DOX loading capacity due to stacking and showed a pH-dependent drug release behavior. However, compared to neat DOX and DOX-GO-Fe_3O_4, Lf-DOX-GO-Fe_3O_4 exhibited efficient intracellular delivery and cytotoxicity against C6 glioma cells. The synthesis of Lf-DOX-GO-Fe_3O_4 is shown in Fig. 8.14.

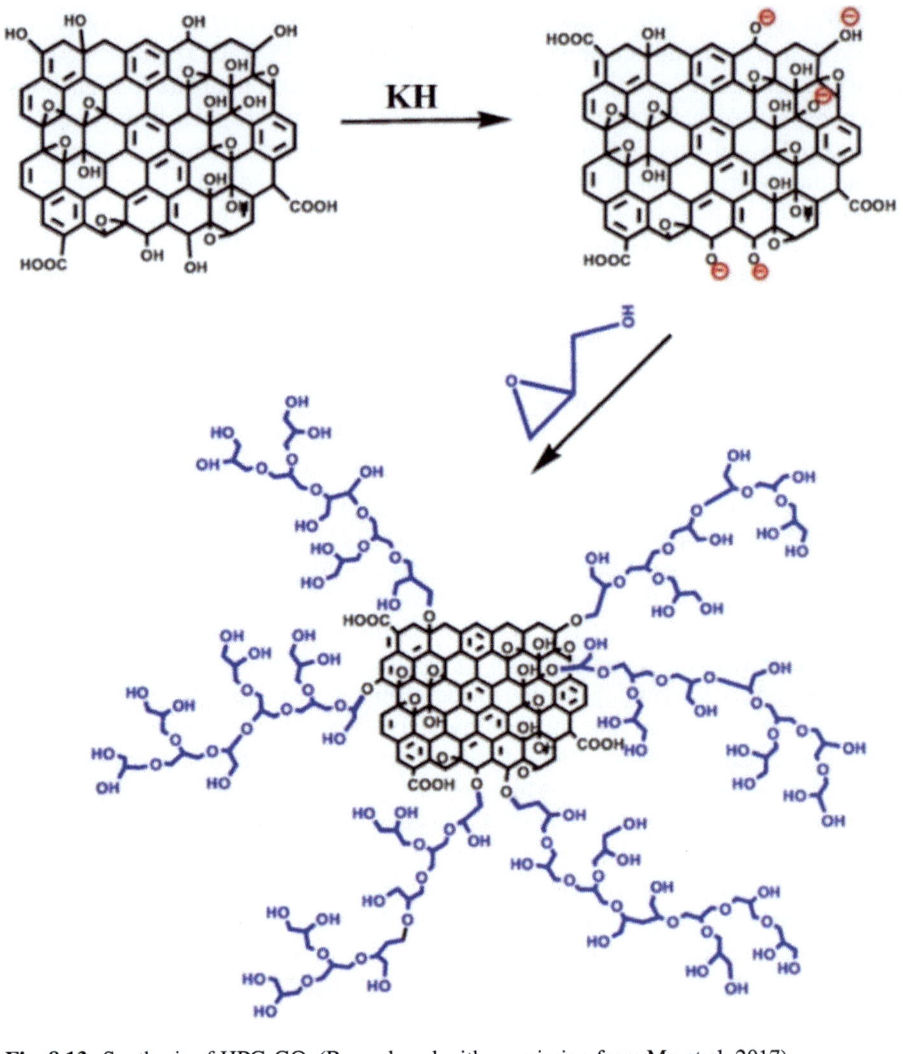

Fig. 8.13 Synthesis of HPG-GO. (Reproduced with permission from Mu et al. 2017)

The planar polymer-graphene oxide nanohybrid was loaded with 5-FU to achieve pH-controlled drug release. In this method, GO sheets and branched poly(ethyleneimine) chains are covalently tethered to tannic acid (Aliabadi et al. 2018). The drug loading capacity was 29%, and in vitro drug release studies at lower pH showed an increased degradation of the carrier and therefore controlled drug release for up to 120 h.

Through steglich esterification, the functionalization of GO with poly(ε-caprolactone) (PCL) is shown in Fig. 8.15 (Weng et al. 2017). Lomefloxacin drug was used in the studies to control the release rate of the drug. However, compared to neat PCL, the GO-PCL blend was able to release the drug at a constant rate and was possible to achieve controlled drug release.

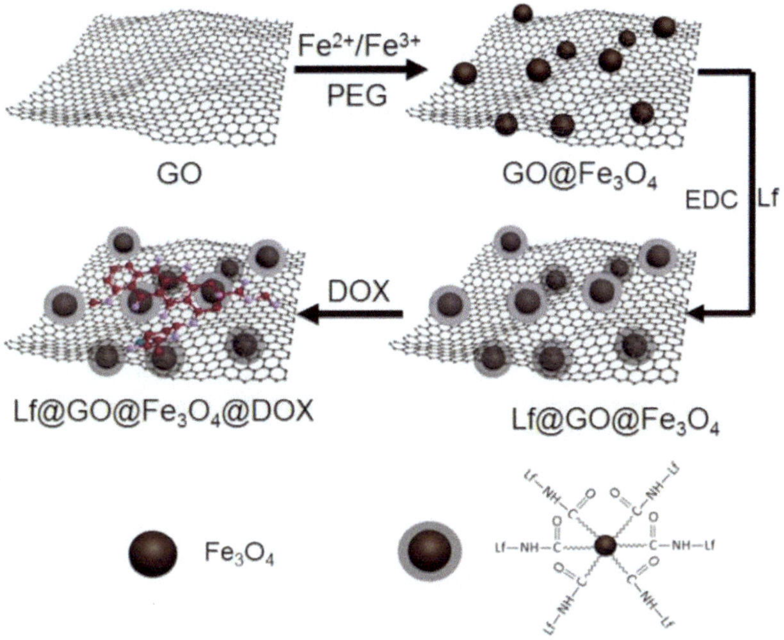

Fig. 8.14 Formation of GO-Fe_3O_4-Lf-GO-Fe_3O_4 and Lf-GO-Fe_3O_4-DOX. (Reproduced with permission from Song et al. 2017)

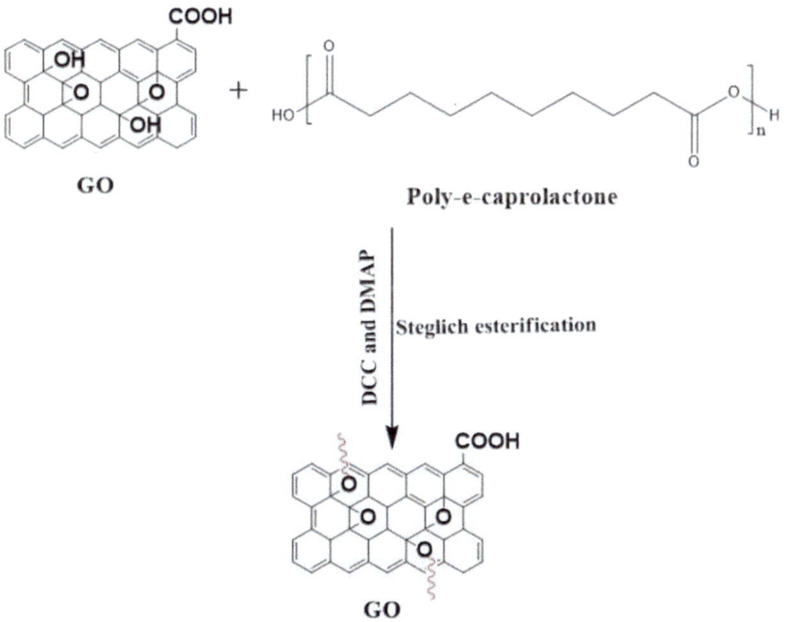

Fig. 8.15 Synthesis of GO-modified poly(ε-caprolactone) (PCL) by steglich esterification. (Reproduced and modified with permission from Weng et al. 2017)

Doping of silver nanoparticles in GO (GO/AgNPs) using glucose as a reducing agent to form a combination of methotrexate MTX-GO/AgNPs has been reported (Thapa et al. 2017). The cell viability and cellular uptake showed targeted and cell-specific uptake. The designed nanodrug carrier improved several properties such as ROS generation, enhanced apoptosis and NIR induced the photothermal effect. Thus, it can be an effective and target-specific drug carrier for cancer treatment. The preparation method for MTX-GO/AgNPs is depicted in Fig. 8.16.

The development of zeolitic imidazolate frameworks and graphene oxide (ZIF)-8/GO nanocrystals was carried out, and it is encapsulated with fluorescein molecules to form a combination of fluorescein-ZIF-8-GO (Tian et al. 2017). This carrier achieved pH-controlled release, and for breast cancer cell line 4T1, it showed less toxicity with easy internalization inside the cancer cells. Fluorescein-ZIF-8-GO showed a distinctive photothermal effect to effectively kill cancer cells. The synthesis of the drug-loaded ZIF-8/GO nanocrystals is shown in Fig. 8.17.

Saeednia et al. reported thermosensitive chitosan-graphene oxide hybrid hydrogels (Saeednia et al. 2017). The loaded drug MTX showed a decrease in drug release and achieved a controlled release due to the presence of GO in chitosan hydrogels. Interestingly, the designed hybrid hydrogels loaded with MTX could inhibit the growth of MCF-7 breast cancer cells. In addition, the designed nanohybrid hydrogels can be delivered through simple injection and can be delivered locally in a sustainable manner.

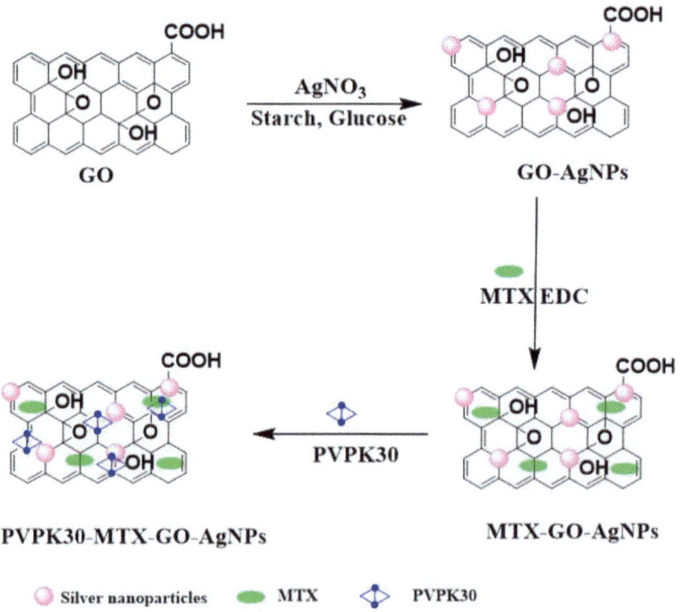

Fig. 8.16 Preparation method for MTX-GO/AgNPs. (Reproduced and modified with permission from Thapa et al. 2017)

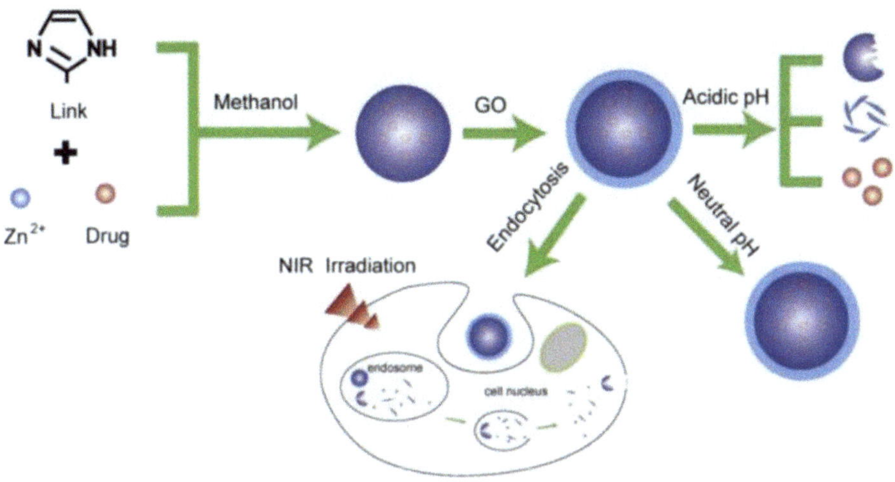

Fig. 8.17 Schematic illustration of the drug-loaded ZIF-8/GO nanocrystal synthesis, synergistic drug delivery, and photothermal therapy. (Reproduced with permission from Ref (Tian et al. 2017))

Gupta et al. tested the combination of both superparamagnetic iron oxides (SPIONs) and reduced GO as a drug carrier for targeted drug delivery applications (Gupta et al. 2018). In this method, the superparamagnetic Fe_3O_4 nanodrug carrier was prepared by one-step co-precipitation using iron salts and GO. DOX was loaded for the different concentrations of nanodrug carriers. At pH 4.3, the acidic microenvironment of cancer cells facilitates the release of drug molecules due to the sensitivity of GO in acidic condition. In contrast, at physiological conditions, i.e., pH 7.4, it reduces the side effects of the chemotherapeutic drug.

Aliabadi and coworkers used SPIONs and functionalized the surface of GO and further coated with polyvinyl alcohol (PVA) (Aliabadi et al. 2017). PVA helps to reduce the circulation time in blood to reach particular tissues. The release of loaded 5-FU drug from PVA-SPIONs-GO nanocomposites depends on the pH of the solutions. However, under acidic conditions, pH 5.8, the drug release rate from the nanocomposites was 91.9%, and at pH 7.4, the drug release rate was decreased to 80.6%.

A novel thermal/pH-sensitive nanogel-based salep modified with GO was designed and achieved with a high drug loading capacity of DOX (Bardajee et al. 2017). The drug was loaded through the interior of nanogels, attaining a loading content of 57%, and the drug loaded through the surface of the designed nanogel exhibited a high loading content of 72%. These systems improved toxicity to HeLa cells while compared to an equivalent dose of the free drug.

GO was pegylated using sodium hydroxide and chloroacetic acid (Deb and Vimala 2018). This converts the hydroxyl group (OH) to carboxyl group (COOH) where the conjugation of acetic acid occurs, producing GO-COOH. The obtained GO-COOH product was activated by adding EDC to form PEG-NH_2 at room temperature. Further, the target ligand folic acid was conjugated with COOH groups. The camptothecin (CPT) drug was loaded to GO-PEG-FA through π-π stacking.

However, the designed carrier showed pH-dependent release and enhanced anticancer activity on MCF-7 breast cancer cell lines.

A flexible nanocomposite hydrogel was designed using GO-quantum dots (QGD) as nanoparticles and carboxymethyl cellulose (CMC) hydrogel as a polymeric matrix (Javanbakht and Namazi 2018). The loaded DOX drug showed pH-sensitive and controlled drug release. The cytotoxicity results showed the DOX-loaded CMC-QGD as a potential high efficient anticancer agent. In addition, CMC-QGD showed nonsignificant toxicity against blood cancer cells (K562).

A novel pH-sensitive drug delivery system was reported by the dispersion of the GO-Ag nanohybrid in a chitosan hydrogel matrix and cross-linking it with sodium tripolyphosphate (STTP) anions (Rasoulzadehzali and Namazi 2018). DOX was used as a drug for the studies, and it showed significant enhancement in the loading capacity. The release time of the drug was increased by the GO-Ag nanoparticles inside the nanocomposite hydrogel beads.

The functionalization of the GO surface with folic acid-polyethylene glycol (FA-PEG) was carried out using the EDS-NHS crosslinking agents. However, the morphology of GO was not altered upon functionalization with FA-PEG (De Sousa et al. 2018). The designed material showed sustained release of the camptothecin drug. The drug release from the nanocarrier was slow in acidic conditions when compared in physiological conditions. Interestingly, the developed system could efficiently internalize by J774 cells, and it intensely adhered to the surface of HepG2 cells. Due to this, it showed tumor cell death by apoptosis.

The chemical structure of drugs, dyes, and pharmaceutical products used in the various studies which are included in this chapter is shown in Fig. 8.18. In another study, GO was coated with cobalt ferrite nanoparticles to form GO-$CoFe_2O_4$ by co-precipitation method (Kooti et al. 2018). The silver nanoparticles were embedded on the surface of GO-$CoFe_2O_4$ to form GO-$CoFe_2O_4$-Ag composites. The developed composites exhibited pH-sensitive drug release in a moderate and controlled manner.

Deb et al. and coworkers used natural polymer chitosan and modified on the surface of GO for codelivery of anticancer drugs (Deb et al. 2018). The two anticancer drugs camptothecin (CPT) and 3,3′-Diindolylmethane (DIM) were used for the drug delivery studies. The difference in the mechanism of action of these two drugs acts as one of the promising candidates that undergo synergistic effect against breast cancer. The chitosan-modified GO was tagged with folic acid, and the two different drugs were loaded which showed delayed and controlled release.

However, few of the studies such as serum analysis, bioavailability and for biodistribution the rats were treated with GO-CS-FA-CPT-DIM when compared with free CPT. The serum analysis studies revealed that the two drugs loaded into the carrier increased anticancer activity and also masked the toxicological effects of free CPT. Further, these data were supported by histological examination. Overall to the nanocomposites the loaded drug showed the effective combinational effect and acts as one of the promising chemotherapeutic agents for breast cancers with improved bioavailability and fewer side effects.

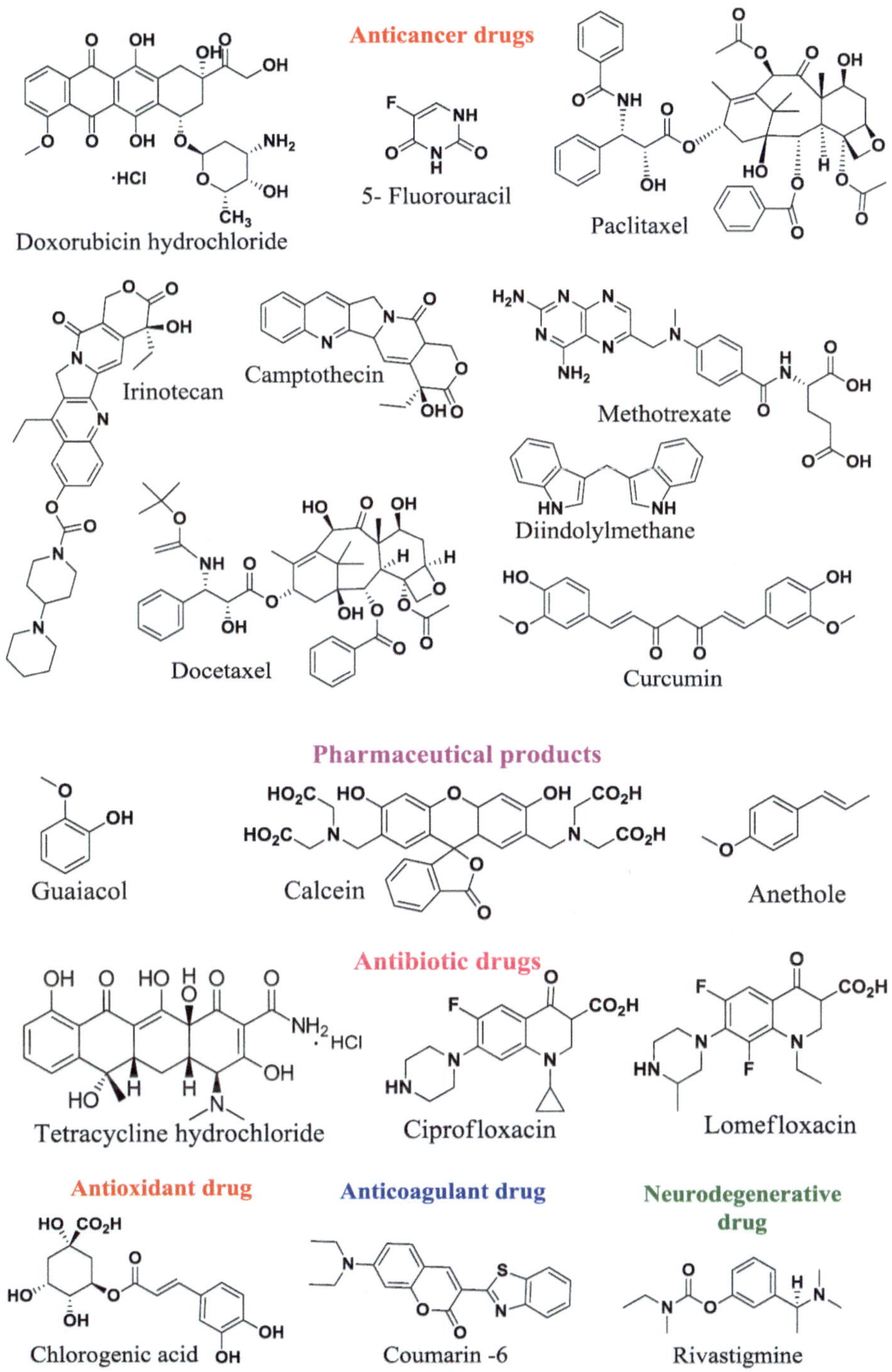

Fig. 8.18 Chemical structure of drugs, dyes, and pharmaceutical products used in the studies

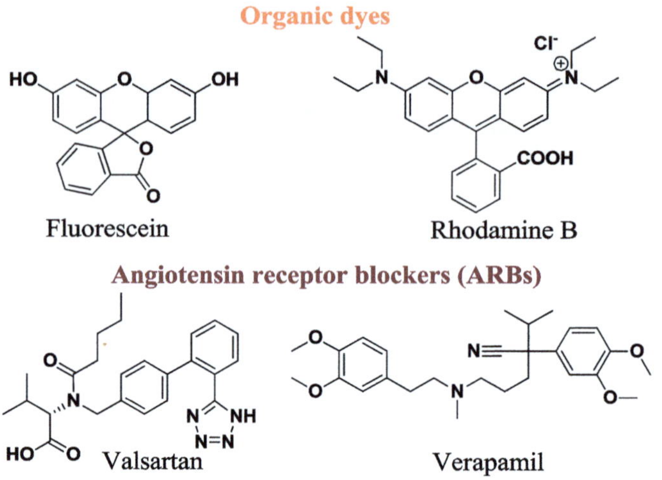

Fig. 18 (continued)

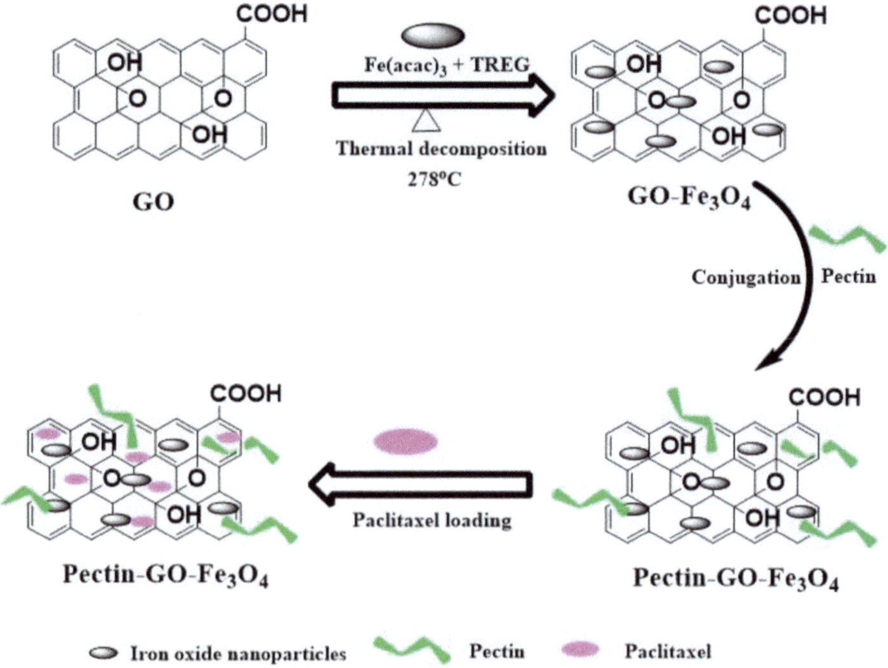

Fig. 8.19 Schematic representation of the synthesis of GO-Fe_3O_4, PEC-GO-Fe_3O_4, and PEC-GO-Fe_3O_4-PAC. (Reproduced and modified with permission from Hussien et al. 2018)

Another study suggested design of pectin-conjugated magnetic GO could be used to deliver paclitaxel drug (Hussien et al. 2018). The material synthesis and drug loading are neatly shown in Fig. 8.19. In this work, GO was conjugated with Fe_3O_4 to form GO-Fe_3O_4 magnetic nanocarrier. Further, the pectin (PEC) natural polymer which has excellent biocompatibility and is nontoxic in nature was conjugated to the magnetic nanocarrier, GO-Fe_3O_4-PEC, by simple mixing. The cytotoxicity studies showed the prepared material has relatively 80% cell viability. The conjugation of PEC enhanced the bioavailability of GO-Fe_3O_4. The loaded drug showed high drug loading and pH-responsive release. Besides, in an endosomal cancer cell medium, the release was more than that in physiological conditions indicating tumor-targeted drug delivery.

8.3 Metal Organic Frameworks (MOFs) and Surface-Modified MOFs as a Drug Carrier

Metal organic frameworks (MOFs) are one of the most exciting architectures in the field of nanotechnology. The MOFs are known to be hybrid self-assemblies of metal ions or clusters (coordination centers) and organic fragments (linkers). They exhibit utmost porosities which made them one of the ideal candidates for drug delivery applications (André and Quaresma 2016). The MOFs reflected as a new class of crystalline nanoporous material family comprising thousands of different structures (Jiang et al. 2017). MOFs possess several advantages over traditional drug carriers. Firstly, their flexible structures provide various surface morphologies, chemical properties, compositions, sizes, etc. Secondly, their high porosities and high surface area enable the MOFs with high drug loading capacity. Thirdly, the most important feature in drug delivery, the weak coordination bonds, indicates MOFs are biodegradable. With these in mind, MOFs are exhibited as high-potential and one of the favorable drug carriers exclusively in drug delivery applications (Wu and Yang 2017).

Park et al. prepared facile nanosized Terbium-MOFs (Tb-MOFs) for drug delivery applications (Park et al. 2016). In this work, Tb-MOF and Pluronic F127 were mixed in water, sonicated, and filtered to obtain Tb-MOF-NPs. The porosity created by MOFs allows to load the DOX drug and deliver to kill cancer cells. However, the facile method created a nanosized, multifunctional biomaterial as a drug carrier for drug delivery applications.

A degradable MOF called zeolitic imidazolate framework-8 (ZIF-8) with uniform size/morphology, well-regulated hollow mesoporous silica nanoparticles (HMSN), and mesoporous blocker as pH-responsive drug delivery system has been designed, as shown in Fig. 8.21 (Zou et al. 2017). A layer of mesoporous silica was coated on ZIF-8 and consequently self-degraded in acidic conditions to obtain HMSN. These MOFs showed two different morphologies such as cubic and dodecahedral. It was noted that the DOX loading was 503.4 μg/mol for HMSN-ZIF-8. The DOX loaded HMSN-ZIF-8 exhibited as efficient drug carrier with pH responsive release which showed enhanced efficiency in killing cancer cells. The synthesis of ZIF-8, MSN coating, and drug loading are shown in Fig. 8.20.

A modified hydrothermal method was used to produce γ-cyclodextrin-MOFs-polyacrylic acid (CD-MOFs-PAA) with a spherical cavity of diameter 1.7 nm and γ-CD with diameter 0.8 nm, which was prepared by an oil-in-oil (s/o/o) emulsifying

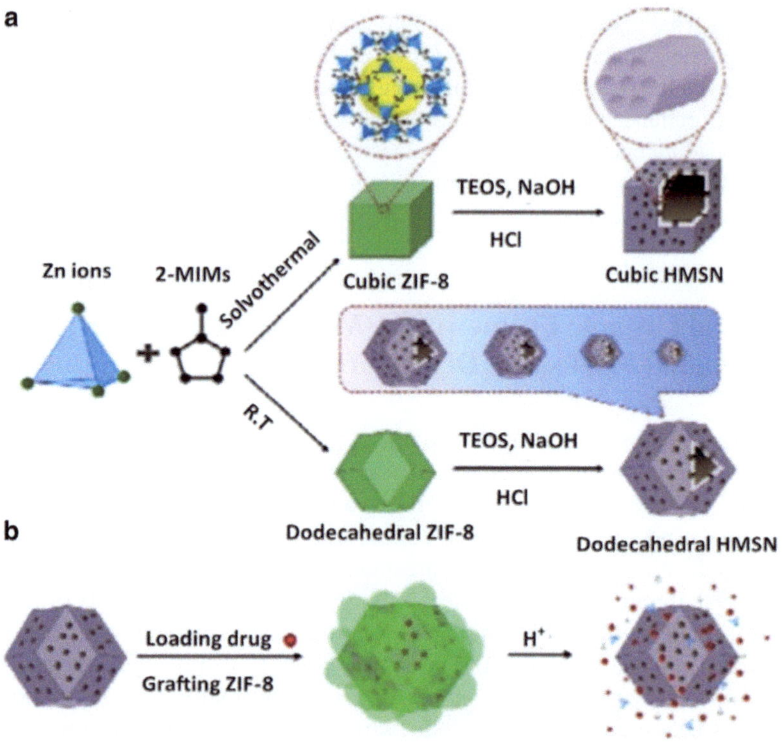

Fig. 8.20 Schematic representation of (**a**) ZIF-8 synthesis and (**b**) MSN coating followed by drug loading for ZIF-8. (Reproduced with permission from Zou et al. 2017)

solvent evaporation method (Li et al. 2017). In this study, insoluble drugs with less stability (such as ibuprofen and lansoprazole) were trapped inside the CD-MOFs. Therefore, for effective drug loading, compared to the impregnation method, the co-crystallization method leads to increase in drug loading, and there was no reduction in CD-MOF during drug loading. However, ibuprofen showed 12% w/w drug loading and 9.4% w/w for the lansoprazole. Interestingly incorporation of CD-MOF inside PAA microspheres leads to control of the drug release in a prolonged way and also helps in the reduction of cell toxicity.

Liu et al. reported MOF-based drug carriers. In this study, two different carriers such as CD-MOF-1 and CD-MOF-2 were reported using β-CD (cyclodextrin) as a building block and cesium (Cs) metal salts as the common backbone (Liu et al. 2017). The templates 1,2,3-triazole-4, 5-dicarboxylic acid (H_3tzdc) and methyl benzene sulfonic acid (TsOH) or ibuprofen (IBU) are used for the preparation of CD-MOF-1 and CD-MOF-2, respectively. The drug 5-Flurouracil showed an adsorption of 1.379 g/g and 1.510 g/g for CD-MOF-1 and CD-MOF-2. However, MTX showed loading contents of 0.689 g/g and 1.217 g/g for CD-MOF-1 and CD-MOF-2, respectively. Interestingly these two CD-MOF-1 and CD-MOF-2 showed good biocompatibility and cytotoxicity assay. Due to the size and shape of the pores, it is possible to achieve high drug loading capacity and also alters the release rate of the anticancer drug molecules.

A MOF (HKUST-1) was synthesized using Cu_3 $(BTC)_2$ and 1,3,5-benzenetricarboxylate (BTC) as a drug carrier and used to release the drug in a controlled manner (Chen et al. 2017). In a typical procedure, Cu $(NO_3)_2{\cdot}3H_2O$ and BTC in 2:1 molar ratio were dissolved in 10 mL of DMF, C_2H_5OH, and H_2O (1:1:1 v/v). The mixture was heated at 80 °C for 24 h which resulted in the formation of a blue product. Further, this product was washed with DMF and C_2H_5OH and dried in air to obtain HKUST-1. Ibuprofen, anethole, and guaiacol are the pharmaceutical products used to adsorb on HKUST-1 with the loading contents of 0.34, 0.38, and 0.40 g/g, respectively. These pharmaceutical products can achieve better controlled drug release properties.

A new class of Zn (11) MOF using $[Zn_6(L)_3(DMA)_4]{\cdot}5DMA$, (H4L [1,1,3,1-terphenyl]-3,3,5,5-tetracarboxylic acid) as a drug carrier was synthesized and used to load 5-FU drug (Aiqing Ma et al. 2017). This material showed 21.1 wt. % of drug loading content and effective drug release in a controlled manner, with 95.8% and 91.6% for around 96 h in acidic and PBS conditions. The cytotoxicity studies showed that there are no specific cytotoxicity issues.

Furthermore, ZIF-8 was encapsulated with aggregated gold nanoclusters (aAuNCs) through a facile in situ method (Fangfang Cao et al. 2013). The purpose of using ZIF-8 is to encapsulate AuNCs, and the AuNCs will aggregate in methanol, forming aAuNCs with improved luminescence property. The prepared ZIF-8-aAuNCs revealed rhombic dodecahedron. However, the designed carrier showed good biocompatibility, and further, this system was explored to load Camptothecin (CPT) drug for controlled release study. The loading efficiency of the drug was 9.4%. The fluorescence of CPT showed strong background emission, thus, the fluorescence selected as a model molecule and the encapsulation efficiency was 12.3%. However, ZIF-8 tends to decompose in acidic solution and is stable in the neutral or alkaline medium. Thus, the release rate at pH 5.0 and 6.0 was faster compared to that at pH 7.4. The delivery of CPT into cancer cells results in growth inhibition and cell death. Subsequently, compared to free CPT, the CPT loaded in ZIF-8-aAuNCs showed higher cytotoxic efficacy results. Thus, it could be one of the promising platforms to transport drug into specific cancer cells.

For the first time, exosome coated on iron-based MOFs known as MIL-88A (composed of iron (111) and fumaric acid) as an efficient carrier for drug delivery was synthesized (Illes et al. 2017). In this study, MIL-88A was synthesized using a microwave technique, loaded with cargo and successively coated with exosomes as represented in Fig. 8.21. In some of the studies, the lipid was coated by the solvent exchange method, which led to a total disassembly of the exosome bilayer. Thus, in order to solve these issues, the exosomes were coated using the fusion method. To study the drug uptake and release, the membrane-impermeable calcein as a model cargo was used and incubated in the exosome-coated MIL-88A, particularly on HeLa cells. There was no release of calcein up to 2 days after incubation. Amazingly, the drug release was observed from the exosome-coated MIL-88A after 3 days of incubation, and, after 4 days, there was an increase in the release. The release of

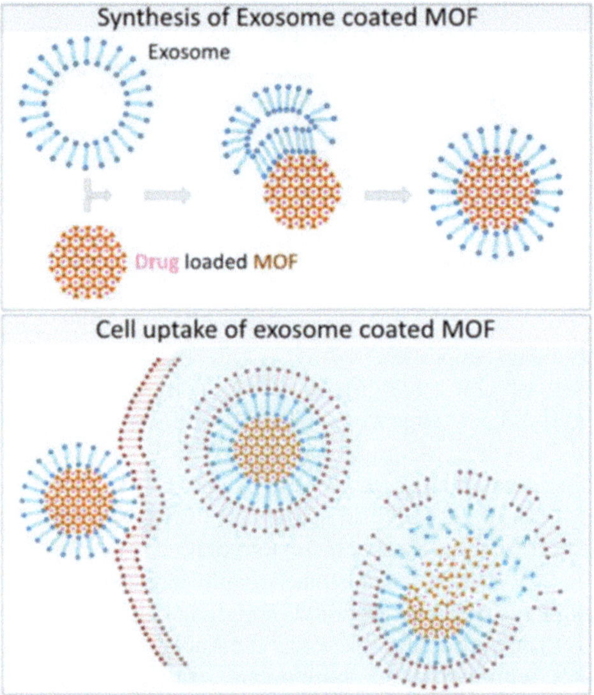

Fig. 8.21 Exosome-coated MOF for cell uptake and designed mechanism for the release of cargo. (Reproduced with permission from Illes et al. 2017)

calcein occurs in the absence of photosensitizers, thus exhibiting as one of the promising drug delivery systems.

In order to solve the common problems in the existing drug carriers such as poor drug loading capacity or premature release from the external surface, Bhattacharjee et al. synthesized Fe_3O_4 -MOF-MIL-100 (Fe) (mixture of trimmers of metal "Fe" octahedral and BTC or trimesic acid) by varying the amount of Fe_3O_4 (Bhattacharjee et al. 2018). This material developed a high surface area which showed 19.4% of loading capacity for DOX drug. Importantly for this material the drug release was up to 20 days without any premature release.

A combination of mixed ligand Cu-MOFs was used to deliver IBU and DOX as shown in Fig. 8.22 (Sun et al. 2017). The MOFs were synthesized by varying mixed ligands BTC (1,3,5-benzenetricarboxylate) and isophthalic acid (IPA) through a hydrothermal method. The variation of the ligand trend is as follows: MOFs-1 (100% IPA), MOFs-2 (40% BTC, 60% IPA), MOFs-3 (70% BTC, 30% IPA), and MOFs-4 (100% BTC, HKUST-1). However, the mixed ligand MOF-1, MOF-2, and MOF-3 exhibited negligible cytotoxicity toward human normal cells (HEK 293A). Noteworthy, the mixed ligand MOFs-2 showed better performance on drug loading and release.

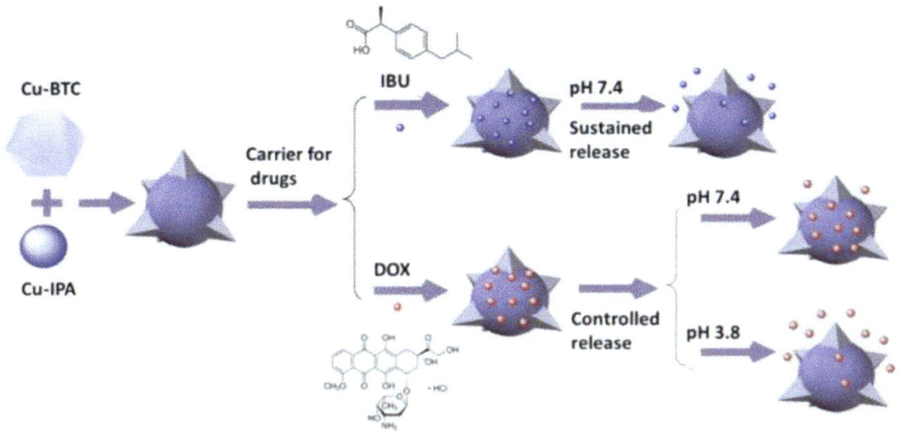

Fig. 8.22 Mixed ligand MOFs and their application as a drug carrier. (Reproduced with permission from Sun et al. 2017)

Mesoporous iron-carboxylate MOF (meso-MOFs) by double template method as a nanocarrier for intratumoral delivery was reported (Wu et al. 2017). In this study, cetyltrimethylammonium bromide (CTAB) and citric acid (CA) were used as a double template. The DOX was loaded and the loading content was up to 55 wt. %. The DOX-loaded meso-MOFs showed better antitumor effect when compared to free DOX. Interestingly, the intratumoral injection of DOX-loaded meso-MOFs exhibited the meso-MOFs could suggestively reduce the systemic toxicity of DOX and maintain an effective DOX concentration for chemotherapy. In addition to this, the designed nanocarrier gave more insight with excellent therapeutic efficacy with reduced side effects for local chemotherapy.

The other study revealed crystalline MOF nanosheets were designed from new bulk MOFs and thereby accumulated by Zn^{2+} metal ions as nodes and IBU as a ligand (Gao et al. 2017). The synthetic surfactant-assisted protocol was followed to control the growth of bulk MOF to obtain MOF nanosheets. However, compared to bulk MOFs, the designed nanosheet MOFs realized better pH-controlled drug release for the ibuprofen drug.

The targeted multidrug delivery approach can be one of the potential strategies to overcome multidrug resistance (MDR). An efficient pH-responsive PEG-FA functionalized multidrug delivery system for targeted drug delivery was reported in order to overcome MDR (Huiyuan Zhang et al. 2017). In this study, two drugs such as DOX and verapamil (VER) were loaded to develop PEG-FA-ZIF-8, which proved with better drug loading efficiency for both the drugs. Strikingly, PEG-FA (DOX + VER)-ZIF-8 exhibited a negligible hemolytic activity and less toxicity to normal organs, indicating their biocompatibility and safe intravenous drug delivery system. In addition, when compared with free DOX and DOX-ZIF-8, PEG-FA/(DOX + VER)-ZIF-8 proved enhanced therapeutic efficiencies. The schematic representation of the synthesis of PEG-FA (DOX + VER)-ZIF-8 is shown in Fig. 8.23.

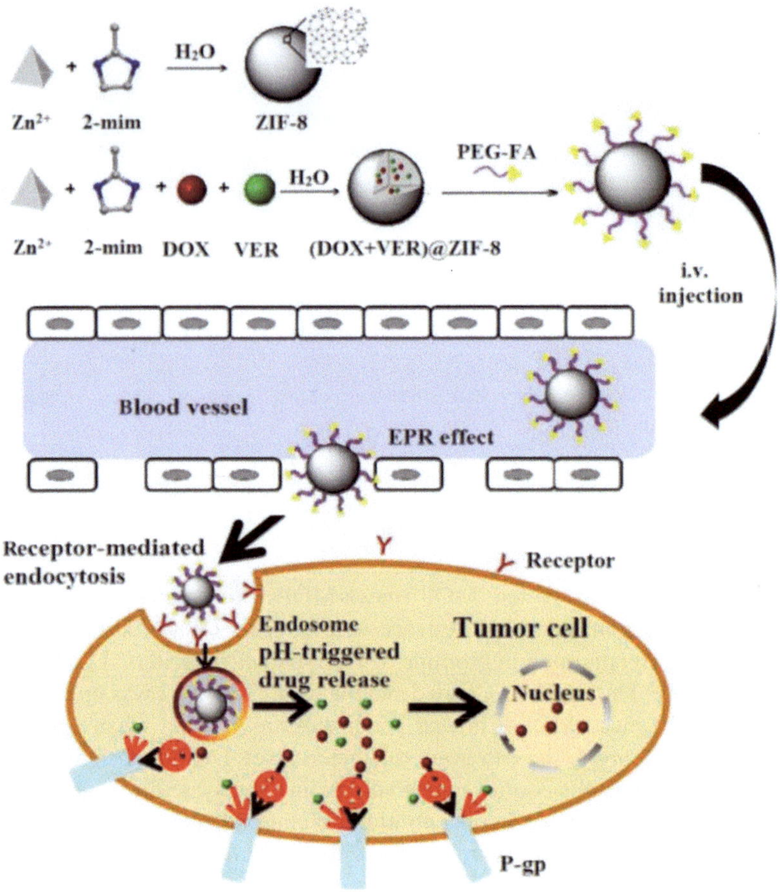

Fig. 8.23 Schematic illustration of pH-responsive ZIF-8 as drug delivery vehicles; the synthesis of PEG-FA/(DOX + VER)-ZIF-8. (Reproduced with permission from Huiyuan Zhang et al. 2017)

Selenium-containing polymer-metal organic framework (P-ZIF-8) as nanocomposites for efficient multiresponsive drug delivery system has been synthesized, as shown in Fig. 8.24 (Zhou et al. 2017). The loaded DOX drug could be selectively released in the presence of external redox agents under the acidic environment. ZIF-8 played an important role in diffusing DOX into ZIF-8 before the release and also helped in a prolonged release for a period of time. The prepared ZIF-8 proved excellent biocompatibility with better drug loading capacity, thereby achieving controlled release of the drug.

ZIF-8-Polyacrylic sodium salt (PAAS) nanocomposites using PAA as a soft template in the range of 20–300 nm size was produced for pH-responsive and improved in vitro/vivo cancer treatment, as shown in Fig. 8.25 (Yan et al. 2017). The developed material exhibited high ultra-drug loading capacity for DOX about 385 wt. %. The release kinetics and pH sensitivity with tremendous tumor inhibition efficacy proved its significant biocompatibility.

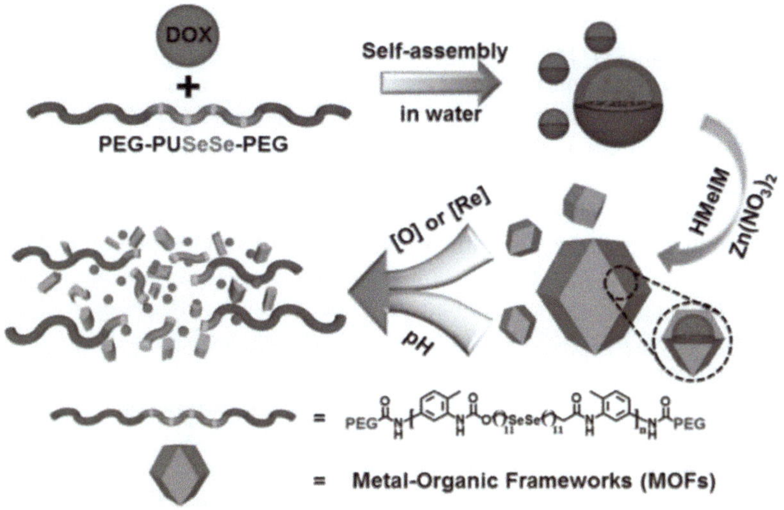

Fig. 8.24 Schematic synthetic procedures of the P-ZIF-8 nanocomposites. (Reproduced with permission from Zhou et al. 2017)

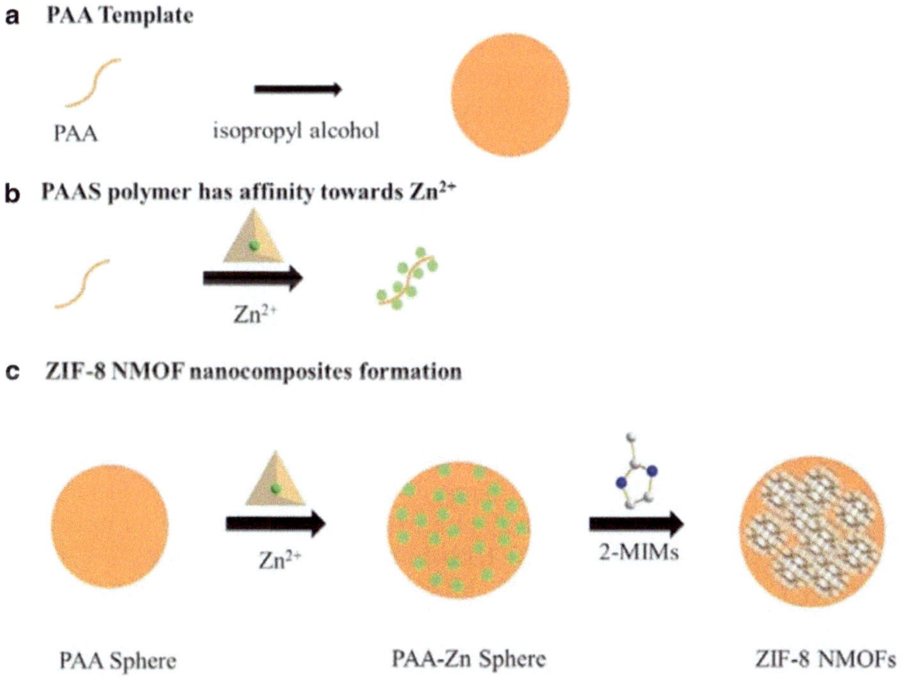

Fig. 8.25 Schematic approach of synthesizing ZIF-8 NMOFs using PAAS as a soft template. (Reproduced with permission from Yan et al. 2017)

Fabrication of ZIF-8 was used to encapsulate the antibiotic drug ciprofloxacin (Nabipour et al. 2017). In this study, the loaded drug proved with a high loading capacity of 21 wt. %, and this acted as a pH-sensitive drug delivery system. The drug release was slow in physiological conditions at pH 7.4. Besides, the drug release was fast in mild acidic phosphate buffer solutions at pH 5.0.

A series of MOFs with hendecahedron cage using [Cu_2 $(COO)_4$] as secondary building unit and H_3L (H_3L = [1,1′;3′,1″]Terphenyl-4,5′,4″-tricarboxylic acid) was synthesized using pyrazine derivatives (Wei et al. 2018). For ibuprofen (IBU), the loading content was 0.33 g/g and 0.37 g/g for guaiacol. Besides the obtained cages, the amino-tethered cages showed controlled release for the IBU drug.

Recently, Zr-MOF-based NU-1000 was used as a carrier for insulin (Yijing Chen et al. 2018). Interestingly, the insulin achieved high loading to the developed carrier. It is noteworthy to mention that in mild conditions, the insulin can diffuse through the MOF framework and easily interact with the pore surface, which leads to rapid encapsulation. Furthermore, the insulin can easily withstand the harsh conditions which mimic the stomach environment. These results suggest NU-1000 can be used as one of the potential insulin carriers for oral delivery.

For instance, the conjugation of hyaluronic acid with MOF as a carrier for targeted drug delivery applications was reported. In this study, the DOX was loaded to ZIF-8, resulting in the formation of DOX-ZIF-8, which was coated with polydopamine and chelated with Fe^{3+} (Shu et al. 2018). This was further conjugated with hyaluronic acid to form a stable nanocarrier. The loading amount of the drug was found to be $8.92 \pm 0.53\%$. The designed system showed sustained and pH-dependent drug release from the designed nanocarrier. Further, DOX-ZIF-8 showed intracellular uptake and enhanced therapeutic efficacy when compared with free DOX.

The MOF DUT-32 was synthesized by ultrasound irradiation using a pyridine modulator (Abazari et al. 2018). Due to the high surface area of DUT-32, there was a favorable DOX release. The release efficiency of 54%–98% after 100 h in pH 7.4–4.5 was achieved. The DOX-loaded DUT-32 was possible to kill cancerous cells similar to free DOX, and DUT-32 showed low cytotoxicity to healthy cells.

A one-pot synthesis of hierarchical-pore Al-MOF (H-MOF) used as a carrier for drug delivery (Xuechuan Gao et al. 2018). In this method, $AlCl_3{\cdot}6H_2O$ and rhodamine B (RhB) are dissolved in water and stirred to form a solution. RhB was added to Al-MOF to form a hierarchical pore structure where it can form weak coordination bonds with Al^{3+}. Consequently, 2-aminoterephthalic acid (NH_2-H_2BDC) was added to the above solution followed by the addition of urea. Importantly, due to hierarchical porous network, excellent loading and release behavior was observed for macromolecular tetracycline hydrochloride (TCH) and small molecule 5-FU drug.

Ibrahim et al. synthesized MOFs by using iron nitrate and 2,6-naphthalenedicarboxylic acid in the presence of DMF (N,N-dimethylformamide) through conventional solvothermal and microwave-assisted technique (Ibrahim et al. 2018). The synthesized carrier has a diameter of 50–80 nm and length of 300–450 nm. However, due to its shape, size, and porosity, it could act as a drug carrier in the near future.

8.4 Diatoms and Surface-Modified Diatoms as a Drug Carrier

As early as 1994, Gordan and Drum anticipated the idea of using diatom frustules in the nanotechnology area (Uthappa et al. 2018a, b). The progress in the field of nanotechnology made the diatoms as admirable candidates in terms of drug delivery. Diatoms are photosynthetic algae covered and composed of the SiO_2 skeleton known as frustule (Monica Terracciano et al. 2017). The diatoms have some remarkable features such as nontoxicity, biocompatibility, chemical stability, and the low cost making them an efficient carrier to replace the existing synthetic silica (Patil et al. 2017). Despite amorphous silica nature, the diatom frustules have been approved by the Food and Drug Administration (FDA) and recognized as a safe material for the food and pharmaceutical products (Uthappa et al. 2018b).

In the year 2010, Losic et al. reported the use of diatoms (*Aulacoseira* sp.) as a drug carrier for the first time (Losic et al. 2010). By the surface modification of diatoms with iron oxide nanoparticles, the drug release profile can be fine-tuned, and also magnetically guided drug delivery systems can be achieved. The magnetized diatoms were synthesized by one-step functionalization through an electrostatically driven self-assembly of dopamine-modified iron oxide nanoparticles (NPs) onto the diatom. Through electrostatic interaction, the cationic magnetic materials were then accumulated onto anionic diatoms, as shown in Fig. 8.26. The model drug indomethacin showed a drug loading capacity of 0.28 ± 0.05 g/g. The drug release from the DOPA/Fe_3O_4/diatom composite in the first phase was about 60%, i.e., initial burst release for 6 h where the drug molecules adhered on the surface, and the remaining 40% of the drug was released in 2 weeks that can be correlated due to the hollow inner pores of diatoms that made the drug molecules to be released in a controlled manner for around 2 weeks.

Kumeria et al. developed a nanohybrid composite made of diatoms and GO by either covalently linking GO nanosheets onto the diatom surface or by electrostatic attachment (Kumeria et al. 2013). The surface modification technique is through covalent attachment of GO onto APTES-modified diatom through NHS-EDC amine coupling resulting in a stable composite. The electrostatic approach is a facile one-step method, where electrostatic interactions occurring between the GO functional groups (COOH) with amine groups (NH_2) on the diatom surface (APTES-DE) can be witnessed. The schematic approach for both the reactions is shown in Fig. 8.27. The indomethacin drug was used to study the release profile from both types of modified diatoms; the drug loading capacity for GO-DE was 28.5 ± 1.1%, which is significantly higher than that of APTES-DE (control), i.e., 15.8 ± 1.8%. The drug release profile showed 14 days of controlled release at pH 3.5 for the control sample. However, for the surface-functionalized GO-DE, the release was controlled up to 37 days, which is a superior evidence for sustained release of drug from GO-modified diatoms. On the other hand, enhancement in drug loading was achieved due to the presence of GO sheets on the surface of diatoms. The GO-modified diatoms showed an enhanced interaction between GO sheets and drug molecules through

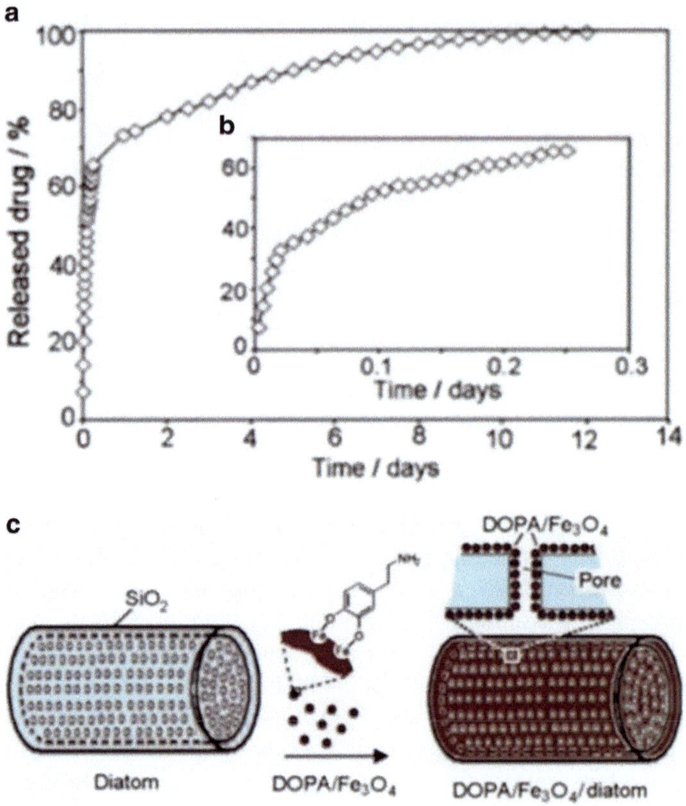

Fig. 8.26 Indomethacin drug release from DOPA/iron oxide: (**a**) 12 days, (**b**) 6 h. (**c**) Schematic approach of diatom-modified iron oxide nanoparticles. (Reproduced with permission from Losic et al. 2010)

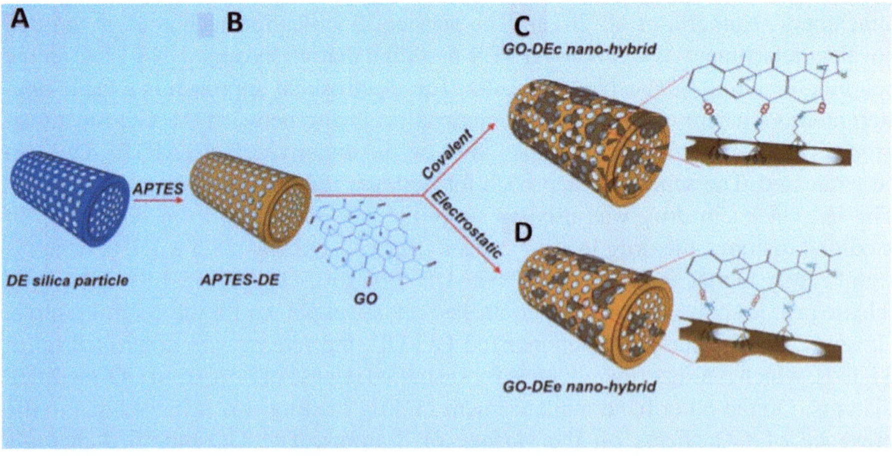

Fig. 8.27 GO-DE nanohybrid: (**a**) DE, (**b**) (3-aminopropyl)triethoxysilane (APTES), (**c**) covalent, and (**d**) electrostatic attachment. (Reproduced with permission from Kumeria et al. 2013)

hydrogen bonding due to the presence of more hydroxyl groups, π-π bonding, and hydrophobic interactions. Besides, for APTES-DE, the drug molecules can interact with only weak van der Waals forces which leads to less drug loading capacity.

For the first time, Rea et al. used porous diatomite nanocarriers for deliverings-mall interfering ribonucleic acid (siRNA) inside the human epidermoid cancer cells (H1355) (Rea et al. 2014). The diatomite nanoparticles (NPs) of diameter < 450 nm were bioconjugated with siRNA for silencing genes linked with a variety of pathogenic conditions. However, the other NPs like gold and liposomes were reported to be not effective in crossing the cell membrane. The diatomite NPs of concentration 300 μg/mL were exposed to cells for around 72 h, and in vitro studies displayed very low cytotoxicity. Further, by using confocal microscopy, localization of diatomite NPs in the cytoplasm confirmed the potent uptake of NPs into H1355 cancer cells. Also, diatomite NP bioconjugates with siRNA have also been used for gene silencing, thus ensuring the downregulation of targeted proteins in cancer cells.

Ruggiero et al. developed surface functionalization of diatomite NPs for drug delivery applications (Immacolata Ruggiero et al. 2014). The purified diatomite NPs were functionalized with APTES and labeled with fluroscent probe, tetramethylrhodamine isothiocyanate. To examine the fluorescent probe, tetramethylrhodamine isothiocyanate was tagged, and to inspect the NP internalization in the lung epidermoid cancer cells (H1355), confocal microscopy was used, which showed effective cellular uptake. This may find applications for targeted drug delivery of anticancer molecules, which might advance delivery of antitumor molecules.

Todd et al. fabricated iron oxide NP-encapsulated diatoms for magnetic controlled transport of small molecules to tumors (Todd et al. 2014). In this study, dye was used to mimic drug molecules and loaded into conjugated diatom-iron-oxide NPs. The advantages of using diatoms with 10 μm in size facilitated to hold more amount of functionalities via physical adsorption. Fluorescence and magnetic resonance imaging were used to examine the movement of tumors in an animal model. Interestingly, diatoms were directed toward tumors due to magnetic properties, and the designed system showed enhancement in tumor retention. The iron oxide and diatoms rendered lesser toxicity and biodegradability, and they have much greater potential in targeted drug delivery applications. The general scheme is depicted in Fig. 8.28.

The diatomite was surface-bioengineered as a nanovector for intracellular uptake and drug delivery (Terracciano et al. 2015). The diatom nanoparticles (DNPs) were functionalized with APTES. Through EDC/NHS chemistry, the carboxyl group of polyethylene glycol (PEG) was reacted with DNP-APTES. Further, it was covalently tethered to the carboxyl group of a cell-penetrating peptide (CPP) by EDC/NHS chemistry. Later poorly water-soluble anticancer drug sorafenib was used in the studies. However, the drug loading capacity followed the trend of DNPs-APTES 10.4 ± 1.1%, DNPs-APTES-PEG 22 ± 2%, and 17 ± 2% in the case of DNPs-APTES-PEG-CPP. It is worthwhile to mention that these biofunctionalized nanovectors have greater intracellular localization in cancer cells; the schematic is shown in Fig. 8.29.

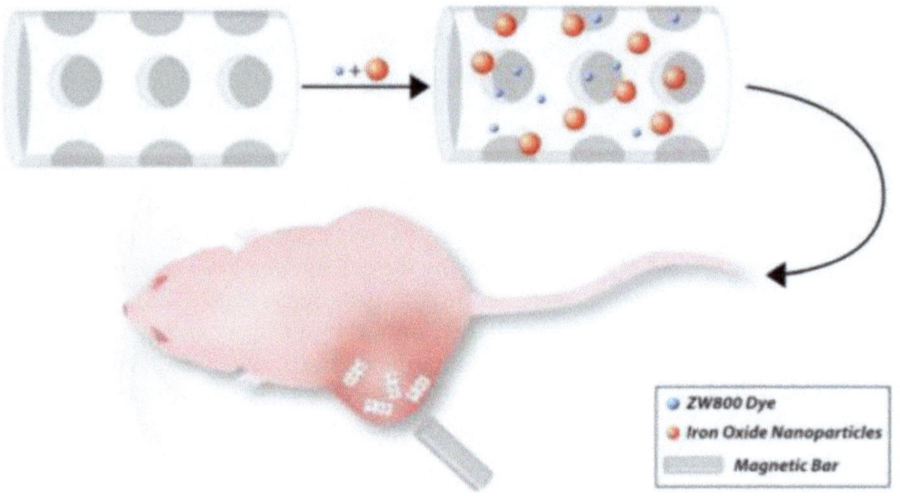

Fig. 8.28 Iron oxide nanoparticle-loaded diatoms for magnetic drug delivery. Iron oxide nanoparticles and small particle drugs loaded into diatom frustules; an external magnetic field was applied, and these diatoms are attracted toward tumor after tail vein injection. (Reproduced with permission from Uthappa et al. 2018a, b)

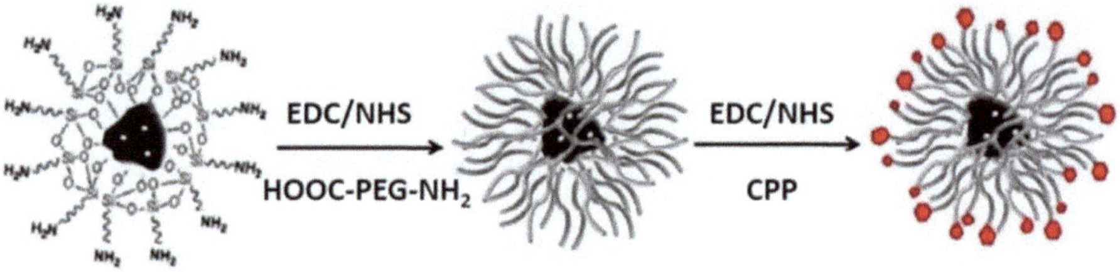

Fig. 8.29 PEGylation of DNP-APTES through EDS/NHS chemistry and bioconjugation of CPP-peptide to DNP-APTES. (Reproduced with permission from Uthappa et al. 2018a, b)

Javalkote et al. (2015) used *Nitzschia* sp. of diatoms for better chemotherapy using curcumin for magnetically guided delivery (Javalkote et al. 2015). In this method, the diatom surface was treated with iron oxide NPs by two different procedures such as curcumin magnetic diatom microparticles ferrofluid (CMDM-F) method and curcumin magnetic diatom microparticle in situ method (CDM-I). The drug loading efficiency was 14.7% in the case of curcumin-loaded diatom microparticles (CDM). However, the loading pattern changed to 8.2% for curcumin magnetic diatom microparticles (CMDM). The loading efficiency was 9.1% for CMDM-F and CDM-I. Similarly, cell viability studies exhibited the cytotoxicity was more about 60.2% for CMDM-F and 59.9% in case of CDM-I. In comparison with the bare drug, which is 42.1%, the bare diatom carrier and curcumin-loaded diatom microparticles showed cytotoxicity of 1.9 and 44.8%, respectively. It was noted that a high cell viability was observed for CMDM-F, which could be the novel route for magnetic hyperthermia applications.

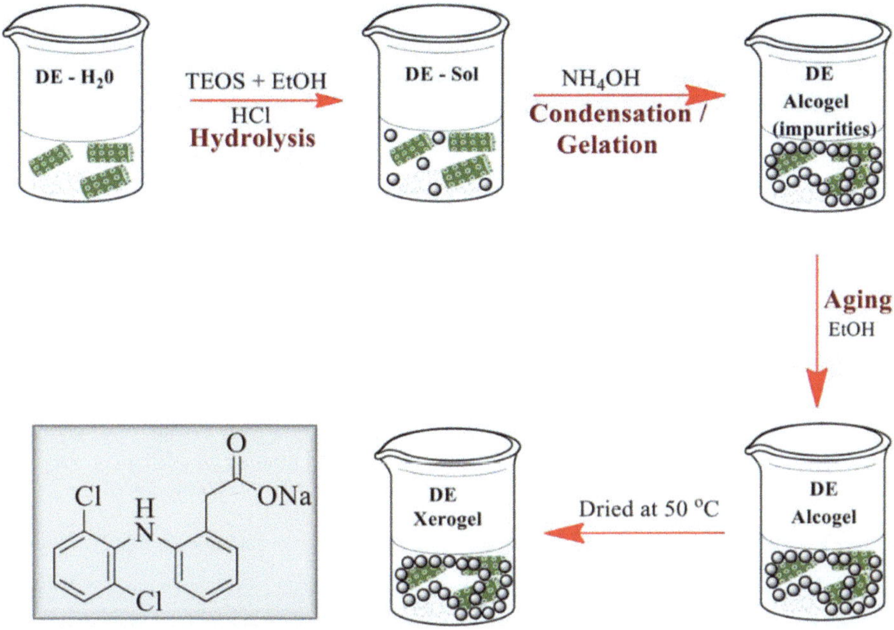

Fig. 8.30 Schematic representation of DE-XER formation: Inset: Chemical structure of the drug. (Reproduced with permission from Uthappa et al. 2018a, b)

Recently a surface-modified naturally available DE (*Aulacoseira* sp.) with high surface area material xerogel particles as a drug carrier was reported (Uthappa et al. 2018a). An anti-inflammatory drug diclofenac sodium (DS) was used in this study. For neat DE, the drug loading capacity was 38.8 ± 1.3%. However, for the surface-modified material DE-XER, the loading capacity was 46.2 ± 1.6%. When compared with the control material, neat DE was able to control the drug release only up to 16 days. Strikingly, the drug release was controlled up to 36 days for DE-XER matrices, suggesting the importance of surface modification. The xerogel formation on the surface of diatoms and drug used in the studies is shown in Fig. 8.30.

8.5 Conclusion

In this chapter, various surface functionalization strategies on the current existing materials such as graphene oxides, metal organic frameworks, and naturally available diatoms have been focused. The use of nanotechnology in developing various nanocarriers for drug delivery applications offers vast potential values. Therefore, surface modification of the existing materials using nanotechnology approach results in the formation of stable hybrid materials or composites. Thus, by using such kind of stable materials, it is possible to tune the surface properties of various materials. Moreover, enhanced drug loading capacity, efficiency, better drug release profiles, targeted delivery, improved biocompatibility, cellular uptake, and cytotoxicity can be achieved. Thus, the designed nanocarriers for drug delivery applications might grow to greater heights in the near future.

Acknowledgment The authors acknowledge the financial support from DST Nanomission, India (SR/NM/NS-20/2014); DST, India (DST-TM-WTI-2K14-213); and SERB-DST, India (YSS/2015/000013) for financial support. They also thank Jain University, India, for providing facilities.

References

Abazari R, Reza Mahjoub A, Slawin AMZ, Carpenter-Warren CL (2018) Morphology- and size-controlled synthesis of a metal-organic framework under ultrasound irradiation: an efficient carrier for pH responsive release of anti-cancer drugs and their applicability for adsorption of amoxicillin from aqueous solution. Ultrason Chem 42:594–608

Aiqing Ma ZL, Chuying G, Li B, Liu J (2017) Cytotoxicity of a metal–organic framework drug delivery. Inorg Chem Commun 77:68–71

Aliabadi M, Shagholani H, Yunessnia Lehi A (2017) Synthesis of a novel biocompatible nanocomposite of graphene oxide and magnetic nanoparticles for drug delivery. Int J Biol Macromol 98:287–291

Aliabadi M, Yunessnia lehi A, Shagholani H, Gerayeli A (2018) Planar polymer-graphene oxide nanohybrid as a 5-fluorouacil carrier in pH-responsive controlled release. J Drug Delivery Sci Technol 43:103–106

Alibolandi M, Mohammadi M, Taghdisi SM, Ramezani M, Abnous K (2017) Fabrication of aptamer decorated dextran coated nano-graphene oxide for targeted drug delivery. Carbohydr Polym 155:218–229

André V, Quaresma S (2016) Bio-inspired metal-organic frameworks in the pharmaceutical world: a brief review. https://doi.org/10.5772/64027

Barahuie F, Saifullah B, Dorniani D, Fakurazi S, Karthivashan G, Hussein MZ, Elfghi FM (2017) Graphene oxide as a nanocarrier for controlled release and targeted delivery of an anticancer active agent, chlorogenic acid. Mater Sci Eng C 74:177–185

Bardajee GR, Hooshyar Z, Farsi M, Mobini A, Sang G (2017) Synthesis of a novel thermo/pH sensitive nanogel based on salep modified graphene oxide for drug release. Mater Sci Eng C 72:558–565

Bhattacharjee A, Gumma S, Purkait MK (2018) $Fe_3 O_4$ promoted metal organic framework MIL-100(Fe) for the controlled release of doxorubicin hydrochloride. Microporous Mesoporous Mater 259:203–210

Chen Q, Chen Q-W, Zhuang C, Tang P-P, Lin N, Wei L-Q (2017) Controlled release of drug molecules in metal–organic framework material HKUST-1. Inorg Chem Commun 79:78–81

de Sousa M, Visani de Luna LA, Fonseca LC, Giorgio S, Alves OL (2018) Folic-acid-functionalized graphene oxide nanocarrier: synthetic approaches, characterization, drug delivery study, and antitumor screening. ACS Appl Nano Mater 1:922–932

Deb A, Vimala R (2018) Camptothecin loaded graphene oxide nanoparticle functionalized with polyethylene glycol and folic acid for anticancer drug delivery. J Drug Delivery Sci Technol 43:333–342

Deb A, Andrews NG, Raghavan V (2018) Natural polymer functionalized graphene oxide for co-delivery of anticancer drugs: in-vitro and in-vivo. Int J Biol Macromol 113:515–525

Fan L, Ge H, Zou S, Xiao Y, Wen H, Li Y, Feng H, Nie M (2016) Sodium alginate conjugated graphene oxide as a new carrier for drug delivery system. Int J Biol Macromol 93:582–590

Fangfang Cao EJ, Liu C, Li W, Zhang Y, Dong K, Liu Z, Ren J, Qu X (2013) Encapsulation of aggregated gold nanoclusters into metal–organic frameworks for real-time monitoring of drug release. Nanoscale 9:4128–4134

Gao X, Cui R, Zhang M, Liu Z (2017) Metal-organic framework nanosheets that exhibit pH-controlled drug release. Mater Lett 197:217–220

Gupta J, Prakash A, Jaiswal MK, Agarrwal A, Bahadur D (2018) Superparamagnetic iron oxide-reduced graphene oxide nanohybrid-a vehicle for targeted drug delivery and hyperthermia treatment of cancer. J Magn Magn Mater 448:332–338

Hemmati K, Sahraei R, Ghaemy M (2016) Synthesis and characterization of a novel magnetic molecularly imprinted polymer with incorporated graphene oxide for drug delivery. Polym J 101:257–268

Huang Y-S, Lu Y-J, Chen J-P (2017) Magnetic graphene oxide as a carrier for targeted delivery of chemotherapy drugs in cancer therapy. J Magn Magn Mater 427:34–40

Huiyuan Zhang WJ, Liu R, Zhang J, Zhang D, Li Z, Luan Y (2017) Rational design of MOF nanocarrier-based co-delivery system of doxorubicin hydrochloride/verapamil hydrochloride for overcoming multidrug resistance with efficient targeted cancer therapy. ACS Appl Mater Interfaces 9:19687–19697

Hussien NA, Isiklan N, Turk M (2018) Pectin-conjugated magnetic graphene oxide nanohybrid as a novel drug carrier for paclitaxel delivery. Artif Cells Nanomed Biotechnol 46:1–10

Ibrahim M, Sabouni R, Husseini GA (2018) Synthesis of metal-organic framework from iron nitrate and 2,6-naphthalenedicarboxylic acid and its application as drug carrier. J Nanosci Nanotechnol 18:5266–5273

Illes B, Hirschle P, Barnert S, Cauda V, Wuttke S, Engelke H (2017) Exosome-coated metal–organic framework nanoparticles: an efficient drug delivery platform. Chem Mater 29:8042–8046

Immacolata Ruggiero MT, Martucc NM, De Stefano L, Migliaccio N, Rosarita Tatè IR, Arcari P, Lamberti A, Rea I (2014) Diatomite silica nanoparticles for drug delivery. Nanoscale Res Lett 9:1–7

Javalkote VS, Pandey AP, Puranik PR, Deshmukh PK (2015) Magnetically responsive siliceous frustules for efficient chemotherapy. Mater Sci Eng C 50:107–116

Javanbakht S, Namazi H (2018) Doxorubicin loaded carboxymethyl cellulose/graphene quantum dot nanocomposite hydrogel films as a potential anticancer drug delivery system. Mater Sci Eng C 87:50–59

Jiang K, Zhang L, Hu Q, Zhang Q, Lin W, Cui Y, Yang Y, Qian G (2017) Thermal stimuli-triggered drug release from a biocompatible porous metal-organic framework. Chem Eur J 23:10215–10221

Jianping Wu KW, Peng Y (2018) Advances in synthesis and application of nanometer drug carriers. Characterization Appl Nanomater 1:12–18

Jung S, Han U, Hong J (2017) Chemical and physical modification of layer-by-layer assembled nanofilms composed of block copolymer micelles and graphene oxide for controlled drug release. J Ind Eng Chem 56:413–421

Khatamian M, Divband B, Farahmand-Zahed F (2016) Synthesis and characterization of Zinc (II)-loaded Zeolite/Graphene oxide nanocomposite as a new drug carrier. Mater Sci Eng C 66:251–258

Kooti M, Sedeh AN, Motamedi H, Rezatofighi SE (2018) Magnetic graphene oxide inlaid with silver nanoparticles as antibacterial and drug delivery composite. Appl Microbiol Biotechnol 102:3607–3621

Kumeria T, Bariana M, Altalhi T, Kurkuri M, Gibson CT, Yang W, Losic D (2013) Graphene oxide decorated diatom silica particles as new nano-hybrids: towards smart natural drug microcarriers. J Mater Chem B 1:6302–6311

Lei H, Xie M, Zhao Y, Zhang F, Xu Y, Xie J (2016) Chitosan/sodium alginate modificated graphene oxide-based nanocomposite as a carrier for drug delivery. Ceram Int 42:17798–17805

Li H, Lv N, Li X, Liu B, Feng J, Ren X, Guo T, Chen D, Fraser Stoddart J, Gref R, Zhang J (2017) Composite CD-MOF nanocrystals-containing microspheres for sustained drug delivery. Nanoscale 9:7454–7463

Liu J, Bao TY, Yang XY, Zhu PP, Wu LH, Sha JQ, Zhang L, Dong LZ, Cao XL, Lan YQ (2017) Controllable porosity conversion of metal-organic frameworks composed of natural ingredients for drug delivery. Chem Comm 53:7804–7807

Losic D, Yu Y, Aw MS, Simovic S, Thierry B, Addai-Mensah J (2010) Surface functionalisation of diatoms with dopamine modified iron-oxide nanoparticles: toward magnetically guided drug microcarriers with biologically derived morphologies. Chem Comm 46:6323–6325

Lv Y, Tao L, Annie Bligh SW, Yang H, Pan Q, Zhu L (2016) Targeted delivery and controlled release of doxorubicin into cancer cells using a multifunctional graphene oxide. Mater Sci Eng C 59:652–660

Ma N, Liu J, He W, Li Z, Luan Y, Song Y, Garg S (2017) Folic acid-grafted bovine serum albumin decorated graphene oxide: an efficient drug carrier for targeted cancer therapy. J Colloid Interface Sci 490:598–607

Mahkam M, Rafi AA, Faraji L, Zakerzadeh E (2014) Preparation of poly (methacrylic acid)–graphene oxide nanocomposite as a pH-sensitive drug carrier through in-situ copolymerization of methacrylic acid with polymerizable graphene. Polym Plast Technol Eng 54:916–922

Maity AR, Chakraborty A, Mondal A, Jana NR (2014) Carbohydrate coated folate functionalized colloidal graphene as a nanocarrier for both hydrophobic and hydrophilic drugs. Nanoscale 6:2752–2758

Masoudipour E, Kashanian S, Maleki N (2017) A targeted drug delivery system based on dopamine functionalized nano graphene oxide. Chem Phys Lett 668:56–63

Monica Terracciano IR, De Stefanoa LC, Santos H (2017) Diatoms: a natural source of nanostructured silica for drug delivery, https://doi.org/10.1039/9781788010160-00201

Mu S, Li G, Liang Y, Wu T, Ma D (2017) Hyperbranched polyglycerol-modified graphene oxide as an efficient drug carrier with good biocompatibility. Mater Sci Eng C 78:639–646

Mukherjee B, Dey NS, Maji R, Bhowmik P, Das PJ, Paul P (2014) Current status and future scope for nanomaterial in drug delivery, pp 526–544. https://doi.org/10.5772/58450

Nabipour H, Sadr MH, Bardajee GR (2017) Synthesis and characterization of nanoscale zeolitic imidazolate frameworks with ciprofloxacin and their applications as antimicrobial agents. New J Chem 41:7364–7370

Nasrollahi F, Varshosaz J, Khodadadi AA, Lim S, Jahanian-Najafabadi A (2016) Targeted delivery of docetaxel by use of transferrin/poly(allylamine hydrochloride)-functionalized graphene oxide nanocarrier. ACS Appl Mater Interfaces 8:13282–13293

Pan Q, Lv Y, Williams GR, Tao L, Yang H, Li H, Zhu L (2016) Lactobionic acid and carboxymethyl chitosan functionalized graphene oxide nanocomposites as targeted anticancer drug delivery systems. Carbohydr Polym 151:812–820

Park KM, Kim H, Murray J, Koo J, Kim K (2016) A facile preparation method for nanosized MOFs as a multifunctional material for cellular imaging and drug delivery. Supramol Chem 29:441–445

Patil P, Bhat MP, Gatti MG, Kabiri S, Altalhi T, Jung H-Y, Losic D, Kurkuri M (2017) Chemodosimeter functionalized diatomaceous earth particles for visual detection and removal of trace mercury ions from water. Chem Eng J 327:725–733

Pereira de Sousa I, Buttenhauser K, Suchaoin W, Partenhauser A, Perrone M, Matuszczak B, Bernkop-Schnurch A (2016) Thiolated graphene oxide as promising mucoadhesive carrier for hydrophobic drugs. Int J Pharm 509:360–367

Prasad R, Pandey R, Varma A, Barman I (2017) Polymer based nanoparticles for drug delivery systems and cancer therapeutics. In: Natural Polymers for Drug Delivery (eds. Kharkwal H and Janaswamy S), CAB International, UK 53–70

Rao Z, Ge H, Liu L, Zhu C, Min L, Liu M, Fan L, Li D (2018) Carboxymethyl cellulose modified graphene oxide as pH-sensitive drug delivery system. Int J Biol Macromol 107:1184–1192

Rasoulzadeh M, Namazi H (2017) Carboxymethyl cellulose/graphene oxide bio-nanocomposite hydrogel beads as anticancer drug carrier agent. Carbohydr Polym 168:320–326

Rasoulzadehzali M, Namazi H (2018) Facile preparation of antibacterial chitosan/graphene oxide-Ag bio-nanocomposite hydrogel beads for controlled release of doxorubicin. Int J Biol Macromol 116:54–63

Rea I, Martucci NM, De Stefano L, Ruggiero I, Terracciano M, Dardano P, Migliaccio N, Arcari P, Tate R, Rendina I, Lamberti A (2014) Diatomite biosilica nanocarriers for siRNA transport inside cancer cells. Biochim Biophys Acta 1840(12):3393–3403

Saeednia L, Yao L, Berndt M, Cluff K, Asmatulu R (2017) Structural and biological properties of thermosensitive chitosan-graphene hybrid hydrogels for sustained drug delivery applications. J Biomed Mater Res A 105:2381–2390

Samadi S, Moradkhani M, Beheshti H, Irani M, Aliabadi M (2018) Fabrication of chitosan/poly(lactic acid)/graphene oxide/TiO2 composite nanofibrous scaffolds for sustained delivery of doxorubicin and treatment of lung cancer. Int J Biol Macromol 110:416–424

Shu F, Lv D, Song X-L, Huang B, Wang C, Yu Y, Zhao S-C (2018) Fabrication of a hyaluronic acid conjugated metal organic framework for targeted drug delivery and magnetic resonance imaging. RSC Adv 8:6581–6589

Singh J, Garg T, Rath G, Goyal AK (2015) Advances in nanotechnology-based carrier systems for targeted delivery of bioactive drug molecules with special emphasis on immunotherapy in drug resistant tuberculosis – a critical review. Drug Deliv 23:1676–1698

Song MM, Xu HL, Liang JX, Xiang HH, Liu R, Shen YX (2017) Lactoferrin modified graphene oxide iron oxide nanocomposite for glioma-targeted drug delivery. Mater Sci Eng C 77:904–911

Sun K, Li L, Yu X, Liu L, Meng Q, Wang F, Zhang R (2017) Functionalization of mixed ligand metal-organic frameworks as the transport vehicles for drugs. J Colloid Interface Sci 486:128–135

Tan J, Meng N, Fan Y, Su Y, Zhang M, Xiao Y, Zhou N (2016) Hydroxypropyl-beta-cyclodextrin-graphene oxide conjugates: carriers for anti-cancer drugs. Mater Sci Eng C 61:681–687

Terracciano M, Shahbazi MA, Correia A, Rea I, Lamberti A, De Stefano L, Santos HA (2015) Surface bioengineering of diatomite based nanovectors for efficient intracellular uptake and drug delivery. Nanoscale 7:20063–20074

Thapa RK, Kim JH, Jeong JH, Shin BS, Choi HG, Yong CS, Kim JO (2017) Silver nanoparticle-embedded graphene oxide-methotrexate for targeted cancer treatment. J Colloid Interface Sci 153:95–103

Tian Z, Yao X, Zhu Y (2017) Simple synthesis of multifunctional zeolitic imidazolate frameworks-8/graphene oxide nanocrystals with controlled drug release and photothermal effect. Microporous Mesoporous Mater 237:160–167

Todd T, Zhen Z, Tang W, Chen H, Wang G, Chuang YJ, Deaton K, Pan Z, Xie J (2014) Iron oxide nanoparticle encapsulated diatoms for magnetic delivery of small molecules to tumors. Nanoscale 6:2073–2076

Uthappa UT, Brahmkhatri V, Sriram G, Jung HY, Yu J, Kurkuri N, Aminabhavi TM, Altalhi T, Neelgund GM, Kurkuri MD (2018a) Nature engineered diatom biosilica as drug delivery systems. J Control Release 281:70–83

Uthappa UT, Sriram G, Brahmkhatri V, Kigga M, Jung H-Y, Altalhi T, Neelgund GM, Kurkuri MD (2018b) Xerogel modified diatomaceous earth microparticles for controlled drug release studies. New J Chem 42:11964–11971

Wang J, Fang J, Fang P, Li X, Wu S, Zhang W, Li S (2017) Preparation of hollow core/shell Fe_3O_4@graphene oxide composites as magnetic targeting drug nanocarriers. J Biomater Sci Polym Ed 28:337–349

Wei L-Q, Li Y, Mao L-Y, Chen Q, Lin N (2018) A series of porous metal–organic frameworks with hendecahedron cage: structural variation and drug slow release properties. J Solid State Chem 257:58–63

Wen Jiang BYK, Rutka JT, Chan WCW (2007) Advances and challenges of nanotechnology-based drug delivery systems. Expert Opin Drug Deliv 4:621–633

Weng F, Yin J, Bao F, Gao J, Ma R, Yan S, Liu Y, Ding H (2017) Preparation and the controlled release effect study of graphene oxide-modified poly(ε-caprolactone). Int J Polym Mater 67:307–312

Wu MX, Yang YW (2017) Metal-organic framework (MOF)-based drug/cargo delivery and cancer therapy. Adv Mater 29:1–20

Wu Z, Hao N, Zhang H, Guo Z, Liu R, He B, Li S (2017) Mesoporous iron-carboxylate metal-organic frameworks synthesised by the double-template method as a nanocarrier platform for intratumoral drug delivery. Biomater Sci 5:1032–1040

Xie M, Lei H, Zhang Y, Xu Y, Shen S, Ge Y, Li H, Xie J (2016) Non-covalent modification of graphene oxide nanocomposites with chitosan/dextran and its application in drug delivery. RSC Adv 6:9328–9337

Xuechuan Gao YW, Ji G, Cuia R, Liu Z (2018) One-pot synthesis of hierarchical-pore metal–organic frameworks for drug delivery and fluorescent imaging. CrystEngComm 20:1087–1093

Yan L, Chen X, Wang Z, Zhang X, Zhu X, Zhou M, Chen W, Huang L, Roy VAL, Yu PKN, Zhu G, Zhang W (2017) Size controllable and surface tunable zeolitic imidazolate framework-8-poly(acrylic acid sodium salt) nanocomposites for pH responsive drug release and enhanced in vivo cancer treatment. ACS Appl Mater Interfaces 9:32990–33000

Yao C, Zhu J, Xie A, Shen Y, Li H, Zheng B, Wei Y (2017) Graphene oxide and creatine phosphate disodium dual template-directed synthesis of GO/hydroxyapatite and its application in drug delivery. Mater Sci Eng C 73:709–715

Yijing Chen PL, Modica JA, Drout RJ, Farha OK (2018) Acid-resistant mesoporous metal–organic framework toward oral insulin delivery: protein encapsulation, protection & release. J Am Ceram Soc 140:1–5

Zhang FGL, Chan JM, Wang AZ, Langer RS, Farokhzad OC (2007) Nanoparticles in medicine: therapeutic applications and developments. Clin Pharmacol Ther 83:761–769

Zhang B, Yang X, Wang Y, Zhai G (2017) Heparin modified graphene oxide for pH-sensitive sustained release of doxorubicin hydrochloride. Mater Sci Eng C 75:198–206

Zhao X, Liu P (2014) Biocompatible graphene oxide as a folate receptor-targeting drug delivery system for the controlled release of anti-cancer drugs. RSC Adv 46:24232–24239

Zhao X, Liu L, Li X, Zeng J, Jia X, Liu P (2014) Biocompatible graphene oxide nanoparticle-based drug delivery platform for tumor microenvironment-responsive triggered release of doxorubicin. Langmuir 30:10419–10429

Zhou W, Wang L, Li F, Zhang W, Huang W, Huo F, Xu H (2017) Selenium-containing polymer@metal-organic frameworks nanocomposites as an efficient multiresponsive drug delivery system. Adv Funct Mater 27:1–8

Zou Z, Li S, He D, He X, Wang K, Li L, Yang X, Li H (2017) A versatile stimulus-responsive metal–organic framework for size/morphology tunable hollow mesoporous silica and pH-triggered drug delivery. J Mater Chem B 5:2126–2132

Chapter 9
Nanotechnology: A Boon for Food Safety and Food Defense

Poorva Sharma, Anil Panghal, Vishwajeet Gaikwad, Shubham Jadhav, Akshay Bagal, Akshay Jadhav, and Navnidhi Chhikara

Contents

9.1 Introduction

Richard Feynman, a famous physicist, first proposed the concept of nanotechnology in 1959. Subsequently, the term "nanotechnology" was coined by Norio Taniguchi in 1974. In 1980, nanotechnology usage became multidisciplinary, and by 2014 the

P. Sharma · A. Panghal · V. Gaikwad · S. Jadhav · A. Bagal · A. Jadhav · N. Chhikara (✉)
Department of Food Technology and Nutrition, Lovely Professional University, Phagwara, Punjab, India

R. Prasad et al. (eds.), *Nanobiotechnology in Bioformulations*, Nanotechnology in the Life Sciences, https://doi.org/10.1007/978-3-030-17061-5_9

global market value of nanotechnology was calculated around US$1 trillion per annum (Handford et al. 2014). The efficacy of nanotechnology is calculated with processing device, application of structure, system, design and by changing the size and shape of nanomaterial (10^{-9} m) of materials (Ravichandran 2009, 2010). Nanotechnology involves a wide range of technological activities for characterization, regulation, and fabrication of a huge range of materials, encompassing chemical applications and processes, electronic sciences, and physical and biological engineering (Handford et al. 2014; Bata-Vidács et al. 2013). It has proved to be a boon for food processing and production of different foodstuffs with an enhanced shelf life. Nanoparticles and nanomaterials are used in various forms such as nanocoatings, nanorods, nanosheets, nanofilms, nanolayers, nanofibers, and nanotubes, with a size between 1 and 100 nm. These nanoparticles are capable of displaying unique chemical, physical, and biological properties that are absent in the bulk forms of the same materials (Cushen et al. 2012). Nanoscale devices are generally made to copy nanodevices that already exists in nature, involving DNA, membranes, proteins, crystalline structures in different starches, cellulose fibrils, molecular building, nanosized plant cell networks, and other biomolecules (Handford et al. 2014; Sozer and Kokini 2009).

Nanoparticles have a large surface area–to–volume ratio, which allows a large fraction of their atoms to be present on their surfaces. This is responsible for their greater stability, strength, and biological and chemical activities, enabling development of new materials with a extensive range of probable applications. Many applications have already been observed in various industries and applications, including sports, cosmetics, medicine, agriculture, food, construction, wastewater treatment, and electronics (Handford et al. 2014; Prasad et al. 2014, 2016, 2017a).

Nanotechnology is an emerging approach in the food industry to ensure the safety of food. The major food safety objective with regard to nanotechnology is to assess risks and safety consequences associated with use of engineered nanoparticles in nanotechnology applications (Prasad et al. 2017b). Food safety risks related to nanotechnology was also introduced in FSMS and considered during risk assessment studies (Panghal et al. 2018a).

9.2 Food Defense

"Food defense" is a term used to define activities and systems that are allied with protection of a country's food supply from intentional and unintentional acts of tampering and contamination (which may also be referred to as "adulteration") (Manning and Soon 2016). It has also been described as actions (mainly related to the safety of human consumers) during the supply of food that protect it from intentional and unintentional contamination, with a focus on safety measures in processed food (Mitenius et al. 2014). Food defense systems also safeguard food and its supply chains from mischievous attacks that could lead to supply failure (Global Food Safety Initiative 2013). Strategies for food defense can be implemented at local or national levels. Manning and Soon (2016) have distinguished between organizational or supply chain management and national risk assessment models. In the USA, the CARVER+ [which stands for "Criticality, Accessibility, Recuperability, Vulnerability, Effect, Recognizability"] shock technique is implemented at the national level.

Intentional food contamination is the one of the global threats of the twenty-first century. Nowadays, food can be used as a tool by terrorists. Threats and fraud in food can have injurious effects on the health of society, consumers, the economy, and national security. Possible threats to the food supply chain are misuse of food for criminal purposes, food shortages due to disruption of the supply chain, and intentional contamination of food with toxic materials that lead to poisoning and death. The US Food and Drug Administration (FDA has developed a personalized food defense plan that includes guidance, tools, and resources in a single application to control food threats and reduce the risk of intentional food adulteration (Manning and Soon 2016).

9.3 Use of Nanotechnology Components in Food Safety and Food Defense

Nanotechnology mainly focuses on operation of biological and nonbiological structures that are smaller than 100 nm. Structures on this scale have distinctive and unique functional properties. Promising advantages of nanotechnology have been identified in different industries, and nanoproducts have been commercially manufactured in the aerospace, pharmaceutical, and microelectronics industries. Development of nanotechnology in industries is determined by research in engineering, materials science, physics, biology, and chemistry. So far, the applications of nanotechnology in the food industry have been limited. However, discoveries and achievements in nanotechnology are becoming more established in the food industry and many related industries, and they have significant influences on various aspects ranging from the safety of food to molecular synthesis of novel food products and their ingredients (Chen et al. 2006). Nanotechnology allows researchers to control, measure, and handle substances at the nanoscale, and to alter their functions and properties in a favorable way.

The unique properties of nanomaterials hold many prospects for the food industry (Cho et al. 2008). Various types of nanostructures are used to build unique structures and establish new functionalities in foods. These include nanoparticles, nanofibers, nanoemulsions, and nanoliposomes. Weiss et al. (2006) described various nanostructures and their potential and actual uses in food processing. Nowadays, nanomaterials used in food applications consist of both organic and inorganic materials. Nanomaterials are classified into three groups: organic engineered nanomaterials (ENMs), inorganic nanomaterials, and surface-functionalized materials (Chaudhary et al. 2008). ENMs are mostly found in nanofood products. Nowadays, nanotechnology is also gaining increased attention in food safety. Nanotechnology-based devices have been reported to enable very quick and specific detection of pathogens to ensure food safety.

9.4 Nanoparticles

The four major categories of nanoparticles are magnetic nanoparticles (MNPs), nonmagnetic nanoparticles, carbon nanotubes (CNTs), and nanowires. They and their applications are discussed further in Sects. 9.4.1, 9.4.2, 9.4.3, and 9.4.4.

9.4.1 *Magnetic Nanoparticles*

Magnetic nanoparticles are <100 nm in size and are influenced by external magnetic fields (EMFs). Applications of MNPs as biosensors in the clinical and pharmaceutical sectors offer remarkable advantages over conventional detection methods. All biological samples manifest higher sensitivity toward magnetic fields; therefore, even a small amount of a sample can be analyzed through this approach (López-Rubio et al. 2012).

MNPs have been reported to be nontoxic, nonhazardous, environmentally friendly, inexpensive, biocompatible, and physiochemically constant. They also exhibit properties of superparamagnetism (Glynn et al. 2006), which is a form of magnetism that appears in small ferromagnetic or ferrimagnetic nanoparticles (Tarui et al. 2009). In the absence of an EMF, superparamagnetic nanoparticles (SPNs) shows some magnetic properties, which increase with escalations in an EMF. One type of magnetic particle that is commonly used is superparamagnetic iron oxide nanoparticles (SPIONs), which are characteristically applied in immunoassays by surface modification with suitable ligands that may bind with a unique selected target.

9.4.1.1 Nuclear Magnetic Resonance–Based Pathogen Detection

Magnetic nanoparticles possess a nuclear magnetic resonance (NMR) property, which is utilized to detect biomarkers and pathogens. When an EMF is applied, a local magnetic dipole field is generated by the MNPs, which act as proximity sensors, disturbing the stability of the spin–spin relaxation time of adjacent water molecules. The NMR based detection platform exploits this property of MNPs to modulate the spin-spin T2 relaxation time of targeted biological samples. Binding of MNPs to biotic or molecular targets reduces the transverse relaxation time, followed by disturbance of the magnetic resonance. In NMR-based assays, as shown in Fig. 9.1, two detection mechanisms are available. Detection of smaller molecules such as proteins and nucleic acids is achieved by magnetic relaxation switching (MRSw), using the properties of MNPs. This form of detection is less time consuming because there is no requirement for free MNPs. For detection of bacteria, functionalized MNPs are required. Zhao et al. (2017) applied this technique to detect *Listeria monocytogenes* in lettuce, milk, and milk products.

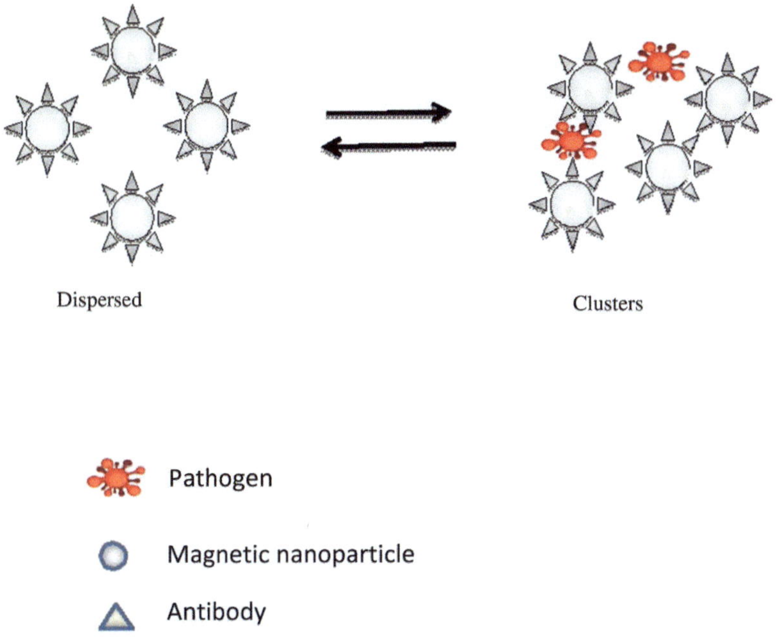

Fig. 9.1 Nuclear magnetic resonance detection platform (Krishna et al. 2018)

9.4.1.2 Search Coil–Based Detection Platform

Magnetic nanoparticles have been reported to possess nonlinear magnetism, and this property is used for immunoassays (Ching et al. 2012; Wang et al. 2014). A coil-based detection system records nonlinear magnetic responses at high and low frequencies. The responses produced are directly proportional to the number of MNPs used. This method has been observed to be better than others, as sample preparation is not required. Orlov (2013) detected toxic shock syndrome toxin (TSST) and staphylococcal enterotoxin A (SEA) in a milk sample with this method, without sample preparation. The sample of untreated milk was filtered via the tip of a pipette for capture of the antigens (i.e., SEA/TSST), followed by cleaning for removal of any free proteins. Finally, specific antibodies, biotinylated antigen, and streptavidin-coated MNPs were distributed in the filter for detection of bound antigens.

9.4.1.3 Giant Magnetoresistance Nanosensors

Magnetic sensors using the giant magnetoresistance (GMR) effect are used to detect the local magnetic dipole field generated by MNPs. The GMR effect is usually examined in multilayers consisting of alternating ferromagnetic and nonmagnetic

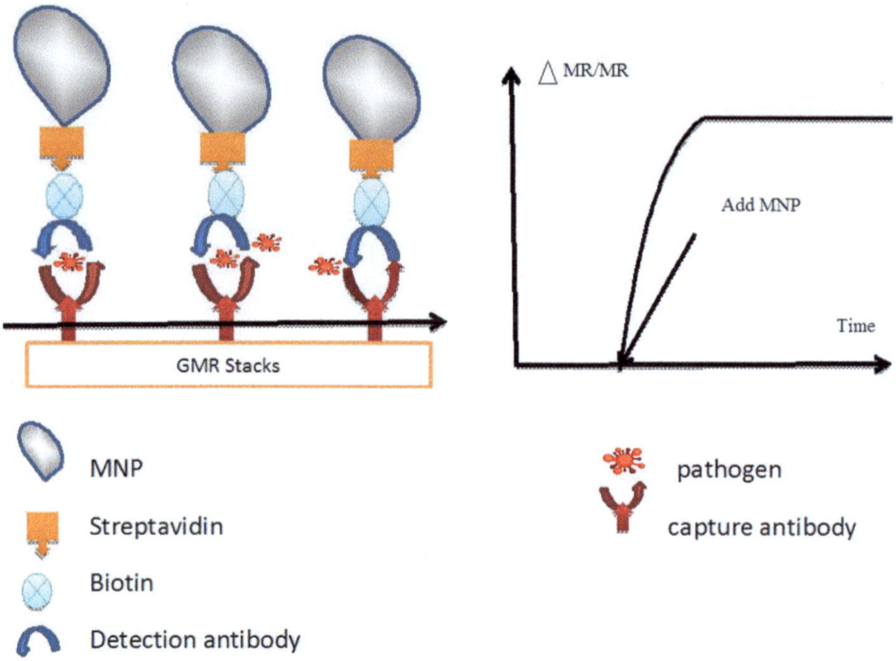

Fig. 9.2 Giant magnetoresistance (*GMR*) sensor surface showing the sandwich structure and time signal change (Wu et al. 2018). MNP magnetic nanoparticle, MR magnetic resonance

conductive layers. Depending on the magnetization alignment of the ferromagnetic layers, the electrical resistance changes; usually, more resistance is present when the layers are in an antiparallel alignment than when they are in a parallel alignment, since electrons are more scattered in the former alignment than in the latter. In standard GMR structures, the magnetic direction of a ferromagnetic layer is in a stationary phase during the deposition and annealing process, and another ferromagnetic layer moves freely under the impact of an applied field. On the basis of this effect, MNPs are used to detect various biological compounds. As shown in Fig. 9.2, the surface of the sensor is functionalized with captured antibodies; MNPs and target antigens are attached with detection antibodies. The number of MNP-tagged antibodies that bind to the antigens is directly proportional to the number of antigens that are present on the surface of the sensors. When affected by an EMF, the magnetic dipole field generated by the MNPs present on the bound antibodies is detected by the GMR sensors, resulting in resistance variation that is directly proportional to the concentration of the antigens (Fig. 9.2). In comparison with other sensing methods, GMR is low in cost, has high sensitivity, and is capable of real-time signal detail. Moreover, it can be used to detect multiple analytes simultaneously.

9.4.2 Nonmagnetic Nanoparticles

9.4.2.1 Localized Surface Plasmon Resonance

This detection method is used for Label free biological sensing. In this method, a beam of light, which is monochromatically polarized, is delivered via a prism and reflected from a film of thin metal on a glass slide, followed by interaction with a test liquid. The surface plasmon resonance (SPR) is attained when the frequency of photon strikes on the metal film matches the normal occurrence of electron fluctuations on the metal surface. A swing of the resonance frequency occurs, which is directly proportional to the concentration of the species adsorbed on the metallic surface; actual measurements of the binding may be attained by obtaining the alteration in the visual reflectivity with respect to time (Yu and Yang 2017). However, although SPR technology is moderately well established, the instrumentation used in this technique is costly and complicated to operate (Krishna et al. 2018).

9.4.2.2 Nanoparticle-Assisted Colorimetry

The simplest method for detection of biological targets is colorimetric testing. The presence of a target analyte can be easily determined either by visualization, by a colorimeter, or by employing a chemical reaction that involves color change. However, the efficacy of color conversion in a normal colorimeter is comparatively low, which lowers the device sensitivity. Kuswandi and Heng (2017) designed nanoparticle-assisted colorimetry to resolve the aforementioned problem. The surface area on which the color precursor molecules are situated is increased considerably by bringing nanoparticles into the sensing matrix.

A sandwich structure was formed on a plate by immobilizing capture probe, during hybridization between capture probes, target sequence and a detection probe by means of functionalized AuNPs. Luo (2014) detected the *invA* gene of *Salmonella* species using a combination of polymerase chain reaction (PCR) and DNAzyme probe self-assembled gold nanoparticles. In the presence of hemin, the detection probe was formed as G-quadruplex/hemin complex, which performed as a catalyst for the oxidation reaction, facilitated by H_2O_2, leading to an intense variation in color. The detection limit was 3×10^3 colony-forming units (CFU) per milliliter. For recognition of *Salmonella typhimurium*, AuNPs have been combined with antibodies (Banerjee et al. 2017).

9.4.2.3 Quantum Dots

Quantum dots (QDs) are widely employed in detection of traditional fluorescence with organic fluorescent dyes, showing corresponding emission peaks and high sensitivity to photobleaching. QDs are nanosized crystalline particles with good resistant properties against chemical degradation and photobleaching (Shao et al. 2012).

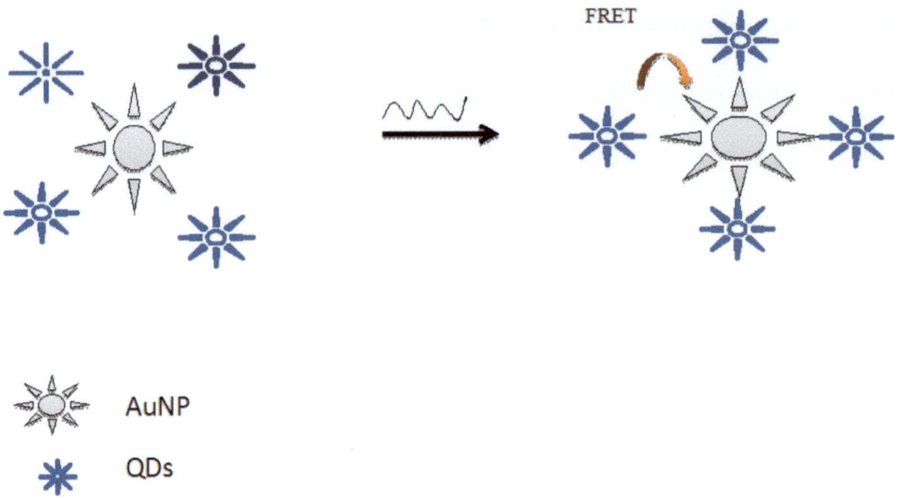

Fig. 9.3 Fluorescence resonance energy transfer (*FRET*) between graphene quantum dots (*QDs*) and gold nanoparticles (*AuNP*). Graphene QDs conjugated with capture probe and AuNPs conjugated with reporter probe (Krishna et al. 2018)

The QD bandgap energy is inversely proportional to their size, resulting in tunable and narrow emission bands, independent of the excitation wavelength (Wu et al. 2015). QDs are available in various sizes with a wide variety of emission wavelengths that allow their use in multiplex analysis. QDs coupled with immunomagnetic separation (IMS) are used to detect *Salmonella enteritidis* and *Escherichia coli* (Landeghem et al. 2009). Magnetic beads and QDs coated with the corresponding antibodies bind the targeted antigens by using a magnet, as shown in Fig. 9.3.

9.4.3 Carbon Nanotubes

Carbon nanotubes are allotropes of a carbon compound with a cylindrical nanostructure. Usually, CNTs measure up to a few millimeters in length, with a diameter of up to few nanometers. These carbon molecules have attracted significant interest because of their exclusive mechanical and electronic characteristics, which make CNTs a capable contender to increase the performance of electrochemical biosensors (Park et al. 2016). Immobilization of ligands for molecular recognition is permitted by the large surface area of CNTs. They function as a sensing component, which allows communication between the conjugated antigen–antibody complex and the underlying electrode.

Single-walled CNTs (SWCNTs) enclosing monoclonal antibodies of *Salmonella* can be applied to modify a surface electrode of glassy carbon (GC) for use as a nanosensor to detect *Salmonella* at a very low concentration (Park et al. 2016). This type of immunosensor results in formation of antigen–antibody multiplexes by estimating changes in the electrical and chemical properties of the sensor Afforded by the insulating properties of bacterial cell membrane.

9.4.4 *Nanowires*

In addition to having a large surface-to-volume ratio, nanowires have great electrical conductance sensitivity to variations on the surface of an electrical field. As a result, they are used in research (Star 2006).

To detect *L. monocytogenes*, TiO_2 nanowire bundles were developed by Wang et al. (2008). Monoclonal antibodies were immobilized on the exterior of the TiO_2 nanowire bundles to capture *L. monocytogenes*. The cell wall of *Listeria* has a neutral pH with a negative charge on its surface; therefore, specific binding of the TiO_2 nanowire bundle and the *Listeria* cell produces substantial changes in the electrical conductance. A TiO_2 nanowire–based activity immunosensor has the capability to detect *L. monocytogenes* at a low concentration (4.7×10^2 CFU/mL) with a detection time of 50 min. For detection of the foodborne pathogen *Bacillus cereus,* Pal et al. (2007) used polyaniline nanowire–based direct-charge transmission biosensors. The working principle is binding of the antigen to its similar antibody, which disturbs the movement of the electron charge in the polyaniline nanowire, resulting in increased resistance. The detection limit for this pattern is specified as 10^1–10^2 CFU/mL, with a detection time of 6 min.

ZnO nanowires are mostly used to detect breast cancer and uric acid levels in patients with Parkinson's disease (Yue et al. 2014). Silicon nanowires are used to detect cancer risk biomarkers, cardiac troponin I (cTnI), and circulating tumor cells.

9.5 Nanotechnology Applications in the Food Supply Chain

Applications of nanotechnology in food contact materials (FCMs) account for the greatest market share of the current and short-term forecasted benefits of nanotechnology in the food sector (Chaudhry et al. 2008). Various characteristics of food and agricultural systems are influenced by nanotechnology. The science and engineering of food and agricultural systems have some important links with nanotechnology—for example, food security, delivery of medical treatments, different tools for use in cellular and molecular biology, pathogen detection with new materials, and environmental protection (Prasad et al. 2014, 2017a, b). Momin et al. (2013) reported that the USA leads the nanofood product sector, followed by Japan and China, although the largest future market for nanofood products may be the Asian countries, which are led by China. An application matrix of nanotechnology is presented in Fig. 9.4.

Examples of nanotechnology as a tool used for accomplishing further improvements in the food industry are:

- Sensors for detecting contaminants and pathogens to increase food security during manufacturing, processing, and distribution of products
- Tracking of individual products to maintain historical records, along with environmental data, with the help of devices

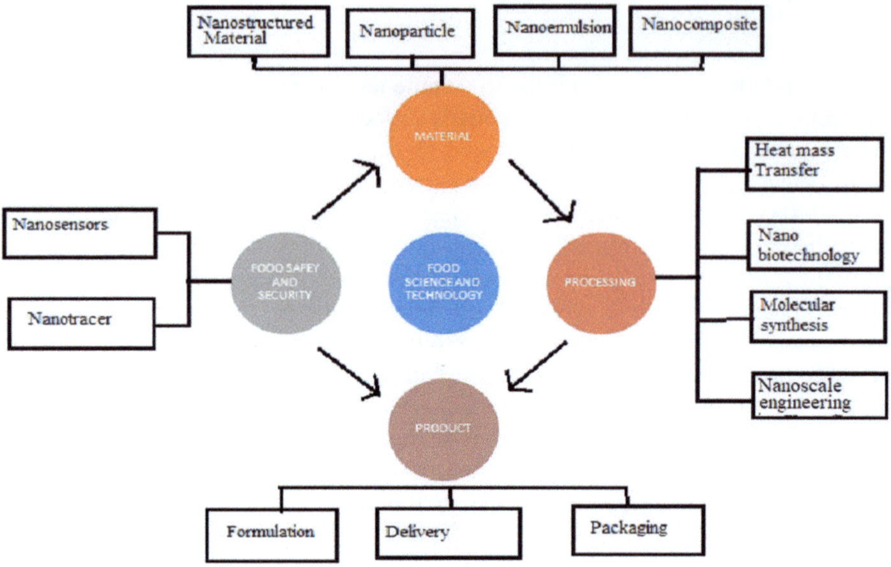

Fig. 9.4 Nanotechnology application matrix in food science (Weiss et al. 2006)

- Increased effectiveness and security of food processing and distribution through remote control of food products with a combination of sensing, localization, and (intelligent) reporting systems
- Protection of functional food ingredients with conservation of their specific modes of action by encapsulation and delivery systems
- Upgraded biosecurity and food safety methods and equipment design, with new product development, microscale and nanoscale processing, and new functional material development, enabling further advancements in the main areas of food production with the help of nanotechnology

9.5.1 Nanotechnology in Food Processing

Nanotechnology has been deployed in food production units and processing areas to increase food quality and food safety. Nanosieves and nanosensors are two basic approaches used in processing units. Nanosieves are a tool by which contaminants are separated from food products, and nanosensors are designed for detection of microorganisms during packaging of food. Silver has been reported to control the growth of bacteria during storage and is widely used in packaging materials (Geys et al. 2008), although an important safety concern is migration of nanoparticles into the food. Inert nanoparticles are used as a system for nanodelivery, in which nano-capsules are prepared that contain bioactive compounds and nutrients. The advantage of this delivery system is that desired active compounds (e.g., pesticides or medicines) can be delivered to their target sites. The use of this system leads to increased absorption and bioavailability (Geys et al. 2008).

9.5.2 Nanoparticles as Food Additives

Chemical reactions occurring in food products, as a result of interaction between food components and outer atmospheric conditions, are the major cause of food spoilage. In the reported literature, numerous nanomaterials have been originated and applied for prevention of these undesirable activities in food (Manke et al. 2013). Pool et al. (2012) studied preparation of polymeric nanoparticles with encapsulation of two bioactive compounds—quercetin and catechin—and their controlled release in food products. Their study indicated that use of poly(lactic-*co*-glycolic acid) (PLGA) encapsulation effectively enhanced the antiradical and chelating characteristics of these bioactive compounds and could be a constructive tool for prolonging the shelf life of lipid-based food. Nanotechnology has also proved to be an effective way to enhance food safety and the nutritional and sensory properties of food products. Rojas-Graü et al. (2009) observed that incorporation of active ingredients into edible coating solutions and films increases the functionality of fruits, and these components work as antimicrobial, antioxidant, antibrowning, and nutraceutical agents. Furthermore, nanoparticles have been reported to improve food characteristics such as appearance and color, but only to a limited extent. The US FDA's permitted limit for TiO_2 coloring additive ingredients is 1% (w/w). Mehrad et al. (2018) reported use of SiO_2 as an anticaking agent, an aroma transmitter, and a means of maintaining the flow properties of dried powder. However, despite their positive attributes, nanomaterials and their metal oxides have been reported to cause generation of reactive oxygen species (ROS), which can result in genotoxicity, inflammation, fibrosis, and carcinogenesis (Manke et al. 2013).

9.5.3 Nanobiosensors for Foodborne Microbial Pathogens

Prokaryotic microorganisms have a hugely dominant influence on the earth and survive in almost all environmental conditions. Microbial pathogens and/or the toxins they produce can contaminate food and cause illness in those who consume it; thus, foods may need to be treated with pasteurization. A vast number of food- and waterborne diseases can ensue when microbial contamination occurs (Malhotra et al. 2014). Microbial analysis can be accomplished with advanced analytical methods such as gas chromatography with mass spectrometry (GC-MS) and high-performance liquid chromatography (HPLC), but these techniques are costly and require technical proficiency to perform (Lee et al. 2012; Valdés et al. 2009). The aforementioned limitations may possibly be overcome with the help of nanotechnology-based biosensing. The many techniques that can produce and manipulate materials in a size range of 1–100 nm are referred to as nanotechnology, which is one of the most promising technologies of the twenty-first century. Microorganisms such as *Salmonella* spp., *Listeria* spp., and *E. coli* are detectable with nanobiosensors. *S. typhimurium* concentrations as low as 1000 CFU/mL can be detected with the use of pathogen-specific antibodies and aptamer-functionalized magnetic particles in an enzyme-linked antibody–aptamer sandwich (nano-ELAAS) assay (Wu et al. 2014).

9.5.4 Nanoparticles in Packaging Materials

Nanotechnology has been reported to have wide applications in the food packaging industry. Incorporation of nanoparticles into packaging materials leads to improvements in their mechanical and heat resistance characteristics, with greater water and gas resistivity (Youssef 2013). Laoutid et al. (2009), Lizundia et al. (2016) reported that nanocomposite polymers covered with silicates have more flame and ultraviolet resistance. Nanotechnology-based packaging materials have been reported to be environmentally friendly because of their biodegradable nature. Rashidi and Khosravi-Darani (2011) described preparation of composite polymers consisting of potato starch and calcium carbonate, and found that this form of packaging was an effective replacement for polystyrene packaging of fast food.

9.6 Limitations of Nanotechnology in Food Safety and Defense

Silver nanoparticles have been reported to affect fibroblasts in the human lung by increasing ROS creation, reducing adenosine triphosphate (ATP) content, and damaging mitochondria and DNA (Kim et al. 2007). Incorporation of nanoparticles into food packaging materials has been reported to cause effects on the skin and lungs (Mills and Hazafy 2009). Morones (2005) verified that nanoparticles could be toxic to plankton and that aluminum nanoparticles could inhibit plant growth. However, the particles present on the equipment or the packaging material directly comes in contact with the food substance that are expected to migrate into food in small amount, these particles are removed from the regulation as food additives if they satisfy certain criteria (US Food and Drug Administration 2014). All these types of pacaking materials thoughout the food suppy chain should be assessed and taken care during food safety studies (Panghal et al. 2018b). Chawengkijwanich and Hayata (2008) reported that different nanoparticles that are known to be harmful to human well-being are employed in food defense nanotechnology.

Cioffi et al. (2005) stated that minerals, proteins, polysaccharides, phospholipids, and surfactants are the most important components of nanoemulsions. Nanoemulsions are made from surfactants and solvents, which can lead to hazardous effects after ingestion at higher concentrations (Cushen et al. 2012). Fujishima et al. (2000) observed that ingestion of large amounts of lipid-containing nanoemulsions can cause obesity and cardiovascular diseases. DNA or RNA can be delivered through the intestinal wall by cochleates, which are lipid-based encapsulates; therefore, a virus or other hazardous components could be transferred through the intestinal wall and cause contamination, which could be a bigger risk factor. There may be some risk associated with the presence of these constituents. Nanosized crystals of lycopene have a tendency to dissolve faster than natural crystals, which could be hazardous. Emulsions containing nanosized components are less stable than those containing larger components; thus, they are absorbed and broken down very quickly (Boom 2011).

Greater knowledge is needed in order to determine the carcinogenicity of these substances, their migration inside the food product through packaging materials and their impact on environment. Determination of the migration patterns of other polymer–nanomaterial compounds resulting from biopolymers requires further research (Chhikara et al. 2018).

9.7 Regulation of Nanotechnology

To date, there has been no regulatory body to formulate specific regulations for use of nanotechnology in food processing. However, the European Parliament and nongovernmental organizations (NGOs) including several stakeholders have suggested some guidelines. The need for such guidelines to evaluate possible risks and provide recommendations for safe use of nanomaterials has been recognized, and numerous bodies are now active in this field, such as the Organization for Economic Co-operation and Development (OECD), the European Union (EU) Scientific Committees and Agencies, the International Standards Organization (ISO), and the US FDA. The input of the EU, the USA, and Australia (which are different jurisdictions) should be sufficient to “capture” the applications of nanotechnology in the food sector within the current framework for food and FCMs. General product safety, water quality, FCMs, chemical safety, specific health claims, novel foods, food additives, and other specific regulations related to general food safety within these jurisdictions cover the use of some chemicals in food production/preservation—for example, pesticides, biocides, and veterinary medicines. For more information on the regulatory features of nanotechnologies, see Chaudhry et al. (2008) and Marchant et al. (2009).

9.8 Future Needs for Nanotechnology

Different nanosystems are being established as capable components of applications in the food industry. Diligent efforts are being made to improve the efficiency of nanocarriers with enhanced bioavailability without affecting the organoleptic properties of food products. Detectors have been designed with incorporation of

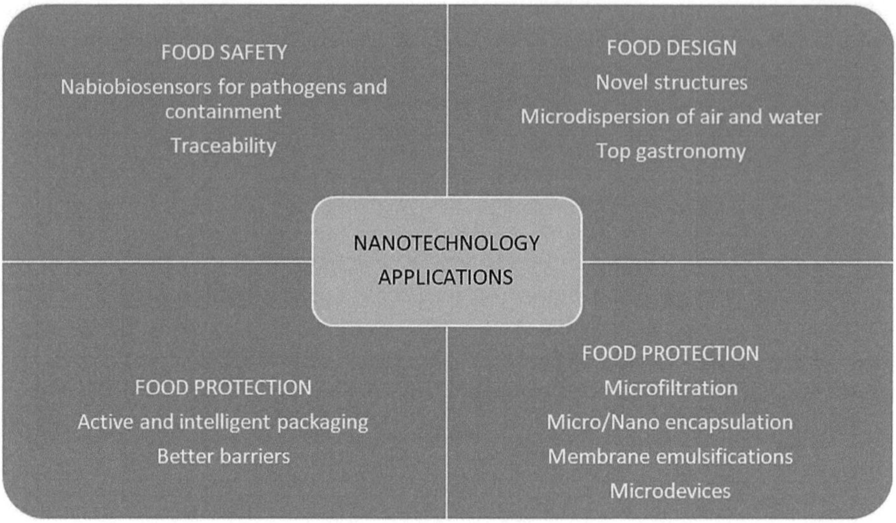

Fig. 9.5 Future needs for nanotechnology applications in the food industry (Rossi et al. 2014)

specific antigen markers to examine the presence of specific microorganisms (Graveland-Bikker and Kruif de 2006). The smart-packaging idea is gradually becoming more appreciated, and research is being done to produce antigen-specific biomarkers for use in food packaging and to create polymeric nanocomposite films containing nanoparticles (Cho et al. 2008). Use of the radiofrequency identification (RFID) concept has also been introduced into the distribution of fast food products with a short shelf life (Farhang 2009).

The field of nanotechnology in product development and commercialization is still lagging behind in India, because the Indian scenario is different from those of developed countries, but it is slowly following world trends and devising new methods. The Indian government has set up a national Institute of Nano Science and Technology for development of nanofoods with use of nanotechnology during cultivation, processing, and packaging of food to improve food safety and food defense.

In the future, nanotechnologies will be used to improve various characteristics of food products, with the use of so-called soft nanomaterials (such as vesicles and micelles) to encapsulate nutrients and deliver them to specific locations in the gastrointestinal tract and to improve the flow and behavior of powdered foodstuffs. Future needs for nanotechnology applications in the food industry are shown in Fig. 9.5.

9.9 Conclusion

Nanotechnology offers pervasive applications in various different areas of the food industry, such as in production, processing, packaging, storage, transportation, traceability, and food security. Numerous advances in nanotechnology in food systems—for example, in food processing and distribution—have been observed in many countries. Nanotechnology also has potential to innovate technology for improvement of functional foods, nutrient delivery systems, food packaging, food hygiene, water decontamination, and shelf life extension. Finally, nanotechnology will enable us to change the existing food system and processing to ensure food safety, construct a healthy food culture, and improve the nutritional quality of food products. However, there is need for better regulations in many countries to ensure proper safety evaluation of products involving nanotechnology in order to protect consumers.

References

Banerjee T, Shelby T, Santra S (2017) How can nanosensors detect bacterial contamination before it ever reaches the dinner table? Future Microbiol 12:97–100

Bata-Vidács I, Adányi N, Beczner J, Farkas J, Székács A (2013) Nanotechnology and microbial food safety. In: Méndez-Vilas A (ed) Microbial pathogens and strategies for combating them: science, technology and education. Formatex Research Center, Badajoz, pp 155–159

Boom RM (2011) Nanotechnology in food production. In: Frewer LJ, Norde W, Fischer A, Kampers F (eds) Nanotechnology in the agri-food sector: implications for the future. Weinheim: Wiley-VCH, pp 39–57

Chaudhry Q, Scotter M, Blackburn J, Ross B, Boxall A, Castle L, Aitken R, Watkins R (2008) Applications and implications of nanotechnologies for the food sector. Food Addit Contam 25(3):241–258

Chawengkijwanich C, Hayata Y (2008) Development of TiO_2 powder–coated food packaging film and its ability to inactivate *Escherichia coli* in vitro and in actual tests. Int J Food Microbiol 123(3):288–292

Chen Z, Meng H, Xing G, Chen C, Zhao Y, Jia G (2006) Acute toxicological effects of copper nanoparticles in vivo. Toxicol Lett 163:109–120

Ching KH, Lin A, McGarvey JA, Stanker LH, Hnasko R (2012) Rapid and selective detection of botulinum neurotoxin serotype-A and-B with a single immunochromatographic test strip. J Immunol Methods 380(1–2):23–29

Chhikara N, Jaglan S, Sindhu N, Anshid V., Charan MVS, Panghal A (2018) Importance of Traceability in Food Supply Chain for Brand Protection and Food Safety Systems Implementation. Annals of Bio 34(2):111–118

Cho YJ, Kim CJ, Kim N, Kim CT, Park B (2008) Some cases in applications of nanotechnology to food and agricultural systems. Biochip J 2(3):183–185

Cioffi N, Torsi L, Ditaranto N, Tantillo G, Ghibelli L, Sabbatini L, Bleve-Zacheo T, D'Alessio M, Zambonin PG, Traversa E (2005) Copper nanoparticle/polymer composites with antifungal and bacteriostatic properties. Chem Mater 17(21):5255–5262

Cushen M, Kerry J, Morris M, Cruz-Romero M, Cummins E (2012) Nanotechnologies in the food industry—recent developments, risks and regulation. Trends Food Sci Technol 24(1):30–46

Farhang B (2009) Nanotechnology and applications in food safety. In: Barbosa-Canovas GV, Mortimer A, Lineback D, Spiess W, Buckle K, Colonna P (eds) Global issues in food science and technology. Academic, San Diego, pp 401–410

Fujishima A, Rao TN, Tryk DA (2000) Titanium dioxide photocatalysis. J Photochem Photobiol C 1(1):1–21

Geys J, Nemmar A, Verbeken E, Smolders E, Ratoi M, Hoylaerts MF, Nemery B, Hoet PH (2008) Acute toxicity and prothrombotic effects of quantum dots: impact of surface charge. Environ Health Perspect 116(12):1607

Global Food Safety Initiative (2013) The Global Food Safety Initiative: GFSI guidance document. Version 6.3. https://www.mygfsi.com/images/mygfsi/gfsifiles/information-kit/GFSI_Guidance_Document.pdf. Accessed 4th October 2018

Glynn B, Lahiff S, Wernecke M, Barry T, Smith TJ, Maher M (2006) Current and emerging molecular diagnostic technologies applicable to bacterial food safety. Int J Dairy Technol 59(2):126–139

Graveland-Bikker JF, Kruif de CG (2006) Food nanotechnology. Trends Food Sci Technol 17(5):196–203

Handford CE, Dean M, Henchion M, Spence M, Elliott CT, Campbell K (2014) Implications of nanotechnology for the agri-food industry: opportunities, benefits and risks. Trends Food Sci Technol 40(2):226–241

Kim JS, Kuk E, Yu KN, Kim JH, Park SJ, Lee HJ, Kim SH, Park YK, Park YH, Hwang CY, Kim YK (2007) Antimicrobial effects of silver nanoparticles. Nanomed Nanotechnol Biol Med 3(1):95–101

Krishna VD, Wu K, Su D, Cheeran MC, Wang JP, Perez A (2018) Nanotechnology: review of concepts and potential application of sensing platforms in food safety. Food Microbiol 75:47–54. https://doi.org/10.1016/j.fm.2018.01.025

Kuswandi BD, Heng LY (2017) Nanosensors for the detection of food contaminants. In: Grumezescu AM, Oprea A (eds) Nanotechnology applications in food. Academic, London, pp 307–333

Landeghem FK, Maier-Hauff K, Jordan A, Hoffmann KT, Gneveckow U, Scholz R, Thiesen B, Brück W, Von Deimling A (2009) Post-mortem studies in glioblastoma patients treated with thermotherapy using magnetic nanoparticles. Biomaterials 30(1):52–57

Laoutid F, Bonnaud L, Alexandre M, Lopez-Cuesta JM, Dubois P (2009) New prospects in flame retardant polymer materials: from fundamentals to nanocomposites. Mater Sci Eng R Rep 63(3):100–125

Lee SH, Pie JE, Kim YR, Lee HR, Son SW, Kim MK (2012) Effects of zinc oxide nanoparticles on gene expression profile in human keratinocytes. Mol Cell Toxicol 8(2):113–118

Lizundia E, Ruiz-Rubio L, Vilas JL, León LM (2016) Poly(l-lactide)/ZnO nanocomposites as efficient UV-shielding coatings for packaging applications. J Appl Polym Sci 133:2

López-Rubio A, Sanchez E, Wilkanowicz S, Sanz Y, Lagaron JM (2012) Electrospinning as a useful technique for the encapsulation of living bifidobacteria in food hydrocolloids. Food Hydrocoll 28(1):159–167

Luo R (2014) A colorimetric assay method for invA gene of *Salmonella* using DNAzyme probe self-assembled gold nanoparticles as single tag. Sensors Actuators B Chem 198:87–93

Malhotra BD, Srivastava S, Ali MA, Singh C (2014) Nanomaterial-based biosensors for food toxin detection. Appl Biochem Biotechnol 174(3):880–896

Manke A, Wang L, Rojanasakul Y (2013) Mechanisms of nanoparticle-induced oxidative stress and toxicity. BioMed Res Int 2013:942916. https://doi.org/10.1155/2013/942916

Manning L, Soon JM (2016) Food safety, food fraud, and food defense: a fast evolving literature. J Food Sci 81(4):823–834. https://doi.org/10.1111/1750-3841.13256

Marchant G, Sylvester D, Abbott KW (2009) Nanotechnology regulation: the United States approach. In: Hodge GA, Bowman D, Ludlow K (eds) New global frontiers in regulation: the age of nanotechnology. Cheltenham: Edward Elgar, pp 189–211

Mehrad B, Ravanfar R, Licker J, Regenstein JM, Abbaspourrad A (2018) Enhancing the physicochemical stability of β-carotene solid lipid nanoparticle (SLNP) using whey protein isolate. Food Res Int 105:962–969

Mills A, Hazafy D (2009) Nanocrystalline SnO_2-based, UVB-activated, colourimetric oxygen indicator. Sensors Actuators B Chem 136(2):344–349

Mitenius N, Kennedy SP, Busta FF (2014) Food defense. In: Motarjemi Y, Lelieveld H (eds) Food safety management: a practical guide for the food industry. Academic, London, pp 937–958

Momin JK, Jayakumar C, Prajapati JB (2013) Potential of nanotechnology in functional foods. Emirates J Food Agric 25(1):10–19

Morones JR (2005) The bactericidal effect of silver nanoparticles. Nanotechnology 16(10):2346–2353

Orlov AV (2013) Magnetic immunoassay for detection of staphylococcal toxins in complex media. Anal Chem 85(2):1154–1163

Pal S, Alocilja EC, Downes FP (2007) Nanowire labeled direct-charge transfer biosensor for detecting *Bacillus* species. Biosens Bioelectron 22(9–10):2329–2336

Panghal A, Chhikara N, Sindhu N, Jaglan S (2018a) Role of Food Safety Management Systems in safe food production: A review. Journal of Food Safety 38(4):e12464

Panghal A, Yadav DN, Khatkar BS, Sharma H, Kumar V, Chhikara N (2018b) Post-harvest malpractices in fresh fruits and vegetables: food safety and health issues in India. Nutrition & Food Sci 48(4):561–578

Park KS, Chung HJ, Khanam F, Lee H, Rashu R, Bhuiyan MT, Qadri F (2016) A magneto-DNA nanoparticle system for the rapid and sensitive diagnosis of enteric fever. Sci Rep 6:32878

Pool H, Quintanar D, Dios Figueroa J, Mano CM, Bechara JEH, Godínez LA, Mendoza S (2012) Antioxidant effects of quercetin and catechin encapsulated into PLGA nanoparticles. J Nanomater 2012:86

Prasad R, Kumar V, Prasad KS (2014) Nanotechnology in sustainable agriculture: present concerns and future aspects. Afr J Biotechnol 13(6):705–713

Prasad R, Pandey R, Barman I (2016) Engineering tailored nanoparticles with microbes: quo vadis. WIREs Nanomed Nanobiotechnol 8:316 330. https://doi.org/10.1002/wnan.1363

Prasad R, Bhattacharyya A, Nguyen QD (2017a) Nanotechnology in sustainable agriculture: recent developments, challenges, and perspectives. Front Microbiol 8:1014. https://doi.org/10.3389/fmicb.2017.01014

Prasad R, Kumar V and Kumar M (2017b) Nanotechnology: Food and Environmental Paradigm. Springer Nature Singapore Pte Ltd. (ISBN 978-981-10-4678-0)

Rashidi L, Khosravi-Darani K (2011) The applications of nanotechnology in food industry. Crit Rev Food Sci Nutr 51(8):723–730

Ravichandran R (2009) Nanoparticles in drug delivery: potential green nanobiomedicine applications. Int J Green Nanotechnol Biomed 1(2):108–130

Ravichandran R (2010) Nanotechnology applications in food and food processing: innovative green approaches, opportunities and uncertainties for global market. Int J Green Nanotechnol Phys Chem 1(2):72–96

Rojas-Graü MA, Soliva-Fortuny R, Martín-Belloso O (2009) Edible coatings to incorporate active ingredients to fresh-cut fruits: a review. Trends Food Sci Technol 20(10):438–447

Rossi M, Cubadda F, Dini L, Terranova ML, Aureli F, Sorbo A, Passeri D (2014) Scientific basis of nanotechnology, implications for the food sector and future trends. Trends Food Sci Technol 40(2):127–148

Shao H, Min C, Issadore D, Liong M, Yoon TJ, Weissleder R, Lee H (2012) Magnetic nanoparticles and microNMR for diagnostic applications. Theranostics 2(1):55

Sozer N, Kokini JL (2009) Nanotechnology and its applications in the food sector. Trends Biotechnol 27(2):82–89

Star A (2006) Label-free detection of DNA hybridization using carbon nanotube network field-effect transistors. Proc Natl Acad Sci U S A 103(4):921–926

Tarui A, Kawasaki H, Taiko T, Watanabe T, Yonezawa T, Arakawa R (2009) Gold-nanoparticle-supported silicon plate with polymer micelles for surface-assisted laser desorption/ionization mass spectrometry of peptides. J Nanosci Nanotechnol 9(1):159–164

US Food and Drug Administration (2014) CFR—code of federal regulations title 21. https://www.accessdata.fda.gov/scripts/cdrh/cfdocs/cfcfr/cfrsearch.cfm. Accessed 05 Nov 2018

Valdés MG, González ACV, Calzón JAG, Díaz-García ME (2009) Analytical nanotechnology for food analysis. Microchim Acta 166(1–2):1–19

Wang R, Ruan C, Kanayeva D, Lassiter K, Li Y (2008) TiO_2 nanowire bundle microelectrode based impedance immunosensor for rapid and sensitive detection of *Listeria monocytogenes*. Nano Lett 8(9):2625–2631

Wang L, Cai J, Wang Y, Fang Q, Wang S, Cheng Q, Liu F (2014) A bare-eye-based lateral flow immunoassay based on the use of gold nanoparticles for simultaneous detection of three pesticides. Microchim Acta 181(13–14):1565–1572

Weiss J, Takhistov P, McClements DJ (2006) Functional materials in food nanotechnology. J Food Sci 71(9):R107–R116

Wu W, Li J, Pan D, Li J, Song S, Rong M, Lu J (2014) Gold nanoparticle-based enzyme-linked antibody–aptamer sandwich assay for detection of *Salmonella typhimurium*. ACS Appl Mater Interfaces 6(19):16974–16981

Wu K, Batra A, Jain S, Ye C, Liu J, Wang JP (2015) A simulation study on superparamagnetic nanoparticle based multi-tracer tracking. Appl Phys Lett 107(17):173701

Youssef AM (2013) Polymer nanocomposites as a new trend for packaging applications. Polym-Plast Technol Eng 52(7):635–660

Yu X, Yang H (2017) Pyrethroid residue determination in organic and conventional vegetables using liquid–solid extraction coupled with magnetic solid phase extraction based on polystyrene-coated magnetic nanoparticles. Food Chem 217:303–310

Yue HY, Huang S, Chang J, Heo C, Yao F, Adhikari S, Li B (2014) ZnO nanowire arrays on 3D hierarchical graphene foam: biomarker detection of Parkinson's disease. ACS Nano 8(2):1639–1646

Zhao Y, Li Y, Jiang K, Wang J, White WL, Yang S, Lu J (2017) Rapid detection of *Listeria monocytogenes* in food by biofunctionalized magnetic nanoparticle based on nuclear magnetic resonance. Food Control 71:110–116

Chapter 10
Microbial Bioformulations: Present and Future Aspects

Usha Rani and Vivek Kumar

Contents

10.1 Introduction

One of the hardest tasks these days is to employ sustainable agricultural practices. Moreover, agricultural practices around the globe are not uniform, but usage of chemical pesticides and fertilizers is common. Around 890 man-made chemically prepared various chemicals are certified that can be used as pesticides and insecticides (Stenersen 2004). These chemicals play a crucial role in improving the crop yield and inhibiting diseases, but they are also leaving their harmful effect on environment (Fenske and Day 2005; Colt et al. 2007). The condition in developing countries is very bad; despite the fact that the usage of agrochemicals in these countries is only 20%, they encounter about 99% of mortality rate due to pesticide poisoning (Kesavachandran et al. 2009). Chief victims are farmers due to high exposure and lack of awareness. About 20,000 workers die because of exposure to pesticides every

U. Rani · V. Kumar (✉)
Himalayan School of Biosciences, Swami Rama Himalayan University, Dehradun, Uttarakhand, India

R. Prasad et al. (eds.), *Nanobiotechnology in Bioformulations*, Nanotechnology in the Life Sciences, https://doi.org/10.1007/978-3-030-17061-5_10

year (Pimentel et al. 1992). According to the WHO, majority of the reports show that pesticides used for agricultural practices have very malignant influence on human well-being; short-term effects include headaches, nausea and vomiting, rashes, respiratory failure, coma, shock, etc. (Moses et al. 1993), whereas long-term effects include reproductive problems, cancer, and neurological disorders, and in serious cases it can cause death (Sanborn et al. 2007). Further, wildlife, aquatic ecosystems, and environment also get disturbed (Berny 2007). Continuous usage of pesticides drastically influences the microbial systems present in soil, and it kills large diversity of microbial population which is beneficial for the agriculture or crop production (Dorigo et al. 2009). Different types of pesticides have different effect on microbial populations and greatly influence the microbial diversity (Johnsen et al. 2001; Spyrou et al. 2009). Chemically synthesized fertilizers are another important factor on which the majority of our agriculture depends. Different studies were conducted to determine the long-term effects of these fertilizers on fertility of soil, and reports show that regular usage can elevate the strength of total nitrogen, organic matter, and different nutrients in soil as compared to the primary values present at the start of experiment (Liang et al. 2013; Mitchell et al. 1991; Mandal et al. 2007). But quality and productivity of soil is gradually deteriorating due to excessive usage of chemical fertilizers. They also influence the microbial diversity and their functions (Nakhro and Dkhar 2010). Formulations based on microbes are known as "bioformulations." In other words, formulations which constitute of several valuable strains of microbes, which are immobilized or trapped on an inert carrier material, that can be employed to enhance plant growth and inhibit plant pathogens and can increase fertility of soil are known as bioformulations (Mendes et al. 2011). Bioformulations are found to be more effective than synthetic chemicals because formulations synthesized by a single microbe can interact with plant pathogens and can have a role in plant growth promotion and disease inhibition (Arora et al. 2010). The abiotic substrates that have the ability to provide protective environment to cells and can deliver cells viably under proper physiological environment are used as carrier molecules. Various types of substrate can be employed as carriers such as inert substances (perlite, vermiculite, polymers), some liquids, and soils (clays, peat, coal) (Bashan et al. 2014). Majority of the studies executed emphasize on the (i) development of better carrier molecules; (ii) search for microbes which can enhance crop yield; and (iii) enhancement of metabolic state and potential of the cells, so that they can be used as intercellular storage devices and can survive within carrier molecules (Kadouri et al. 2005). Bioformulations have to go through various stressful conditions during storage and production, where microbes have to survive through different situations like desiccation, hot conditions, etc. Microbes should retain high survival rates and sustain their capabilities to enhance plant development for extended durations. Various strategies are used by microbes for their survival like formation and aggregation of polyhydroxyalkanoates (PHA) or osmolytes. Higher level of tolerance toward desiccation is shown by microbes which are osmoadapted and osmolytes such as glycine betaine or trehalose are aggregated by them, in comparison to the cells which are non-osmoadapted (Bonaterra et al. 2005). Those cells which have higher PHA levels have more survival potential than those cells which have lower PHA levels; this is due to the fact that PHA imparts caliber to the cells so that they can withstand against the unfavorable physical and chemical strains (Morel et al. 2012). The most remarkable microbial-plant synergism is seen in the diazotrophic microbial relationship with plants. Diazotrophs may be symbiotic or free-living microbes which have potential to fix as well as reduce atmospheric nitrogen into ammonia; some examples of diazotrophs include rhizobia, *Azotobacter*, *Gluconacetobacter diazotrophicus*, *Azospirillum*, and *Azoarcus*. Certain phytohormones, ACC deaminase, phosphate-solubilizing molecules, iron-sequestering siderophores, and other molecules are produced by some plant growth-promoting bacteria (PGPB) and diazotrophs. *Bacillus* and *Pseudomonas* come under non-diazotrophic, plant growth-promoting bacteria (Morel and Castro-Sowinski 2013). The most commonly studied symbiotic relationship includes legume-rhizobia association for agricultural improvement and productivity enhancement. Such type of associations provides enough evidence that unification of different valuable microbes, exhibiting distinct routes of plant growth enhancement, has cumulative and synergic influence on plant development and crop yield (Morel et al. 2012). Several reports also suggest that combo of secondary metabolites produced by plants with bioformulations may amplify the agricultural productivity. However, nature-friendly bioformulations are not so popular in agro-market because of some limitations associated with them (Morel et al. 2015). The constraints include sustenance of microbial biota and vigor, unpredictable field performance, inconsistency in quality, and inadequate shelf life. There are some queries which are yet to be answered so as to gain trust of consumers and to make agricultural practices chemical free (Arora et al. 2010). To answer the questions, it is essential to ascertain the work done in the previous reports and also to know the idea in which mainstream research is going so as to determine the future aspects for the development of superior bioformulations. This chapter deals with the current and future aspects of bioformulations.

10.2 Current Situation of Bioformulations

For the elimination of plant pathogens, microbial-based formulations are being utilized all around the globe, but the supportive information about its usage all over the world is very limited (Leggett et al. 2011; Naderifar and Daneshian 2012). One of the major reasons is difference in terminology. Majority of the developing countries use the term "biofertilizers," whereas in developed countries the term "bioinoculant" for crop yield and improvement is used, but in both of the cases either compounds are isolated from living organisms or whole organism is employed for the enhancement in nutrient uptake by plants so as to improve crop yield and soil quality (Vessey 2003; Chen et al. 2006; Prasad et al. 2018). Many producers (farmers) around the globe regularly employ biofertilizers and biopesticides into their different types of crops. The most advanced and prevalent market for formulations is European biofertilizer market as compared to all other domains, and growth from \$2566.4 million in 2012 to \$4582.2 million was observed in 2017, at an annual growth rate of 12.3% from 2012 to 2017 (PRWEB 2014). In 2012, biofertilizer market was highest in North America and is expected to develop at the rate of 14.4%

in the duration of 2013–2018 (Micro Market Monitor 2015). China is the chief grower of wheat and rice along with cabbage and onions as well as promotes the use of biofertilizers (Grand View Research 2015). In India, there are around 151 biofertilizer manufacturing units which are run by government and nongovernment agencies (Mahajan and Gupta 2009). Mainly nitrogen-fixing biofertilizers were used in maximum as compared to all other biofertilizers, and in 2012 their worldwide demand increased over 78% (Agro news 2014). In controlling plant diseases, *Bacillus thuringiensis* (BT)-based biopesticides are more popular and constitute for 95% of the overall microorganisms employed (Bravo et al. 2011). Around 322 BT products generate a revenue of $210 million annually (CAB International Centre 2010). Usage of fungal and non-BT biopesticides is also growing. Various agencies have carried out market research survey, but outcomes of these kinds of surveys are not very reliable and are questionable. This is because criteria employed in market research survey may deviate since many firms and agro-based industries involve subcategories such as biochemicals, plant-induced protectants (PIPs), microbes, plant growth regulators, pheromones, insect growth regulators, and essential oil in the name "biopesticide," while others use only the products which are microbial in origin (Gelernter 2007). Around $672 million was biopesticide turnover in 2005 (description of category was not included) (Thakore 2006), and worldwide market of biopesticide was about $280 million in 2007 (for true microbial agents) (Harwood et al. 2007). CPL (2006) and BCC (2010) are business consultancies which are vigorously conducting direct market survey and generating a reliable data, and their reports show that biopesticide market is enhancing at a rate of 10% every year, globally. Biopesticide market was expected to exceed by $2.5 billion by 2015, via global industry analysis 2015. Other research surveys conducted by BCC on biopesticide suggested that total sale of biopesticide in 2008 was $1.2 billion and in 2009 it was $ 1.6 billion. And it was expected to increase in 2014 to around $3.3 billion and in 2017 around $10 billion (Marrone 2007). Region-wise research reports highlight that the United States is the largest region of biopesticide globally, while Europe is the fast-growing regional market for biopesticide and represents an average annual growth rate of 15.0%. Asia Pacific is also an emerging market for biopesticides, where sales were expected to be around $362 million in 2012. Latin America has shown very little increase as compared to other regions. In 2005, market was about $70 million, and in 2010 it reached only $88 million, with an average annual growth rate of 5.0% (Industrial Equipment News 2011). Table 10.1 shows the types of microbes, main crops, and areas in world utilizing microbes as bioinoculants/bioformulations.

10.3 Consortia- or Inoculant-Based Bioformulations

Majority of the bioformulations available are mostly composed of single strain or mixed cultures or are in co-inoculations with other microbes. Use of such co-inoculations helps in overall development and plant growth promotion. When mycorrhiza co-inoculated with rhizobia, it displayed enhanced performance with legumes. This association helps to enhance the nutritional value of most nodulated plants and

Table 10.1 Types of crop, microbes, and agricultural market of bioinoculants

Market by type	Market by mode of application	Market by microbial type	Market by type of crop	Market by type of geography
Biocontrol agents	Seed inoculant	Bacteria: rhizobacteria N_2 fixing P-solubilizing Cyanobacteria Others	Cereals and grains	North America: USA, Canada, Mexico, other North American countries
Plant resistance stimulants	Soil inoculant	Fungi: *Trichoderma* sp., AM fungi *Aspergillus*	Oil seed and pulses	Europe: Germany, France, Spain, Italy, Denmark, other European countries
Plant growth-promoting microbes	Foliar spray	Others: Azolla-Anabaena	Fruits and vegetables	Asia Pacific: Australia, China India, Japan, other countries
Plant stress manager	Other inoculants		Other crop types	South America: Brazil, Chile, Argentina. LAMEA countries (Latin America, Middle East, and African countries)

also raises the tolerance toward drought and other osmotic stress in pigeon pea (Bhattacharjee and Sharma 2012), lucerne (Ardakani et al. 2009), soybean (Gao et al. 2012), broad bean (Jia et al. 2004), and chickpea (Tavasolee et al. 2011). Several reports also suggest that plant growth is stimulated after the employment of nodule-forming bacteria with Phosphate solubilizing bacteria (PSB) in leguminous plants (Messele and Pant 2012). Currently, several studies have been conducted related to consortia formulation development by different researchers and patents have also been filed (Paikray and Malik 2010). Maiyappan et al. (2010) conducted a study in which a bioformulation consortium was prepared (as a wettable powder) which involved nine strains of the following genera—*Frauteria*, *Bacillus*, *Azotobacter*, and *Streptomyces*, and this formulation was found to be useful for black gram. Similarly, consortium bioformulation was prepared using *Burkholderia* species MSSP plus three other plant growth-promoting bacteria and was examined for the development of *Cajanus cajan*, by employing different carriers like sawdust, rock phosphate, bagasse, wheat bran, cocoa peat, charcoal, rice husk, and paneer whey, and this consortium was found to be effective in enhancement of pigeon plant growth, when used as a formulation (Pandey and Maheshwari 2007). Studies conducted by Tajini et al. (2012) demonstrated that in bean plants, when arbuscular mycorrhizal (AM) fungi and rhizobia are inoculated in combination, they help in enhancing uptake of nitrogen and potassium in plants as compared to single inoculation. Further, according to some researchers, the consortia can be prepared using *Azotobacter*, microalgae, and cyanobacteria, which can be employed as biofertilizer and bio-stimulator (Zayadan et al. 2014). Figure 10.1 shows the basic comparison of plant growth promoting rhizobacteria (PGPR) and bioformulations used in agroecosystem.

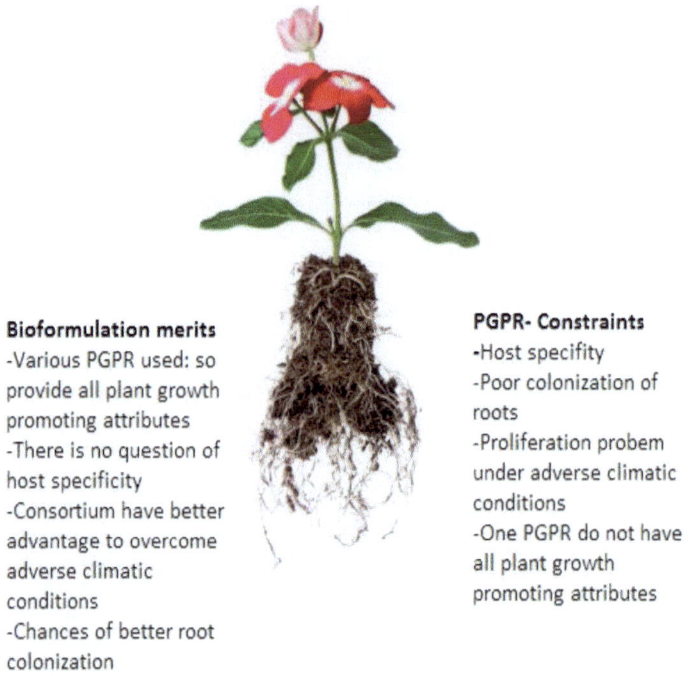

Fig. 10.1 Some comparisons of bioformulations with PGPR under agro-climatic conditions

10.3.1 Inoculums Production Approaches

10.3.1.1 Inoculants Based on Carriers

A carrier is usually locally available and reasonably affordable material that has efficient water-holding capacity and is competent enough to slowly liberate its viable cells into soil. Agro-industrial residues, peat, compost, vermiculture, charcoal mixed with soil, bentonite, and perlite are some of the commonly available and economically friendly organic materials that are utilized for making microbial inoculants. After a suitable carrier is selected, it is sterilized and amalgamated with definite microbes such as phosphate-solubilizing or nitrogen-fixing microbes under sterile environment. Such carrier-based bioformulations can be stored only for 3–4 months and have high cell density. When stored at room temperature, cell density decreases at a very high rate (Trivedi et al. 2005).

10.3.1.2 Inoculants Based on Clays

The most widely utilized formulation in agricultural domain is "clay." It may be employed in a number of forms such as powder, suspension, or granules. Use of clays can increase the shelf life of microbial strains as they have large pore size and surface area. It offers suitable conditions for microbes to flourish (Smith 1992).

10.3.1.3 Liquid Formulations

Liquid bioformulations are made by using specific microbes which are prepared in a mineral-based medium, broth-based medium, or organic oil-based suspension (Schisler et al. 2004). When these liquid formulations are employed, the microbial inoculants come in direct contact with soil or seeds and perform their action; however, protective environment for microbial strains is absent. This kind of formulation is not suitable for long-term storage and is highly susceptible to contamination during storage. Suitable environmental conditions are required for maintenance so that microbial strains do not lose their viability and efficacy (Brar et al. 2006).

10.3.1.4 Encapsulation

In bio-encapsulation procedure, active strains are incorporated into solid or liquid matrix, where materials like polystyrene, alginate, agarose, polyacrylamides, agar-agar, polyurethane, carrageenan, synthetic polymers, etc. are utilized as matrix (John et al. 2011). The stabilization is done by chemical polymerization. It can be stored as long as 5 years.

10.3.1.5 Use of Biofilms

Biofilms are based on microbial cell aggregates. Biofilm involves four stages in its development which are (a) primary adherence, (b) irrevocable immobilization by generation of exo-polysaccharide, (c) prematuration stage, and (d) maturation stage. Biofilms prepared by employing fungal and bacterial strains have been found adequate (Seneviratne et al. 2008).

10.3.2 Outcome of Inoculants/Formulations

10.3.2.1 Mycorrhizal Fungi Inoculants

Arbuscular mycorrhizal fungi belonging to phylum *Glomeromycota* have potential to form symbiotic relationship with majority of the land plants (Schubler et al. 2001; Prasad et al. 2017a). They use their host as a carbon source and help the host plant by promoting its growth by providing nutrients, water, and minerals. AMF influence the soil microbes and form mycorrhizospheric zone in the soil (Linderman 1988). The AMF present in mycorrhizosphere may have positive (Albertsen et al. 2006) or negative (Cavagnaro et al. 2006) or no effects (Olsson et al. 1996) at all on the growth of microbes or microbial biomass. Several studies show that certain bacterial species behave differently in the vicinity of specific AMF; this suggests that bacteria and AMF might have high degree of specificity between them (Artursson et al. 2006). Therefore, AMF associated with specific bacteria may prove to be useful for plant development by providing essential nutrients, enhancing branching of roots (Barea 1997). AMF also provide protective benefits by inhibiting the proliferating pathogens by forming the bacterial populations which limits the invasion of pathogens (St-Arnaud and Vujanovic 2007). *Glomus intraradices* shows beneficial effects on the development of bacterial fauna and saprotrophic fungal biomass (Albertsen et al. 2006).

10.3.2.2 *Azospirillum* Inoculants

Agricultural benefits of *Azospirillum* are well known (Okon and Labandera-Gonzalez 1994). *Azospirillum* inoculation highly influences the root exudation and development; the utilization of these phyto-stimulatory plant growth promoters can also affect the microbial fauna present in rhizospheric zone (Dobbelaere et al. 2001). In some studies, *Azospirillum brasilense* sp 245 was found to be useful in plant growth, as it helps in generation of auxins, gibberellins, and cytokinins (Steenhoudt and Vanderleyden 2000). Naiman's study suggested that when *Azospirillum* and *Pseudomonas* are co-inoculated on wheat fields, they displayed different effects on the bacterial communities flourishing on rhizospheric zone of the wheat. This inoculation altered the carbon source usage of soil microbial communities (Naiman et al. 2009). Carbon source utilization is directly related to the total microbial population capable to utilize every carbon source, and their growth reflects the functional capabilities of the microbial community. The two *A. brasilense* strains (42M and 40M) isolated from roots of maize when inoculated altered the physiological profiles of the microbial flora and fauna linked to rice (De Salamone et al. 2010).

10.3.2.3 Rhizobia Inoculants

As per several reports, rhizobia have enhanced effect on crop yield and plant development. They help in nitrogen fixation and uptake of nutrients like iron and phosphorus; stimulate plant hormones; promote the growth of favorable bacteria and fungi; and check fungal and bacterial diseases, insects, and pests. The entire rhizospheric biodiversity can be seen in the form of different functional groups which is mostly influenced by the alterations in the residual nitrogen rather than the effects of inoculation (Antoun and Prévost 2005). Further, the exact mechanisms which play key role in these changes are still not clearly known and need to be explored (Saharan and Nehra 2011).

10.3.2.4 Biocontrol Agents

Majority of rhizobacterial products have been employed as biological control for plant disease prevention instead of effecting plant nutrition uptake or dealing with abiotic stress factors (Berg 2009). Variety of microorganisms like *Bacillus subtilis*

(Dawar et al. 2010), *Trichoderma harzianum* (Mohiddin et al. 2010), and *Pseudomonas fluorescens* (Peighami-Ashnaei et al. 2009) have antagonistic effect against the diseases caused by *Sclerotium* species, *Pythium* species, *Rhizoctonia* species, and *Fusarium* species, which leads to high yield or plant growth promotion. *P. fluorescens* (Pal et al. 2000), *Pochonia chlamydosporia* (Kerry 2000), and *B. subtilis* (Khan et al. 2001) when employed are found to be effective against the diseases triggered by nematodes. Some *Pseudomonas* species are antagonist against the activity of tomato plant pathogen *Ralstonia solanacearum* (Kozdroj et al. 2004). Biocontrol agents like *Corynrbacterium glutamicumin* (Vahjen et al. 1995), *P. fluorescens* (Natsch et al. 1998), and *Streptomyces melanosporofaciens* (Prevost et al. 2006) when applied show transient effects on soil ecosystem and fungal flora and fauna, and it designates that the efficacy of the biocontrol agent may be for limited duration.

10.3.3 Effects of Co-inoculations vs Mono-inoculations

In most of the inoculations, usually single strain is applied which has displayed inconsistent results in the fields. This issue can be solved by employing different strains or different species of valuable microorganisms in the single formulation of microbes. By applying consortium of bioinoculates, there is no need of genetic engineering (Janisiewicz 1996) as different strains involve different working mechanisms to enhance the plant development, promise the efficiency and reliability that it will have positive effects on crops (Marimuthu et al. 2002). It was reported that when PGPR or AMF were inoculated, they induced slight modification on the bacterial community structure present in wheat rhizosphere (Roesti et al. 2006). The type of PGPR consortium utilized had additional effect on the bacterial community structure as compared to the AMF. Further PGPR strains employed produce an antibiotic, i.e., 2-4-diacetylphloroglucinol, which has antifungal effect; however, it does not affect the growth of AMF, and associative or synergic outcome of PGPR and AMF co-inoculation was seen.

When *Pinus pinea* was inoculated with two different strains of *Bacillus*, i.e., *Bacillus pumilus* CECT105 and *Bacillus licheniformis* CECT5160, both of these *Bacillus* strains encourage the seedling development of *P. pinea* (Probanza et al. 2002). But this positive effect was not observed when both the *Bacillus* strains were co-inoculated, maybe as a result of competition effect (Probanza et al. 2002). Further, the combo of *A. brasilense* and *B. subtilis* did not display any synergic or associative effects on tomato plants in comparison to their single inoculations. Therefore, it may be concluded that when inoculants are employed in combination, they may not necessarily show synergic effects, instead they may display competitive effects, and as a result the growth and development could be minimum or gradually disappear. Similarly, the effects on microbial flora and fauna present in soil are also uncertain or unpredictable.

10.4 Problems, Challenges, and Approaches

The current researchers are becoming more aware and focusing their work to deal with growing problems like increasing urbanization and growing population. It is expected that by 2050, about 10 billion people would inhabit the planet. With increase in population, many issues will also arise, putting more pressure to produce food, fiber, and energy resources with simultaneous sustainable approach. The growing demands cannot be avoided; however, this will put immense risk to nonrenewable resources like fossil fuels, water, energy resources, agricultural soil, etc. Further, with more and more expanding industries, contamination is also at its highest peak (Browne et al. 2013). Emission of greenhouse gases is causing the rise in earth's temperature, disturbing the environmental stability, and giving rise to various stressful scenarios which affect both agriculture and natural systems (Duarte et al. 2006). Problems like salinity of soil, drought, nutrition deficiency, diseases, soil erosion, pests, crop destruction due to natural calamities, loss of biodiversity, deforestation, landscape fragmentation, use of chemicals, etc. are affecting humans either directly or indirectly (Vitousek et al. 1997).

In view of these issues, sustainable agricultural practices and various approaches are being implemented to meet the demands without hampering the natural ecosystems; this can be achieved only when there is balance between the three major interacting domains, i.e., environment, society, and economy. The balanced interaction cycle between these three domains can finally give the true meaning to the "sustainable development" approach (Altieri 2004). In the concept of sustainability, the vital issues are regarding depletion of nonrenewable resources, controlling pests and pathogens, suitable methods for recycling soil nutrients, dealing with abiotic stress, maintaining the vitality of soil microbes (which depends on soil microbes), etc. for the global ecosystems and human welfare (Zancarini et al. 2013; Prasad et al. 2014, 2017b). Majority of these issues can be resolved by using microbial services (Zolla et al. 2013). Microorganisms can be exploited after the identification of their beneficial functions/features in terms of both cost-effective and ecological sustainability.

10.5 Conclusions

The discussion in this chapter regarding development and uses of some novel bioformulations will definitely be useful in sustainable agroecosystem. The application of microbial consortium is an essential constituent of agroecological practice, which is a reliable technology whose time has come to sustain the soil and fulfill the requirement of food and feed. These microbes in the bioformulations have been successfully employed in several parts of the globe and with encouraging results, that with time this notion of utilizing microbial consortium as bioformulations will certainly grow. In the case of underdeveloped, developing, and developed world, where agricultural inputs are synthetic and quite expensive, the application of bioformulations conquers an insignificant but developing role in organic agriculture development. Additionally, the microbial consortium in bioformulations also acts as biocontrol and stress manager and in many phytoremediation approaches. The cost-effectiveness and easy availability of bioformulations have made it a choice of farmers and scientists, since this is also a step toward organic agriculture. According to one survey carried in the USA, it was found that both organic and conventional growers are taking interest in growing and consuming organic products, since such products are not having side effects on health as caused by chemically growing agriproducts. Rzewnicki (2000) suggested that in the coming years the bioformulations demand will have huge market potential. Globally, the sale of organic products grew by 8% in the year 2010, which is expected to increase by 27% by the year 2020 (Komorowska 2014). In relation to organic production using organic manure and bioformulations, encouraging growth has been seen and still continued in the main European markets and in the case of the USA, and the viewpoint regarding bioformulations for coming years is very positive. Interestingly, China's organic market has grown four times in the last 5 years only. Moreover, in the case of Brazil, the organic produce has shown the growth rate of 40%, which is quite amazing. Regarding Asia, the market analysts forecast that organic sales using bioformulations is expected to grow by 20% a year over the next coming 3 years. All over the globe, 37 mh of land are now farmed organically, and most of the land also utilizes the microbes in one form or another. India has to work hard to achieve the good growth of organic farming using microbial bioformulations. The agricultural universities and institutions working on bioformulations development can work on these aspects using the help of state agricultural departments.

References

Agro news (2014) Biofertilizers market–global industry analysis, size, share, growth, trends and forecast, 2013–2019. Available online http://news.agropages.com/News/NewsDetail-11612–e.htm

Albertsen A, Ravnskov S, Green H, Jensen DF, Larsen J (2006) Interactions between the external mycelium of the mycorrhizal fungus *Glomus intraradices* and other soil microorganisms as affected by organic matter. Soil Biol Biochem 38(5):1008–1014

Altieri MA (2004) Linking ecologists and traditional farmers in the search for sustainable agriculture. Front Ecol Environ 2(1):35–42

Antoun H, Prévost D (2005) Ecology of plant growth promoting rhizobacteria. In: PGPR: biocontrol and biofertilization. Springer, Dordrecht, pp 1–38

Ardakani MR, Pietsch G, Moghaddam A, Raza A, Friedel JK (2009) Response of root properties to tripartite symbiosis between lucerne (*Medicago sativa* L.), rhizobia and mycorrhiza under dry organic farming conditions. Am J Agric Biol Sci 4:266–277

Arora NK, Khare E, Maheshwari DK (2010) Plant growth promoting rhizobacteria: constrains in bioformulation, commercialization, and future strategies. In: Maheshwari DK (ed) Plant growth and health promoting bacteria microbiology. Springer, Berlin, pp 97–116

Artursson V, Finlay RD, Jansson JK (2006) Interactions between arbuscular mycorrhizal fungi and bacteria and their potential for stimulating plant growth. Environ Microbiol 8:61–70

Barea JM (1997) Mycorrhiza/bacteria interactions on plant growth promotion. In: Plant growth-promoting rhizobacteria, present status and future prospects. OECD, Paris, pp 150–158

Bashan Y, de-Bashan LE, Prabhu SR, Hernandez JP (2014) Advances in plant growth-promoting bacterial inoculant technology: formulations and practical perspectives (1998–2013). Plant Soil 378:1–33

BCC Research (2010) Biopesticides: the global market report CHM029C, Wellesley

Berg G (2009) Plant–microbe interactions promoting plant growth and health: perspectives for controlled use of microorganisms in agriculture. Appl Microbiol Biotechnol 84(1):11–18

Berny P (2007) Pesticides and the intoxication of wild animals. J Vet Pharmacol Ther 30:93–100

Bhattacharjee SG, Sharma D (2012) Effect of dual inoculation of Arbuscular Mycorrhiza and *Rhizobium* on the chlorophyll, nitrogen and phosphorus contents of pigeon pea (*Cajanus cajan* L.). Adv Microbiol 2:561–564

Bonaterra A, Camps J, Montesinos E (2005) Osmotically induced trehalose and glycine betaine accumulation improves tolerance to desiccation, survival and efficacy of the postharvest biocontrol agent *Pantoea agglomerans* EPS125. FEMS Microbiol Lett 250:1–8

Brar SK, Verma M, Tyagi RD, Valéro JR (2006) Recent advances in downstream processing and formulations of *Bacillus thuringiensis* based biopesticides. Process Biochem 41(2):323–342

Bravo A, Likitvivatanavong S, Gill SS, Sobero NM (2011) *Bacillus thuringiensis*: a story of a successful bioinsecticide. Insect Biochem Mol Biol 41:423–431

Browne P, Barret M, Morrissey JP, O'Gara F (2013) Molecular based strategies to exploit the inorganic phosphate-solubilization ability of *Pseudomonas* in sustainable agriculture. Mol Microbial Ecol Rhizosphere 1:615–628

CAB International Centre (2010) The 2010 world wide biopesticides market summary. CAB International Centre, Wallingford

Cavagnaro TR, Jackson LE, Six J, Ferris H, Goyal S, Asami D, Scow KM (2006) Arbuscular mycorrhizas, microbial communities, nutrient availability, and soil aggregates in organic tomato production. Plant Soil 282(1–2):209–225

Chen YP, Rekha PD, Arun AB, Shen FT, Lal WA, Young CC (2006) Phosphate solubilizing bacteria from subtropical soil and their tricalcium phosphate solubilizing abilities. Appl Soil Ecol 34:33–41

Colt JS, Cyr MJ, Zahm SH, Tobias GS, Hartge P (2007) Inferring past pesticide exposures: a matrix of individ-ual active ingredients in home and garden pesticidesused in past decades. Environ Health Perspect. 115:248–254

CPL (2006) Biopesticides 2007. CPL Business Consultants, Wallingford

Dawar S, Wahab S, Tariq M, Zaki MJ (2010) Application of *Bacillus* species in the control of root rot diseases of crop plants. Arch Phytopathol Plant Protect 43(4):412–418

De Salamone IE, Di Salvo LP, Ortega JS, Sorte PM, Urquiaga S, Teixeira KR (2010) Field response of rice paddy crop to *Azospirillum* inoculation: physiology of rhizosphere bacterial communities and the genetic diversity of endophytic bacteria in different parts of the plants. Plant Soil 336(1–2):351–362

Dobbelaere S, Croonenborghs A, Thys A, Ptacek D, Vanderleyden J, Dutto P, Labandera-Gonzalez C, Caballero-Mellado J, Aguirre JF, Kapulnik Y, Brener S (2001) Responses of agronomically important crops to inoculation with *Azospirillum*. Funct Plant Biol 28(9):871–879

Dorigo U, Lefranc M, Leboulanger C, Montuelle B, Humbert JF (2009) Spatial heterogeneity of periphytic microbial communities in a small pesticide–polluted river. FEMS Microbiol Ecol 67:491–501

Duarte CM, Alonso S, Benito G, Dachs J, Montes C, Pardo Buendía M, Ríos AF, Simó R, Valladares F (2006) Global change. Impact of human activity on the Earth system. CSIC Superior Council of Scientific Investigations, Madrid

Fenske RA, Day EW Jr (2005) Assessment of exposure for pesticide handlers in agricultural, residential and institutional environments. In: Franklin CA, Worgan JP (eds) Occupational and residential exposure assessment for pesticides. Wiley, Chichester, pp 13–43

Gao X, Lu X, Wu M, Zhang H, Pan R, Tian J, Li S, Liao H (2012) Co-inoculation with rhizobia and AMF inhibited soybean red crown rot: from field study to plant defense-related gene expression analysis. PLoS One 7(3):e33977. https://doi.org/10.1371/journal.pone.0033977

Gelernter WD (2007) Microbial control in Asia: a bellwether for the future? J Invertebr Pathol 95:161–167

Global Industry Analysts (2015) Global biopesticides market to reach US$2.8 billion by 2015, according to a new report by global industry analysts, Inc. Available online http://prweb.com/printer/8041130.htm

Grand View Research (2015) Biofertilizers market analysis by product (nitrogen fixing, phosphate solubilizing), by application (seed treatment, soil treatment) and segment forecasts to 2022. Available online. http://www.grandviewresearch.com/industry–analysis/biofertilizers–industry

Harwood RWJ, Lee MSK, Lisansky SG, Quinlan R (2007) Current worldwide markets for biopesticides and success factors for the business. In: Proceedings of the XVIth international plant protection congress/BCPC international congress. Crop Science and Technology, Glasgow, pp 598–599

Industrial Equipment News (2011) Biopesticides market to reach $1 billion in 2010. Available online http://www.ien.com/article/biopesticides–market–to/8648

Janisiewicz W (1996) Ecological diversity, niche overlap, and coexistence of antagonists used in developing mixtures for biocontrol of postharvest diseases of apples. Phytopathology 86:473–479

Jia Y, Gray VM, Straker CJ (2004) The influence of Rhizobium and arbuscular mycorrhizal fungi on nitrogen and phosphorus accumulation by *Vicia faba*. Ann Bot 94:251–258

John RP, Tyagi RD, Brar SK, Surampalli RY, Prévost D (2011) Bio-encapsulation of microbial cells for targeted agricultural delivery. Crit Rev Biotechnol 31(3):211–226

Johnsen K, Jacobsen CS, Torsvik V (2001) Pesticide effects on bacterial diversity in agricultural soils: a review. Biol Fertil Soils 33:443–453

Kadouri D, Jurkevitch E, Okon Y, Castro-Sowinski S (2005) Ecological and agricultural significance of bacterial polyhydroxyalkanoates. Crit Rev Microbiol 31:55–67

Kerry BR (2000) Rhizosphere interactions and the exploitation of microbial agents for the biological control of plant-parasitic nematodes. Annu Rev Phytopathol 38(1):423–441

Kesavachandran C, Pathak MK, Fareed M, Bihari V, Mathur N, Srivastava AK (2009) Health risks of employees working in pesticide retail shops: an exploratory study. Indian J Occup Environ Med 13:121–126

Khan MR, Khan N, Khan SM (2001) Evaluation of agricultural materials as substrate for mass culture of fungal biocontrol agents of fusarial wilt and root-knot nematode diseases. Ann Appl Biol (TAC-21 Suppl) 22:50–51

Komorowska M (2014) Innovative bioformulations for seed treatment. Preliminary assessment of functional properties in the initial plant growth phase. Przemysl chemiczny. 93:959–963

Kozdroj J, Trevors JT, Van Elsas JD (2004) Influence of introduced potential biocontrol agents on maize seedling growth and bacterial community structure in the rhizosphere. Soil Biol Biochem 36(11):1775–1784

Leggett M, Leland J, Kellar K, Epp B (2011) Formulation of microbial biocontrol agents–an industrial perspective. Can J Plant Pathol 33:101–107

Liang LZ, Zhao X, Yi XY, Chen ZC, Dong XY, Chen RF, Shen RF (2013) Excessive application of nitrogen and phosphorus fertilizers induces soil acidification and phosphorus enrichment during vegetable production in Yangtze River Delta, China. Soil Use Manag 29:161–168

Linderman RG (1988) Mycorrhizal interactions with the rhizosphere microflora: the mycorrhizosphere effect. Phytopathology 78(3):366–371

Mahajan A, Gupta RD (2009) Bio-fertilizers: their kinds and requirement in India. In: Mahajan A, Gupta RD (eds) Integrated nutrient management (INM) in a sustainable rice-wheat cropping system. Springer, Dordrecht, pp 75–100

Maiyappan S, Amalraj ELD, Santhosh A, Peter AJ (2010) Isolation, evaluation and formulation of selected microbial consortia for sustainable agriculture. J Biofertil Biopestic 2:109–121

Mandal A, Patra AK, Singh D, Swarup A, Masto RE (2007) Effect of long-term application of manure and fertilizer on biological and biochemical activities in soil during crop development stages. Bioresour Technol 98:3585–3592

Marimuthu S, Subbian P, Ramamoorthy V, Samiyappan R (2002) Synergistic effect of combined application of *Azospirillum* and *Pseudomonas fluorescens* with inorganic fertilizers on root rot incidence and yield of cotton. J Plant Dis Protec 109(6):569–577

Marrone PG (2007) Barriers to adoption of biological control agents and biological pesticides. CAB Rev: Perspect Agri Vet Sci Nutr Natur Resour, CAB International, Wallingford, pp 2–51

Mendes R, Kruijt M, de Bruijn I, Dekkers E, van der Voort M, Schneider JH, Piceno YM, DeSantis TZ, Andersen GL, Bakker PA, Raaijmakers JM (2011) Deciphering the rhizosphere microbiome for disease-suppressive bacteria. Science 332(6033):1097–1100

Messele B, Pant LM (2012) Effects of inoculation of *Sinorhizobium ciceri* and phosphate solubilizing bacteria on nodulation, yield and nitrogen and phosphorus uptake of chickpea (*Cicer arietinum* L.) in Shoa Robit Area. J Biofertil Biopestic 3:1000129

Micro Market Monitor (2015) North America biofertilizer market by application (cereals & grains, fruits & vegetables, pulses & oilseeds), by type (nitrogen fixing biofertilizers, phosphate solubilizing biofertilizers, potash mobilizing biofertilizers), by source, by geography –analysis and forecast to 2019. Available online http://www.micromarketmonitor.com/market/north–america–bio–fertilizer–5250154124.html

Mitchell CC, Westerman RL, Brown JR, Peck TR (1991) Overview of long-term agronomic research. Agron J 83:24–25

Mohiddin FA, Khan MR, Khan SM, Bhat BH (2010) Why *Trichoderma* is considered super hero (super fungus) against the evil parasites? Plant Pathol J 9(3):92–102

Morel MA, Castro-Sowinski S (2013) The complex signaling network in microbe-plant interaction. In: Arora NK (ed) Plant microbe symbiosis: fundamentals and advances. Springer, New Delhi, pp 134–149

Morel MA, Braña V, Castro-Sowinski S (2012) Legume crops, importance and use of bacterial inoculation to increase the production. In: Goyal A (ed) Crop plant. InTech, Rijeka, pp 217–240

Morel MA, Cagide C, Minteguiaga MA, Dardanelli MS, Castro-Sowinski S (2015) The pattern of secreted molecules during the co-inoculation of alfalfa plants with *Sinorhizobium meliloti* and *Delftia* sp. strain JD2: an interaction that improves plant yield. Mol Plant-Microbe Interact 28:134–142

Moses M, Johnson ES, Anger WK, Burse VW, Horstman SW, Jackson RJ (1993) Environmental equity and pesticide exposure. Toxicol Health 9:913–959

Naderifar M, Daneshian J (2012) Effect of different nitrogen and biofertilizers effect on growth and yield of *Brassica napus* L. Int J Agric Crop Sci 4:478–482

Naiman AD, Latrónico A, de Salamone IE (2009) Inoculation of wheat with *Azospirillum brasilense* and *Pseudomonas fluorescens*: impact on the production and culturable rhizosphere microflora. Eur J Soil Biol 45(1):44–51

Nakhro N, Dkhar MS (2010) Impact of organic and inorganic fertilizers on microbial populations and biomass carbon in paddy field. Soil J Agron 9:102–110

Natsch A, Keel C, Hebecker N, Laasik E, Défago G (1998) Impact of *Pseudomonas fluorescens* strain CHA0 and a derivative with improved biocontrol activity on the culturable resident bacterial community on cucumber roots. FEMS Microbiol Ecol 27(4):365–380

Okon Y, Labandera-Gonzalez CA (1994) Agronomic applications of *Azospirillum*: an evaluation of 20 years worldwide field inoculation. Soil Biol Biochem 12:1591–1601

Olsson PA, Bååth E, Jakobsen I, Söderström B (1996) Soil bacteria respond to presence of roots but not to mycelium of arbuscular mycorrhizal fungi. Soil Biol Biochem 28(4–5):463–470

Paikray S, Malik V (2010) Microbial formulation for widespread used in agricultural practices: google patents

Pal KK, Tilak KV, Saxena AK, Dey R, Singh CS (2000) Antifungal characteristics of a fluorescent *Pseudomonas* strain involved in the biological control of *Rhizoctonia solani*. Microbiol Res 155(3):233–242

Pandey P, Maheshwari DK (2007) Bioformulation of *Burkholderia* sp. MSSP with a multispecies consortium for growth promotion of *Cajanus cajan*. Can J Microbiol 53:213–222

Peighami-Ashnaei S, Sharifi-Tehrani A, Ahmadzadeh M, Behboudi K (2009) Interaction of different media on production and biocontrol efficacy of *Pseudomonas fluorescens* P-35 and *Bacillus subtilis* B-3 against grey mould of apple. J Plant Pathol 1:65–70

Pimentel D, Acquay H, Biltonen M, Rice P, Silva M, Nelson J (1992) Environmental and economic costs of pesticide use. Bioscience 42:750–760

Prevost K, Couture G, Shipley B, Brzezinski R, Beaulieu C (2006) Effect of chitosan and a biocontrol Streptomycete on field and potato tuber bacterial communities. BioControl 51(4):533–546

Probanza A, Garcıa JL, Palomino MR, Ramos B, Mañero FG (2002) *Pinus pinea* L. seedling growth and bacterial rhizosphere structure after inoculation with PGPR *Bacillus* (*B. licheniformis* CECT 5106 and *B. pumilus* CECT 5105). Appl Soil Ecol 20(2):75–84

Prasad R, Kumar V, Prasad KS (2014) Nanotechnology in sustainable agriculture: present concerns and future aspects. Afr J Biotechnol 13(6):705–713

Prasad R, Bhola D, Akdi K, Cruz C, Sairam KVSS, Tuteja N and Varma A (2017a) Introduction to mycorrhiza: Historical development. In: Mycorrhiza (eds. Varma A, Prasad R and Tuteja N) Springer International Publishing AG 1–7

Prasad R, Bhattacharyya A, Nguyen QD (2017b) Nanotechnology in sustainable agriculture: Recent developments, challenges, and perspectives. Front Microbiol 8:1014. doi: 10.3389/fmicb.2017.01014

Prasad R, Gill SS, Tuteja N (2018) Crop Improvement Through Microbial Biotechnology. Elsevier (ISBN: 9780444639882)

PRWEB (2014) Europe bio fertilizer market is expected to reach $4,582.2 million in 2017 new report by MicroMarket Monitor. Available online http://www.micromarketmonitor.com/market/europe–bio–fertilizer–4637178345.html

Roesti D, Gaur R, Johri BN, Imfeld G, Sharma S, Kawaljeet K, Aragno M (2006) Plant growth stage, fertiliser management and bio-inoculation of arbuscular mycorrhizal fungi and plant growth promoting rhizobacteria affect the rhizobacterial community structure in rain-fed wheat fields. Soil Biol Biochem 38(5):1111–1120

Rzewnicki P (2000) Ohio organic producers: final survey results. Online. Ohio State University Extension, College of Food Agricultural and Environmental Sciences. Bulletin, Special Circular 174

Saharan BS, Nehra V (2011) Plant growth promoting rhizobacteria: a critical review. Life Sci Med Res 21(1):30–39

Sanborn M, Kerr KJ, Sanin LH, Cole DC, Bassil KL, Vakil C (2007) Non cancer health effects of pesticides: systematic review and implications for family doctors. Can Fam Physician 53:1712–1720

Schisler DA, Slininger PJ, Behle RW, Jackson MA (2004) Formulation of *Bacillus* spp. for biological control of plant diseases. Phytopathology 94:1267–1271

Schubler A, Schwarzott D, Walker C (2001) A new fungal phylum, the Glomeromycota: phylogeny and evolution. Mycol Res 105(12):1413–1421

Seneviratne G, Zavahir JS, Bandara WM, Weerasekara ML (2008) Fungal-bacterial biofilms: their development for novel biotechnological applications. World J Microbiol Biotechnol 24(6):739

Smith RS (1992) Legume inoculant formulation and application. Can J Microbiol 38(6):485–492

Spyrou IM, Karpouzas DG, Menkissoglu-Spiroudi U (2009) Do botanical pesticides alter the structure of the soil microbial community. Microb Ecol 58:715–727

St-Arnaud M, Vujanovic V (2007) Effect of the arbuscular mycorrhizal symbiosis on plant diseases and pests. In: Mycorrhizae in crop production. Haworth, New York, pp 67–122

Steenhoudt O, Vanderleyden J (2000) *Azospirillum*, a free-living nitrogen-fixing bacterium closely associated with grasses: genetic, biochemical and ecological aspects. FEMS Microbiol Rev 24(4):487–506

Stenersen J (2004) Chemical pesticides: mode of action and toxicology. CRC Press, Boca Raton

Tajini F, Trabelsi M, Drevon JJ (2012) Combined inoculation with *Glomus intraradices* and *Rhizobium tropici* CIAT899 increases phosphorus use efficiency for symbiotic nitrogen fixation in common bean (*Phaseolus vulgaris* L.). Saudi J Biol Sci 19:157–163

Tavasolee AN, Aliasgharzad G, Salehijouzani MM, Asgharzadeh A (2011) Interactive effects of arbuscular mycorrhizal fungi and rhizobial strains on Chickpea growth and nutrient content in plant. Afr J Microbiol 10:7585–7591

Thakore Y (2006) The biopesticide market for global agricultural use. Ind Biotechnol 2:194–208

Trivedi P, Pandey A, Palni LM (2005) Carrier-based preparations of plant growth-promoting bacterial inoculants suitable for use in cooler regions. World J Microbiol Biotechnol 21:941–945

Vahjen W, Munch JC, Tebbe CC (1995) Carbon source utilization of soil extracted microorganisms as a tool to detect the effects of soil supplemented with genetically engineered and non-engineered *Corynebacterium glutamicum* and a recombinant peptide at the community level. FEMS Microbiol Ecol 18(4):317–328

Vessey JK (2003) Plant growth promoting rhizobacteria as biofertilizers. Plant Soil 255:571–586

Vitousek PM, Mooney HA, Lubchenco J, Melillo JM (1997) Human domination of Earth's ecosystems. Science 277(5325):494–499

Zancarini A, Lépinay C, Burstin J, Duc G, Lemanceau P, Moreau D, Munier-Jolain N, Pivato B, Rigaud T, Salon C, Mougel C (2013) Combining molecular microbial ecology with ecophysiology and plant genetics for a better understanding of plant–microbial communities' interactions in the rhizosphere. Mol Microbial Ecol Rhizosphere 1:69–86

Zayadan BK, Matorin DN, Baimakhanova GB, Bolathan K, Oraz GD, Sadanov AK (2014) Promising microbial consortia for producing biofertilizers for rice fields. Microbiology 83:391–397

Zolla G, Bakker MG, Badri DV, Chaparro JM, Sheflin AM, Manter DK, Vivanco J (2013) Understanding root-microbiome interactions. Mol Microbial Ecol Rhizosphere 1:743–754

Chapter 11
Interaction of Microorganisms with Nanomaterials as a Basis for Creation of High-Efficiency Biotechnological Preparations

I. K. Kurdish

Contents

11.1 Introduction

In natural conditions, microorganisms function in close interaction with solid materials, including interaction with nanoparticles of natural minerals, which are significant components of the soil (Costerton et al. 1985; Zehnder et al. 1996; Mishra and Kumar 2009; Nannipieri et al. 2017). A relevant role in this interaction is played by particles of silicon dioxide and clay minerals. The properties of these particles may change significantly as the soil particle size decreases to the nanoscale (Zhang et al. 2017). Clay minerals have considerable effects on the physical, chemical, and biological processes of soils, as well as the physiological and biochemical activities of microorganisms in different taxonomic groups (Kurdish 2010; Zhang et al. 2018).

I. K. Kurdish (✉)
Department of Microbiological Processes on Solid Surface, Zabolotny Institute of Microbiology and Virology, National Academy Sciences of Ukraine, Kiev, Ukraine

R. Prasad et al. (eds.), *Nanobiotechnology in Bioformulations*, Nanotechnology in the Life Sciences, https://doi.org/10.1007/978-3-030-17061-5_11

The properties of the surface are important factors, determining the interaction of microorganisms and nanoparticles. Nanoparticles of different origins are characterized by a large specific surface area with a negative charge and different functional groups thereon, which determine the specificities of the interactions between microorganisms and these particles considerably. Regardless of the total negative charge, the surface of cells and particles of solid materials may contain positively charged and hydrophobic sites, which promote contact interaction between microorganisms and particles of solid materials and other objects (Rutter et al. 1984). It has been demonstrated that nanoparticles of natural soil montmorillonite and kaolinite are efficient sorbents for organic compounds (He et al. 2015).

The interaction of different kinds of microorganisms and particles of clay minerals promotes the survival of cells in soil conditions (van Veen et al. 1997), including the impact of protozoa (Heijnen et al. 1988), coexistence with which affects both the number of microorganisms and the content of biologically active substances in the environment (Pogorelova et al. 2012).

11.2 Influences of Nanomaterials on Growth, Activity, and Viability of Microorganisms

By use of electronic microscopy and microelectrophoresis, it was established that after the introduction of silica nanoparticles (Aerosil A-300) or nanoparticles of clay minerals (bentonite, montmorillonite, or palygorskite) into a suspension of bacteria, they start to interact; as a result, the cells get covered with particles of these materials (Fig. 11.1) (Gordienko et al. 1993).

The interaction between bacteria and nanomaterials has considerable impacts on the physiological and biochemical activities of microbial populations. It was demonstrated by us that during cultivation of the bacteria *Methylomonas rubra* 15sh

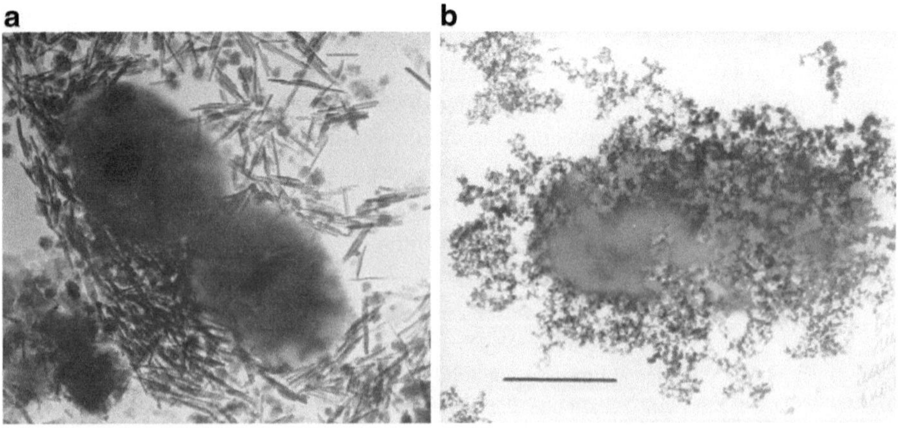

Fig. 11.1 Cells of *Methylomonas rubra* 15sh after interaction with nanoparticles of palygorskite (**a**) and silicon dioxide (**b**)

or *Methylococcus capsulatus* BCB-874 in a culture medium containing 50–200 mg/l of nanoparticles of silicon dioxide or palygorskite, the methane-oxidizing and growth activities of these microorganisms increased considerably (Kurdish and Kigel 1997).

Introduction of silica nanoparticles into the culture medium was accompanied by an increase in the growth activity of the yeast *Saccharomyces cerevisiae* (race XII). The maximal accumulation of biomass in conditions of periodic cultivation of the yeast was observed with a 0.05–0.1% content of this nanomaterial in the medium (Fig. 11.2). An increase in its concentration resulted in a decrease in the biomass gain, and at a content exceeding 0.3% the level of biomass accumulation was lower than that seen in the control.

A further increase in the content of silica nanoparticles was accompanied by more vividly expressed inhibition of yeast growth (Kurdish et al. 1991a). A similar regularity regarding the impact of these nanoparticles was observed during cultivation of the yeast *Candida tropicalis* K-41 (Tsimberg et al. 1991).

A similar effect of these particles was observed during cultivation of *Azotobacter* bacteria. For instance, during cultivation of *Azotobacter chroococcum* 20 in Ashby's medium containing 0.05% silica nanoparticles, the number of these bacteria doubled in comparison with the control. The kinematic viscosity of the culture liquid was also increased by accumulation of a considerable amount of polysaccharide therein (Table 11.1).

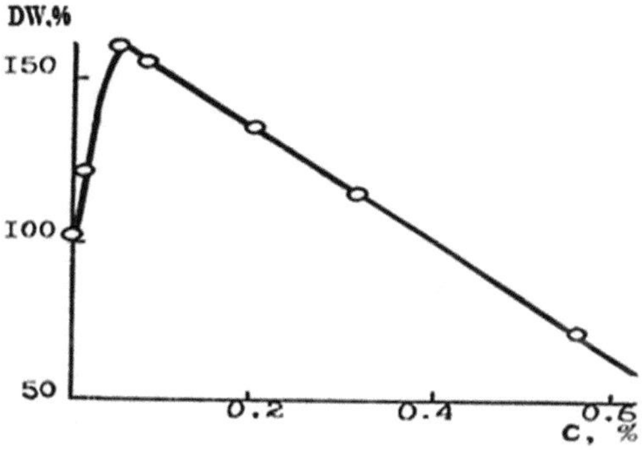

Fig. 11.2 Dependence of biomass accumulation (calculated per total dry weight (*DW*) of yeast biomass) of *S. cerevisiae* on the content (*C*) of silica nanoparticles in the medium

Table 11.1 Effects of silicon dioxide (*SD*) nanoparticles and their modified forms on the growth of *Azotobacter chroococcum* 20 in Ashby's medium and on the kinematic viscosity of the suspension

	Viable bacteria		Kinematic viscosity of the suspension	
Type of nanoparticles	CFU · 10^8/ml	% versus control	mm/cm^2	% versus control
No nanoparticles	0.24 ± 0.05	100.0	6.1 ± 0.1	100.0
SD nanoparticles	0.48 ± 0.06	200.0	17.3 ± 0.3	283.6
SD-AE nanoparticles	0.55 ± 0.06	229.3	20.4 ± 0.4	334.4
SD-Al nanoparticles	0.97 ± 0.17	404.0	84.7 ± 5.0	1388.5

The content of nanoparticles in the medium was 0.05%

CFU Colony-forming units, *SD-AE* silicon dioxide particles modified with aminoethoxy groups, *SD-Al* silicon dioxide particles modified with aluminum oxide

In a medium containing 0.05% silica dioxide nanoparticles modified with aminoethoxy groups (SD-AE) or aluminum oxide (SD-Al), these indicators were higher (Kurdish et al. 1993a). Cultivation of *A. chroococcum* 20 or *A. vinelandii* 56 in a medium containing 0.05% silicon dioxide or alumo-Aerosil resulted in considerable increases in thiamine (vitamin B_1) and pyridoxine (vitamin B_6) content therein (Titova et al. 1994).

Cultivation of *Agrobacterium radiobacter* 10 or *A. radiobacter* 204 in a medium containing 0.05% silica nanoparticles was accompanied by a 32% increase in the growth of bacteria in comparison with the control. The numbers of *A. radiobacter* 10 were increased by 71.5% and 78.6% with addition of 1% montmorillonite or palygorskite nanoparticles, respectively, in comparison with the control. However, cultivation of these bacteria with kaolinite particles resulted in a 28.6% decrease in the number of bacterial cells (Kurdish and Titova 2001) (Table 11.2).

Cultivation of the legume bacteria *Bradyrhizobium japonicum* 634b in a medium with clay mineral particles was also accompanied with a considerable increase in their growth activity (Table 11.3). After 5 days of cultivation in medium without clay minerals, the number of bacteria increased up to 5.0×10^{10} cells/ml. The same index was an order higher in the presence of 1 g/l of nanoparticles of palygorskite, montmorillonite, or bentonite in the medium (Kurdish and Melnykova 2011).

We bred highly active strains of the phosphate-mobilizing bacteria *Bacillus subtilis* IMV B-7023 (Kurdish and Roy 2003) and the nitrogen-fixing bacteria *A. vinelandii* IMV B-7076 (Kurdish and Bega 2006a), promoting the growth and development of plants considerably (Kurdish 2010). It was established that cultivation of *A. vinelandii* IMV B-7076 in Berk's medium containing 0.1–5.0 g/l of silica nanoparticles was accompanied by a considerable increase in the growth activity of these bacteria (Fig. 11.3). With a silica nanoparticle content of 0.5 g/l in

Table 11.2 Effects of clay mineral nanoparticles on the growth of *Agrobacterium radiobacter* 10

Type of nanoparticles	Viable bacterial cells, %
No nanoparticles (control)	100.0 ± 6.2
Palygorskite	178.6 ± 5.4
Montmorillonite	171.5 ± 16.0
Kaolinite	128.6 ± 11.3

The mineral content of the medium was 1 g/l

Table 11.3 Effects of clay mineral nanoparticles on the growth of *Bradyrhizobium japonicum* 634b

Type of nanoparticles	Viable bacterial cells/ml of culture medium		
	At 1 day	At 3 days	At 5 days
No nanoparticles (control)	$(8.4 \pm 1.6) \times 10^6$	$(3.1 \pm 0.2) \times 10^8$	$(5.0 \pm 0.4) \times 10^{10}$
Palygorskite	$(2.4 \pm 0.4) \times 10^6$	$(2.3 \pm 0.1) \times 10^9$	$(8.2 \pm 0.2) \times 10^{11}$
Montmorillonite	$(3.1 \pm 0.6) \times 10^6$	$(5.3 \pm 1.0) \times 10^9$	$(2.5 \pm 0.6) \times 10^{11}$
Bentonite	$(1.5 \pm 0.1) \times 10^6$	$(1.2 \pm 0.2) \times 10^9$	$(6.3 \pm 0.6) \times 10^{11}$

The mineral content of the medium was 1 g/l

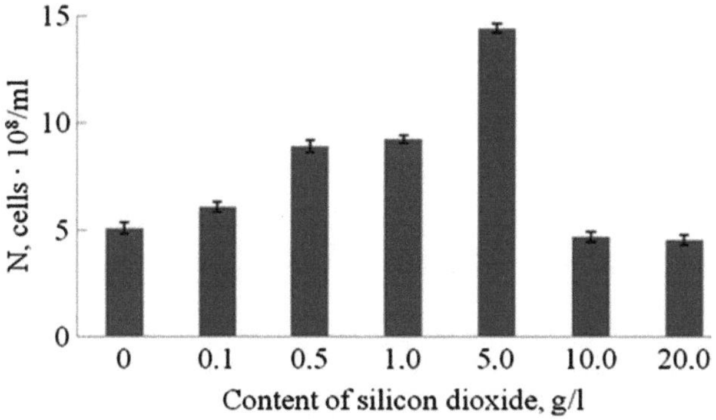

Fig. 11.3 Numbers of viable cells of *Azotobacter vinelandii* IMV B-7076 during cultivation in Berk's medium with silica nanoparticles

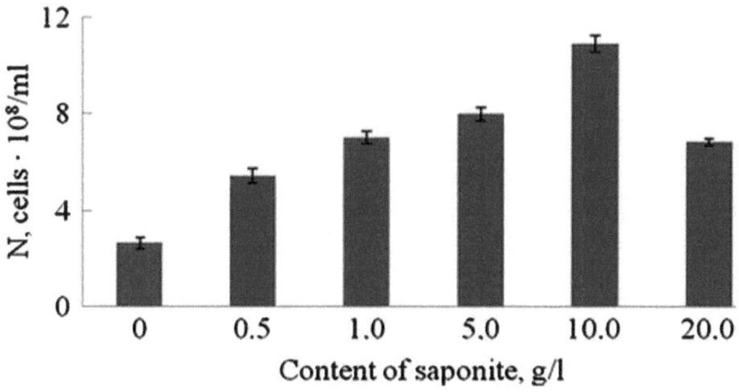

Fig. 11.4 Dependence of the numbers of viable cells of *Azotobacter vinelandii* IMV B-7076 on the content of saponite particles in the culture medium

the culture medium, the number of bacteria therein increased by 71% in comparison with the control. The maximal growth activity was obtained after addition of 5 g/l of silica nanoparticles to the culture medium (Chobotarov et al. 2010a). In these conditions, the number of *A. vinelandii* IMV B-7076 cells was 2.8 times that in the control conditions after 48 h of cultivation in the culture medium (Fig. 11.3). At the same time, the viscosity of the culture medium increased 55 times. Data in the scientific literature have confirmed that silica nanoparticles increase the content of bacteria in soil conditions considerably, promoting germination of corn seeds (Karunakaran et al. 2012).

Saponite particles had a stimulating effect on the growth of *A. vinelandii* IMV B-7076. As a result of introduction of 0.5 g/l of this mineral into the culture medium, the number of bacteria almost doubled in comparison with the control (Fig. 11.4). An increase in the content of saponite particles led to enhanced growth activity of these microorganisms. Their greatest increased occurred at a concentration of 10.0 g/l of this mineral in the medium. Under such conditions the number of cells in the experimental variants was almost four times that in the control. A further

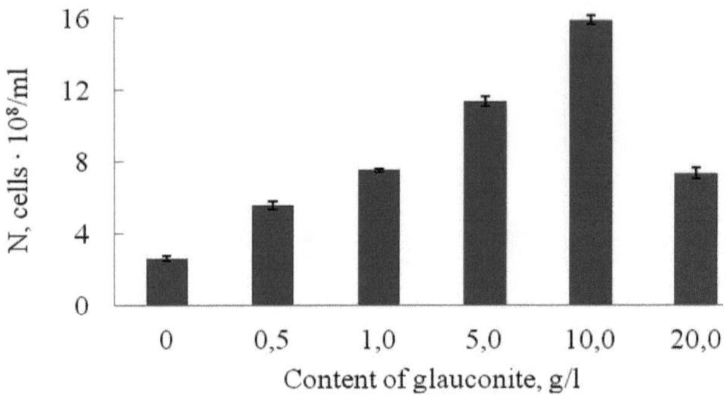

Fig. 11.5 Dependence of the numbers of viable cells of *Azotobacter vinelandii* IMV B-7076 on the content of glauconite particles in the culture medium

increase in the content of saponite in the medium was accompanied by a decrease in its stimulating effect on the growth of these bacteria (Chobotarov et al. 2010a).

Addition of glauconite to the culture medium significant affected the growth of the nitrogen-fixing bacteria *A. vinelandii* IMV B-7076. After introduction of 0.5 g/l of glauconite particles, the number of these bacteria doubled in comparison with the control (Chobotarov et al. 2010a). The greatest growth of bacteria was observed during their cultivation in a medium containing 5.0 or 10.0 g/l of these mineral particles (Fig. 11.5). Under these conditions the numbers of *Azotobacter* were 4.3 and 6.0 times that in the control conditions, respectively. A further increase in the content of this mineral in the medium was accompanied by a decrease in its stimulating effect on bacterial growth (Chobotarov et al. 2010a) (Fig. 11.5).

Cultivation of this *Azotobacter* strain in a medium with phosphorite particles also stimulated the growth activity of these bacteria. The maximal stimulating effect was observed during cultivation of the bacteria in a medium containing 5 g/l of mineral particles. Under these conditions, the number of viable cells was five times that seen in the control conditions (Chobotarov et al. 2010a).

It was established that cultivation of *B. subtilis* IMV B-7023 in a medium containing saponite particles had an obvious stimulating effect on the growth activity of these bacteria. At a content of 0.5 g/l of saponite particles in the medium, the number of these bacterial cells increased by 80% in comparison with the control. The stimulating effect of saponite was enhanced with an increase in its content in the medium. The maximal growth of cells was observed in media containing 5 g/l and 10 g/l of saponite; under these conditions the numbers of bacteria were 2.3 and 2.4 times that in the control medium, respectively (Chobotarov et al. 2010b, 2013). Therefore, introduction of nanoparticles of silicon dioxide, saponite, glauconite, and phosphorite into the culture medium considerably increased the growth activity of *A. vinelandii* IMV B-7076 and *B. subtilis* IMV B-7023.

It was demonstrated by us that *B. subtilis* IMV B-7023 is capable of absorbing phosphorus from both its organic (Roy et al. 2001) and inorganic compounds, including the poorly soluble calcium phosphate $Ca_3(PO_4)_2$ (Bulavenko et al. 2000). A considerable influence on growth was the presence of particles of clay minerals in a medium containing $Ca_3(PO_4)_2$ as the single source of phosphorus (Kurdish and

Bega 2006a). Their stimulating effect on the growth of these bacteria depended on the size of the particles of these minerals. Cultivation of *B. subtilis* IMV B-7023 in a medium containing 0.2% free-flowing montmorillonite increased the number of these bacteria by 67%. A more noticeable increase in bacterial growth activity was observed during cultivation of this strain with nanoparticles of this mineral obtained after its ultrasonic processing. In these conditions, the number of bacteria was increased by 72% in comparison with the control. The maximal values of growth activity were obtained during cultivation of *B. subtilis* IMV B-7023 in a medium containing 0.5% montmorillonite particles. A further increase in the content of this mineral in the medium was accompanied by a decrease in the growth activity of the bacteria (Kurdish and Bega 2006a).

The presence of palygorskite particles had a more obvious stimulating effect on the growth activity of *B. subtilis* IMV B-7023. Cultivation of bacteria in a medium containing 0.2% free-flowing palygorskite particles resulted in an increase in the number of bacteria by 43%. In the presence of nanoparticles of this mineral the number of cells was enhanced by 56% in comparison with the control. The most noticeable stimulating effect on the growth activity of the bacilli was observed during cultivation in a medium containing 1% palygorskite nanoparticles. In this variant, the number of bacteria was increased by 263% in comparison with the control. The increase in the content of nanoparticles in the cultivation medium was accompanied by a decrease in their stimulating effect on bacterial growth (Kurdish and Bega 2006a).

It was demonstrated that the growth activity of methanotrophic bacteria is due to the concentration of the substrates and the ratios between the bacterial biomass and the specific concentrations of the substrates (Kurdish et al. 1988, 1990). To assess the influence of mineral nanoparticles on the growth activity of *B. subtilis* IMV B-7023, we investigated the features of its growth by cultivating it for 20 h in a culture medium containing calcium phosphate and montmorillonite particles (1%) and different initial densities of the population of bacteria. In the control variant, the cultivation of *B. subtilis* was conducted without the presence of the clay mineral.

The results of the study (Table 11.4) demonstrated that when *B. subtilis* was cultivated in a medium not containing particles of the clay mineral, the highest specific rate of bacterial growth (0.29 h^{-1}) was obtained in the presence of the lowest

Table 11.4 Dependence of the growth of *Bacillus subtilis* IMV B-7023 on montmorillonite content in the culture medium and the initial density of the bacterial population

Montmorillonite content, %	Viable bacterial cells/ml of culture medium		Specific growth rate, μ	Culture medium pH
	At the start of cultivation	At the end of cultivation		
0	$(8.42 \pm 0.68) \cdot 10^5$	$(2.87 \pm 0.41) \cdot 10^8$	0.29	5.30
0	$(8.61 \pm 0.40) \cdot 10^6$	$(2.83 \pm 0.32) \cdot 10^8$	0.19	5.28
0	$(4.13 \pm 0.54) \cdot 10^7$	$(3.71 \pm 0.78) \cdot 10^8$	0.11	5.40
1	$(8.42 \pm 0.68) \cdot 10^5$	$(3.53 \pm 0.46) \cdot 10^8$	0.30	5.08
1	$(8.61 \pm 0.40) \cdot 10^6$	$(5.26 \pm 0.75) \cdot 10^8$	0.21	5.07
1	$(4.13 \pm 0.54) \cdot 10^7$	$(6.54 \pm 0.60) \cdot 10^8$	0.14	5.08

The cultivation period was 20 h

initial density of the bacterial population (8.42×10^5 cells per milliliter) (Kurdish and Bega 2006a). With an increase in the latter, the specific rate of growth of the microorganisms decreased considerably in the medium with an initial density of 4.13×10^7 bacteria per milliliter; the specific rate of growth amounted to 0.11 h^{-1} (Table 11.4).

During cultivation of *B. subtilis* IMV B-7023 in a medium with montmorillonite particles, the specific rates of growth were higher than in the absence of these particles. However, a regularity concerning the dependence of the growth of the bacteria on the initial content of biomass, similar to the aforementioned regularity, was observed in this experimental variant as well (Kurdish and Bega 2006a). Therefore, the obtained results demonstrate that under such conditions the growth rate of *B. subtilis* is determined by the concentrations of both nutrient substrates and mineral particles in the medium and their ratios to the number of bacteria—that is, their specific concentrations.

A stimulating effect of the interaction between microorganisms and nanomaterials on the physiological activity of microbial populations may be conditioned by a number of factors. One of the mechanisms of the stimulating effect that nanoparticles of the studied minerals have on the growth activity of microorganisms may be found in the enhanced mass transfer of oxygen at their introduction into the medium used for cultivation of the bacteria. Stirring of the medium with mineral nanoparticles results in formation of highly turbulent microzones around the aforementioned particles, which are capable of conditioning the increase in the mass transfer of oxygen (Kurdish 2001). It was demonstrated by us that introduction of 1% and 4% powdered palygorskite into the medium resulted in increases in the mass transfer of oxygen by 6% and 15.5%, respectively. The introduction of nanoparticles of this mineral and silicon dioxide caused a noticeable effect on this indicator. At a content of 1% palygorskite in the medium, the mass transfer of oxygen therein increased by 16.9% (Kurdish 2001).

As one of the relevant factors for aerobic microorganism functioning is the supply of oxygen to them, an impact of this mechanism is possible. At the same time, it should be noted that an increase in the content of mineral nanoparticles by >4% resulted in an increase in the medium viscosity, accompanied by a decrease in the mass transfer of oxygen. This may cause a negative effect on the growth activity of aerobic microorganisms (Kurdish 2001).

Taking the above into consideration, we were interested in studying the impact of different concentrations of oxygen on the growth of these bacteria. It was demonstrated that with an increase of oxygen mass transfer into the medium from 0.41 to 1.83 g $O_2 \cdot l \cdot h^{-1}$, the growth activity of *B. subtilis* IMV B-7023 was boosted, whereas that of *A. chroococcum* 21 was decreased.

To study the effect of mineral nanoparticles on the growth activity of these bacteria with different values of mass transfer of oxygen, they were cultivated in hermetic vials with artificial gas mixtures (Kisten et al. 2006). After 13 h of incubation at a content of about 3% O_2 in the gas phase, the number of bacteria that grew was 57.1% greater in the medium with palygorskite than in the control medium (with no palygorskite). The increase in the oxygen content up to 6% resulted in the increase in the number of bacteria by 148.3%, and at 20.7% O_2 by 487.5% compared against

Table 11.5 Effects of the clay mineral palygorskite on accumulation of *Azotobacter chroococcum* 21 cells at different concentrations of O_2 in the gas phase

	Incubation in culture medium				
	For 0 h	For 13 h		For 24 h	
Palygorskite content, mass %	O_2, vol. %	O_2, vol. %	Viable bacterial cells/ml	O_2, vol. %	Viable bacterial cells/ml
0	20.7	10.6	$(1.6 \pm 0.5) \cdot 10^8$	8.6	$(2.1 \pm 0.8) \cdot 10^8$
1.0	20.7	6.7	$(9.4 \pm 0.7) \cdot 10^8$	4.0	$(1.2 \pm 0.2) \cdot 10^9$
0	12.6	5.7	$(2.0 \pm 0.3) \cdot 10^8$	5.2	$(2.7 \pm 1.2) \cdot 10^8$
1.0	12.6	3.8	$(9.5 \pm 1.2) \cdot 10^8$	1.8	$(1.3 \pm 0.2) \cdot 10^9$
0	6.9	3.3	$(2.9 \pm 0.5) \cdot 10^8$	2.3	$(3.1 \pm 0.6) \cdot 10^8$
1.0	6.8	2.0	$(7.2 \pm 0.8) \cdot 10^8$	1.2	$(8.9 \pm 1.6) \cdot 10^8$
0	3.5	1.4	$(2.1 \pm 0.4) \cdot 10^8$	1.4	$(2.8 \pm 0.4) \cdot 10^8$
1.0	3.8	1.1	$(3.3 \pm 0.3) \cdot 10^8$	1.0	$(5.3 \pm 1.3) \cdot 10^8$

vol. Volume

the control (Table 11.5). Cultivation of *A. chroococcum* in the aforementioned conditions for 24 h resulted in rapid decreases in the concentrations of glucose and phosphate and in the content of O_2 in the gas phase of the hermetic vials. Regardless of this fact, the clay mineral particles stimulated the growth of these bacteria.

Therefore, the obtained results demonstrated that the most traceable stimulating effect of palygorskite particles on the growth activity of *Azotobacter* was shown at high concentrations of oxygen. It was shown that during cultivation of *A. chroococcum* 21 bacteria in Ashby's medium, palygorskite protects cells from the toxic effect of high concentrations of oxygen to some degree.

A relevant impact on the growth of microorganisms may be caused by ion exchange processes occurring in the medium upon the introduction of nanomaterials. It is known that protein may be adsorbed on the surface of silica nanoparticles and other mineral particles (Kisten et al. 2006). It was demonstrated by us that introduction of saponite particles into a *B. subtilis* IMV B-7023 cultivation medium led to an evident increase in the concentrations of the cations Mg^{2+}, Ca^{2+}, Na^+, and K^+ (by between 63 and 191 mg/l, respectively) (Chobotarov et al. 2010b). Therefore, the increase in the physiological activity of microorganisms upon their interaction with nanomaterials of different natures may be conditioned by the impacts of many factors, and determining their actions is a relevant prerequisite for accomplishment of biotechnological tasks.

It was demonstrated that interaction of the bacteria *M. rubra* 15sh, *Pseudomonas aureofaciens* UKM-111, *A. radiobacter* 204, and other microorganisms with nanoparticles of the clay minerals montmorillonite, palygorskite, and bentonite had a considerable stimulating effect on the viability of cells in long-term storage and during exposure to increased temperature (Gerasymenko and Kurdish 2015; Kurdish et al. 1993a, b, 1999; Kurdish and Antonyuk 1999). For instance, incubation of an *A. radiobacter* 204 suspension at 50 °C for 15 min (as a control treatment) was accompanied by a 40% decrease in the number of viable cells. However, in the case of previous introduction of 1% palygorskite particles into such a suspension, the number of viable cells decreased by only 20% (Kurdish and Antonyuk 1999).

A similar effect on the viability of *B. japonicum* 634b was caused by particles of this mineral (Kurdish et al. 1999). When suspensions of these bacteria were incubated at 45 °C for 15 min in the absence of clay mineral nanoparticles, 32.4–40.6% of cells were viable (Table 11.6). Introduction of montmorillonite nanoparticles into the suspension of this strain was accompanied by an increase in the resistance of the cells to the effect of supraoptimal temperatures. For instance, in the presence of 1 g/l of montmorillonite particles, the percentage of viable cells after warming of the suspension was increased to 69%, and in the presence of 10 g/l of these nanoparticles, the percentage was increased to 82.7%. A similar effect on the influence of supraoptimal temperatures on *B. japonicum* 634b was observed with introduction of palygorskite nanoparticles into the suspension (Table 11.6).

Therefore, it was demonstrated that the interaction of the investigated species of bacteria and clay mineral particles considerably increases the resistance of cells to the effect of increased temperatures on them.

It was shown (Kurdish and Antonyuk 1999) that the interaction of *A. chroococcum* 20 and clay mineral particles also considerably increases their resistance to supraoptimal temperatures. In the absence of minerals, the cells of these bacteria are very sensitive to the effect of higher temperatures. After incubation of the bacteria at 45 °C for 10 min without clay minerals, only 12.7% of cells in the suspension were viable, but after interaction between these microorganisms and palygorskite (0.5%), the proportion of viable cells was increased to 34.5% (Table 11.7). Montmorillonite particles had even a more evident effect on the viability of *Azotobacter* (Kurdish and Antonyuk 1999).

It should be noted that this effect of mineral particles was enhanced with an increase in the content of montmorillonite in the medium. At a concentration of 10 g/l the proportion of viable bacteria was increased to 68%—more than three

Table 11.6 Effects of clay mineral nanoparticles on resistance of *Bradyrhizobium japonicum* 634b to warming at 45 °C for 15 min

Clay mineral particle content, g/l	Viable bacterial cells, % versus control (accepted as 100%)	
	With montmorillonite	With palygorskite
0	40.6 ± 3.9	40.6 ± 3.9
1.0	69.0 ± 4.2	67.1
5.0	64.8 ± 5.1	ND
10.0	82.7 ± 8.1	ND

ND Not determined

Table 11.7 Effects of clay mineral nanoparticles on resistance of *Azotobacter chroococcum* 20 to warming at 45 °C for 10 min

Clay mineral particle content, g/l	Viable bacterial cells, % versus control (accepted as 100%)	
	With montmorillonite	With palygorskite
0	12.7 ± 1.4	12.7 ± 0.8
0.1	28.5 ± 1.8	28.4 ± 1.8
0.5	59.4 ± 3.7	34.5 ± 2.1
1.0	68.0 ± 4.2	32.4 ± 2.0

times that in the control conditions. Therefore, the interaction between *Azotobacter* and clay mineral particles increases not only the growth, nitrogen-fixing activity, and synthesis of B vitamins by these microorganisms, but also the viability of these bacteria with exposure to supraoptimal temperatures. The technology of using nanoparticles of natural minerals, developed by us to promote survival of bacteria in case of their exposure to extreme environmental factors, is a promising approach for improving the storage of collection strains of microorganisms and for creating new forms of bacterial preparations (Kurdish 2001).

Taking into consideration the protective impact of clay mineral particles on the survival of microorganisms, we developed biotechnology to enhance the yield of viable bacteria in preparations (Kurdish and Titova 2000) produced by the method of spray drying (Gordienko et al. 1990). For this purpose, 10 g/l of palygorskite particles was introduced into the suspension of bacteria. The mixture was stirred, and after 15 min of interaction between the cells and the mineral particles, we introduced a protective medium (dry skim milk), and then the composite underwent further spray drying. This method allowed the yield of viable *Streptococcus faecium* cells and other species of bacteria to be increased by >60% during manufacturing of their preparations by the aforementioned technology in comparison with the control treatment (without mineral particles, but with dry milk) (Gordienko et al. 1990; Kurdish et al. 1991b). The aforementioned biotechnology was implemented in the production of the Litosyl preparation at the Kiev Pharmaceutical Drugs Plant.

11.3 Influences of Nanoparticles on Physiological and Biochemical Activities of Microorganisms

It is known that the physiological and biochemical activities of microorganisms depend on their cultivation conditions, which determine the accumulation of a number of metabolites in the medium that are capable of affecting other components of the biota (Niste et al. 2013). We demonstrated that the energy potential of cells was increased by interaction between bacteria and mineral nanoparticles. During cultivation of *B. subtilis* IMV B-7023 in a culture medium containing 5 g/l of exfoliated vermiculite, the dehydrogenase activity of the bacteria was increased by 34% in comparison with the control (Gerasymenko and Kurdish 2015).

During cultivation of these bacteria in a medium with addition of 0.5 g/l of silica nanoparticles, the dehydrogenase activity of *B. subtilis* IMV B-7023 was increased by only 6–7% in comparison with the control. Silica nanoparticles had an insignificant effect on the dehydrogenase activity of *A. vinelandii* IMV B-7076. However, during cultivation of this strain in a medium containing vermiculite nanoparticles, the indices of the dehydrogenase activity of *Azotobacter* were higher; when 5.0 g/l of vermiculite nanoparticles were added to the medium, the dehydrogenase activity was 40% higher in comparison with the control (Fig. 11.6). Therefore, the introduction of vermiculite nanoparticles into the medium was accompanied by an increase in the dehydrogenase activity of the investigated bacteria (Gerasymenko and Kurdish 2015; Kurdish et al. 2014).

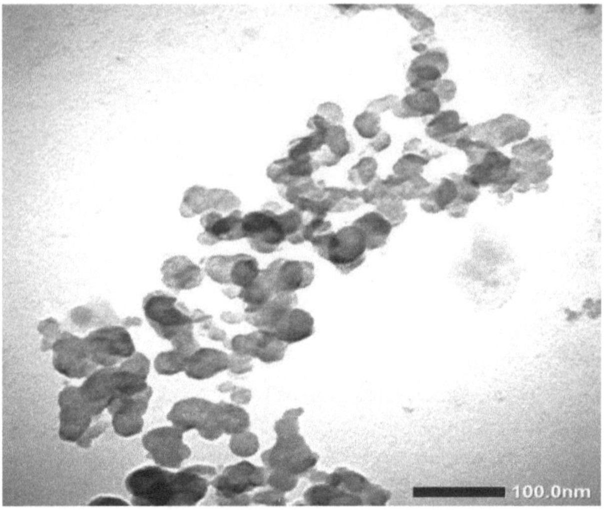

Fig. 11.6 Vermiculite particles after processing with an ultrasonic disintegrator

A relevant property of microorganisms (as components of different ecosystems and strains) that is promising for biotechnological applications, including plant production, is their capability to synthesize biologically active substances that can affect other representatives of the biota, such as plants, by influencing their growth and development and protecting them against the impacts of negative factors.

It was demonstrated by us that during cultivation of the phosphate-mobilizing bacteria *B. subtilis* IMV B-7023 in Menkina medium with calcium glycerophosphate and glucose, accumulation of various amino acids (arginine, isoleucine, valine, proline, and phenylalanine) and organic acids occurred (Table 11.8), with acetic acid (up to 32.5 μg/ml) and pelargonic acid (up to 6 μg/ml) being the most abundant (Tsercovniak et al. 2009b; Skorochod et al. 2013).

During cultivation of *B. subtilis* in a medium with addition of 5.0 g/l of titanium dioxide nanoparticles, the total content of amino acids in the culture medium increased by 27.7% in comparison with the control (Table 11.8). Under these conditions the content of phenylalanine was 7.85 μg/ml, histidine 2.63 μg/ml, and valine 1.59 μg/ml. Insignificant concentrations of glutamic acid, asparagine, and methionine were also accumulated in the culture medium. No traces of isoleucine was found.

The total amounts of free amino acids accumulated in the culture media of *B. subtilis* IMV B-7023 with addition of particles of the natural minerals glauconite or saponite were 6.80 and 3.46 μg/ml, respectively—1.7 and 3.4 times lower, respectively, than the total amino acid content accumulated with the control treatment (Table 11.8). These findings may have been conditioned by the adsorption of a certain part of these substances on the surface of the mineral nanoparticles.

Cultivation of *A. vinelandii* IMV B-7076 in Ashby's medium was accompanied by accumulation of an insignificant amount of amino acids therein, the total amount of which was up to 2.78 μg/ml (Table 11.9). An evident stimulating impact on the

Table 11.8 Free amino acid content in a culture medium containing *Bacillus subtilis* IMV B-7023, depending on the type of mineral particle content

Amino acid	Amino acid content, μg/ml			
	With no mineral particles (control)	With titanium dioxide[a]	With glauconite[a]	With saponite[a]
Alanine	ND	ND	ND	0.29 ± 0.02
Arginine	2.90 ± 0.08	ND	ND	ND
Asparagine	ND	0.37 ± 0.08	ND	ND
Valine	1.04 ± 0.04	1.59 ± 0.04	1.00 ± 0.01	1.25 ± 0.08
Histidine	ND	2.63 ± 0.08	2.77 ± 0.03	ND
Glutamic acid	ND	1.73 ± 0.01	ND	ND
Isoleucine	4.11 ± 0.05	ND	0.69 ± 0.08	ND
Methionine	ND	0.90 ± 0.01	ND	0.37 ± 0.09
Proline	2.12 ± 0.06	ND	ND	ND
Serine	ND	ND	0.67 ± 0.04	ND
Tryptophan	ND	ND	0.30 ± 0.06	ND
Phenylalanine	1.63 ± 0.03	7.85 ± 0.03	1.37 ± 0.05	1.55 ± 0.07
Total amino acid content	11.80 ± 0.08	15.07 ± 0.08	6.80 ± 0.05	3.46 ± 0.09

ND Not determined
[a]The content of mineral particles in the medium was 5.0 g/l
[b]$p < 0.01$

Table 11.9 Free amino acid content in a culture medium containing *Azotobacter vinelandii* IMV B-7076, depending on the type of mineral particle content

Amino acid	Amino acid content, μg/ml			
	With no mineral particles (control)	With titanium dioxide[a]	With glauconite[a]	With saponite[a]
Alanine	ND	ND	0.13 ± 0.02	ND
Arginine	ND	ND	6.98 ± 0.05	ND
Asparagine	ND	1.58 ± 0.06	ND	0.32 ± 0.01
Valine	0.58 ± 0.08	ND	0.56 ± 0.04	1.42 ± 0.09
Histidine	ND	1.23 ± 0.09	1.91 ± 0.03	ND
Glycine	ND	ND	0.14 ± 0.03	ND
Glutamic acid	ND	ND	1.15 ± 0.02	ND
Isoleucine	ND	ND	ND	0.34 ± 0.08
Methionine	ND	ND	ND	0.74 ± 0.07
Proline	ND	ND	ND	9.55 ± 0.10
Tryptophan	ND	ND	0.57 ± 0.01	0.86 ± 0.06
Phenylalanine	2.18 ± 0.08	ND	3.16 ± 0.02	3.16 ± 0.10
Cysteine	ND	ND	ND	0.29 ± 0.09
Total amino acid content	2.76 ± 0.08	2.81 ± 0.07	14.58 ± 0.03	16.67 ± 0.10

ND Not determined
[a]The content of mineral particles in the medium was 5.0 g/l
[b]$p < 0.01$

synthesis of amino acids by these bacteria was caused by their cultivation in a medium containing nanoparticles of some natural minerals. During cultivation of these bacteria in a medium with glauconite particles, eight amino acids were found, the total content of which was up to 14.4 μg/ml. Arginine, phenylalanine, and histidine were the most abundant. In a medium containing 5 g/l of saponite, the total accumulation of amino acids exceeded 16 μg/ml (which was 6 times greater than in the control) (Table 11.9).

The highest accumulated concentrations were noted for proline (9.55 μg/ml), phenylalanine (3.16 μg/ml), and valine (1.42 μg/ml) (Kurdish et al. 2014; Chobotarov 2015). Therefore, particles of the natural minerals saponite and glauconite have an evident stimulating effect on accumulation of free amino acids in a culture medium of *A. vinelandii* IMV B-7076.

One of the most important properties of bacteria is their ability to stimulate and improve plant growth and development by producing phytohormones. About 95% of soil microorganisms can produce different hormonal compounds that regulate physiological and biochemical reactions in plants (including stress responses) and play an essential role in plant growth and development (Davies 2004).

Auxins have a considerable impact on the growth and development of plants. They accelerate the process of root formation and affect the processes of photosynthesis, growth, tropism, blossoming, and fruit bearing of plants. As a rule, L-tryptophan should be added to the medium for synthesis of indole acetic acid (IAA), the source of which in soil conditions may be found in root excretions (Shakirova 2001).

It has been established that *B. subtilis* IMV B-7023 and *A. vinelandii* IMV B-7076 bacteria are capable of synthesizing a number of substances of a phytohormonal nature. Cultivation of *A. vinelandii* IMV B-7076 in a medium with L-tryptophan was accompanied by accumulation of 140 ng/ml of free IAA therein, and about 160 ng/ml of this auxin was in a bound state (Tsercovniak et al. 2009a). During cultivation of these bacteria without L-tryptophan, much lower concentrations of this phytohormone were accumulated.

The total amount of IAA in the medium containing *A. vinelandii* IMV B-7076 amounted to 67.1 ng/ml. After cultivation of these bacteria in a medium with nanoparticles of silicon dioxide or vermiculite, the total amount of this phytohormone in the medium was much smaller (Chobotarov et al. 2017a, b).

Cultivation of *B. subtilis* IMV B-7023 in a glucose–mineral medium was accompanied by accumulation of 46 ng/ml of IAA therein. However, during cultivation of these bacteria in a medium with silica nanoparticles, the total accumulation of this phytohormone was almost doubled. There was a particular increase in the content of IAA during cultivation of these bacteria in a medium with vermiculite: 4.6 times the amount observed in the absence of these particles (Chobotarov et al. 2017a).

Among phytohormones, abscisic acid (ABA) plays an important role in plants. ABA accumulates in plants upon their exposure to stress factors and is involved in bud differentiation, fruit development, and formation of additional roots (Kulaeva 1973). The data obtained have revealed that *A. vinelandii* IMV B-7076 accumulates ABA in its culture medium (Table 11.10). The ABA content not associated with other organic compounds was 8.1 ng/ml, while the content of its bound form was 21.5 ng/ml (Chobotarov et al. 2017a, b).

Table 11.10 Effects of nanoparticles on the ability of *A. vinelandii* IMV B-7076 and *B. subtilis* IMV B-7023 to accumulate abscisic acid (*ABA*) in the culture medium

Type of nanoparticles	ABA form	ABA accumulation, ng/ml By *A. vinelandii* IMV B-7076	By *B. subtilis* IMV B-7023
No nanoparticles (control)	Free	8.1 ± 0.4	41.4 ± 1.1
	Bound	21.5 ± 1.0	ND
Nano-SiO_2	Free	10.8 ± 0.5	10.5 ± 0.5
	Bound	9.8 ± 0.5	14.7 ± 0.7
Vermiculite	Free	48.6 ± 1.4	65.0 ± 1.8
	Bound	20.3 ± 1.0	4.2 ± 0.2

ND Not determined

Table 11.11 Effects of nanoparticles on accumulation of cytokinins in a culture medium of *A. vinelandii* IMV B-7076

Phytohormone	Phytohormone content, ng/ml With no nanoparticles (control)	With silica[a]	With vermiculite[a]
Zeatin	53.9 ± 1.7	129.2 ± 1.5	53.6 ± 1.7
Zeatin riboside	49.2 ± 1.5	176.8 ± 1.3	113.6 ± 1.7
Zeatin glycoside	104.3 ± 1.2	107.4 ± 1.3	83.3 ± 1.2

[a]The content of mineral particles in the medium was 5.0 g/l

Addition of 5.0 g/l of nano-SiO_2 stimulated biosynthesis of the free form of ABA in a culture medium of *A. vinelandii* IMV B-7076 by 1.3 times, whereas the amount of the bound form of ABA decreased rapidly. With addition of 5.0 g/l of vermiculite nanoparticles to a culture medium of *A. vinelandii* IMV B-7076, a substantial increase in ABA synthesis was observed, with the total ABA content being 2.3 times that seen in the control conditions. The most noticeable increase, resulting in an ABA content six times that seen in the control conditions, was observed for the free ABA form not associated with other organic compounds.

B. subtilis IMV B-7023 is also capable of ABA accumulation (41.4 ng/ml) in a culture medium. The total amount of ABA in a culture medium of *B. subtilis* IMV B-7023 was lower with addition of SiO_2 nanoparticles than in the control; the content of the free and bound ABA forms was 10.5 ng/ml and 14.7 ng/ml, respectively. Addition of vermiculite to a culture medium of *B. subtilis* IMV B-7023 increased synthesis of the free ABA form to 65.0 ng/ml, which was 1.4 times that observed with the control, whereas the concentration of its bound forms was 3.5 times lower than in the variants with addition of nano-SiO_2 (Chobotarov et al. 2017a).

Cytokinins play an important role in regulation of plant growth and development (Shakirova 2001; Giron et al. 2013). It was established that *A. vinelandii* IMV B-7076 can accumulate different compounds of a cytokinin nature in its culture medium (Table 11.11). The amounts of zeatin, zeatin riboside, and zeatin glycoside were 53.9, 49.2, and 104.3 ng/ml, respectively (Chobotarov et al. 2017a, b). Cultivation of these bacteria in a medium with SiO_2 nanoparticles increased accumulation of zeatin by 2.4 times, zeatin riboside by 3.6 times, and zeatin glycoside by 3% (Table 11.11).

Addition of vermiculite particles to a culture medium of *A. vinelandii* stimulated synthesis of zeatin riboside, resulting in a content 2.3 times that seen in the control, but did not change the content of zeatin and actually reduced the content of zeatin glycoside by 20% (Table 11.11).

B. subtilis IMV B-7023 bacteria are also capable of cytokinin production (Chobotarov et al. 2017a). Addition of vermiculite and silica dioxide particles to a culture medium of *B. subtilis* resulted in a significant increase in cytokinin production (Kurdish et al. 2014). Thus, *A. vinelandii* IMV B-7076 and *B. subtilis* IMV B-7023 bacteria are capable of plant hormone synthesis. Cultivation of bacteria in a culture medium with silica or vermiculite nanomaterials promotes accumulation of phytohormones.

A relevant role among those of the biologically active compounds of cells, including microorganisms, is attributed to metabolites of a phenolic nature. They participate in initiation of interactions between bacteria and plants (Long 2001), and they are important components of the cell, capable of protecting them from reactive oxygen intermediates (Skorochod and Kurdish 2018) and phytopathogenic microorganisms (Kiprushkina and Kolodyaznaya 2014). It was demonstrated by us that during cultivation of *B. subtilis* IMV B-7023 in Menkina medium, these bacteria accumulated a number of phenolic compounds, with phenylacetic and 4-hydroxyphenyl-acetic acids being the most abundant (Tserkovniak and Kurdish 2009). It was established that 4-hydroxyphenyl-acetic acid is capable of stimulating the development of plants and inhibiting the growth of the phytopathogenic micromycetes *Fusarium culmorum*, *Fusarium solani*, *Alternaria alternata*, whose zones of growth inhibition were 21–31 mm.

It was shown that accumulation of phenolic compounds in a culture medium of *A. vinelandii* IMV B-7076 depends on the type of carbon source and its concentration in the medium. The greatest concentration of phenolic compounds in the culture medium (223 μg/ml) was observed during cultivation of the bacteria with 30 g/l of glucose (Ocheretyanko et al. 2016). It was established that during cultivation of the strain *A. vinelandii* IMV B-7076 with 0.05–0.1 g/l of bentonite nanoparticles, the content of phenolic compounds in the culture medium exceeded that in the control by 2–16%. However, when *A. vinelandii* IMV B-7076 was cultivated with silica nanoparticles, an increase in phenolic compound content in the culture medium was observed only in variants containing 0.05–0.1 g/l of nano-SiO_2. After introduction of 0.5 g/l of silica nanoparticles into the culture medium, the phenolic compound content decreased (Skorochod and Kurdish 2018).

It is known that different microbial metabolites may protect cells against reactive oxygen species (Skorochod and Kurdish 2014). Among these are enzymes (catalase, peroxidase, and superoxide dismutase (SOD)) (Labas et al. 2010), low molecular weight antioxidants of the thiol redox system (e.g., glutathione) (Filomeni et al. 2002), phenolic antioxidants, and other bioactive substances (Shtarkman et al. 2008). Phenolic antioxidants effectively inhibit peroxide, alkoxy radicals, hydroxyl radicals, superoxide anion radicals, and singlet oxygen (He et al. 2015). In addition, some phenolic compounds in microorganisms selectively inhibit the functioning of individual species of phytopathogenic micromycetes (Kohen and Nyska 2002).

It was established that *B. subtilis* IMV B-7023 and *A. vinelandii* IMV B-7076 bacteria are characterized by a high degree of antioxidant and antiradical protection. Cultivation of *B. subtilis* IMV B-7023 in a medium containing low concentrations of silica nanoparticles was accompanied by activation of antioxidant protection of the cells, whereas at a silica nanoparticle content of 1 g/l, antioxidant activity was decreased by 11.8%, hydroxyl radical scavenging was decreased by 17.6%, and oxidation activity was increased by 26.9% (Skorochod et al. 2016). A considerable impact on these indices was made by cultivation of *B. subtilis* IMV B-7023 in a medium with vermiculite nanoparticles (Skorochod and Kurdish 2013).

The first link in the chain of protection of living cells from reactive oxygen intermediates is found in such enzymes as catalase, peroxidase, and SOD (Labas et al. 2010). We studied the dependence of the activity of these enzymes in *B. subtilis* IMV B-7023 and *A. vinelandii* IMV B-7076 bacteria on the content of mineral particles in the culture medium.

It was shown that with addition of 1.5 or 2.5 g/l of vermiculite nanoparticles into the culture medium, the peroxidase activity of *B. subtilis* IMV B-7023 tripled. However, with a higher content of vermiculite nanoparticles (5 g/l), the extracellular peroxidase activity was lower (Table 11.12).

It was established that during cultivation of *B. subtilis* IMV B-7023 in a medium containing 0.05 g/l of silica nanoparticles, the extracellular peroxidase activity of the bacteria increased by 43.8% and the intracellular activity by 74.2% (Skorochod and Kurdish 2013). These indicators increased with further increases in the content of these nanoparticles in the medium to 0.1 and 0.5 g/l. However, the extracellular and intracellular peroxidase activity decreased after addition of 1 g/l of silica nanoparticles to the medium. Silica and vermiculite nanoparticles did not have any substantial effect on the extracellular and intracellular catalase activity of *B. subtilis* IMV B-7023 or on the intracellular peroxidase activity (Skorochod and Kurdish 2013).

Some effect of particles of these natural minerals on SOD activity of *A. vinelandii* IMV B-7076 was observed. We demonstrated that cultivation of these bacteria in a medium containing 1 g/l of bentonite or saponite nanoparticles was accompanied by boosted SOD activity of the bacteria. A more evident increase in this index was observed during cultivation of *Azotobacter* with bentonite particles together with addition of 0.5 mM of manganese ions to the medium. SOD activity was 17.5% higher in this medium than in the medium containing bentonite alone and 31.5% higher than during cultivation without these ions and nanoparticles (Chobotarov et al. 2017a, b). Cultivation of this strain in a medium with both saponite nanoparticles and Mn^{2+} ions had a less considerable impact on SOD activity of these bacteria.

Table 11.12 Effects of vermiculite nanoparticles on antioxidant enzyme activity in a culture medium of *B. subtilis* IMV B-7023

Vermiculite nanoparticle content, g/l	Catalase activity, mmol of H_2O_2/min/mg of protein	Peroxidase activity, mmol of indigo carmine/min/mg of protein
0	6.98 ± 0.12	0.75 ± 0.03
1.5	7.61 ± 0.17	1.92 ± 0.89
2.5	7.69 ± 0.17	2.09 ± 0.95
5.0	7.79 ± 0.18	1.61 ± 0.65

It was established that cultivation of *A. vinelandii* IMV B-7076 in a medium containing 0.25 mM of Fe^{2+} ions stimulated the SOD activity of this strain to some degree. However, an increase in the content of these cations in the medium was accompanied by inhibition of SOD activity. The obtained results demonstrate a considerable effect of saponite and bentonite nanoparticles on the SOD activity of the investigated bacteria.

Therefore, *A. vinelandii* IMV B-7076 and *B. subtilis* IMV B-7023 synthesize a number of organic acids, amino acids, phenol compounds, and phytohormones that are capable of improving the growth and development of plants. Cultivation of these strains in culture media containing nanoparticles of natural materials stimulates the growth of these bacteria and their accumulation of biologically active substances in the media.

11.4 Impacts of Interaction Between Bacteria and Nanomaterials on Chemotaxis

Along with physical and chemical factors, a significant role in the interaction between microorganisms and other surfaces, including nanomaterials, is attributed to mobility of cells (van der Mei et al. 2001), which is affected by hydrodynamic forces, sedimentation (Marshall 1985), the sensor properties of cells, and their ability to have taxis. Taxis of microorganisms it has a remarkable role in the process of their interaction with other objects, including solid surfaces (Begonia and Kremer 1994). Cells may react to different factors: a number of chemical factors, light, the pH of the medium, magnetic fields, etc. (Begonia and Kremer 1994; Bashan and Holguin 1997). Through chemotaxis, bacteria may move to places where they interact with solid particles, especially at the surface of plant roots, as considerable quantities of plant metabolites are released into the space near the roots, which may be attractive for bacteria (Kravchenko et al. 2003).

The bacterial examples *B. japonicum* 634b, *B. subtilis* IMV B-7023, and *A. vinelandii* IMV B-7076 were used by us to demonstrate that bacteria demonstrate chemotaxis regarding a wide range of carbohydrates, amino acids, and organic acids (Kurdish et al. 2001, 2010; Chuiko and Kurdish 2004, 2017; Chuiko et al. 2006). We investigated the effects of a number of factors, including nanoparticles of different natures, on the chemotaxis properties of these microorganisms. It was demonstrated that interaction of bacteria with nanoparticles of silicon dioxide or clay minerals promotes the mobility of the cells considerably. However, the chemotaxis of bacilli and nodule bacteria was decreased (Chuiko and Kurdish 2004; Chuiko et al. 2006). After interaction with montmorillonite in a concentration of 0.2 g/l, the mobility of *B. japonicum* 634b increased by 25% (Chuiko and Kurdish 2004).

A similar increase in mobility was observed in the strain *B. japonicum* 604k. After introduction of montmorillonite in a concentration of 0.1 g/l into a suspension of these bacterial cells, the quantitative indicators of their mobility increased

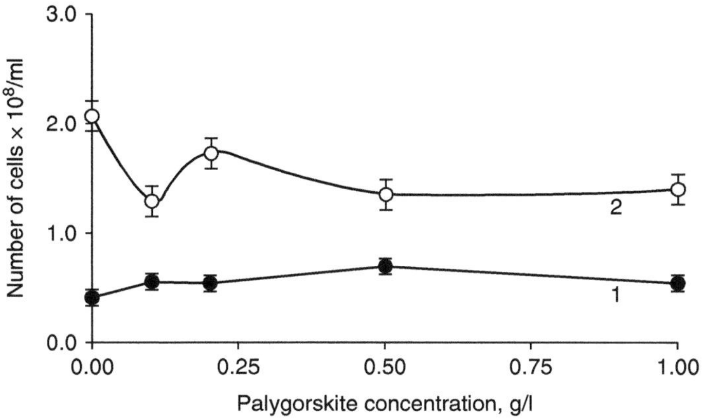

Fig. 11.7 Effects of palygorskite on the mobility and chemotaxis properties of *Bradyrhizobium japonicum* 634b. *1* Number of bacterial cells in capillaries containing phosphate buffer, *2* number of bacterial cells in capillaries containing 5.6×10^{-2} M of glucose

by 27%, and at a mineral concentration of 0.5 g/l, their mobility increased by 105% (Chuiko and Kurdish 2004).

Interaction of bacteria and palygorskite particles also increased the mobility of the investigated strains of *B. japonicum*, but somewhat less than their interaction with montmorillonite (Chuiko and Kurdish 2004). The chemotaxis of bacteria decreased at the content of palygorskite nanoparticles. With addition of 0.1 g/l of these nanoparticles to a suspension of *B. japonicum* 634b, the chemotaxis of the bacteria decreased by 38% (Fig. 11.7).

Similar dependence of the mobility of cells on the content of silica and clay mineral nanoparticles in their suspension was also observed in *B. subtilis* IMV B-7023 and *A. vinelandii* IMV B-7076 (Kurdish et al. 2010; Chuiko and Kurdish 2017). Introduction of 0.05–0.2 g/l of saponite particles into the phosphate buffer did not affect the mobility and chemotaxis of *B. subtilis* IMV B-7023. However, with an increase in the saponite content to 0.5–1.0 g/l, the mobility of these bacteria increased by 121–124% and their chemotaxis to glucose decreased by 2.2–3.7 times. A stimulating impact on the mobility of these bacteria was caused by introduction of 0.05–0.5 g/l of silicon dioxide into the phosphate buffer. In this case, the content of bacilli in capillaries containing the phosphate buffer was 11–46% higher than that in the absence of the nanoparticles. The chemotaxis of these bacteria was decreased the most by a content of 1 g/l of these nanoparticles.

A decrease in the chemotaxis of bacilli was also observed during their contact with the polysaccharide complex of *Azotobacter*, which was adsorbed on the cell surfaces of *B. subtilis* IMV B-7023; this was accompanied by an increase in the zeta potential of the bacilli from −35.8 mV to −46 mV, which is notable for *Azotobacter* cells (Chuiko et al. 2013).

During interaction of *A. vinelandii* IMV B-7076 with saponite particles (0.1–1.0 g/l), 5–22% more cells entered capillaries with phosphate buffer, which demonstrated stimulation of chaotic mobility of these bacteria by the aforementioned particles. However, the number of bacterial cells in capillaries containing glucose along with mineral particles differed from that seen in the previous variant considerably (Chuiko and Kurdish 2017). This may have been influenced by the fact that during interaction of nanoparticles and the surface of the bacilli, these particles block chemotaxis receptors, whereas in *A. vinelandii* IMV B-7076 they may be under the polysaccharide layer, protecting these receptors from interaction with nanoparticles and blocking their function. Therefore, the interaction of the investigated bacteria with silica, montmorillonite, and palygorskite particles promoted an increase in the mobility of their cells. This effect may have been influenced by an increase in the energetic metabolism of the bacteria during their interaction with the investigated nanomaterials. This phenomenon could promote distribution of these microorganisms in natural conditions.

It was demonstrated by us that the increase in the mobility of *B. subtilis* IMV B-7023 and *A. vinelandii* IMV B-7076 with the impact of some factors correlates with their enhanced adhesion to plant roots (Kurdish et al. 2008a, b, c). It was established that bacteria that were selected in the phase of their logarithmic growth (24 h) had the highest adhesive activity of *Azotobacter* to the roots of Konkurent cucumber plants. An evident decrease in the chemotaxis of bacteria was observed in the stationary phase of growth (after 72 h of cultivation). The bacterial cells adhered to the surface of the plant roots in much smaller quantities, and after their cultivation for 96 h the number of adhered cells on this surface was only a tenth of that seen after cultivation of the bacteria for 24 h. A considerable decrease in the adhesion of these bacteria was influenced by high indices of the negative charge of their surface and the loss of cell mobility during their long-term cultivation (Kurdish et al. 2008a, b, c).

The decrease in the chemotaxis of bacteria upon their interaction with nanoparticles of natural minerals demonstrates the possibility of blocking chemotaxis receptors on the surface of bacteria by the investigated nanoparticles and the polysaccharide complex of *Azotobacter* (Chuiko and Kurdish 2004, 2017; Chuiko et al. 2006, 2013). Therefore, the results of these investigations demonstrate that interaction between the investigated species of bacteria and nanoparticles of many minerals is accompanied by an increase in the mobility of cells and a reduction in their chemotaxis.

11.5 Interaction of Plant Growth–Promoting Rhizobacteria and Natural Mineral Nanomaterials as a Basis for Highly Efficient Preparations for Plant Production

Taking into consideration the stretch properties of clay minerals, their stimulating effects on the physiological and biochemical activity of bacteria, the protective effect of nanoparticles on the viability of cells, their antioxidant properties, and the interaction of our selected highly active strains of the nitrogen-fixing bacteria *A. vinelandii* IMV B-7076 and the phosphate-mobilizing bacteria *B. subtilis* IMV

Table 11.13 Numbers of viable cells of *A. vinelandii* IMV B-7076 and *B. subtilis* IMV B-7023 in the granulated complex bacterial preparation Azogran during storage at different temperatures (Kurdish and Bega 2006b)

	Viable bacterial cells/g			
	At room temperature		At 4 °C	
Storage period, months	*A. vinelandii* IMV B-7076	*B. subtilis* IMV B-7023	*A. vinelandii* IMV B-7076	*B. subtilis* IMV B-7023
0	$(7.1 \pm 0.3) \cdot 10^8$	$(3.8 \pm 0.2) \cdot 10^8$	$(7.0 \pm 0.3) \cdot 10^8$	$(3.5 \pm 0.2) \cdot 10^8$
3	$(4.6 \pm 0.3) \cdot 10^8$	$(3.4 \pm 0.2) \cdot 10^8$	$(5.1 \pm 0.4) \cdot 10^8$	$(3.5 \pm 0.3) \cdot 10^8$
6	$(1.8 \pm 0.3) \cdot 10^8$	$(2.8 \pm 0.1) \cdot 10^8$	$(2.1 \pm 0.2) \cdot 10^8$	$(3.4 \pm 0.1) \cdot 10^8$

B-7023 with particles of bentonite, we created a granulated complex bacterial preparation, named Azogran, for use in plant production (Kurdish 2010), containing $>10^8$ viable cells of each species of bacteria per gram of the preparation. This preparation is stable during long-term storage. It is registered in Ukraine and used in plant production. The number of bacteria in the granulated preparation is somewhat dependent on the storage temperature (Kurdish 2010). After 6 months of storage at room temperature, the number of viable cells of *A. vinelandii* IMV B-7076 therein was >25% of the initial number, and after 6 months of storage at 4°C, it was 30% of the initial number. Under these conditions of storage of the preparation, the number of viable *B. subtilis* IMV B-7023 cells was decreased to some degree (Table 11.13).

This preparation improves nitrogen and phosphorus nutrition of plants, and it stimulates their growth and development through synthesis of many biologically active substances by the bacteria (including substances of a phytohormonal nature) and the ability to inhibit plant damage by phytopathogens and phytophages. It was established that *B. subtilis* IMV B-7023 is an antagonist of many strains of phytopathogenic bacteria and micromycetes (Roy et al. 2005). This strain inhibits the growth of phytopathogenic bacteria (*Pseudomonas syringae* pv. *syringae*, *Pseudomonas syringae* pv. *atrofaciens*, *Erwinia carotovora*, *Clavibacter michiganensis*, and *Agrobacterium tumefaciens*) and many species of phytopathogenic micromycetes (*Fusarium graminearum*, *Fusarium oxysporum*, *Fusarium solani*, *Fusarium sambucinum*, *Bipolaris sorokiniana*, *Alteromonas alternate*, and *Gliocladium roseum*) (Roy et al. 2005). *B. subtilis* IMV B-7023 bacteria inhibit spreading of the agent of bacterial cancer of tomatoes considerably (Roy et al. 2012).

It was observed that presowing treatment of Beliy Naliv tomato seeds with a suspension of the phytopathogen *Clavibacter michiganensis* subsp. *michiganensis* was accompanied by a reduction in plant growth. However, combined inoculation of the seeds with this phytopathogen and *B. subtilis* IMV B-7023 enhanced their germination by 14.6% and plant growth by 16–18%; the development of bacterial cancer in these plants was not determined. However, when control plants were treated with this phytopathogen (their seeds were treated with water), they developed classic symptoms of bacterial cancer (Roy et al. 2012).

A relevant factor in the efficiency of use of the complex bacterial preparation Azogran in plant production is its capability to inhibit distribution of both phytopathogens and phytophages in agroecosystems. It was established by us that epiphytical treatment of plants using a suspension of the bacteria that are components of the preparation decreases distribution of many species of phytophages in agroecosystems (Roy et al. 2014; Zubko and Kurdish 2017) and improves the growth and development of plants considerably. It was determined that treatment of flowering plants (*Coleus* and *Pelargonium*, planted in a greenhouse) with a suspension of *B. subtilis* IMV B-7023 considerably decreased the numbers of greenhouse whitefly (*Trialeurodes vopårariorum*) and green peach aphid (*Aulacorthum circumflexus*) phytophages by 50–70%. It was shown that subsequent treatment of the plants with a suspension of the studied bacteria provided effective biocontrol of phytophages in greenhouse conditions (Zubko and Kurdish 2017).

Taking into consideration the fact that the interaction of *B. subtilis* IMV B-7023 and *A. vinelandii* IMV B-7076 with vermiculite particles stimulates the physiological and biochemical activities of these bacteria considerably and improves their survival during long-term storage (Table 11.14), we also created a free-flowing form of the complex bacterial preparation Azogran on the basis of the interaction between these strains and particles of exfoliated vermiculite, which is convenient for bacterization of plant seeds (Kurdish and Roy 2014).

Creation of a free-flowing complex bacterial preparation requires optimization of the culture conditions used for production. We optimized the composition of culture media for cultivation of the individual bacterial monocultures and mixed bacterial cultures that are components of the complex preparation. It was established that the best medium for growth of the mixed bacterial culture is a liquid culture medium with the following composition: treacle 30.0 g/l, corn steep extract 2.0 g/l, $K_2HPO_4 \cdot 3H_2O$ 0.25 g/l, KH_2PO_4 0.25 g/l, $MgSO_4 \cdot 7H_2O$ 0.3 g/l, NaCl 0.3 g/l, and $CaCO_3$ 3.0 g/l (pH 7.0–7.2). Cultivation of mixed cultures of *A. vinelandii* IMV B-7076 and *B. subtilis* IMV B-7023 in this medium for 24 h yields a culture liquid with more than 10^9 cells of each strain per milliliter (Table 11.14) (Kurdish et al. 2015).

The process used for manufacturing this free-flowing complex bacterial preparation based on the highly active strains *A. vinelandii* IMV B-7076 and *B. subtilis* IMV B-7023 was optimized. Because the bacilli were characterized as having a higher specific growth rate than *A. vinelandii* IMV B-7076, the inoculation of the

Table 11.14 Numbers of *A. vinelandii* IMV B-7076 and *B. subtilis* IMV B-7023 cells during incubation in a preparation containing powdered vermiculite

	Viable bacterial cells/g		
Bacterial species	At 0 days	At 1 day	At 2 days
A. vinelandii	$(4.0 \pm 0.1) \cdot 10^7$	$(1.9 \pm 0.1) \cdot 10^9$	$(1.7 \pm 0.1) \cdot 10^9$
B. subtilis	$(1.4 \pm 0.1) \cdot 10^7$	$(3.8 \pm 0.2) \cdot 10^9$	$1.0 \pm 0.1) \cdot 10^{10}$
A. vinelandii	$(3.7 \pm 0.2) \cdot 10^7$	$(1.6 \pm 0.1) \cdot 10^9$	$(1.8 \pm 0.1) \cdot 10^9$
B. subtilis	$(1.9 \pm 0.1) \cdot 10^6$	$(1.1 \pm 0.1) \cdot 10^9$	$(3.7 \pm 0.4) \cdot 10^9$
A. vinelandii	$(3.8 \pm 0.3) \cdot 10^7$	$(2.6 \pm 0.2) \cdot 10^9$	$(2.1 \pm 0.1) \cdot 10^9$
B. subtilis	$(1.2 \pm 0.1) \cdot 10^5$	$(3.0 \pm 0.1) \cdot 10^8$	$(6.0 \pm 0.2) \cdot 10^8$

vermiculite-mixed suspension of bacteria in the optimized culture medium amounted to about 10^7 colony-forming units (CFU) per milliliter of *A. vinelandii* IMV B-7076 and 10^6 CFU/ml of *B. subtilis* IMV B-7023 in a 3:1 mass ratio to this carrier. Thus, incubation with vermiculite for 24 h at 28 °C ensured high quality of the resulting complex bacterial preparation containing >10^9 CFU of *A. vinelandii* IMV B-7076 and *B. subtilis* IMV B-7023 per gram (Table 11.14) (Kurdish et al. 2015).

It was established by us that the obtained free-flowing preparation Azogran based on the bacteria *A. vinelandii* IMV B-7076 and *B. subtilis* IMV B-7023 could be stored well at room temperature (20–23 °C) for a long time (Table 11.15).

During storage for 6 months, both strains of these bacteria were defined in the preparation to the ninth power. After 1 year of storage, the number of *B. subtilis* cells exceeded 10^9 cells/g, while the number of *A. vinelandii* had decreased to 6.6×10^8 cells/g. Thus, it was established that Azogran, a granulated and free-flowing complex bacterial preparation, was characterized by compositional stability during long-term storage (for up to 6 months).

The complex bacterial preparation Azogran causes a more evident stimulating effect on the growth and development of plants than use of monoculture preparations (Kurdish 2010). Introduction of the complex bacterial preparation Azogran into agroecosystems considerably improves the growth and development of lawn grass, many species of decorative plants (*Chlorophytum*, dragon tree, jade tree, boxwood, and *Thuja*), flowering plants (*Begonia*, roses, and others), and seedlings and young plants of pine and fir trees. It also enhances yields of technical crops, cereals, and vegetables by 16–37% (Kurdish 2010; Kurdish et al. 2008a, b, c; Skorokhod et al. 2012).

Introduction of one granule of the complex bacterial preparation Azogran into the root zone of Ilius roses increased the number of inflorescence shoots on the plants by 45% (Kurdish 2010; Kurdish et al. 2008a, b, c). However, its stimulating effect decreased with introduction of two granules of this preparation (Table 11.16). A less evident impact was made by introduction of one granule of the Azogran preparation into the root zone of Grand Prix roses (Table 11.16). In this case, the number of inflorescence shoots increased by 22.6%, but after introduction of two granules, the stimulating action of the preparation, again, decreased.

Table 11.15 Dependence of the numbers of viable cells of *B. subtilis* IMV B-7023 and *A. vinelandii* IMV B-7076 in a free-flowing complex bacterial preparation on the storage duration at 20–23 °C

Storage period, months	Bacterial species	Viable bacterial cells/g
0	*A. vinelandii* *B. subtilis*	$(4.3 \pm 0.2) \times 10^9$ $(6.2 \pm 0.1) \times 10^9$
2	*A. vinelandii* *B. subtilis*	$(2.5 \pm 0.2) \times 10^9$ $(6.0 \pm 0.4) \times 10^9$
6	*A. vinelandii* *B. subtilis*	$(2.0 \pm 0.2) \times 10^9$ $(2.4 \pm 0.3) \times 10^9$
12	*A. vinelandii* *B. subtilis*	$(6.6 \pm 0.2) \times 10^8$ $(1.4 \pm 0.05) \times 10^9$

Table 11.16 Effects of a granulated complex bacterial preparation (Azogran) on the numbers of inflorescence shoots in Ilius and Grand Prix roses

Preparation	Cut inflorescence shoots			
	Ilius		Grand Prix	
	N	%	*N*	%
No preparation (control)	471	100.0	610	100.0
One granule	683	145	748	122.6
Two granules	587	125	658	107.9
$HIP_{0.5}$	94.5	–	97.4	–

Table 11.17 Effects of granulated bacterial preparations of *A. vinelandii* and *B. subtilis* IMV B-7023 on the yield of Chervona Strila tomato species

Preparation	Yield, kg	Yield gain	
		kg	%
No preparation (control)	2556.5	–	100.0
A. vinelandii 56	3037.1	480.5	118.8
A. vinelandii IMV B-7976	3264.5	708.0	127.7
A. vinelandii 56 + *B. subtilis* IMV B-7023	3351.6	795.1	131.1
A. vinelandii IMV B-7076 + *B. subtilis* IMV B-7023	3502.4	945.9	137.0

It was established that introduction of one granule of the complex granulated preparation Azogran into the root zone of young coniferous plants was accompanied by intensification of their growth. The most evident impact of this preparation was observed in *Thuja*, European spruce, and juniper plants. After 4.5 months of growth, *Thuja* plants treated with the preparation were 17% taller, young European spruce trees were 15.0% taller, and juniper plants were 22% taller than their respective controls (Chuiko et al. 2010).

It was demonstrated that introduction of two granules (0.5 g) of Azogran preparation into the root zone of Olexandria sugar beet improved the yield of seeds considerably. After introduction of two granules of a preparation based on one strain of *A. vinelandii* IMV B-7076, the yield of seeds increased by 21.3%. After introduction of two granules of a complex bacterial preparation based on both these bacteria and *B. subtilis*, this index increased by 37.8% (Kurdish et al. 2005).

It was established that introduction of two granules of a monoculture (the mass of one granule is 0.2 g) based on the nitrogen-fixing bacteria *A. vinelandii* 56 during planting of Chervona Strila tomato species increased the yield of the plants by 18.8% (Table 11.17). With use of a monoculture based on *A. vinelandii* IMV B-7076, the yield of the plants increased by 27.7%. However, the most marked stimulating impact on the investigated plants was seen with introduction of the complex bacterial preparation Azogran containing strains of *A. vinelandii* IMV B-7076 and *B. subtilis* IMV B-7023. In this case, the yield of tomatoes increased by 37% (Table 11.17).

It was determined that processing of Tsarivna winter wheat seeds with the free-flowing complex bacterial preparation Azogran increased the grain yield up to 0.57–0.67 t/ha and also increased the content of crude protein and fiber in the grain up to 0.6–1.0% and 1.1–1.3%, respectively. Moreover, the occurrences of root rot lesions and *Septoria* leaf spot on the wheat were decreased significantly (Kurdish et al. 2015). Seed treatment of Nabat spring barley with the free-flowing complex bacterial preparation increased the yield of the grain up to 0.35–0.43 t/ha, increased its content of crude protein up to 0.4–0.6%, and reduced the occurrence of dark brown spot lesions on the plant leaves (Kurdish et al. 2014, 2015).

Thus, the technology for creation of the free-flowing complex bacterial preparation Azogran has been developed for crop growing. This preparation is stable during storage and convenient for use in cereal agroecosystems. Positive impacts of application of this free-flowing complex bacterial preparation to crop seeds of winter wheat and spring barley have been shown. The preparation promotes a significantly increased yield and improves the quality of grain.

It has been determined that in the rhizosphere soil of cereals whose seeds have been treated with the complex bacterial preparation, significant changes are observed in microbial biocenosis, with increases in the total number of bacteria and in the content of oligotrophic bacteria, phosphate-mobilizing bacteria, and other physiologically trophic groups of microorganisms (Kurdish et al. 2014, 2015).

One of the relevant factors that determine the efficiency of using the complex bacterial preparation Azogran in plant production is its capability to inhibit distribution of both phytopathogens and phytophages in agroecosystems. It was established by us that epiphytic treatment of plants, using a suspension of these bacteria, decreases distribution of many species of phytophages in agroecosystems (Kurdish et al. 2008a, b, c; Skorokhod et al. 2012) and improves the growth and development of the plants greatly.

11.6 Conclusion and Future Prospects

It has been demonstrated that the interaction of many species of microorganisms with nanomaterials of different natures has an evident stimulating effect on the physiological and biochemical activity of microbial populations, protecting them from the effects of negative environmental factors. Interactions of *Azotobacter vinelandii* IMV B-7076 and *Bacillus subtilis* IMV B-7023 with bentonite and vermiculite particles have been used as the basis for creation of a granulated and free-flowing complex bacterial preparation, Azogran, for use in plant production. This preparation improves the growth and development of decorative, flowering, and other plants considerably, and increases the yield of technical crops, cereals, and vegetables by 16–37%. The obtained results may be used as the basis for creation of novel biotechnologies.

References

Bashan J, Holguin G (1997) *Azospirillum*—plant relationships: environmental and physiological advances (1990–1996). Can J Microbiol 43(2):103–121

Begonia MF, Kremer RJ (1994) Chemotaxis of deleterious rhizobacteria to velvetleaf seed and seedlings. FEMS Microbiol Ecol 15(3–4):227–236. https://doi.org/10.1111/j.1574-6941.1994.tb00246.x

Bulavenko LV, Bega ZT, Kurdish IK (2000) Phosphorus mobilization from poorly soluble non-organophosphates by some microorganisms. Bull Agric Microbiol (6):55–56

Chobotarov AY (2015) Physiological activity of *Azotobacter vinelandii* IMV B-7076 and *Bacillus subtilis* IMV B-7023 in condition of disperse materials activity. PhD thesis. IMV NAS Ukraine

Chobotarov A, Volkogon M, Voytenko L, Kurdish I (2017a) Accumulation of phytohormones by soil bacteria *Azotobacter vinelandii* and *Bacillus subtilis* under the influence of nanomaterials. J Microbiol Biotechnol Food Sci 7(3):271–274. https://doi.org/10.15414/jmbfs.2017/18.7.3.271-274

Chobotarov AY, Gordienko AS, Kurdish IK (2010a) Influence of natural minerals on growth of *Azotobacter vinelandii* IMV B-7076. Mikrobiol Z 72(5):27–31

Chobotarov AY, Gordienko AS, Samchuk AI, Kurdish IK (2010b) Influence of silicon dioxide and saponite on growth of *Bacillus subtilis* IMV B-7023. Mikrobiol Z 72(4):33–39

Chobotarov AY, Gordienko AS, Kurdish IK (2013) Growth peculiarities of *Bacillus subtilis* and streptomycin resistant mutant in the medium with saponite. Mikrobiol Z 75(5):62–66

Chobotarov AY, Skorochod IO, Kurdish IK (2017b) Effect of some nanomaterials and cations on superoxide dismutase activity of *Azotobacter vinelandii* IMV B-7076//Theses of reports of XV congress of the Vinogradskyi Society of Microbiologists of Ukraine. September 11–15, 2017, Odesa: 157

Chobotarov A, Volkogon M, Voytenko L, Kurdish I (2017a) Accumulation of phytohormones by soil bacteria *Azotobacter vinelandii* and *Bacillus subtilis* under the influence of nanomaterials. J Microbiol Biotechnol Food Sci 7(3):271–274

Chuiko NV, Kurdish IK (2004) Chemotactic properties of *Bradyrhizobium japonicum* in the presence of natural fine-grained minerals. Microbiology 73(3):364–367

Chuiko NV, Kurdish IK (2017) The influence of high-dispersity saponite and silicone dioxide on chemotaxis of *Azotobacter vinelandii* IMV B-7076 and *Bacillus subtilis* IMV B-7023. Mikrobiol Z 79(3):36–43

Chuiko NV, Gordienko AS, Kurdish IK (2006) Chemotaxis and growth of *Bradyrhizobium japonicum* in presence of fine-dispersed silica. Mikrobiol Z 75(1):44–47

Chuiko NV, Bega ZT, Bulavenko LV, Kurdish IK (2010) Influence of bacterial preparation of complex action on decorative plants growth. Microbiol Biotech 2:43–50

Chuiko NV, Gordienko AS, Kurdish IK (2013) Chemotaxis of *Azotobacter vinelandii* and *Bacillus subtilis* in mixed culture. Mikrobiol Z 82(2):186–189

Costerton JW, Marrie MJ, Cheng KJ (1985) Phenomena of bacterial adhesion: mechanisms and physiological significance. In: Savage DC, Fletcher M (eds) Bacterial adhesion: mechanism and physiological significance. Plenum Press, London/New York, pp 3–43

Davies PJ (2004) Plant hormones—biosynthesis, signal transduction, action. Springer, Berlin, p 750

Filomeni G, Rotilio G, Ciriolo R (2002) Cell signalling and the glutathione redox system. Biochem Pharmacol 64(5–6):1057–1064

Gerasymenko IO, Kurdish IK (2015) Influence of vermiculite and silicon dioxide on dehydrogenase activity of *Bacillus subtilis* IMV B-7023 and *Azotobacter vinelandii* IMV B-7076. Mikrobiol Z 77(1):20–24

Giron D, Frago E, Glevarec G, Pieterse CMJ, Dicke M (2013) Cytokinins as key regulators in plant–microbe–insect interactions: connecting plant growth and defence. Funct Ecol 27(3):599–609

Gordienko AS, Kurdish IK, Krasnobrizhy NY (1990) Effect of clay mineral palygorskite on survivability of bacteria under their dewatering. Mikrobiol Z 52(5):75–78

Gordienko AS, Zbanatskaja IV, Kurdish IK (1993) Change in electrosurface properties of *Methylomonas rubra* cells at contact interaction with particles of silicon dioxide. Can J Microbiol 39(9):902–905
He Y, Zeng F, Lian Z, Xu J, Brookes PC (2015) Natural soil mineral nanoparticles are novel sorbents for pentachlorophenol and phenanthrene removal. Environ Pollut 205:43–51
Heijnen CE, van Elsas JD, Kuikman PJ, van Veen JD (1988) Dynamics of *Rhizobium leguminosarum* biovar *trifolii* introduced in soil: the effect of bentonite clay on predation by protozoa. Soil Biol Biochem 20(4):483–488
Karunakaran G, Suriyaprabha R, Manivasakan P et al (2012) Effect of nanosilica and silicon sources on plant growth promoting rhizobacteria, soil nutrients and maize seed germination. IET Nanobiotechnol:1–8. https://doi.org/10.1049/iet-nbt.2012.0048
Kiprushkina EI, Kolodyaznaya VS (2014) Dynamics of phenolic compounds content during storage of potatoes treated with biological preparation. Processes and Vehicles of Food Production series. Sci J St. Petersburg National Research University of information Technologies, Mechanics and Optics 1:1–6
Kisten AG, Kurdish IK, Bega ZT, Tsarenко IY (2006) The effect of several factors on the growth of pure and mixed culture of *Azotobacter vinelandii* and *Bacillus subtilis*. Appl Biochem Microbiol 42(3):278–283
Kohen R, Nyska A (2002) Oxidation of biological systems: oxidative stress phenomena, antioxidants, redox reactions and methods for their quantification. Toxicol Pathol 30(6):620–650
Kravchenko LV, Azarova TS, Leonova-Erko EI et al (2003) Root exudates of tomato plants and their effect on the growth and antifungal activity of *Pseudomonas* strains. Microbiology 72(1):48–53
Kulaeva ON (1973) Cytokinins, their structure and functions. Nauka, Moscow, p 253
Kurdish IK (2001) Granulated microbial preparation for plant-growing: science and practice. KVITs, Kiev, p 141
Kurdish IK (2010) Introduction of microorganisms in agroecosystems. Naukova Dumka, Kiev, p 253
Kurdish IK, Antonyuk TS (1999) Effect of clayey minerals on viability of some bacteria at high temperatures. Mikrobiol Z 61(3):3–8
Kurdish IK, Bega ZT (2006a) Effect of argillaceous minerals on growth of phosphate-mobilising bacteria *Bacillus subtilis*. Appl Biochem Microbiol 42(4):388–391
Kurdish IK, Bega ZT (2006b) Strain of bacteria *Azotobacter vinelandii* for bacterial fertilizer obtaining for plant-growing. Ukrainian Patent 72856, Bulletin 8
Kurdish IK, Kigel NF (1997) Effect of high-dispersed materials on physiological activity of methanotrophic bacteria. Mikrobiol Z 59(2):29–36. https://doi.org/10.15407/microbiolj78.06.029
Kurdish IK, Roy AO (2003) Strain of bacteria *Bacillus subtilis* for bacterial fertilizer obtaining for plant-growing. Ukrainian Patent 54923A, Bulletin 3
Kurdish IK, Roy AO (2014) Method for obtaining a free-flowing complex bacterial preparation for plant growing. Ukrainian Patent 106135, Bulletin 14, 20 July 2014
Kurdish IK, Titova LV (2000) Granular preparation of *Azotobacter* containing clay minerals. Appl Microbial Biochem 36(4):484–487
Kurdish IK, Muchnik FV, Malashenko YR, Dementieva OA (1988) Specific concentration of nutrients as a criterion controlling microorganism cultivation. Mikrobiol Z 50(3):13–16
Kurdish IK, Muchnik FV, Malashenko JR (1990) The mathematical model to control the microbial cultivation process by the specostat principle. In: Blazej A, Ottova A (eds) Progress in biotechnology, vol 6. Elsevier, Amsterdam, pp 175–182
Kurdish IK, Bikhunov VL, Tsimberg EA, Elchits SV, Vygovskaya EL, Chuiko AA (1991a) Effect of dispersed silicon dioxide-aerosol A-300 on the growth of yeast *Saccharomyces cerevisiae*. Mikrobiol Z 53(2):41–44
Kurdish IK, Krasnobrizhy NY, Gordienko AS (1991b) Method for obtaining dry bacterial preparation. Ukrainian Patent 14654A, Bulletin 5
Kurdish IK, Titova LV, Tsimberg EA, Antipchuk AF, Tantsyurenko EV (1993a) Effect of Aerosil on the growth of *Azotobacter chroococcum*. Mikrobiol Z 55(1):38–42

Kurdish IK, Kigel NF, Bortnik SF (1993b) Stabilization of physiological activity of methanotroph *Methylomonas rubra* 15sh under storage. Mikrobiol Z 55(4):37–43

Kurdish IK, Roy AA, Garagulya AD, Kiprianova EA (1999) Survival and antagonistic activity of *Pseudomonas aureofaciens* UKM B-111 stored in fine materials. Microbiology 68(3):387–391

Kurdish IK, Antonyk TS, Chuiko NV (2001) Influence of environmental factors on the chemotaxis of *Bradyrhizobium japonicum*. Microbiology 70(1):91–95

Kurdish IK, Tserkovniak LS, Tsvey YP, Chernata DM (2005) Perspectives and introduction problems of microbial preparation in agricenosis. Sci Bull Chernivtsy Univ 252:126–131

Kurdish IK, Bega ZT, Gordienko AS, Dyrenko ДІ (2008a) The effect of *Azotobacter vinelandii* on plant seed germination and adhesion of these bacteria to cucumber roots. Appl Biochem Microbiol 45(4):400–404

Kurdish IK, Chuiko NV, Bulavenko LV, Direnko DI (2008b) Efficiency of the introduction of granular bacterial preparations into the agroecosystem of flowering plants. Collection of the Uman Agrarian University. The basis for the formation of productivity of agricultural crop with intensive cultivation technologies, pp 186–192

Kurdish I, Roy A, Chuiko N, Belogubova O, Bulavenko L, Bega Z, Dyrenko D, Chobotarjov A (2008c) The application of granulated bacterial preparation complex action in a plant-grower. Agroecological J:141–142

Kurdish IK, Chuiko NV, Bega ZT (2010) Chemotactic and adhesive properties of *Azotobacter vinelandii* and *Bacillus subtilis*. Appl Biochem Microbiol 46(1):51–56

Kurdish IK, Roy AA, Skorochod IO, Chobotarov AY, Herasimenko IO (2014) Elaboration of a free-flowing complex bacterial preparation for cereal crops. Nanosize system and nanomaterials: studies in Ukraine. Akademperiodika, Kiev, pp 610–614

Kurdish I, Roy A, Skorochod I, Chobotarov A, Herasimenko I, Plotnikov V, Gylchuk V, Korniychuk A (2015) Free-flowing complex bacterial preparation for crop and efficiency of its use in agroecosystems. J Microbiol Biotechnol Food Sci 4(6):527–531

Kurdish IK, Melnykova NM (2011) Influence of clay minerals on growth and nodulation activity of *Bradyrhizobium japonicum*. Mikrobiol Z 73(4):36–40

Kurdish IK, Titova LV (2001) Use of high-dispersed materials for culturing and obtaining granular *Agrobacterium radiobacter* preparation. Appl Biochem Microbiol 37(3):318–321

Labas YA, Gordeeva AV, Deryabina YI, Deryabin AN, Isakova EP (2010) Regulator role of reactive oxygen species: from bacteria to the man. Successes Mod Biol 130(4):323–335

Long SR (2001) Genes and signals in the *Rhizobium*–legume symbiosis. Plant Physiol 125:69–72

Marshall KC (1985) Mechanisms of bacterial adhesion at solid–water interfaces. In: Savage DC, Fletcher M (eds) Bacterial adhesion mechanisms and physiological significance. Plenum, New York, pp 133–161

Mishra VK, Kumar A (2009) Impact of metal nanoparticles on the plant growth promoting rhizobacteria. Dig J Nanomater Biostruct 4(3):587–592

Nannipieri P, Ascher J, Ceccherini MT, Landi L, Pietramellara G, Renella G (2017) Microbial diversity and soil functions. Eur J Soil Sci 68(1):12–26

Niste M, Vidican R, Pop R, Rotar I (2013) Stress factors affecting symbiosis activity and nitrogen fixation by *Rhizobium* cultured in vitro. ProEnviron Promediu 6:42–45

Ocheretyanko A, Roy A, Skorocod I, Kurdish I (2016) Influence of physical and chemical factors on the accumulation of phenolic compounds nitrogen-fixing bacteria *Azotobacter vinelandii* IMV B-7076. Acta Velit 3(1):49–55

Pogorelova VV, Bega ZT, Kurdish IK (2012) Interrelations of infusoria with *Azotobacter* and their influence on plants. Mikrobiol Z 74(5):48–54

Roy AA, Bulavenko LV, Kurdish IK (2001) New strains of soil bacilli mineralizing organic phosphorus compounds. Mikrobiol Z 63(4):9–15

Roy AA, Zaloilo OV, Chernova LS (2005) Antagonistic activity of phosphate-mobilizing bacilli to phytopathogenic fungi and bacteria. Agroecol J (1):50–55

Roy AA, Pasichnyk LA, Tserkovniak LS, Chodos SF, Kurdish IK (2012) Influence of bacteria of *Bacillus subtilis* on the agent of bacterial cancer of tomatoes. Mikrobiol Z 74(5):74–80

Roy AA, Matselyukh OV, Zubko PD, Varbanets LD, Kurdish IK (2014) Proteolitic activity of phosphorous mobilizing bacteria of *Bacillus* genus and their influence on some phytophages. Agric Microbiol 20:66–73

Rutter PR, Dazzo FB, Fletcher M et al (1984) In: Marshall KC, Characklis WG, Filip Z, Fletcher M, Hirsch P, Jones GW, Mitchell R, Pethica BA, Rose AH, Calleja GB (eds) Microbial adhesion and aggregation: report of the Dahlem Workshop on Microbial Adhesion and Aggregation Berlin 1984, January 15–20. Springer, Berlin, pp 5–19

Shakirova FM (2001) Non-specific resistance of plants to stress factors and its regulation. Gilem, Ufa

Shtarkman IN, Gudkov SV, Chernikov AV, Bruskov VI (2008) Formation of hydrogen peroxide and hydroxyl radicals in aqueous solutions of L-amino acids by the action of x-rays and heat. Biophysics 53(1):5–13

Skorochod IO, Kurdish IK (2013) Influence of nanoparticles of silica and vermiculite on activity of enzymes of antioxidant defense. Microbiol Biotechnol 1:59–67

Skorochod IO, Kurdish IK (2014) The low-molecular antioxidants of microorganisms. Mikrobiol Z 76(3):48–59

Skorochod IO, Kurdish IK (2018) Influence of bentonite nanoparticles and nano-SiO_2 on total content of phenolic compounds in the cultural medium of bacteria *Azotobacter vinelandii* IMV B-7076. In: Proceedings of the Ukrainian conference with international participation "Chemistry, Physics and Technology of Surface", 23–24 May 2018, Kiev, p 151

Skorochod IO, Roy AA, Melentiev AI, Kurdish IK (2013) Influence of bioactive substances of phosphate-mineralizing strains of genus *Bacillus* on plant seeds affected by oxidative stress. Microbiol Biotechnol (2):41–51

Skorochod IO, Roy AO, Kurdish IK (2016) Influence of silica nanoparticles on antioxidant potential of *Bacillus subtilis* IMV B-7023. Nanoscale Res Lett 11:139

Skorokhod IO, Tserkovniak LS, Kurdish IK, Plotnikov VV, Gylchuk VG, Korniychuk OV (2012) Influence of granulated bacterial preparation complex action on the growth and yield of barley. Mikrobiol Z 74(3):23–28

Titova LV, Antipchuk AF, Kurdish IK, Skochinskaya NN, Tantsyurenko EV (1994) Effect of high-dispersed materials on physiological activity of *Azotobacter* bacteria. Mikrobiol Z 56(3):60–65

Tsercovniak LS, Bega ZT, Ostapchuk AN, Kuz'min VE, Kurdish IK (2009a) Production of biologically active substances of indol nature by bacteria of *Azotobacter* genus. Ukr Biochem J 81(3):122–128

Tsercovniak LS, Roy AA, Kurdish IK (2009b) Synthesis of amino acids of *Bacillus subtilis* IMV B-7023 in the medium with glicerophosphates. Mikrobiol Z 71(5):18–23

Tserkovniak LS, Kurdish IK (2009) Phosphate-mobilizing bacteria *Bacillus subtilis* as phenolic producers. Appl Biochem Microbiol 45(3):311–317

Tsimberg EA, Titova LV, Kurdish IK (1991) Effect of highly-dispersed materials on the growth of *Candida* yeast. Mikrobiol Z 53(4):55–58

van der Mei HC, van der Belt-Gritter B, Doyle RJ (2001) Cell surface analysis and adhesion of chemically modified streptococci. J Colloid Interface Sci 241(2):327–332

van Veen JA, van Overbeek LS, van Elsas JD (1997) Fate and activity of microorganisms introduced into soil. Microbiol Mol Biol Rev 61(2):121–135

Zehnder AJB, Harms H, Bosma TNP (1996) Proceedings of the International Symposium on Subsurface Microbiology, 15–21 September 1996, Davos, p 42

Zhang ZY, Huang L, Liu F (2017) The properties of clay in soil minerals participles from two ultisols, China. Clays Clay Miner 65(4):273–285

Zhang ZY, Huang L, Liu F (2018) Transformation of clay minerals in nanoparticles of several zonal soils in China. J Soils Sediments. https://doi.org/10.1007/s11368-018-2013-4

Zubko PD, Kurdish IK (2017) Features of the spatial distribution of bacteria-components of the complex preparation Azogran in the root zone of cucumber. In: Proceedings of the XVth Congress of Vinogradskyi, Society of Microbiologists of Ukraine, 11–15 September 2017, Odessa

Chapter 12
Use of Metallic Nanoparticles and Nanoformulations as Nanofungicides for Sustainable Disease Management in Plants

Imran Ul Haq and Siddra Ijaz

Contents

I. Ul Haq (✉)
Department of Plant Pathology, Faculty of Agriculture, University of Agriculture, Faisalabad, Pakistan

S. Ijaz
Centre of Agricultural Biochemistry and Biotechnology (CABB), Faculty of Agriculture, University of Agriculture, Faisalabad, Pakistan

R. Prasad et al. (eds.), *Nanobiotechnology in Bioformulations*, Nanotechnology in the Life Sciences, https://doi.org/10.1007/978-3-030-17061-5_12

12.1 Introduction

Nanotechnology is gaining much popularity due to its broad applications in agriculture (Jha et al. 2011; Chowdappa and Gowda 2013). Nanobiotechnology has a key status among other disease management strategies for early disease detection, putative fungicides (nanofungicides), and efficient systems for delivery of fungicides to plants (Rai and Ingle 2012; Satalkar et al. 2016; Mishra and Singh 2015). This is a revolutionary science, which has turned the green revolution into a green nanobiorevolution (Khan and Rizvi 2014). This is based on two aspects: synthesis and application of nanosized materials.

Application of this technology provides real-time monitoring of crop plants for precision farming, which leads to maximum output while requiring minimum input (Scott and Chen 2003; Sharma et al. 2010). Extensive application of pesticides and fungicides results in ecotoxicity, as well as evolution of new phytopathogens that are resistant to them (Dzhavakhiya et al. 2012; Alghuthaymi et al. 2015; Chen et al. 2015). Hence, there is a great need to find alternative ways to manage plant pathogens and microbes (Vu et al. 2015).

There is an urgent global need for the use of eco-friendly approaches that generate less hazardous waste. This scenario has sensitized scientists to adopt and develop "green synthesis/biosynthesis" methods and strategies. Biosynthesis of nanoparticles (NPs) as a green synthesis approach helps to reduce the production of harmful waste by using nontoxic and environmentally safe resources. Therefore, in green chemistry, use of biological agents (plants and microbes) for synthesis of nanoparticles is a novel concept that opens up new avenues for exploring a broad array of biological species (Sharma et al. 2010; Chowdappa et al. 2013; Prasad et al. 2018a).

Plant extract–based bioreduction reactions for nanoparticle formation involve different biomolecules (phytochemicals) such as proteins, polysaccharides, tannins, organic compounds, plant resins, and pigments, as well as redox enzymes (Huang and Yang 2004; Nam et al. 2008; Wei and Qian 2008; Sanghi and Verma 2009; Prasad 2014). Microbe-assisted synthesis of nanoparticles is the branch of green chemistry that bridges microbial biotechnology and nanotechnology. For nanoparticle synthesis, microbes accumulate intracellular and extracellular inorganic compounds and execute bioreduction of different metals such as silver, platinum, copper, gold, and silica (Meyer 2008; Rai et al. 2009; Prasad et al. 2016).

Nanoparticles play pivotal roles in providing better food by promoting sustainable agriculture (Gruère 2012). A diverse range of microbes damage crop plants, ornamental plants, and trees, leading to major losses in the economy of a country (Tournas 2005). Some of them exert hazardous effects even on the health of human beings. The world's food demand is expected to double in the next half century, and that poses a big challenge for food production to feed the people (Tilman et al. 2002).

As has been documented, only a minute quantity of fungicides and pesticides (<0.1%) reaches the target site of action, because of depletion during application, photodegradation, and off-target deposition; these losses ultimately have effects on the ecosystem and increase the costs of production (Castro et al. 2013). When a fungicide/pesticide/bactericide agrochemical is applied to target pathogens, they may change their population into a new species or strain by genome recombination; thereby, a new species evolves that has resistance against that particular fungicide or pesticide (Schaller et al. 2004; Hettiarachchi and Wickramarachchi 2011; Chowdappa et al. 2013).

In the current scenario, use of nanoparticles in disease management, disease detection, and precise and controlled distribution of functional molecules (Scott and Chen 2003; Johnston 2010) is the best way to tackle this problem. These nanosized particles target specific issues in agriculture regarding crop protection (disease management) and improvement (Ghormade et al. 2011). The characteristic of a high surface-to-volume ratio makes nanoparticles more reactive and biochemically active (Dubchak et al. 2010). They bind to the cell walls of pathogens, resulting in deformation of cell membranes due to high-energy transfer, which leads to the death of the pathogen (Schaller et al. 2004).

These nanoparticles and nanoparticle-based formulations mediate a strong nanoscale system, which provides entrapment and encapsulation of agrochemicals for slow and targeted delivery of their active compounds and to reduce agrochemical runoff into the environment (Chen and Yada 2011; Gruère 2012). Thereby, this emerging science may be the key player for sustainable agriculture globally. This chapter considers the prospects for this putative field and comprehensively discusses the importance, synthesis, and characteristics of nanoparticles (particularly metallic nanoparticles), as well as their role as nanofungicides for sustainable disease management in plants.

12.2 Nanoparticles and Their Synthesis

Nanoparticles are very tiny in size in comparison with bacterial and viral cells (Wang et al. 2011). These particles may be rod shaped, spherical, polyhedral, etc. (Dubchak et al. 2010; Wang et al. 2011), and possess a high surface area–to–volume ratio (Satalkar et al. 2016). Nanoparticles show differential actions based on their size and thereby exhibit new aspects that differ from the characteristics of their bulk form. For example, silver nanoparticles possess antimicrobial action, while their bulk form does not (Sofi et al. 2012). Nanoparticles are also known for controlling genotoxicity, oxidative stress, and apoptosis responses (Kuppusamy et al. 2014).

Different techniques have been established and employed for synthesis of nanomaterials. Thus, nanoparticle synthesis methods are categorized into two broad groups (Fig. 12.2):

1. Top-down approaches
2. Bottom-up approaches

In top-down approaches, nanoparticles are prepared from massive and bulk materials by cutting them into nanosized materials, whereas in the case of bottom-up approaches, atoms are built into nanoparticles (Mazhar et al. 2017), as shown in Fig. 12.1.

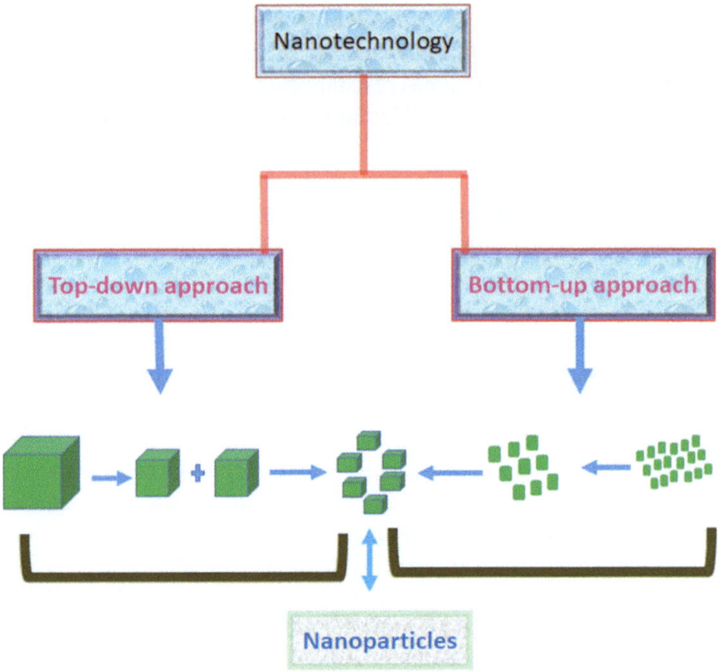

Fig. 12.1 Top-down and bottom-up approaches. Top-down approaches are based on reduction of the size of massive entities to nanosized materials, whereas bottom-up approaches are based on aggregation and build-up of atomic-state entities into nanomaterials

The different methods used for nanoparticles synthesis (such as biological, chemical, and physical methods, as listed in Table 12.1) affect their properties and efficiency (Narayanan and Sakthivel 2008; Rai and Yadav 2013). However, chemical and physical methods of nanoparticle synthesis are not cost effective and even require lethal and toxic compounds. It is a well-established fact that these methods have deleterious impacts on human or environment health through harmful radiation and the presence of synthetic reductants in concentrated forms and stabilizing agents (Pileni 1997; Joerger et al. 2000; Panigrahi et al. 2004; Oliveira et al. 2005; Gan et al. 2012). Conversely, biological methods based on plants and microbes have greater efficacy, are more cost effective and eco-friendly (Kumar et al. 2012; Bonde et al. 2012), and involve only a one-step bioreduction process (Sathishkumar et al. 2009; Iravani 2011), as shown in Fig. 12.2.

Plant secondary metabolites (phytochemicals) and microbial enzymes are utilized extensively in nanoparticle formulations because of their reducing actions (Bawaskar et al. 2010; Dar et al. 2013). The major loopholes of biogenic approaches are that it is hard to attain monodispersity and there is no control over the shape and size of the nanoparticles (Li et al. 2007; Nayak et al. 2011). However, with adjustment and optimization of the reaction medium and the metal concentration, the biosynthesis reaction may be controlled to influence the size and shape of the nanoparticles (Chandran et al. 2006; Shameli et al. 2012).

Table 12.1 Chemical, physical, and biogenic methods used for synthesis of nanoparticles

Chemical methods	Physical methods	Biological methods
Colloidal method	Electrochemical method	Plant extract–based synthesis
Chemical reduction method	Microwave method	Bacterial system–based synthesis
Sonochemical method	Solvothermal decomposition	Fungal system–based synthesis
Sol-gel method	Laser ablation	Microalgae system–based synthesis
Chemical solution deposition	Plasma arcing	
Chemical vapor deposition	Ball milling	
Catalytic route	Lithographic method	
Hydrolysis	Sputter deposition	
Langmuir–Blodgett method	Layer-by-layer growth	
Electrodeposition	Pulsed laser desorption	
Soft chemical method	Spray pyrolysis	
Wet chemical method	Molecular beam épistaxis	
Coprecipitation method	Ultra thin film	
	Diffusion flame–based synthesis	

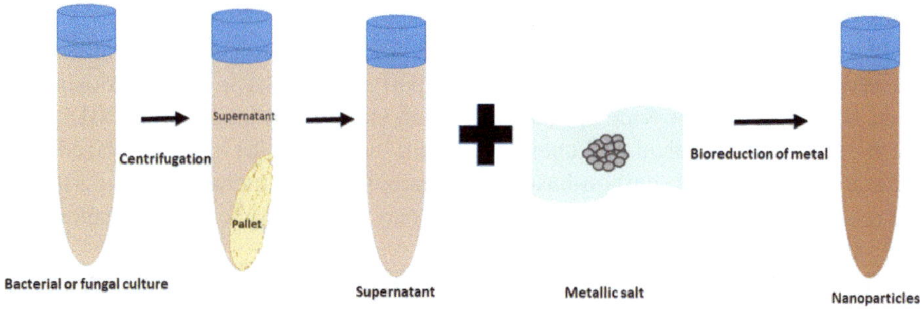

Fig. 12.2 Bioreduction process of metallic nanoparticle synthesis. This biogenic method of nanoparticle formation involves either a plant extract, a bacterial culture, or a fungal culture as a biological medium, which provides extracellular biological compounds for reduction of metals to metallic nanoparticles

In plant-assisted nanoparticle synthesis, plant extracts are used, comprising various biomolecules such as saponins, flavonoids, alkaloids, phenolic acids, and terpenoids. These secondary metabolites mediate the redox reaction and perform the reduction of the metals to nanoparticles (Aromal and Philip 2012; Prasad 2014; Prasad et al. 2018a).

In mycosynthesis of nanoparticles, the potential of fungi (especially filamentous fungi) is exploited because of their rapid growth on available substrates and metabolite production. Fungal-based extracellular synthesis of nanoparticles includes three mechanisms: (1) use of nitrate reductase action, (2) use of electron shuttle quinones, and (3) use of both of them (Sastry et al. 2010; Dhillon et al. 2012). Polysaccharides in the fungal cell wall are the main players in metal ion reduction (Sastry et al. 2003).

At the first step of the bioreduction process, metal ions are trapped and an interface is established between the metal ions and the cell surface of the fungus, which could be due to the electrostatic interaction among functional groups of enzymes (in the mycelial cell wall) that possess a positive charge. The next step involves enzyme-mediated reduction of metal ions, which leads to synthesis of nanoparticles (Meyer 2008; Dhillon et al. 2012; Prasad 2016, 2017; Prasad et al. 2018b).

12.3 Structure and Physicochemical Properties of Nanoparticles

Nanoparticles are particulate materials with dimensions measuring <100 nm (Laurent et al. 2010; Tiwari et al. 2012). They have attracted massive interest in multidisciplinary fields because of their exceptional attributes. The size of nanoparticles is the key factor that influences their chemical and physical attributes (Khan et al. 2017a, b). These nanomaterials exhibit sporadic biological, chemical, and physical properties that are entirely distinct and diverse from those of their bulk forms (Li et al. 2001). Variations in the physical characteristics of nanoparticles, such as their size and shape, lead to changes in their other physicochemical attributes (Dreaden et al. 2012; Barrak et al. 2016).

Physicochemical properties (e.g., chemical reactivity, mechanical strength, optical properties, and a large surface area) impart uniqueness to nanoparticles and make them suitable for wide applications (Wan et al. 2009; Gupta et al. 2013). On the basis of their physical and chemical nature, nanoparticles are categorized into different classes: (1) carbon-based nanoparticles, (2) metallic nanoparticles, (3) semiconductor nanoparticles, (4) ceramic nanoparticles, (5) polymer nanoparticles, and (6) lipid-based nanoparticles.

Nanoparticles are small particles but not simple ones. They comprise three discrete layers: a surface layer, a shell layer, and a core region (Shin et al. 2016); however, the core region is preferentially referred to as the nanoparticle itself (Khan et al. 2017a, b). The structural characterization of a substance is of prime importance in exploration of its composition and bonding nature (Ullah et al. 2017). Different techniques—such as x-ray diffraction (XRD), scanning electron microscopy (SEM), transmission electron microscopy (TEM), zeta sizing, infrared spectroscopy (IR), x-ray photoelectron spectroscopy (XPS), and energy-dispersive x-ray (EDX)—are employed to reveal the structural characteristics of nanoparticles (Ingham 2015).

Identification of single and multiphase nanoparticles, as well as their crystallinity, is achieved using an XRD approach (Ingham 2015; Ullah et al. 2017). SEM and TEM provide insights into and estimations of the size of nanoparticles. Estimation of the elemental composition of nanoparticles is done using EDX, because they are constituted from elements that emit characteristic energy x-rays. The intensity of a specific x-ray is in direct proportion to an explicit element concentration in a nanoparticle (Avasare et al. 2015). The XPS technique is widely considered the most sensitive approach to determine the elemental ratio and the nature of the bonds between elements in nanoparticles (Mansha et al. 2016).

12.4 Metallic Nanoparticles

Metals have been used to cure different diseases in plants, animals, and human beings since ancient times. Metallic nanoparticles with remarkable physicochemical attributes—such as nanoscale size, a high surface-to-volume ratio, structural stability, and target affinity—are used as antimicrobial agents and as the best alternative to synthetic fungicides (Kumar et al. 2010; Aziz et al. 2014, 2015, 2016). Bioreduction of metals to stable metallic nanoparticles through a green route is eco-friendly and safe (Kumar and Yadav 2009). As mentioned earlier, microbes and plants are attractive candidates for this green nanotechnology because of their nontoxic and cost-effective attributes (Prasad et al. 2016, 2018a). These metal-based nanoparticles have been shown to be effective weapons against phytopathogens; thus they will supersede synthetic fungicides, pesticides, and other agrochemicals, as a better option (Jo et al. 2009; Medici et al. 2015; Ismail et al. 2017; Gupta et al. 2018; Abd-Elsalam and Prasad 2018).

Several plant species have been designated as hyperaccumulators of metals. They accumulate metals in high concentrations and then assimilate them as nanoparticles (Dubey et al. 2009). Plant-based reduction of metals to nanoparticles involves phytochemicals (aldehydes, amides, carboxylic acids, flavonoids, ketones, terpenoids, quinones, etc.) (Bali et al. 2006; Ali et al. 2011). Besides plant-assisted synthesis, microbes (fungi and bacteria) have also emerged as suitable eco-friendly candidates for nanoparticle synthesis (Mandal et al. 2006; Ingle et al. 2009; Golinska et al. 2014; Tiwari et al. 2014; Prasad et al. 2016; Abdel-Aziz et al. 2018). Among the commonly used nanoparticles are silver-, gold-, silica-, copper-, and zinc-based nanoparticles.

12.4.1 Silver Nanoparticles

Silver nanoparticles possess antimicrobial properties in their ionic form, as well as in their nanosized form. Silver exhibits manifold inhibitory modes against microorganisms such as plant bacterial and fungal pathogens (Clement and Jarrett 1994; Kim et al. 2006; Wei et al. 2009). It has been shown in different in vitro and in planta assays that silver (in both its ionic and nanosized forms) inhibits colony formation by affecting spore and germ tube viability, and reduces disease progression (Kim et al. 2006; Gul et al. 2014). The reduction of silver nitrate to silver nanoparticles is due to the involvement of different metabolites and proteins in leaf tissues. A leaf extract provides a medium to synthesize and stabilize nanoparticles by acting as a reducing and capping agent (Singh et al. 2010; Jha and Prasad 2010). Silver nanoparticles have demonstrated highly significant inhibition of fungal phytopathogens in disease outbreaks under field conditions (Aguilar-Mendez et al. 2011; Gupta et al. 2018).

12.4.2 Silica Nanoparticles

Silica nanoparticles strengthen plants by enhancing their resistance against diseases and stimulating their physiological mechanisms (Carver et al. 1998; Brecht et al. 2004).

12.4.3 Copper Nanoparticles

Copper nanoparticles act as fungicides by generating highly reactive hydroxyl radicals, which may damage cellular materials (DNA, proteins, lipids, and other biomolecules) in fungal pathogens, leading to their death (EstebanTejeda et al. 2009; Brunel et al. 2013). Use of copper-based nanoparticles has been shown to be an effective control measure against bacterial blight in rice and leaf spot in mung bean (Gogoi et al. 2009).

12.4.4 Zinc Nanoparticles

Upon application as nanofungicides, zinc nanoparticles produce hydroxyl and superoxide radicals, which cause deformity of fungal cell walls and result in cellular death due to high-energy transfer (Patra et al. 2012). These nanoparticles interrupt the electron transfer chain and thereby disrupt related biological processes (Xia et al. 2008). Zinc nanoparticles deform fungal hyphae, impede conidiophores and conidial development, and eventually cause the death of fungal hyphae (Borkow and Gabbay 2005).

12.4.5 Gold Nanoparticles

The toxic effects of gold nanoparticles on *Salmonella* spp. were determined by Wang et al. (2011), who documented that they exhibited more toxic effects than gold in its bulk form.

12.4.6 Iron Nanoparticles

Iron nanoparticles establish a direct interface with fungal cell surfaces as a result of electrostatic interactions, and affect membrane permeability (Corredor et al. 2009; Parveen et al. 2018). When entering fungal cells, these nanoparticles generate oxidative stress by producing high levels of reactive oxygen species (ROS). Thereby, these nanoparticles inhibit growth, resulting in cell death (Yanping et al. 2011).

Because of their ability to rapidly permeate through the microbial cell membrane, metallic nanoparticles disorganize the cell's polymeric subunits, interrupt its protein synthesis mechanism, and thereby arrest the cell cycle (Sondi and Salopek-Sondi 2004; Kasthuri et al. 2009). Moreover, a pit appears on the cell wall, resulting in cell lysis. Coagulation of metallic nanoparticles on the microbial cell membrane results in increased permeability of the plasma membrane, which causes cellular content leakage (Sondi and Salopek-Sondi 2004; Panácek et al. 2006). The presence of nanoparticles inside microorganisms has been noted in different reports, revealing their interactions with sulfur- and phosphorus-containing compounds (Panácek et al. 2006; Raffi et al. 2008; Kasthuri et al. 2009).

Biosynthesized metallic nanoparticles are considered to have a stronger fungicidal mechanism than synthetic fungicides. Metallic nanoparticles damage fungal membranes and intercellular modules, and destroy cell functioning. As a result of nanoparticle activity, both fungal spore formation and fungal growth are restricted (Gardea-Torresdey et al. 2002; Marambio-Jones and Hoek 2010). The consensus on the antifungal mechanism of nanoparticles is that when a fungal cell takes up these metallic nanoparticles, upon entry they interrupt the process of adenosine triphosphate (ATP) synthesis in the cell and halt its DNA replication mechanism. Because of this, excess ROS are generated, disorganizing the integrity of the cellular membrane and causing development of pits on the surface of the membrane, which leads to cellular death (Logeswari et al. 2012; Prabhu and Poulose 2012; Reidy et al. 2013), as shown in Fig. 12.3.

As far as the effect of nanoparticles on bacterial cells is concerned, it is believed that a bacterial cell uses an enzyme for oxygen metabolization, which is required to sustain its life. Silver nanoparticles cripple this enzyme and inhibit oxygen metabolism, resulting in suffocation and ultimately leading to the death of the bacterium (Alvarez-Puebla et al. 2004; Raffi et al. 2008). The mechanism of the antibacterial

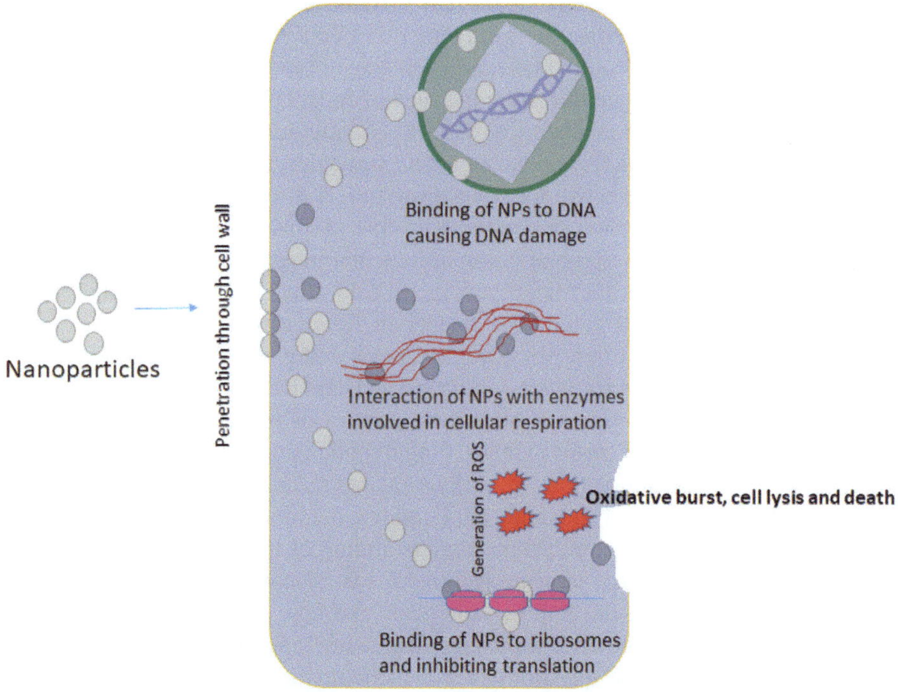

Fig. 12.3 Antifungal effects of nanoparticles (*NPs*) on a fungal cell and its functioning through interaction with the cellular machinery, resulting in cellular pathway arrest and generation of reactive oxygen species (*ROS*), which leads to an oxidative burst and fungal death

effect of metallic nanoparticles is believed to be their interaction with protein functional groups, particularly thiol groups of cysteine residues.

As a result of the oxidation reaction, a disulfide bond is established between thiol groups, which results in protein folding and structural alteration. Thus, a change in protein conformation results in inhibition of enzyme activity and, ultimately, inactivation of the bacterial cell (Liau et al. 1997; Schierholz et al. 1998). Each metal has different target sites in a cell; for example, Na^+-translocating reduced nicotinamide adenine dinucleotide (NADH) ubiquinone oxidoreductase (an enzyme in the bacterial respiratory chain) is the target site of silver. Thus, interaction of this enzyme with silver results in inhibition of the enzyme NADH dehydrogenase (Gupta et al. 1998; Holt and Bard 2005; Prasad et al. 2016).

12.5 Use of Metallic Nanoparticles as Nanofungicides for Sustainable Disease Management in Plants

Fungi are responsible for about 70% of diseases in crop plants such as cereals, fiber crops, pulses, and fruits (Agrios 2005). Pre- and postharvest losses due to fungal diseases have been reported to exceed €200 billion, and more than US$600 million is spent on fungicides annually in the USA (González-Fernández et al. 2010). Therefore, for sustainable disease management, effective and well-organized crop protection strategies are required because each stage of the fungus life cycle differs (Dhekney et al. 2007) depending upon the type of fungal pathogen.

For this purpose, different chemical controls (fungicides) and biological controls have been devised as plant disease management strategies. In general, chemical control is considered more effective for controlling fungal disease. Nevertheless, application of fungicides has some nonspecific impacts and causes ecological disturbance by destroying beneficial microbial communities that inhabit the rhizosphere of the crop plants afflicted with the fungal pathogen (Zaki et al. 1998; Manczinger et al. 2002). Injudicious use of fungicides results in fungicidal resistance in existing pathogens and creation of new physiological races and pathotypes, and even more virulent strains of fungal pathogens, which are resistant to fungicides.

In the case of agrochemical applications, control of fungal diseases is mostly influenced by fungicidal resistance in fungal pathogens. Fungicidal resistance develops by complex interactions of various factors such as the mode of action of the fungicide, the pattern of fungicide applications, the biology of the fungal pathogens, and the cropping system. Proper understanding of the biological philosophy of fungicide resistance (through unraveling of two major aspects—namely, how does fungicide resistance develop in pathogens, and how can it be managed?) is a prerequisite for sustainable disease management using fungicides. However, finding the exact answers to these questions is still difficult for scientists because of the multifactorial complexity of the mechanisms involved. Hence, they have begun to attempt alternative strategies and have now started to use nanomaterials to manage fungal diseases in plants (Kim et al. 2012).

After considering loopholes in chemical control, the scientific community has tried to open up new avenues in the form of nanoparticle synthesis and applications for plant disease management. Different protocols and practices have been implemented to evaluate and determine the efficacy of nanoparticles and to find the best alternatives to use of agrochemicals against different microbial pathogens, especially fungi (Jo et al. 2009; Rai et al. 2009).

Metallic nanoparticles are now becoming popular and accepted alternatives to agrochemicals. They have potential to eliminate unwanted and lethal microbes from soils, from plants, and even from hydroponic systems (Park et al. 2006; Sharma et al. 2012). Therefore, to determine the status and impact of metallic nanoparticles for disease management in plants, their effects can be addressed in two ways (Khan and Rizvi 2014): (1) direct application of nanoparticles to phytopathogens, and (2) use of nanoparticles in formulating fungicides. In both ways, nanoparticles are applied as nanofungicides.

These nanoparticles are applied as foliar sprays to kill pathogens that cause different plant diseases. In addition, the nanoparticles may even stimulate plant growth (Agrawal and Rathore 2014). Lower concentrations of these nanoparticles are recommended for effective control of plant diseases (Nel et al. 2003; Park et al. 2006). Nanoparticle administration is also effective for those microbes that possess less or even no sensitivity to antimicrobial agents because of poor penetration by the antimicrobial compounds through the cell membrane (Samuel and Guggenbichler 2004). Microscopic studies have revealed that metallic nanoparticles damage cell walls of fungal hyphae, resulting in hyphal plasmolysis (Min et al. 2009).

Different theories have been put forward by different scientists regarding the mechanism of action of nanoparticles. Those most widely accepted are the following (Zeng et al. 2008; Prabhu and Poulose 2012; Lemire et al. 2013):

1. They bind to sulfur groups of proteins, prevent their functioning in the cellular membrane, and thereby affect membrane permeability.
2. They have a genotoxic effect and cause DNA damage.
3. They disrupt protein oxidation and the electron transport mechanism in the cell.
4. They generate ROS, which mediate cellular damage.
5. They hinder proper uptake of nutrients.

These mechanisms are interlinked, which illustrates the multitargeted action of nanoparticles in effectively combating phytopathogens (Alghuthaymi et al. 2015; Abd-Elsalam and Prasad 2018).

As discussed earlier in this chapter, bioreduction of different metals (such as silver, gold, zinc, copper, and iron) alone, or in combinations, has been evaluated in metallic nanoparticle synthesis. Moreover, different in vitro and in vivo assays have been documented in the literature regarding investigations into the antimicrobial effects of these metallic nanoparticles against various plant pathogens. Among them, silver nanoparticles have been shown to be more toxic to pathogens, and that is why they are generally known as nanoweapons (Alghuthaymi et al. 2015; Mishra and Singh 2015).

Different researchers around the globe have tested the toxicity of nanoparticles to pathogens and their safety for nontargeted organisms (such as plants, animals, human beings, and even other beneficial microbes present in the microflora of candidate plants) at low concentrations. They have suggested that application of nanoparticles for controlling and managing plant diseases is a relatively safer approach than use of synthetic agrochemicals (Thomas and McCubbin 2003; Zeng et al. 2008).

The fungicidal effect of metallic nanoparticles on *Raffaelea* species (a fungal pathogen causing oak wilt) was evaluated in vitro by Woo et al. (2009). They documented fungal growth inhibition, damaged hyphae, and restricted conidial germination. Likewise, their antifungal effect was tested against *Magnaporthe grisea* and *Bipolaris sorokiniana* (which cause cereal diseases) by Jo et al. (2009), who observed that the progress and severity of disease were inhibited in both cases.

Among metallic nanoparticles, silver nanoparticles are most commonly and widely used in biosystems. The antimicrobial effect of silver nanoparticles is due to their oligodynamic action, which inactivates enzymes that are key players in metabolic pathways in microorganisms (Thomas and McCubbin 2003). Silver nanoparticles are thus detrimental to microbial pathogens and cause cellular damage and dysfunction of the fungal ion efflux transport system (Morones et al. 2005). The disruption of ion efflux causes silver ion accumulation and thereby interrupts cellular processes such as respiration and metabolism. Upon entry into the cell, nanosilver rapidly produces ROS by reacting with oxygen molecules and damages biomolecules such as DNA, RNA, protein, lipids, and polysaccharides (Hwang et al. 2008; Aziz et al. 2016, 2019). Thus, DNA replication is halted, ribosomal protein is inactivated, other proteins and enzymes involved in ATP synthesis are degraded, and dysfunction of cell membrane–bounded enzymes takes place (Thomas and McCubbin 2003; Kim et al. 2012). Hence, silver nanoparticles are widely accepted as a nanofungicide and considered a potential agrochemical replacement. Moreover, numerous patents have been filed for use of silver nanoparticles in the treatment of plant diseases (Sharon et al. 2010).

The effectiveness of nanosilver depends on its physical characteristics (i.e., the size and shape of the particles). The efficacy decreases with an increase in the particle size (Cioffi et al. 2004; Duhan et al. 2017). The "-cidal" (i.e., lethal) effect of the nanoparticles is also influenced by their shape. Nanoparticles of a truncated triangular shaped exhibit more -cidal effects than rod-shaped and spherical ones (Cioffi et al. 2004; Kim et al. 2012).

The fungicidal properties of silver nanoparticles have also been evaluated against *Sclerotinia minor*, *Sclerotinia sclerotiorum*, and *Rhizoctonia solani*. Min et al. (2009) reported strong inhibition in germination of sclerotia, as well as inhibition of fungal growth. In the case of powdery mildew in cucurbits and cucumber, silver nanoparticles exhibited inhibitory and phytotoxic effects on conidial growth, as well as on fungal hyphae. Moreover, a good disease prognosis was observed upon application of nanoparticles in field conditions. Likewise, application of silver nanoparticles was shown to have a strong antifungal effect on the fungal pathogen of powdery mildew in roses (Sharon et al. 2010).

A nanocomposite constituted from pullulan and silver nanoparticles was tested by Pinto et al. (2013) against a phytopathogen, *Aspergillus niger*, and they reported sporulation inhibition and disruption of spore cells of this fungus. Chowdappa et al. (2013) applied a chitosan–nanosilver composite against *Colletotrichum gloeosporioides* and observed inhibited germination of conidia and conidiophores.

Considering the strong antimicrobial action of nanosilver, Park et al. (2006) created a nanosilica–nanosilver composite for evaluation as a nanofungicide under both greenhouse and field conditions. When administered in a dose of 10 parts per million (ppm), it showed high efficacy and effectivity against powdery mildew in pumpkin and inhibited 100% of fungal growth; within 3 days after its application, the pathogen had disappeared from the infected plant parts and thereafter the treated plants were healthy. In other studies, silica–silver nanocomposites were found to achieve 100% control of powdery mildew in cucurbits under field conditions (Brecht et al. 2003; Banik and Sharma 2011; Patel et al. 2014).

Sulfur nanoparticles have shown efficacy in preventing early blight and wilt diseases in tomato (caused by *Fusarium solani*), as well as apple scab (caused by *Venturia inaequalis*) (Rao and Paria 2013; Boxi and Paria 2015). The fungicidal action of these nanoparticles is due to their deposition on the cell wall of the fungus and its subsequent lysis (Jampílek and Kráľová 2015). Silver nanoparticles of cylindrical and spherical shapes have been shown to significantly reduce the total lipid content of fungal cells upon their application to *Aspergillus niger* isolates. The expression of desaturase enzymes was downregulated, saturated fatty acid was accumulated in high levels, and lipid layer depletion was also observed with silver nanoparticle–mediated fungistasis. These nanoparticles have also been shown to significantly reduce the phospholipid content of *Fusarium oxysporum* cells (Choudhury et al. 2011, 2012; Chhipa 2017). Similarly, nanosized copper has been reported to exhibit a strong antifungal action against bacterial blight in pomegranate (Hezave and Esmaeilzadeh 2010).

12.6 Use of Nanoformulations as Nanofungicides for Sustainable Disease Management in Plants

Agricultural crops all over the world are affected by fungal diseases, which pose a considerable threat to their yield. A reduction in crop yield has a significant impact on the economy of a country. Different fungicides of narrow and broad spectra are prepared in order to combat fungal diseases in plants. However, this control measure is not widely and adequately effective. Therefore, there is a trend toward use of nanofungicides that consist of nanoparticles or contain an active nanoformulation compound (Jampílek and Kráľová 2015; Bhattacharyya et al. 2016). Broad application of fungicides causes environmental pollution, biodiversity loss, and evolution of new pathogens (Rai et al. 2015). To resolve these issues, nanoformulations are being applied, which more effectively facilitate site specificity for targeted and controlled delivery of fungicidal compounds, avoiding collateral damage (Nikhil and Bharat 2004).

Extremely small particle size and a large surface area are the features of core importance for permeation through cellular membranes and for carriage and transport of compounds. Generally, nanosystems comprise two core components: an active material and a nanocarrier. Nanocarriers facilitate transportation and site-specific distribution of the active ingredients of fungicides/pesticides by stabilizing them. In nanoformulations, nanoparticles are combined with active chemical compounds and with other organic and inorganic compounds, all of which must be of a nanometric size (Jampílek and Kráľová 2015).

Nanoformulations are prepared in order to enhance stability and effectivity but with lesser amounts of fungicidal compounds (Zachariah et al. 1995; Parham and Saeed 2014; Kumar et al. 2014). Therefore, these nanobased fungicidal formulations are smart delivery systems for progressive farmers, and they assist growers to minimize their use of fungicides (Sarlak et al. 2014; Shyla et al. 2014). Moreover, high reactivity of materials at the nanoscale (in comparison with their bulk counterparts) support use of lesser quantities of nanoformulations, with improved impacts on crop protection (Badami 2008; Debnath et al. 2011). Thus, use of nanoparticle-based nanoformulations promotes safer administration of fungicides at low doses (Kuzma and VerHage 2006) by decreasing their toxicity while increasing their efficiency (Mousavi and Rezaei 2011).

The rate of release of nanoparticles loaded with fungicides is influenced by environmental factors (Lauterwasser 2005). Nanofungicide formulations improve the solubility of active compounds and facilitate their gradual release at the target sites. Thereby, the bioavailability of agrochemicals with poor solubility in water is increased (Kah and Hofmann 2014).

Combination of nanoparticles of diverse metal types with fungicides increases their activity (Zielińska-Jurek et al. 2012; Lopes et al. 2013). Combination of silica nanoparticles with chlorfenapyr has been reported to double its pesticidal activity. Likewise, combination of these nanoparticles with calcium carbonate have also shown controlled and effective release of the -cidal component for a longer time span (Sonawane and Dongare 2006; Türk and Bolten 2010).

Application of nanoformulations prior to pathogen attacks and disease outbreaks is also very helpful, because of their slow release pattern. Their presence in the plant root zone at the initial developmental stages of crop growth strengthen and protect the plants from invasive pathogens, as well as keeping pathogen populations below the threshold level (Bhattacharyya et al. 2010; Castro et al. 2013; Khan et al. 2014). The presence of nanofungicides prior to fungal pathogen invasion of host plants is effective because of the greater persistence and slow release (of active compounds) by the nanoformulations, which enhance their effectiveness against pathogens (Khan et al. 2011). These attributes reduce the net amounts of fungicides required to control diseases (Khan and Jairajpuri 2012), as well as decreasing their concomitant environmental menace.

As discussed earlier in this chapter, nanofungicides are small structures that provide fungicidal properties and/or formulations with active ingredients of fungicides in a nanoform. The stability and gradual release of the active -cidal ingredients for longer periods make them eco-friendly in comparison with agrochemicals. Scientists have experimented with preparation of a variety of nanofungicides in different forms such as nanocapsules (nanoencapsulated formulations), metallic nanoparticles, metal oxide–based nanoparticles, nanospheres, nanogels, and nanoemulsions (NEs) (Yan et al. 2005). They are all nanofungicides and delivery systems with a large surface area and increased target affinity, and have been shown to be effective in plant protection strategies (Lyons and Scrinis 2009; Bordes et al. 2009; Bergeson 2010). Nanomaterials exhibit various favorable properties—such as solubility, crystallinity, enhanced permeation, stability, stiffness, and biodegradability—that are required to formulate nanofungicides and/or nanopesticides (Yan et al. 2005; Bouwmeester et al. 2009).

12.6.1 *Nanoemulsions*

Use of nanoemulsions is a better approach for nanofungicides or nanofungicide delivery systems because of their small size, optical transparency, low viscosity, and greater kinetic stability (Tice 2001; Senturk et al. 2013; Bernardes et al. 2014). Nanoemulsions improve the solubility of active agents of agrochemicals and thereby enhance the bioavailability of the active ingredients (Xu et al. 2010). According to the literature, nanoemulsion preparation by dispersion into liquid phases enhances the solubility and distribution ability of fungicides many times. The characteristics of nanoemulsions—such as wettability, spreadability, and better mechanical stability—make them helpful for low volatilization and lesser degradation of active compounds (Guillette and Iguchi 2012; Mason et al. 2006; Anton et al. 2008).

A surfactant-based nanoemulsion of β-cypermethrin was developed by Wang et al. (2007), and nanoemulsions of neem oil and permethrin were made by Anjali et al. (2010, 2012). According to their reports, the droplet size affected the activity and effectiveness of the nanoemulsions (Jiang et al. 2012; Díaz-Blancas et al. 2016). Droplets of a nanoemulsion can be used to encapsulate active compounds of agrochemicals in a formulation with less degradation of the functional ingredients (McClements and Decker 2000).

12.6.2 *Metallic Nanoparticles*

Use of metallic nanoparticles as nanofungicides has been discussed comprehensively in this chapter. These metallic nanoparticles are also used as part of nanoformulations. Because of their toxic effects on pathogens, silver nanoparticles are used in nanoformulations (Sondi and Salopek-Sondi 2004; Retchkiman-Schabes et al. 2006). A mixture comprising silver nanoparticles with macromolecules that have an amphiphilic hyperbranched conformation was shown to be an effective surface coating with antimicrobial effects on a diverse range of pathogens (Aymonier et al. 2002; Gu et al. 2003; Ahmad et al. 2005; Lead and Wilkinson 2006; Gong et al. 2007). Silver nanoparticles in association with fluconazole (a triazole fungicide) showed significant antifungal activity against *Trichoderma* spp., *Phoma glomerate*, and *Candida albicans* (Gajbhiye et al. 2009). Silver nanoemulsions showed strong growth inhibition of *Sclerotium rolfsii* in mung bean, and this nanoemulsion was also reported to have a strong inhibitory action against *Magnaporthe grisea* and *Bipolaris sorokiniana* (Agrawal and Rathore 2014).

12.6.3 Nanogels

Nanogels of chitosan with copper and pheromone were evaluated against *Fusarium graminearum* and fruit pests, respectively, and showed increased antifungal activity due to their synergistic effects (Brunel et al. 2013; Bhagat et al. 2013).

12.6.4 Nanocapsules

Nanocapsules are a nanosystem in which the active ingredient of a fungicide is placed within the core, surrounded by a membrane. Nanoencapsulation has potential scope for use in nanofungicide formulations. Polymeric and solid lipid nanocapsules loaded with tebuconazole and carbendazim have been created for use as nanofungicides (Campos et al. 2015).

12.6.5 Nanospheres

Nanospheres (in a monolithic system) comprise irregular spherical nanoscale particles in which -cidal compounds or active agents of fungicides are dispersed and/or dissolved in polymeric matrices (Sotthivirat et al. 2007). Various polymers such as natural polymers (proteins and polysaccharides), synthetic polymers (polyamide, polyacrylamide, polystyrene, polyesters, etc.) and inorganic compounds (zoolites, silica, ceramics, glass beads, inorganic oxides, etc.) have been tested to explore their potential in nanofungicide formulations for crop protection (Shukla et al. 1992; Chuan et al. 2013).

These nanofungicides can be formulated in easy and cost-effective ways. In this section, some examples are mentioned. Size reduction of the active ingredients or functional compounds of existing fungicides to the nanoscale and subsequent nanoencapsulation have been used to develop different nanofungicides. Syngenta have made nanofungicide formulations containing nanoparticles—for example, Banner MAXX™ (which contains the active chemical compound propiconazole), Apron MAXX™ (in which the active ingredient is fludioxonil), and Primo MAXX™ (containing cyclopropyl, a derivative of cyclohexanone) (Gogoi et al. 2009; Abd-Elsalam 2012).

Other nanoproducts have also been launched; for example, Nano-5 has been developed for use against different phytopathogens, and Nano-Gro—a nanoproduct from Agro Nanotechnology Corp. (Miami, FL, USA)—has been certified as organic nanomaterial with no harmful impact. This nanofungicide has reported to be effective against *Magnaporthe grisea*, for eliminating rice blast disease (Gogoi et al. 2009).

12.7 Effects of Nanoparticles on Ecosystems: Challenges and Prospects

The application of nanomaterials has been gaining importance and popularity; thus, assessment of toxic effects of nanoparticles in ecosystems is indispensable for their downstream applications. As their chemical and physical features determine the property of nanoparticles, accurate assessments of their physicochemical properties are required to investigate toxic manifestations of each aspect of nanoparticles. Hence, it could be said that the toxicity of nanoparticles depends upon their physicochemical characteristics. Different studies have been conducted over a long period to assess the ecotoxicological impacts of nanoparticles, but large knowledge gaps still exist, particularly regarding nanoparticle-related biological concerns (Antisari et al. 2011; Banik and Sharma 2011; Alghuthaymi et al. 2015).

A reactive interface develops between nanoparticles and their surrounding environment because of their high surface-to-volume ratio (Orts-Gil et al. 2011). Because materials of identical chemistry may vary to a significant extent on the basis of their size (Murdock et al. 2008), these materials need to be critically characterized in different physiological contexts to estimate the correlation of their biological impacts with their colloidal characteristics. A multitude of vital characteristics of nanoparticles should be investigated for their proper characterization, such as their size, shape, surface area, functional groups, size distribution, crystal structure, porosity, chemical composition, and charge (Oberdörster et al. 2005).

According to different reports published in the literature, metallic nanoparticles generate stress in plants, which disturbs the equilibrium between generation of ROS and their removal. Because of this, ROS start to accumulate in plant cells and affect photosynthesis and other metabolic and biosynthetic mechanisms (Barazzouk et al. 2005; Bujak et al. 2011; Keller et al. 2013). Nanoparticles have been found to be phytotoxic by altering photosynthesis processes, the quantum yield, and other processes in plant cells (Peralta-Videa et al. 2011; Olejnik et al. 2013).

As discussed earlier in this chapter, nanoparticles have enormous applicability for plant disease management. However, their broad applications may be the major cause of their accumulation in the environment, increasing their lethal and toxic effects in ecosystems. It has been shown in different studies that excessive amounts of nanoparticles beyond a certain limit lead to negative effects on the environment. For example, different metallic oxides (ZnO and TiO_2) have displayed negative effects on wheat biomass, as well as its growth, and repressed the biological activities of different enzymes such as catalases, proteases, and peroxidases (Du et al. 2011).

The first report on soil-based toxicity of nanoparticles (nanotoxicity) to plants was published by Yang and Watts (2005). They reported that aluminum oxide (Al_2O_3) in conjugation with phenanthrene, and even alone, showed adverse effects on root elongation and germination in different crop plants such as maize, soybean, carrot, cucumber, and different brassica species (Lin and Xing 2007). Likewise, application of titanium dioxide (TiO_2) caused a reduction in water use efficiency in maize and altered the apoplastic pathway (Asli and Neumann 2009). Palladium nanoparticle accumulation has been observed in barley leaves (Battke et al. 2008), and iron oxide (Fe_2O_3) nanoparticles have been shown to accumulate in pumpkin tissues (Zhu et al. 2008; Lin et al. 2009).

All of these reports in the literature suggest that application of nanoparticles has environmental consequences because the presence of nanoparticles in the soil may affect and disturb the microflora, and even plants themselves, by absorption through the soil. From plants, they may be transferred to animals and human beings who consume and uptake food from them; thereby, they may affect the entire food chain. Hence, there is a dire need to explore strategies to set and standardize optimal criteria for nanoparticle synthesis and applications, and for their impacts on ecosystems. Thus, there is a need for determination of physicochemical properties of nanoparticles that affect plant diseases by targeting specific pathogens without having adverse impacts on ecosystems. Despite to standardize physicochemical attributes of nanoparticles, only doses of them that are known to be safe should be administered, in order to avoid detrimental effects of them on food chain.

In the field of agriculture, for sustainable disease management, most patents that have been filed for nanoparticle-based pesticides involve nanosilver. The increase in the popularity of nanopesticides and nanofungicides necessitates some regulations regarding their usage. Hence, in 2008, the International Center for Technology Assessment (ICTA) submitted a petition to the US Environmental Protection Agency (EPA) to regulate use of silver nanoparticles in the creation of nanoproducts (such as nanopesticides) under the Federal Insecticide, Fungicide, and Rodenticide Act (FIFRA) (Baier-Anderson 2009).

12.8 Conclusions

Nanotechnology is merely alteration of the size and shape of particles to the nanoscale. Their small size imparts extraordinary and miraculous properties to these particles. The exploitation of their potential has moved on from basic research to their use in applied technologies. Nanotechnology, in juxtaposition with biotechnology, has extended the capability of nanomaterials in crop plant production and protection at a meaningful level. For plant disease management, use of nanoparticles for controlled delivery of agrochemicals holds great promise in the field of agriculture. Nanoparticles, nanoemulsions, nanoencapsulations, and other nanobiotechnology approaches light the path for targeted delivery of fungicides, pesticides, etc., in an efficient and environmentally friendly manner to tackle epidemic diseases in plants. Advancements in nanobiotechnology by application of green chemistry for synthesis of nanoparticles, using living cells and plant extracts, offers assurances of ecoprotection. The myriad potential of nanoparticles includes not only their utility as vehicles for targeted delivery of antimicrobial compounds but also their innate antimicrobial effects and characteristics, both of which demonstrate their value for use as nanopesticides or nanofungicides against plant pathogens. The epitome of the nanobiotechnology for the synthesis of nanoparticles and nanoemulsions and their use as nanofungicides or nanopesticides revolves around the prospects of its applications for plant disease management.

References

Abdel-Aziz SM, Prasad R, Hamed AA, Abdelraof M (2018) Fungal nanoparticles: A novel tool for a green biotechnology? In: Fungal Nanobionics: Principles and Applications (eds. Prasad R, Kumar V, Kumar M and Wang S), Springer Singapore Pte Ltd. 61–87

Abd-Elsalam KA (2012) Nanoplatforms for plant pathogenic fungi management. Fungal Genomics Biol 2:107

Abd-Elsalam KA, Prasad R (2018) Nanobiotechnology applications in plant protection. Springer, Cham. https://doi.org/10.1007/978-3-319-91161-8

Agrawal S, Rathore P (2014) Nanotechnology pros and cons to agriculture: a review. Int J Curr Microbiol App Sci 3(3):43–55

Agrios GN (2005) Plant pathology. Elsevier Academic, San Diego

Aguilar-Mendez MA, San Martin Martinez E, Ortega Arroyo L, Cobian Portillo G, Sánchez Espindola E (2011) Synthesis and characterization of silver nanoparticles: effect on phytopathogen *Colletotrichum gloesporioides*. J Nanopart Res 13(6):2525–2532

Ahmad A, Senapati S, Khan MI, Kumar R, Sastry M (2005) Extra intracellular biosynthesis of gold nanoparticles by an alkalotolerant fungus, *Trichothecium* sp. J Biomed Nanotechnol 1(1):47–53

Alghuthaymi MA, Almoammar H, Rai M, Said-Galiev E, Abd-Elsalam KA (2015) Myconanoparticles: synthesis and their role in phytopathogens management. Biotechnol Biotechnol Equip 29(2):221–236

Ali EA, Yuri NG, Ray HB (2011) Mirage effect from thermally modulated transparent carbon nanotube sheets. Nanotechnology 22(43):435704

Alvarez-Puebla RA, Dos Santos DS Jr, Aroca RF (2004) Surface-enhanced Raman scattering for ultrasensitive chemical analysis of 1 and 2-naphthalenethiols. Analyst 129(12):1251–1256

Anjali CH, Khan SS, Margulis-Goshen K, Magdassi S, Mukherjee A, Chandrasekaran N (2010) Formulation of water-dispersible nanopermethrin for larvicidal applications. Ecotoxicol Environ Saf 73(8):1932–1936

Anjali CH, Sharma Y, Mukherjee A, Chandrasekaran N (2012) Neem oil (*Azadirachta indica*) nanoemulsion: a potent larvicidal agent against *Culex quinquefasciatus*. Pest Manag Sci 68(2):158–163

Antisari LV, Carbone S, Ferronato C, Simoni A, Vianello G (2011) Characterization of heavy metals atmospheric deposition for assessment of urban environmental quality in the Bologna city (Italy). EQA Int J Environ Qual 7(7):49–63

Anton N, Benoit JP, Saulnier P (2008) Design and production of nanoparticles formulated from nano-emulsion templates: a review. J Control Release 128:185–199

Aromal SA, Philip D (2012) Green synthesis of gold nanoparticles using *Trigonella foenum-graecum* and its size-dependent catalytic activity. Spectrochim Acta A Mol Biomol Spectrosc 97:1–5

Asli S, Neumann PM (2009) Colloidal suspensions of clay or titanium dioxide nanoparticles can inhibit leaf growth and transpiration via physical effects on root water transport. Plant Cell Environ 32:577–584

Avasare V, Zhang Z, Avasare D, Khan I, Qurashi A (2015) Room-temperature synthesis of TiO_2 nanospheres and their solar driven photoelectrochemical hydrogen production. Int J Energy Res 39(12):1714–1719

Aymonier C, Schlotterbeck U, Antonietti L, Zacharias P, Thomann R, Tiller JC, Mecking S (2002) Hybrids of silver nanoparticles with amphiphilic hyperbranched macromolecules exhibiting antimicrobial properties. Chem Commun 24:3018–3019

Aziz N, Fatma T, Varma A, Prasad R (2014) Biogenic synthesis of silver nanoparticles using *Scenedesmus abundans* and evaluation of their antibacterial activity. J Nanopart 2014:689419. https://doi.org/10.1155/2014/689419

Aziz N, Faraz M, Pandey R, Sakir M, Fatma T, Varma A, Barman I, Prasad R (2015) Facile algae-derived route to biogenic silver nanoparticles: synthesis, antibacterial and photocatalytic properties. Langmuir 31:11605–11612. https://doi.org/10.1021/acs.langmuir.5b03081

Aziz N, Pandey R, Barman I, Prasad R (2016) Leveraging the attributes of *Mucor hiemalis*–derived silver nanoparticles for a synergistic broad-spectrum antimicrobial platform. Front Microbiol 7:1984. https://doi.org/10.3389/fmicb.2016.01984

Aziz N, Faraz M, Sherwani MA, Fatma T, Prasad R (2019) Illuminating the anticancerous efficacy of a new fungal chassis for silver nanoparticle synthesis. Front Chem 7:65. doi: 10.3389/fchem.2019.00065

Badami BV (2008) Concept of green chemistry. Resonance 13(11):1041–1048

Baier-Anderson C (2009) Regulating nanosilver as a pesticide. Environmental Defense Fund. http://blogs.edf.org/health/2009/02/12/regulating-nano-silver-as-a-pesticide/. Accessed

Bali R, Razak N, Lumb A, Harris AT (2006) The synthesis of metallic nanoparticles inside live plants. In: Proceedings of the 2006 international conference on nanoscience and nanotechnology, Brisbane, 3–7 July 2006. https://doi.org/10.1109/ICONN.2006.340592

Banik S, Sharma P (2011) Plant pathology in the era of nanotechnology. Indian Phytopathol 64(2):120–127

Barazzouk S, Kamat PV, Hotchandani S (2005) Photoinduced electron transfer between chlorophyll and gold nanoparticles. J Phys Chem B 109:716–723

Barrak H, Saied T, Chevallier P, Laroche G, M'nif A, Hamzaoui AH (2016) Synthesis, characterization, and functionalization of ZnO nanoparticles by N-(trimethoxysilylpropyl) ethylenediamine triacetic acid (TMSEDTA): investigation of the interactions between phloroglucinol and ZnO@TMSEDTA. Arab J Chem. https://doi.org/10.1016/j.arabjc.2016.04.019

Battke F, Leopold K, Maier M, Schmidhalter U, Schuster M (2008) Palladium exposure of barley: uptake and effects. Plant Biol 10:272–276

Bawaskar M, Gaikwad S, Ingle A, Rathod D, Gade A, Duran N, Rai M (2010) A new report on mycosynthesis of silver nanoparticles by *Fusarium culmorum*. Curr Nanosci 6(4):376–380

Bergeson LL (2010) Nanosilver pesticide products: what does the future hold? Environ Qual Manag 19(4):73–82

Bernardes PC, Nélio JDA, Nilda de Fátima FS (2014) Nanotechnology in the food industry. Biosci J 30(6):1919–1932

Bhagat D, Samanta SK, Bhattacharya S (2013) Efficient management of fruit pests by pheromone nanogels. Sci Rep 3:1294

Bhattacharyya A, Bhaumik A, Rani PU, Mandal S, Epidi TT (2010) Nanoparticles: a recent approach to insect pest control. Afr J Biotechnol 9(24):3489–3493

Bhattacharyya A, Duraisamy P, Govindarajan M, Buhroo AA, Prasad R (2016) Nano-biofungicides: emerging trend in insect pest control. In: Prasad R (ed) Advances and applications through fungal nanobiotechnology. Springer, Cham, pp 307–319

Bonde SR, Rathod DP, Ingle AP, Ade RB, Gade AK, Rai MK (2012) *Murraya koenigii*–mediated synthesis of silver nanoparticles and its activity against three human pathogenic bacteria. Nanosci Methods 1(1):25–36

Bordes P, Pollet E, Avérous L (2009) Nano-biocomposites: biodegradable polyester/nanoclay systems. Prog Polym Sci 34(2):125–155

Borkow G, Gabbay J (2005) Copper as a biocidal tool. Curr Med Chem 12(18):2163–2175

Bouwmeester H, Dekkers S, Noordam MY, Hagens WI, Bulder AS, De Heer C, Sips AJ (2009) Review of health safety aspects of nanotechnologies in food production. Regul Toxicol Pharmacol 53(1):52–62

Boxi SS, Paria S (2015) Visible light induced enhanced photocatalytic degradation of organic pollutants in aqueous media using Ag doped hollow TiO_2 nanospheres. RSC Adv 5(47):37657–37668

Brecht M, Datnoff L, Nagata R, Kucharek T (2003) The role of silicon in suppressing tray leaf spot development in St. Augustine grass. University of Florida, Gainesville

Brecht MO, Datnoff LE, Kucharek TA, Nagata RT (2004) Influence of silicon and chlorothalonil on the suppression of gray leaf spot and increase plant growth in St. Augustine grass. Plant Dis 88(4):338–344

Brunel F, El Gueddari NE, Moerschbacher BM (2013) Complexation of copper(II) with chitosan nanogels: toward control of microbial growth. Carbohydr Polym 92(2):1348–1356

Bujak N, Czechowski D, Piatkowski R (2011) Fluorescence enhancement of light-harvesting complex 2 from purple bacteria coupled to spherical gold nanoparticles. Appl Phys Lett 99:173701–173703

Campos EVR, De Oliveira JL, Da Silva CMG, Pascoli M, Pasquoto T, Lima R, Fraceto LF (2015) Polymeric and solid lipid nanoparticles for sustained release of carbendazim and tebuconazole in agricultural applications. Sci Rep 5:13809

Carver TLW, Thomas BJ, Robbins MP, Zeyen RJ (1998) Phenylalanine ammonia-lyase inhibition, autofluorescence, and localized accumulation of silicon, calcium and manganese in oat epidermis attacked by the powdery mildew fungus *Blumeria graminis* (DC) Speer. Physiol Mol Plant Pathol 52(4):223–243. https://doi.org/10.1006/pmpp.1998.0148

Castro ML, Ojeda C, Cirelli A (2013) Advances in surfactants for agrochemicals. Environ Chem Lett 1–11

Chandran SP, Chaudhary M, Pasricha R, Ahmad A, Sastry M (2006) Synthesis of gold nanotriangles and silver nanoparticles using *Aloe vera* plant extract. Biotechnol Prog 22(2):577–583

Chen H, Yada R (2011) Nanotechnologies in agriculture: new tools for sustainable development. Trends Food Sci Technol 22(11):585–594

Chen L, Song Y, Tang B, Song X, Yang H, Li B, Zhao Y, Huang C, Han X, Wang S, Li Z (2015) Aquatic risk assessment of a novel strobilurin fungicide: a microcosm study compared with the species sensitivity distribution approach. Ecotoxicol Environ Saf. https://doi.org/10.1016/j.ecoenv.2015.06.027

Chhipa H (2017) Nanopesticide: current status and future possibilities. Agric Res Technol Open Access J 5:1

Choudhury SR, Ghosh M, Mandal A, Chakravorty D, Pal M, Pradhan S, Goswami A (2011) Surface-modified sulfur nanoparticles: an effective antifungal agent against *Aspergillus niger* and *Fusarium oxysporum*. Appl Microbiol Biotechnol 90(2):733–743

Choudhury SR, Pradhan S, Goswami A (2012) Preparation and characterisation of acephate nano-encapsulated complex. Nanosci Methods 1(1):9–15

Chowdappa P, Gowda S (2013) Nanotechnology in crop protection: status and scope. Pest Manag Hortic Ecosyst 19(2):131–151

Chowdappa P, Kumar NB, Madhura S, Kumar MS, Myers KL, Fry WE, Cooke DE (2013) Emergence of 13_A2 blue lineage of *Phytophthora infestans* was responsible for severe outbreaks of late blight on tomato in South-West India. J Phytopathol 161(1):49–58

Chuan L, He P, Pampolino MF, Johnston AM, Jin J, Xu X, Zhou W (2013) Establishing a scientific basis for fertilizer recommendations for wheat in China: yield response and agronomic efficiency. Field Crop Res 140:1–8

Cioffi N, Torsi L, Ditaranto N, Sabbatini L, Zambonin PG, Tantillo G, Traversa E (2004) Antifungal activity of polymer-based copper nanocomposite coatings. Appl Phys Lett 85(12):2417–2419

Clement JL, Jarrett PS (1994) Antibacterial silver. Metal-Based Drugs 1(5–6):467–482

Corredor E, Testillano PS, Coronado MJ, Gonzalez Melendi P, Fernandez Pacheco R, Marquina C, Risueno MC (2009) Nanoparticle penetration and transport in living pumpkin plants: in situ subcellular identification. BMC Plant Biol 9(1):45

Dar MA, Ingle A, Rai M (2013) Enhanced antimicrobial activity of silver nanoparticles synthesized by *Cryphonectria* sp. evaluated singly and in combination with antibiotics. Nanomedicine 9(1):105–110

Debnath N, Das S, Seth D, Chandra R, Bhattacharya SC, Goswami A (2011) Entomotoxic effect of silica nanoparticles against *Sitophilus oryzae* (L.). J Pest Sci 84(1):99–105

Dhekney S, Li A, Anaman M, Dutt M, Tattersall J, Gray D (2007) Genetic transformation of embryogenic cultures and recovery of transgenic plants in *Vitis vinifera*, *Vitis rotundifolia* and *Vitis* hybrids. Acta Hortic 738:743–748

Dhillon GS, Brar SK, Kaur S, Verma M (2012) Green approach for nanoparticle biosynthesis by fungi: current trends and applications. Crit Rev Biotechnol 32(1):49–73

Díaz-Blancas V, Medina DI, Padilla-Ortega E, Bortolini-Zavala R, Olvera-Romero M, Luna-Bárcenas G (2016) Nanoemulsion formulations of fungicide tebuconazole for agricultural applications. Molecules 21(10):1271

Dreaden EC, Alkilany AM, Huang X, Murphy CJ, El Sayed MA (2012) The golden age: gold nanoparticles for biomedicine. Chem Soc Rev 41(7):2740–2779

Du W, Sun Y, Ji R, Zhu J, Wu J, Guo H (2011) TiO_2 and ZnO nanoparticles negatively affect wheat growth and soil enzyme activities in agricultural soil. J Environ Monit 13(4):822–828

Dubchak S, Ogar A, Mietelski JW, Turnau K (2010) Influence of silver and titanium nanoparticles on arbuscular mycorrhiza colonization and accumulation of radiocaesium in *Helianthus annuus*. Span J Agric Res 8(1):103–108. https://doi.org/10.5424/sjar/201008S1-1228

Dubey M, Bhadauria S, Kushwah BS (2009) Green synthesis of nanosilver particles from extract of *Eucalyptus hybrida* (Safeda) leaf. Dig J Nanomater Biostruct 4(3):537–543

Duhan JS, Kumar R, Kumar N, Kaur P, Nehra K, Duhan S (2017) Nanotechnology: the new perspective in precision agriculture. Biotechnol Rep 15:11–23

Dzhavakhiya V, Shcherbakova L, Semina Y, Zhemchuzhina N, Campbell B (2012) Chemosensitization of plant pathogenic fungi to agricultural fungicides. Front Microbiol 3:1–9

EstebanTejeda L, Malpartida F, Esteban Cubillo A, Pecharroman C, Moya JS (2009) Antibacterial and antifungal activity of a soda-lime glass containing copper nanoparticles. Nanotechnology 20(50):505701

Gajbhiye M, Kesharwani J, Ingle A, Gade A, Rai M (2009) Fungus-mediated synthesis of silver nanoparticles and their activity against pathogenic fungi in combination with fluconazole. Nanomedicine 5(4):382–386

Gan PP, Ng SH, Huang Y, Li SFY (2012) Green synthesis of gold nanoparticles using palm oil mill effluent (POME): a low-cost and eco-friendly viable approach. Bioresour Technol 113:132–135

Gardea-Torresdey JL, Parsons JG, Gomez E, Peralta Videa J, Troiani HE, Santiago P, Yacaman MJ (2002) Formation and growth of Au nanoparticles inside live alfalfa plants. Nano Lett 2(4):397–401

Ghormade V, Deshpande MV, Paknikar KM (2011) Perspectives for nano-biotechnology enabled protection and nutrition of plants. Biotechnol Adv 29(6):792–803

Gogoi R, Dureja P, Singh PK (2009) Nanoformulations—a safer and effective option for agrochemicals. Indian Farming 59(8):7–12

Golinska P, Wypij M, Ingle AP, Gupta I, Dahm H, Rai M (2014) Biogenic synthesis of metal nanoparticles from actinomycetes: biomedical applications and cytotoxicity. Appl Microbiol Biotechnol 98:8083–8097

Gong P, Li H, He X, Wang K, Hu J, Tan W, Yang X (2007) Preparation and antibacterial activity of Fe3O4@Ag nanoparticles. Nanotechnology 18(28):285604

González-Fernández R, Prats E, Jorrín-Novo JV (2010) Proteomics of plant pathogenic fungi. J Biomed Biotechnol 2010:932527. https://doi.org/10.1155/2010/932527

Gruère GP (2012) Implications of nanotechnology growth in food and agriculture in OECD countries. Food Policy 37(2):191–198

Gu H, Ho PL, Tong E, Wang L, Xu B (2003) Presenting vancomycin on nanoparticles to enhance antimicrobial activities. Nano Lett 3(9):1261–1263

Guillette LJ, Iguchi T (2012) Life in a contaminated world. Science 337(6102)):1614–1615

Gul HT, Saeed S, Khan FZA, Manzoor SA (2014) Potential of nanotechnology in agriculture and crop protection: a review. Appl Sci Bus Econ 1(2):23–28

Gupta A, Maynes M, Silver S (1998) Effects of halides on plasmid-mediated silver resistance in *Escherichia coli*. Appl Environ Microbiol 64(12):5042–5045

Gupta K, Singh RP, Pandey A, Pandey A (2013) Photocatalytic antibacterial performance of TiO_2 and Ag-doped TiO_2 against *S. aureus*, *P. aeruginosa* and *E. coli*. Beilstein J Nanotechnol 4:345

Gupta N, Upadhyaya CP, Singh A, Abd-Elsalam KA, Prasad R (2018) Applications of silver nanoparticles in plant protection. In: Nanobiotechnology Applications in Plant Protection (eds. Abd-Elsalam K and Prasad R), Springer International Publishing AG 247–266

Hettiarachchi MA, Wickramarachchi PASR (2011) Synthesis of chitosan stabilized silver nanoparticles using gamma ray irradiation and characterization. J Sci 6:65–75

Hezave AZ, Esmaeilzadeh F (2010) Crystallization of micro particles of sulindac using rapid expansion of supercritical solution. J Cryst Growth 312(22):3373–3383

Holt KB, Bard AJ (2005) Interaction of silver(I) ions with the respiratory chain of *Escherichia coli*: an electrochemical and scanning electrochemical microscopy study of the antimicrobial mechanism of micromolar Ag^{+}. Biochemistry 44(39):13214–13223
Huang H, Yang X (2004) Synthesis of polysaccharide-stabilized gold and silver nanoparticles: a green method. Carbohydr Res 339(15):2627–2631
Hwang ET, Lee JH, Chae YJ, Kim YS, Kim BC, Sang BI, Gu MB (2008) Analysis of the toxic mode of action of silver nanoparticles using stress-specific bioluminescent bacteria. Small 4(6):746–750
Ingham B (2015) X-ray scattering characterisation of nanoparticles. Crystallogr Rev 21(4):229–303
Ingle A, Rai M, Gade A, Bawaskar M (2009) *Fusarium solani*: a novel biological agent for the extracellular synthesis of silver nanoparticles. J Nanopart Res 11(8):2079
Iravani S (2011) Green synthesis of metal nanoparticles using plants. Green Chem 13(10):2638–2650
Ismail M, Prasad R, Ibrahim AIM, Ahmed ISA (2017) Modern prospects of nanotechnology in plant pathology. In: Prasad R, Kumar M, Kumar V (eds) Nanotechnology. Springer, Singapore, pp 305–317
Jampílek J, Kráľová K (2015) Application of nanotechnology in agriculture and food industry, its prospects and risks. Ecol Chem Eng S 22:321–361
Jha AK, Prasad K (2010) Green synthesis of silver nanoparticles using *Cycas* leaf. Int J Green Nanotechnol Phys Chem 1(2):110–117
Jha Z, Behar N, Sharma SN, Chandel G, Sharma DK, Pandey MP (2011) Nanotechnology: prospects of agricultural advancement. Nano Vision 1(2):88–100
Jiang LC, Basri M, Omar D, Rahman MBA, Salleh AB, Rahman RNZRA, Selamat A (2012) Green nano-emulsion intervention for water-soluble glyphosate isopropylamine (IPA) formulations in controlling *Eleusine indica* (*E. indica*). Pestic Biochem Physiol 102(1):19–29
Jo YK, Kim BH, Jung G (2009) Antifungal activity of silver ions and nanoparticles on phytopathogenic fungi. Plant Dis 93(10):1037–1043
Joerger R, Klaus T, Granqvist CG (2000) Biologically produced silver–carbon composite materials for optically functional thin-film coatings. Adv Mater 12(6):407–409
Johnston CT (2010) Probing the nanoscale architecture of clay minerals. Clay Miner 45(3):245–279
Kah M, Hofmann T (2014) Nanopesticide research: current trends and future priorities. Environ Int 63:224–235
Kasthuri J, Kathiravan K, Rajendiran N (2009) Phyllanthin-assisted biosynthesis of silver and gold nanoparticles: a novel biological approach. J Nanopart Res 11(5):1075–1085
Keller AA, McFerran S, Lazareva A, Suh S (2013) Global life cycle releases of engineered nanomaterials. J Nanopart Res 15(6):1692
Khan MR, Rizvi TF (2014) Nanotechnology: scope and application in plant disease management. Plant Pathol J 13(3):214–231
Khan MR, Majid S, Mohidin FA, Khan N (2011) A new bioprocess to produce low cost powder formulations of biocontrol bacteria and fungi to control fusarial wilt and root-knot nematode of pulses. Biol Control 59(2):130–140
Khan MR, Jairajpuri MS (2012) Nematode infestation in horticultural crops, national scenario. In: Khan MR, Jairajpuri MS (eds) Nematode infestation part III: horticultural crops. National Academy of Sciences, India, pp 1–30
Khan MR, Ashraf S, Rasool F, Salati KM, Mohiddin FA, Haque Z (2014) Field performance of *Trichoderma* species against wilt disease complex of chickpea caused by *Fusarium oxysporum* f. sp. *ciceri* and *Rhizoctonia solani*. Turk J Agric For 38(4):447–454
Khan I, Abdalla A, Qurashi A (2017a) Synthesis of hierarchical WO_3 and Bi_2O_3/WO_3 nanocomposite for solar-driven water splitting applications. Int J Hydrog Energy 42(5):3431–3439
Khan I, Yamani ZH, Qurashi A (2017b) Sonochemical-driven ultrafast facile synthesis of SnO_2 nanoparticles: growth mechanism structural electrical and hydrogen gas sensing properties. Ultrason Sonochem 34:484–490
Kim YH, Lee DK, Cha HG, Kim CW, Kang YC, Kang YS (2006) Preparation and characterization of the antibacterial Cu nanoparticle formed on the surface of SiO_2 nanoparticles. J Phys Chem B 110(49):24923–24928

Kim S, Lee S, Lee I (2012) Alteration of phytotoxicity and oxidant stress potential by metal oxide nanoparticles in *Cucumis sativus*. Water Air Soil Pollut 223(5):2799–2806

Kumar V, Yadav SK (2009) Plant-mediated synthesis of silver and gold nanoparticles and their applications. J Chem Technol Biotechnol 84(2):151–157

Kumar R, Sharon M, Choudhary AK (2010) Nanotechnology in agricultural diseases and food safety. J Phytology 2:8392

Kumar RR, Priyadharsani KP, Thamaraiselvi K (2012) Mycogenic synthesis of silver nanoparticles by the Japanese environmental isolate *Aspergillus tamarii*. J Nanopart Res 14(5):860

Kumar S, Bhanjana G, Sharma A, Sidhu MC, Dilbaghi N (2014) Synthesis, characterization and on field evaluation of pesticide loaded sodium alginate nanoparticles. Carbohydr Polym 101:1061–1067

Kuppusamy P, Yusoff MM, Maniam GP, Ichwan SJ, Soundharrajan I, Govindan N (2014) Nutraceuticals as potential therapeutic agents for colon cancer: a review. Acta Pharm Sin 4(3):173–181

Kuzma J, VerHage P (2006) Nanotechnology in agriculture and food production: anticipated applications. Woodrow Wilson International Center for Scholars, The Project on Emerging Nanotechnologies, Washington, D.C.. http://www.nanotechproject.org/process/assets/files/2706/94_pen4_agfood.pdf. Accessed

Laurent S, Forge D, Port M, Roch A, Robic C, Vander LE, Muller RN (2010) Magnetic iron oxide nanoparticles: synthesis, stabilization, vectorization, physicochemical characterizations, and biological applications. Chem Rev 110:2574–2574

Lauterwasser C (2005) Small sizes that matter: opportunities and risks of nanotechnologies. Report in co-operation with the OECD International Futures Programme. Allianz Center for Technology, Munich. http://www.oecd.org/chemicalsafety/nanosafety/44108334.pdf. Accessed

Lead JR, Wilkinson KJ (2006) Aquatic colloids and nanoparticles: current knowledge and future trends. Environ Chem 3(3):159–171

Lemire JA, Harrison JJ, Turner RJ (2013) Antimicrobial activity of metals: mechanisms, molecular targets and applications. Nat Rev Microbiol 11(6):371

Li LS, Hu J, Yang W, Alivisatos AP (2001) Band gap variation of size- and shape-controlled colloidal CdSe quantum rods. Nano Lett 1(7):349–351

Li S, Shen Y, Xie A, Yu X, Qiu L, Zhang L, Zhang Q (2007) Green synthesis of silver nanoparticles using *Capsicum annuum* L. extract. Green Chem 9(8):852–858

Liau SY, Read DC, Pugh WJ, Furr JR, Russell AD (1997) Interaction of silver nitrate with readily identifiable groups: relationship to the antibacterial action of silver ions. Lett Appl Microbiol 25(4):279–283

Lin D, Xing B (2007) Phytotoxicity of nanoparticles: inhibition of seed germination and root growth. Environ Pollut 150(2):243–250

Lin S, Reppert J, Hu Q, Hudson JS, Reid ML, Ratnikova TA, Rao AM, Luo H, Ke PC (2009) Uptake translocation and transmission of carbon nanomaterials in rice plants. Nano Micro Small 5:1128–1132

Logeswari P, Silambarasan S, Abraham J (2012) Synthesis of silver nanoparticles using plant extracts and analysis of their antimicrobial activity. J Saudi Chem Soc 4:23–45

Lopes CM, Fernande JR, Lopes PM (2013) Application of nanotechnology in the agro-food sector. Fac Health Sci 51(2):183–197

Lyons K, Scrinis G (2009) Under the regulatory radar? Nanotechnologies and their impacts for rural Australia. In: Merlan F, Raftery D (eds) Tracking rural change: community, policy and technology in Australia, New Zealand and Europe. Australian National University E Press, Canberra, pp 151–171

Manczinger L, Antal Z, Kredics L (2002) Ecophysiology and breeding of mycoparasitic *Trichoderma* strains. Acta Microbiol Immunol Hung 49(1):1–14

Mandal D, Bolander ME, Mukhopadhyay D, Sarkar G, Mukherjee P (2006) The use of microorganisms for the formation of metal nanoparticles and their application. Appl Microbiol Biotechnol 69(5):485–492

Mansha M, Qurashi A, Ullah N, Bakare FO, Khan I, Yamani ZH (2016) Synthesis of In_2O_3/graphene heterostructure and their hydrogen gas sensing properties. Ceram Int 42(9):11490–11495
Marambio-Jones C, Hoek EM (2010) A review of the antibacterial effects of silver nanomaterials and potential implications for human health and the environment. J Nanopart Res 12(5):1531–1551
Mason TG, Wilking JN, Meleson K, Chang CB, Graves SM (2006) Nanoemulsions: formation, structure, and physical properties. J Phys Condens Matter 18(41):635
Mazhar T, Shrivastava V, Tomar RS (2017) Green synthesis of bimetallic nanoparticles and its applications: a review. J Pharm Sci Res 9(2):102
McClements DJ, Decker EA (2000) Lipid oxidation in oil-in-water emulsions: impact of molecular environment on chemical reactions in heterogeneous food systems. J Food Sci 65(8):1270–1282
Medici S, Massimiliano P, Valeria MN, Joanna I, Lachowicz G, Crisponi M, Antonietta Z (2015) Noble metals in medicine: latest advances. Coord Chem Rev 284:329–350
Meyer V (2008) Genetic engineering of filamentous fungi progress, obstacles and future trends. Biotechnol Adv 26(2):177–185
Min JS, Kim KS, Kim SW, Jung JH, Lamsal K, Kim SB, Lee YS (2009) Effects of colloidal silver nanoparticles on sclerotium-forming phytopathogenic fungi. Plant Pathol J 25(4):376–380
Mishra S, Singh HB (2015) Biosynthesized silver nanoparticles as a nanoweapon against phytopathogens: exploring their scope and potential in agriculture. Appl Microbiol Biotechnol 99(3):1097–1107
Morones JR, Elechiguerra JL, Camacho A, Holt K, Kouri JB, Ramirez JT, Yacaman MJ (2005) The bactericidal effect of silver nanoparticles. Nanotechnology 16(10):2346
Mousavi SR, Rezaei M (2011) Nanotechnology in agriculture and food production. J Appl Environ Biol Sci 1:414–419
Murdock RC, Braydich-Stolle L, Schrand AM, Schlager JJ, Hussain SM (2008) Characterization of nanomaterials dispersion in solution prior to in vitro exposure using dynamic light scattering technique. Toxicol Sci 101(2):239–253
Nam KT, Lee YJ, Krauland EM, Kottmann ST, Belcher AM (2008) Peptide-mediated reduction of silver ions on engineered biological scaffolds. ACS Nano 2(7):1480–1486
Narayanan KB, Sakthivel N (2008) Coriander leaf mediated biosynthesis of gold nanoparticles. Mater Lett 62(30):4588–4590
Nayak R, Pradhan N, Behera D, Pradhan K, Mishra S, Sukla L, Mishra B (2011) Green synthesis of silver nanoparticle by *Penicillium purpurogenum* NPMF: the process and optimization. J Nanopart Res 13:3129–3137
Nel A, Xia T, Madler L, Li N (2003) Toxic potential of materials at the nano level. Science 311(5761):622–627
Nikhil ST, Bharat B (2004) Scale dependence of micro/nano-friction and adhesion of MEMS/NEMS materials, coatings and lubricants. Nanotechnology 15(11):1561
Oberdörster G, Oberdörster E, Oberdörster J (2005) Nanotoxicology: an emerging discipline evolving from studies of ultrafine particles. Environ Health Perspect 113(7):823–839
Olejnik M, Krajnik B, Kowalska D (2013) Imaging of fluorescence enhancement in photo-synthetic complexes coupled to silver nanowires. Appl Phys Lett 102:083703–083707
Oliveira MM, Ugarte D, Zanchet D, Zarbin AJ (2005) Influence of synthetic parameters on the size, structure, and stability of dodecanethiol-stabilized silver nanoparticles. J Colloid Interface Sci 292(2):429–435
Orts-Gil G, Natte K, Drescher D et al (2011) Characterisation of silica nanoparticles prior to in vitro studies: from primary particles to agglomerates. J Nanopart Res 13(4):1593–1604
Panácek A, Kvitek L, Prucek R, Kolar M, Vecerova R, Pizurova N (2006) J Phys Chem B 110:16248–16253. https://doi.org/10.1021/jp063826h
Panigrahi S, Kundu S, Ghosh S, Nath S, Pal T (2004) General method of synthesis for metal nanoparticles. J Nanopart Res 6(4):411–414
Parham H, Saeed S (2014) Ultrasound-assisted solid phase extraction of nitro- and chloro-(phenols) using magnetic iron oxide nanoparticles and Aliquat 336 ionic liquid. J Chromatogr A 1336:34–42. https://doi.org/10.1016/j.chroma.2014.02.012

Park HJ, Kim SH, Kim HJ, Choi SH (2006) A new composition of nanosized silica-silver for control of various plant diseases. Plant Pathol J 22(3):295–302
Parveen S, Wani AH, Shah MA, Devi HS, Bhat MY, Koka JA (2018) Preparation, characterization and antifungal activity of iron oxide nanoparticles. Microb Pathog 115:287–292
Patel N, Desai P, Patel N, Jha A, Gautam HK (2014) Agronanotechnology for plant fungal disease management: a review. Int J Curr Microbiol Appl Sci 3(10):71–84
Patra P, Mitra S, Debnath N, Goswami A (2012) Biochemical-, biophysical-, and microarray-based antifungal evaluation of the buffer-mediated synthesized nano zinc oxide: an in vivo and in vitro toxicity study. Langmuir 28(49):16966–16978
Peralta-Videa JR et al (2011) Nanomaterials and the environment: a review for the biennium 2008–2010. J Hazard Mater 186(1):1–15
Pileni MP (1997) Nanosized particles made in colloidal assemblies. Langmuir 13(13):3266–3276
Pinto RJ, Almeida A, Fernandes SC, Freire CS, Silvestre AJ, Neto CP, Trindade T (2013) Antifungal activity of transparent nanocomposite thin films of pullulan and silver against *Aspergillus niger*. Colloids Surf B: Biointerfaces 103:143–148
Prabhu S, Poulose EK (2012) Silver nanoparticles: mechanism of antimicrobial action, synthesis, medical applications, and toxicity effects. Int Nano Lett 2(1):32
Prasad R (2014) Synthesis of silver nanoparticles in photosynthetic plants. J Nanopart 2014:963961. https://doi.org/10.1155/2014/963961
Prasad R (2016) Advances and applications through fungal nanobiotechnology. Springer, Cham
Prasad R (2017) Fungal nanotechnology: applications in agriculture, industry, and medicine. Springer, Singapore
Prasad R, Pandey R, Barman I (2016) Engineering tailored nanoparticles with microbes: quo vadis. WIREs Nanomed Nanobiotechnol 8:316–330. https://doi.org/10.1002/wnan.1363
Prasad R, Jha A and Prasad K (2018a) Exploring the Realms of Nature for Nanosynthesis. Springer International Publishing (ISBN 978-3-319-99570-0) https://www.springer.com/978-3-319-99570-0
Prasad R, Kumar V, Kumar M, Wang S (2018b) Fungal nanobionics: principles and applications. Springer, Singapore
Raffi M, Hussain F, Bhatti TM, Akhter JI, Hameed A, Hasan MM (2008) Antibacterial characterization of silver nanoparticles against *E. coli* ATCC15224. J Mater Sci Technol 24(2):192–196
Rai M, Ingle A (2012) Role of nanotechnology in agriculture with special reference to management of insect pests. Appl Microbiol Biotechnol 94(2):287–293
Rai M, Yadav A (2013) Plants as potential synthesiser of precious metal nanoparticles: progress and prospects. IET Nanobiotechnol 7(3):117–124
Rai M, Yadav A, Gade A (2009) Silver nanoparticles as a new generation of antimicrobials. Biotechnol Adv 27(1):76–83
Rai M, Ribeiro C, Mattoso L, Duran N (eds) (2015) Nanotechnologies in food and agriculture. Springer, Cham
Rao KJ, Paria S (2013) Use of sulfur nanoparticles as a green pesticide on *Fusarium solani* and *Venturia inaequalis* phytopathogens. RSC Adv 3(26):10471–10478
Reidy B, Andrea H, Luch A, Dawson K, Lynch I (2013) Mechanisms of silver nanoparticle release, transformation and toxicity: a critical review of current knowledge and recommendations for future studies and applications. Materials 6:2295–2350
Retchkiman-Schabes PS, Canizal G, Becerra-Herrera R, Zorrilla C, Liu HB, Ascencio JA (2006) Biosynthesis and characterization of Ti/Ni bimetallic nanoparticles. Opt Mater 29:95–99
Samuel U, Guggenbichler JP (2004) Prevention of catheter-related infections: the potential of a new nano-silver impregnated catheter. Int J Antimicrob Agents 23:75–78
Sanghi R, Verma P (2009) Biomimetic synthesis and characterisation of protein capped silver nanoparticles. Bioresour Technol 100(1):501–504
Sarlak N, Taherifar A, Salehi F (2014) Synthesis of nanopesticides by encapsulating pesticide nanoparticles using functionalized carbon nanotubes and application of new nanocomposite for plant disease treatment. J Agric Food Chem 62(21):4833–4838
Sastry M, Ahmad A, Khan MI, Kumar R (2003) Biosynthesis of metal nanoparticles using fungi and actinomycete. Curr Sci 85(2):162–170

Sastry RK, Rashmi H, Rao NH, Ilyas SM (2010) Integrating nanotechnology into agri-food systems research in India: a conceptual framework. Technol Forecast Soc Chang 77(4):639–648
Satalkar P, Elger BS, Shaw DM (2016) Defining nano, nanotechnology and nanomedicine: why should it matter? Sci Eng Ethics 22(5):1255–1276
Sathishkumar M, Sneha K, Won SW, Cho CW, Kim S, Yun YS (2009) Cinnamon zeylanicum bark extract and powder mediated green synthesis of nano-crystalline silver particles and its bactericidal activity. Colloids Surf B: Biointerfaces 73(2):332–338
Schaller M, Laude J, Bodewaldt H, Hamm G, Korting HC (2004) Toxicity and antimicrobial activity of a hydrocolloid dressing containing silver particles in an ex vivo model of cutaneous infection. Skin Pharmacol Physiol 17(1):31–36
Schierholz JM, Lucast LJ, Rump A, Pulverer G (1998) Efficacy of silver coated medical devices. J Hosp Infect 40(4):257–262
Scott N, Chen H (2003) Nanoscale science and engineering for agriculture and food systems. In: Proceedings of the USDA National Planning Workshop, Washington, DC, 18–19 November 2002
Senturk A, Yalcın B, Otles S (2013) Nanotechnology as a food perspective. J Nanomater Mol Nanotechnol 2:6
Shameli K, Bin AM, Jaffar AEA, Ibrahim NA, Shabanzadeh P, Rustaiyan A, Zidan M (2012) Green biosynthesis of silver nanoparticles using *Callicarpa maingayi* stem bark extraction. Molecules 17(7):8506–8517
Sharma S, Ahmad N, Prakash A, Singh VN, Ghosh AK, Mehta BR (2010) Synthesis of crystalline Ag nanoparticles (AgNPs) from microorganisms. Mater Sci Appl 1:1–7
Sharma K, Sharma R, Shit S, Gupata S (2012) Nanotechnological application on diagnosis of a plant disease. In: Proceedings of the 2012 international conference on advances in biological and medical sciences, Singapore, 15–16 July 2012
Sharon M, Choudhary AK, Kumar R (2010) Nanotechnology in agricultural diseases and food safety. J Phytology 2(4):83
Shin WK, Cho J, Kannan AG, Lee YS, Kim DW (2016) Cross-linked composite gel polymer electrolyte using mesoporous methacrylate-functionalized SiO_2 nanoparticles for lithium-ion polymer batteries. Sci Rep 6:26332
Shukla DD, Lauricella R, Ward CW (1992) Serology of potyviruses: current problems and some solutions. In: Barnett OW (ed) Potyvirus taxonomy. Springer, Vienna, pp 57–69
Shyla KK, Natarajan N, Nakkeeran S (2014) Antifungal activity of zinc oxide, silver and titanium dioxide nanoparticles against *Macrophomina phaseolina*. J Mycol Plant Pathol 44:268–273
Singh A, Jain D, Upadhyay MK, Khandelwal N, Verma HN (2010) Green synthesis of silver nanoparticles using *Argemone mexicana* leaf extract and evaluation of their antimicrobial activities. Dig J Nanomater Biostruct 5(2):483–489
Sofi W, Gowri M, Shruthilaya M, Rayala S, Venkatraman G (2012) Silver nanoparticles as an antibacterial agent for endodontic infections. BMC Infect Dis 12(1):60
Sonawane RS, Dongare MK (2006) Sol–gel synthesis of Au/TiO_2 thin films for photocatalytic degradation of phenol in sunlight. J Mol Catal A 243:68–76
Sondi I, Salopek-Sondi B (2004) Silver nanoparticles as antimicrobial agent: a case study on *E. coli* as a model for Gram-negative bacteria. J Colloid Interface Sci 275(1):177–182
Sotthivirat S, Lubach JW, Haslam JL, Munson EJ, Stella VJ (2007) Characterization of prednisolone in controlled porosity osmotic pump pellets using solid-state NMR spectroscopy. J Pharm Sci 96(5):1008–1017
Thomas S, McCubbin P (2003) A comparison of the antimicrobial effects of four silver-containing dressings on three organisms. J Wound Care 12(3):101–107
Tice CM (2001) Selecting the right compounds for screening: does Lipinski's rule of 5 for pharmaceuticals apply to agrochemicals? Pest Manag Sci 57(1):3–16
Tilman D, Cassman KG, Matson PA, Naylor R, Polasky S (2002) Agricultural sustainability and intensive production practices. Nature 418:671–677

Tiwari JN, Tiwari RN, Kim KS (2012) Zero-dimensional, one-dimensional, two-dimensional and three-dimensional nanostructured materials for advanced electrochemical energy devices. Prog Mater Sci 57(4):724–803

Tiwari M, Narayanan K, Thakar MB, Jagani HV, Venkata Rao J (2014) Biosynthesis and wound healing activity of copper nanoparticles. IET Nanobiotechnol 8(4):230–237

Tournas VH (2005) Spoilage of vegetable crops by bacteria and fungi and related health hazards. Crit Rev Microbiol 31(1):33–44

Türk M, Bolten D (2010) Formation of submicron poorly water-soluble drugs by rapid expansion of supercritical solution (RESS): results for naproxen. J Supercrit Fluids 55(2):778–785

Ullah AR, Joyce HJ, Tan HH, Jagadish C, Micolich AP (2017) The influence of atmosphere on the performance of pure-phase WZ and ZB InAs nanowire transistors. Nanotechnology 28(45):454001

Vu HT, Keough MJ, Long SM, Pettigrove VJ (2015) Effects of the boscalid fungicide Filan® on the marine amphipod *Allorchestes compressa* at environmentally relevant concentrations. Environ Toxicol Chem. https://doi.org/10.1002/etc.3247

Wan D, Chen HL, Lin YS, Chuang SY, Shieh J, Chen SH (2009) Using spectroscopic ellipsometry to characterize and apply the optical constants of hollow gold nanoparticles. ACS Nano 3(4):960–970

Wang L, Li X, Zhang G, Dong J, Eastoe J (2007) Oil-in-water nanoemulsions for pesticide formulations. J Colloid Interface Sci 314(1):230–235

Wang S, Lawson R, Ray PC, Yu H (2011) Toxic effects of gold nanoparticles on *Salmonella typhimurium* bacteria. Toxicol Ind Health 27(6):547–554

Wei D, Qian W (2008) Facile synthesis of Ag and Au nanoparticles utilizing chitosan as a mediator agent. Colloids Surf B: Biointerfaces 62(1):136–142

Wei D, Sun W, Qian W, Ye Y, Ma X (2009) The synthesis of chitosan-based silver nanoparticles and their antibacterial activity. Carbohydr Res 344(17):2375–2382

Woo KS, Kim KS, Lamsal K, Kim YJ, Kim SB, Mooyoung J, Sim SJ, Kim HS, Chang SJ, Kim JK, Lee YS (2009) An in vitro study of the antifungal effect of silver nanoparticles on oak wilt pathogen *Raffaelea* sp. J Microbiol Biotechnol 19(8):760–764

Xia T, Kovochich M, Liong M, Maadler L, Gilbert B, Shi H, Nel AE (2008) Comparison of the mechanism of toxicity of zinc oxide and cerium oxide nanoparticles based on dissolution and oxidative stress properties. ACS Nano 2(10):2121–2134

Xu L, Liu Y, Bai R, Chen C (2010) Applications and toxicological issues surrounding nanotechnology in the food industry. Pure Appl Chem 82(2):349–372

Yan J, Huang K, Wang Y, Liu S (2005) Study on anti-pollution nano-preparation of dimethomorph and its performance. Chin Sci Bull 50(2):108–112

Yang L, Watts DJ (2005) Particle surface characteristics may play an important role in phytotoxicity of alumina nanoparticles. Toxicol Lett 158:122–132

Yanping X, Yiping H, Peter LI, Tony J, Xianming S (2011) Antibacterial activity and mechanism of action of zinc oxide nanoparticles against *Campylobacter jejuni*. Appl Environ Microbiol 77:2325–2331

Zachariah MR, Aquino MI, Shull RD, Steel EB (1995) Formation of superparamagnetic nanocomposites from vapor phase condensation in a flame. Nanostruct Mater 5(4):383–392

Zaki K, Misaghi IJ, Heydari A, Shatla MN (1998) Control of cotton seedling damping-off in the field by *Burkholderia* (*Pseudomonas*) *cepacia*. Plant Dis 82(3):291–293

Zeng H, Li XF, Zhang GY, Dong JF (2008) Preparation and characterization of beta cypermethrin nanosuspensions by diluting O/W microemulsions. J Dispers Sci Technol 29:358–361

Zhu H, Han J, Xiao JQ, Jin Y (2008) Uptake, translocation, and accumulation of manufactured iron oxide nanoparticles by pumpkin plants. J Environ Monit 10(6):713–717

Zielińska-Jurek A, Reszczyńska J, Grabowska E, Zaleska A (2012) Nanoparticles preparation using microemulsion systems. InTechOpen. https://doi.org/10.5772/36183

Chapter 13
Nanotechnology Application in Agricultural Sector

Mahmoud Nasr

Contents

13.1 Introduction

Nanotechnology has experienced profound applications in handling various environmental and engineering issues (Nasr et al. 2015). Nanotechnology enables the production of microscopic particles at the nanoscale size having dimensions between 1 and 100 nanometers (nm) being in either unbound or agglomerated state (Hussein et al. 2005). Nanoparticles retain unique physical and chemical properties, and they can be used in various optical, biomedical, and electronic applications (Paret et al. 2013). Previous studies attempted to synthesize nanoparticles using several chemical routes such as N,N-dimethylformamide utilization, surface-enhanced Raman scattering, colorimetric detection, and photo-induced conversion (Rai et al. 2008). However, these methods have experienced a series of

M. Nasr (✉)
Sanitary Engineering Department, Faculty of Engineering, Alexandria University, Alexandria, Egypt

R. Prasad et al. (eds.), *Nanobiotechnology in Bioformulations*, Nanotechnology in the Life Sciences, https://doi.org/10.1007/978-3-030-17061-5_13

environmental issues due to the utilization of toxic, hazardous, and flammable elements (Mattos and Magalhaes 2016). Alternatively, phytosynthesis is used to fabricate nanoparticles from plant parts (e.g., stems, leaves, and roots) and their extracts via intracellular and extracellular methods (Ocsoy et al. 2013). These techniques have the advantages of being stable, environmentally benign, and cost-effective, and hence, phytosynthesized nanoparticles have found several applications in sensors manufacturing, drug delivery, pharmaceutics, and cancer treatment (Mittal et al. 2013; Prasad 2014; Prasad et al. 2018).

Farming practices have resulted in the generation of huge amounts of agricultural wastes (Fawzy et al. 2018). For instance, European countries generate over 90×10^6 tons of agricultural wastes per year (Marchiol 2018). These wastes include high fractions of organic matters that are economically feasible, socially acceptable, and environmentally efficient for their conversion into energy using biological routes (Abdelsalam et al. 2017). Anaerobic digestion of agricultural biomass is a viable and sustainable option to generate biogas that can cope with the intensive consumption of fossil-based fuels (Abdelsalam et al. 2016). The use of nanotechnology in biological processes attempts to (a) maintain enzyme stability, (b) improve electron exchange rate, (c) enhance the conversion efficiency of lignocellulosic components, and (d) offer beneficial trace elements to the microbial communities (Chandra et al. 2012). Hence, nanotechnology can be employed as a promising approach to enhance the anaerobic digestion performance and to increase the biogas productivity from agricultural wastes.

Agricultural practices utilize several attributes including water, fertilizers, energy, manpower, and land. Most plant species require trace elements such as Mn, Zn, and Mo with concentrations of 0.50, 0.05, and 0.01 mg/L, respectively, in the soil solution for healthy growth (Chhipa 2017). Nanoparticles of Mn, Zn, and Mo can improve the performance and yield of some crops and cause little environmental risks compared with the commercially available salts. Mg nanoparticles have higher mobility in soil than conventional Mg salts, and they are used to improve the uptake of Mg by plant roots (Liu and Lal 2015). Similarly, Fe nanoparticles can be used to reduce the iron deficiency chlorosis in plants. Nanoparticles can penetrate and be transported into various cells and tissues of the plant and improve the utilization efficiency of nutrients and water (Rai et al. 2018). Moreover, nanoparticles can act as micro/macro nutrient carriers and nanofertilizers to increase crops production. Liu and Lal (2015) reported that nanofertilizers could (a) ensure plant growth and yields, (b) improve the fertilizer utilization efficiency, (c) limit leaching losses of nutrients, and (d) minimize the negative environmental risks caused by N and P applications. Nanoparticles can also be employed as nanopesticides to protect crops against pests (e.g., insects and weeds), diseases, and climatic changes (Kah and Hofmann 2014; Bhattarcharyya et al. 2016). Nanopriming has also been used to enhance seed germination and seedling growth (Duhan et al. 2017). In this context, the utilization of nanoparticles in the agricultural field has become an essential point of research that attempts to maintain food safety and security and reduce environmental pollution.

This chapter represents the recent utilization of nanotechnology in the agricultural field for an efficient bioconversion of farming residues and the improvement of agricultural practices. The first section describes the phytosynthesis of nanoparticles using plant extracts. The second part demonstrates the use of nanotechnology to enhance the biological conversion of plant biomass into biogas. Other parts of the chapter depict the recent utilization of nanomaterial, i.e., nanofertilizers, carriers of macro-/micronutrients, nanopesticides, and nanopriming, in the sector of agriculture. Finally, the nanotechnology limitations and future prospects regarding the plant-nanotechnology nexus have been listed.

13.2 Agriculture Waste

Agricultural waste, also known as agro-waste, is defined as any residues obtained from the farming practices (Noonari et al. 2019). Agricultural waste includes a food-based fraction of plants (e.g., sugar beet, corn, wheat, and soybean), non-edible part of crops (e.g., leaves, stems, stalks, and roots), and animal residues (Prasad et al. 2014; Sangeetha et al. 2017). It contains sustainable and renewable sources of carbon that can be used to meet the growing demand for energy (Chandra et al. 2012). Agricultural waste is composed of five categories (Marchiol 2018).

(a) Sugars and starches: Agricultural biomass containing sugar and starch components is suitable for biological conversion into CH_4 gas.
(b) Lignocellulosic feedstock: It is a woody, herbaceous, and clean material that comprises high fractions of cellulose, hemicellulose, and lignin elements. It represents a renewable, abundant, and non-edible part of the plant biomass, which can be used as second-generation biofuels.
(c) Bio-oils: Some crops such as *Jatropha*, pennycress, rapeseed, soybean, and canola are good resources of oil, which can be converted into biodiesel.
(d) Solid wastes: Agricultural solid waste is generated during plant harvesting, crop production, or raising of animals.
(e) Other waste including livestock residues, poultry manure, and landfill gases and leachate.

13.3 Nanotechnology in Agricultural Sector

Recently, nanotechnology has been recognized as an attractive solution for various engineering and environmental issues. Particles at the nanoscale size range of 1–100 nm are known as nanoparticles (Mahakham et al. 2016). This nano-size offers the ability to penetrate biological barriers, pass through cell tissues, and interact with immune systems. Nanomaterial is also characterized by a large specific surface area, distinctive physical and chemical properties, high reactivity, and adequate adsorption and uptake capabilities (Mittal et al. 2013). Recently, nanotechnology has found several applications in the agricultural sector to stimulate the plant physiological properties and promote the crop and vegetation sustainability (Ocsoy et al. 2013). Nanotechnology can contribute to the attainment and improvement of agricultural practices to cope with the current food security. For instance, by adding sufficient dosages of carbon nanotubes, plant growth can be induced via enhancing seed germination rate, shoot and root elongations, and biomass fresh and dry weights (Vithanage et al. 2017).

Nanoparticles are introduced to the plant body to (a) improve the water uptake efficiency, (b) transport DNA into the plant tissues through a bombardment system, (c) supply herbicides to crops, causing a stable release of the active ingredients, and (d) activate the reproductive system and adapt the gene functions and expressions (Delfani et al. 2014; Rico et al. 2014; Lahiani et al. 2015). Moreover, nanoparticles can promote the transport of nutrients, pesticides, and fertilizers from the rhizosphere into the plant body (Yearla and Padmasree 2016). Nanoparticles having a diameter smaller than the pore size of the cell wall can penetrate the plant membranes and serve as nutrient carriers. This trend allows for a "smart fertilizer" approach to control the mobility of N and P species. However, a high dosage of nanoparticles may be toxic to the plant physiological properties, and it can inhibit the seedling performance and alter the soil microbial diversity.

Kalteh et al. (2014) revealed that silicon (Si) nanoparticles could be used to reduce the negative impact of high salinity stress on the basil growth and development. Xiumei et al. (2005) found that $CaCO_3$ nanoparticles enhanced the cultivation and development of peanut and promoted the adsorption capacity of shoot and root toward nutrient species (e.g., Ca, N, and P). Rico et al. (2014) investigated the impact of soil amendment with cerium oxide nanoparticles ($n$$CeO_2$) on the cultivation of wheat (*Triticum aestivum* L.). Their study (Rico et al. 2014) depicted that the use of $n$$CeO_2$-H enhanced plant growth by 9.0%, grain yield by 36.6%, and shoot biomass by 12.7%, compared with the control. Lahiani et al. (2015) examined the influence of single-walled carbon nanohorns (SWCNHs) on the biological response of several plant species. Their study (Lahiani et al. 2015) demonstrated that SWCNHs of 100 μg/ml enhanced seed germination of switchgrass (*Panicum virgatum*), rice, barley, and corn, whereas 25 μg/ml SWCNHs improved tomato seed germination. In addition, the growth of tobacco cells increased by 78% compared with control due to SWCNHs exposure. The improvement mechanism of SWCNHs includes overexpression of aquaporins (water channels) in plant bodies, leading to enhance the uptake of water by plant organs and cells (Lahiani et al. 2015). Haghighi and Teixeira da Silva (2014) found that 10–40 mg/L of carbon nanotubes improved the growth of onion and tomato with germination percentages of 95% and 8%, respectively. Yatim et al. (2015) found that multiwalled carbon nanotubes (MWNTs) could carry urea fertilizer and enhance nutrient uptake, reduce nitrogen loss from the plant/soil system, and improve the utilization efficiency of water and oxygen.

13.4 Phytosynthesis of Nanoparticles

Nanoparticles material exhibits several advantages compared with other particulate matter and bulk particles. These benefits include (a) adequate adsorption capacity, (b) large specific surface area, (c) existence of active sites and functional groups on the surface, (d) advanced physical, electrical, chemical, and optical characteristics, and (e) high stability and broad functionalities (Giannousi et al. 2013; Marchiol 2018). The new age of nanotechnology attempts to synthesize nanomaterial using simple, safe, economic, and controlled techniques. The chemical synthesis of nanomaterial can cause adverse health and environmental impacts due to the utilization of harmful reagents (Mahakham et al. 2016). Moreover, the washing step during nanoparticles fabrication results in the release of toxic by-products, which tend to accumulate in the soil, water, and air systems.

Plant biomass and its extracts can be used as an environmentally benign tool for the green synthesis of nanoparticles. Plant parts contain essential proteins, carbohydrates, and carboxylic and amino groups that have main functions in the reduction process and the creation of metallic nanoparticles (Rai et al. 2008; Prasad 2014). The phytosynthesis (or biosynthesis) of nanoparticles aims at avoiding severe threats to human, plant, and animal. Phytosynthesized nanoparticles can be employed in the agro-ecosystem, pharmaceutical industry, and biomedical field (e.g., disease diagnostics, biosensors, and drug development and delivery) to ensure human safety (Raut et al. 2009). Moreover, phytosynthesized nanoparticles are used in food packaging due to their antimicrobial properties that inhibit nutrition spoilage. The advantages of applying phytosynthesized nanoparticles in agriculture include enhancement of crop yield, improvement of enzymatic activities in plant parts, enhancement of chlorophyll pigment, and an increase of beneficial element contents in tissues (Husen and Siddiqi 2014).

Shanmugam et al. (2016) used the leaf extract of *Aristolochia indica* L. to synthesize a silver nanoparticles material, which could act as an oxidation catalyst, disinfectant agent, and free radical scavenger. Gopinath et al. (2014) used fruit extract of *Terminalia arjuna* to synthesize gold nanoparticles, which could be used to improve the seed germination of endangered crop types. A review article by Mittal et al. (2013) represented the use of plant extracts as reducing agents and stabilizing agents to synthesize metallic nanoparticles. Their study (Mittal et al. 2013) reported that metal ions could be reduced to nanoparticles using biomolecules in plant extracts. Raut et al. (2009) used a leaf broth of *Gliricidia sepium* to prepare Ag nanoparticles, which acted as antimicrobial agents against *Klebsiella pneumonia*, *Escherichia coli*, *Staphylococcus aureus*, and *Pseudomonas aeruginosa*.

13.5 Nanotechnology for Anaerobic Digestion of Agricultural Wastes

Anaerobic digestion is recognized as an effective and well-established technique for the biological conversion of agricultural waste into methane gas (Nasr et al. 2015). Biogas can be used as a renewable source of energy for cooking, heating, and electricity generation, providing a practical mean of energy saving. The estimated biogas yields from cattle manure, chicken manure, sweet sorghum, corn silage, and forage beet have been reported as 45, 80, 108, 202, and 111 m^3 per tons of fresh feed, respectively (Ganzoury and Allam 2015). The utilization of biogas as an alternative energy source to fossil fuels offers several advantages including (a) the achievement of waste management target, (b) the reduction of pollution problems from organic wastes, (c) the use of renewable sources without inputs of oil or natural gas, and (d) no release of greenhouse gases to the atmosphere. The degradation and conversion of agricultural waste into biogas undergoes four successive stages, which can be illustrated as follows (Chandra et al. 2012; Ganzoury and Allam 2015):

1. Hydrolysis: Hydrolysis phase is the breakdown of lipids (fats), proteins, and complex carbohydrates (polysaccharides) into fatty acids, amino acids, and simple sugars (monosaccharides) by the actions of hydrolytic bacteria.
2. Acidogenesis: In acidogenesis, the formed monomers are converted into organic acids, alcohols, hydrogen, and carbon dioxide by diverse facultative and obligate acidogenic bacteria.
3. Acetogenesis: During the acetogenic phase, the products obtained from the acidification stage are converted into acetates and hydrogen by the aid of acetogenic bacteria.
4. Methanation: Finally, CH_4 gas can be produced in the methanogenic step via three routes, as follows: (a) Acetate is converted into $CH_4 + CO_2$ by acetoclastic methanogenesis, (b) H_2 and CO_2 are converted into CH_4 by hydogenotrophic methanogenesis, and (c) Methanol is converted into CH_4 and H_2O by methyltrophic methanogenesis.

Some trace elements such Fe, Co, Cu, Zn, and Ni are required at appropriate concentrations during anaerobic digestion to stimulate the enzymatic activities and metabolic pathways of methanogenic bacteria (Ganzoury and Allam 2015). Moreover, various macronutrients have a positive impact on the stability of the biogas performance, and they tend to decrease the lag phase to improve the cumulative biogas volume. For example, zero valance iron (Fe^0) has found a considerable influence on the anaerobic digestion process because it acts as an electron donor for reducing CO_2 into CH_4 (Eq. 13.1). However, high concentrations of chemical additives can inhibit biogas productivity due to the generation of toxic substances such as free radicals.

$$8H^+ + 4Fe^0 + CO_2 = CH_4 + 4Fe^{2+} + 2H_2O \tag{13.1}$$

The utilization of trace metals in the form of nanoparticles represents an attractive option to improve the biogas yield and methane percent compared with conventional additives. For instance, Abdelsalam et al. (2016) investigated the utilization of various trace metals in the form of nanoparticles for enhancing the anaerobic digestion of livestock manure. Their study (Abdelsalam et al. 2016) indicated that Co of 1 mg/L, Ni of 2 mg/L, Fe of 20 mg/L, and Fe_3O_4 of 20 mg/L improved the biomethane yield by 1.7, 1.8, 1.5, and 1.7 folds, respectively, compared with the control. Noonari et al. (2019) examined the influence of Fe_3O_4 nanoparticles on the anaerobic co-digestion of plant wastes with buffalo dung. Their work (Noonari et al. 2019) demonstrated that the optimum dosages of 0.81 mg for canola straw with buffalo dung and 0.5 mg for banana waste with buffalo dung attained methane yields of 256.0 and 202.3 mL-CH_4/g-VS, respectively. The addition of Fe_3O_4 nanoparticles tended to stimulate the microbial activities and improve methane production compared with the control experiment. Abdelsalam et al. (2017) examined the enhancement effects of iron nanoparticles on biomethane production from anaerobic digestion of manure. The study (Abdelsalam et al. 2017) indicated that Fe of 20 mg/L and Fe_3O_4 of 20 mg/L, respectively, improved the biogas volume by 1.45- and 1.66-fold and the methane volume by 1.59- and 1.96-fold compared with the control test.

13.6 Nanofertilizers

Fertilizers are chemical substances added during plant cultivation to enhance the crop yield and food production. However, the excess implementation of conventional chemical fertilizers has caused groundwater contamination, human health issues, eutrophication in aquatic ecosystems, and several environmental destabilizations (Chhipa 2017). Nanofertilizers are fertilizers synthesized in the form of nano-sized material, and they provide one or multiple species of nutrients for plant cultivation (Dissanayake and Chandrajith 2009). Nanofertilizers are categorized into (a) macronutrients containing N, P, K, Mg, S, and Ca elements, (b) micronutrients composed of trace elements (e.g., Fe, Mn, Zn, Cu, and Mo), and (c) nanoparticulate fertilizers including CNTs, TiO_2, and SiO_2. The benefits of nanofertilizers are fourfold (Duhan et al. 2017; Liu and Lal 2015): (a) enhance the nutrient uptake efficiency by plants and control the release of nutrients, (b) improve the vegetation resistance against pests and pathogens, (c) synchronize plant assimilation, and (d) reduce the negative influences of conventional chemical fertilizers on the environment.

Nanofertilizers can dissolve in aqueous solutions to release nutrient(s) in the form of soluble ions. This tendency can be clarified by Eqs. 13.2, 13.3, and 13.4.

$$2Cu(NPs) + 2H_2O + O_2 \Leftrightarrow 2Cu^{2+} + 4OH^- \tag{13.2}$$

$$Ca_5(PO_4)_3OH(NPs) \Leftrightarrow 5Ca^{2+} + 3PO_4^{3-} + OH^- \tag{13.3}$$

$$ZnO(NPs) + H_2O \Leftrightarrow Zn^{2+} + 2OH^- \tag{13.4}$$

Nanoparticles of apatite, $Ca_5(PO_4)_3OH$, are used as P fertilizer to improve the agronomical yield and minimize the occurrence of eutrophication (Dissanayake and Chandrajith 2009). Moreover, nanoparticles of calcite, $CaCO_3$, are applied as a fertilizer to provide the plant with the Ca source (Liu and Lal 2015). Delfani et al. (2014) investigated the effect of Fe and Mg (in the forms of nano- and regular- particles) on the physiological responses of a black-eyed pea. Their study (Delfani et al. 2014) found that the combination of 0.5 g/L nano-Fe + nano-Mg provided a yield of 232 g/m^2 compared with 93 g/m^2 when employing 0.5 g/L regular-Fe + regular-Mg.

Parveen et al. (2016) found that the utilization of bio-nanogold could enhance seed germination and seedlings growth of Pearl millet, as well as, stimulate the plant nutrition and growth. Taran et al. (2014) depicted that molybdenum (Mo) nanoparticles at a colloidal solution of 8 ppm were suitable for augmenting the number and mass of nodules per Chickpea (*Cicer arietinum*). Their work (Taran et al. 2014) revealed that the efficient penetration of nanoelements into the Chickpea tissues could improve the plant resistance toward opposed environmental factors. When plants involve a deficient Fe condition, TiO_2 nanoparticles can be employed to improve plant growth by promoting the uptake and utilization of Fe (Lyu et al. 2017). Ti nanoparticles can also enhance the plant photosynthetic efficiency, stimulate the activity of enzymes in crops, and promote nutrient uptake. Nutrient-augmented-zeolite fertilizers can be used to strengthen the soil's ability to reduce the leaching of nutrient species into groundwater (Prasad et al. 2014, 2017).

13.7 Nanopesticides

Pesticides are chemical compounds applied to protect crops against pests, including fungi, insects, and rodents, that cause leaf senescence, decay, rot, and various plant diseases (Qian et al. 2011). However, the excess and uncontrolled applications of chemical pesticides in crop systems have resulted in the reduction of food quality and the loss of soil fertility (Chhipa 2017). Nanopesticide technology is the modification of traditional pesticide with metal or polymer nanoparticles to maintain the optimum utilization of pesticides and preserve the soil health and geo-biological cycle. Moreover, nanopesticides experience a better spatial distribution on leaf surfaces, and they provide positive impacts on the control of vegetation pest and disease (Yearla and Padmasree 2016). These advantages tend to adopt the conventional practices of farming, leading to minimize the environmental deterioration problems.

Mattos and Magalhaes (2016) prepared a nanocomposite material using nanosilica and nanofibrillated cellulose, which was used to control the release rate of tebuconazole biocide. Their study (Mattos and Magalhaes 2016) depicted that tebuconazole entrapped onto nanocomposite experienced a release rate of 30–45% compared with 95% for the conventional biocide. Qian et al. (2011) investigated the application, release behavior, and stability of $CaCO_3$ nanoparticles as a delivery carrier for the pesticide validamycin. Their study (Qian et al. 2011) demonstrated that validamycin wrapped into nanoparticles achieved a better germicidal ability against *Rhizoctonia solani* after 1 week compared with regular validamycin. Paret et al. (2013) investigated the applications of nano-TiO_2, nano-TiO_2/Ag, and nano-TiO_2/Zn to protect tomato against *Xanthomonas perforans* for in vitro, greenhouse, and field conditions. Their work (Paret et al. 2013) demonstrated that the treatment of crops with 500–800 mg/L of TiO_2/Zn considerably reduced bacterial spot disease and provided positive impacts on tomato yield. Hussein et al. (2005) studied the release of an herbicide, namely, 2,4-dichlorophenoxyacetate (2,4-D), from a zinc-aluminum-layered double hydroxide nanocomposite (Zn-Al-24D). Their work

(Hussein et al. 2005) indicated that the release mechanism was controlled by ion exchange between the 2,4-D anion encapsulated in Zn-Al-2,4-D and carbonate or chloride anions in the aqueous medium. Yearla and Padmasree (2016) investigated the loading of herbicide (i.e., diuron) within a subabul stem lignin matrix to prepare a diuron nanoformulation (DNF), which tended to improve the release rate of herbicide. Their work (Yearla and Padmasree 2016) depicted that the seedlings of canola (*Brassica rapa*) augmented with DNF had earlier indications of leaf chlorosis and mortality compared with those supplemented with conventional diuron formulation. Saharan et al. (2013) indicated that several formulations of nanoparticles, viz., chitosan, chitosan-saponin, and Cu-chitosan, could protect crops, particularly against phytopathogenic fungi. It was found that the Cu-chitosan nanomaterial with 0.1% concentration achieved growth inhibitions of 60.1% for *Rhizoctonia solani*, 89.5% for *Alternaria alternate*, and 63.0% for *Macrophomina phaseolina* in an in vitro model. This observation was due to the high surface charge density of Cu-chitosan nanoparticles, leading to high zeta potential and great binding affinity to the negatively charged fungal membrane. Ocsoy et al. (2013) found that a composite of DNA-directed Ag nanoparticles developed on graphene oxide limited the existence of *Xanthomonas perforans* in culture and on tomatoes. Moreover, the plants subjected to nanocomposite provided several advantages including effective antibacterial capability, higher adsorption properties, and no phytotoxicity compared with untreated plants. Ho et al. (2015) demonstrated that Ag core-shell nanoclusters inhibited the growth of plant-pathogenic fungi of *Phytophthora colocasiae*, *Phytophthora capsici*, and *Phytophthora nicotianae*. Rajesh et al. (2012) found that Ag nanoparticles prepared from *Ulva fasciata* crude ethyl acetate could deactivate the growth and reproduction of *Xanthomonas campestris* pv. *Malvacearum*. Giannousi et al. (2013) demonstrated that copper-based nanoparticles, viz., CuO, Cu_2O, and Cu/Cu_2O, could be employed to protect tomato against *Phytophthora infestans*.

13.8 Nanopriming

Nanopriming is the use of nanotechnology to enhance seed germination, growth performance, and seedling vigor in plant species. Mahakham et al. (2016) prepared gold nanoparticles using galanga rhizome extracts (GNPs), which were employed as a nanopriming agent to endorse the germination and growth of maize seedlings. Their work (Mahakham et al. 2016) depicted that the emergence percentage was 83% using 5 ppm of phytosynthesized GNPs compared with 56% and 43% for chemically hydroprimed and unprimed groups, respectively. Mahakham et al. (2017) investigated the application of Ag nanoparticles synthesized using kaffir lime leaf extract to improve the germination of aged rice seeds. The phytosynthesized nanocomposite was used as nanopriming agent (Mahakham et al. 2017), leading to (a) enhance water uptake, (b) support seedlings growth, (c) improve α-amylase activity, (d) produce more reactive oxygen species, and (e) induce starch hydrolysis to maintain embryo growth. Guha et al. (2018) investigated the application of nanoscale zero-valent iron (nZVI) as nanopriming agent for the growth and development of *Oryza sativa* cv. Gobindabhog. Their work (Guha et al. 2018) depicted that rice seeds primed with 20 mg/L of nZVI could (a) maintain high hydrolytic and antioxidant enzymes activities, (b) increase root and shoot lengths, (c) avoid membrane damage, and (d) increase photosynthetic pigment content. Sayedena et al. (2018) examined the impact of nanopriming using multiwall carbon nanotubes (MWCNT) on the growth of mountain ash. Their work (Sayedena et al. 2018) indicated that the nanopriming treatment increased the seedling moisture and oxygen infiltration into seeds, leading to disrupt seed dormancy and develop seed germination. Raja et al. (2019) studied the implementation of ZnO and Cu nanoparticles for seed priming of blackgram (*Vigna mungo*) under in vitro conditions. Their work (Raja et al. 2019) depicted that ZnO and Cu nanoparticles at 600 and 400 mg/L, respectively, in aqueous solutions for 3 h could enhance the seed germination quality.

13.9 Future Perspectives

The application of nanotechnology in the agricultural sector has revealed profound impacts on food production, environmental protection, and energy saving. However, various future perspectives should be considered for sustainable and eco-friendly implementation of nanomaterials:

(a) Endorse the social awareness and perception toward the benefits of nanotechnology.
(b) Ensure the integration between nanoscience, engineering, environmental, and public approaches.
(c) Establish a long-term collaboration between funding agencies and scientific organizations to validate the impact of phytosynthesized nanoparticles on the environment and human health.
(d) In-depth investigations on the nanoparticles roles to support water and nutrients uptake by plants should be intensively provided.
(e) Long-term studies should be conducted to illustrate the relationship between nanofertilizers and soil structure and ecology.
(f) Additional researches on the functions of nanopesticides to accumulate in plant tissues and disrupt microbial cell walls and membranes should be illustrated.

13.10 Conclusions

This chapter represents an adequate assessment of the application of nanotechnology for efficient utilization of agricultural waste and improvement of farming practices. It is found that plants and their extracts can be used as reducing agents and stabilizing agents in the preparation of nanoparticles. Various trace elements can be utilized in the form of nanoparticles to improve the biogas yield and methane percent during the anaerobic digestion of agricultural waste. Nanoparticles can be used to promote the transport of nutrients, fertilizers, and pesticides to the plant body. The application of nanotechnology in agriculture can improve the water uptake efficiency, transport DNA into the plant tissues through a bombardment system, supply herbicides to crops, activate the reproductive system, and adapt the gene functions and expressions. These advantages tend to stimulate the plant physiological properties and promote crop and vegetation sustainability. Nanofertilizers are employed to enhance seed germination rate, shoot and root elongations, and biomass fresh and dry weights. Nanopesticide technology is the modification of conventional pesticides using metal or polymer nanoparticles to limit the vegetation pests and diseases. Moreover, several literature studies have reported that nanotechnology can be applied to enhance seed germination, growth performance, and seedling vigor in plant species via nanopriming. In-depth investigations should be conducted to comprehensively describe (a) the role of nanoparticles in water and nutrients uptake by plants, (b) the relationship between nanofertilizers and soil structure and ecology, and (c) the disruption of microbial cell walls and membranes via nanopesticides.

References

Abdelsalam E, Samer M, Attia YA, Abdel-Hadi MA, Hassan HE, Badr Y (2016) Comparison of nanoparticles effects on biogas and methane production from anaerobic digestion of cattle dung slurry. Renew Energy 87:592–598

Abdelsalam E, Samer M, Attia YA, Abdel-Hadi MA, Hassan HE, Badr Y (2017) Influence of zero valent iron nanoparticles and magnetic iron oxide nanoparticles on biogas and methane production from anaerobic digestion of manure. Energy 120:842–853

Bhattacharyya A, Duraisamy P, Govindarajan M, Buhroo AA, Prasad R (2016) Nano-biofungicides: emerging trend in insect pest control. In: Prasad R (ed) Advances and applications through fungal nanobiotechnology. Springer International Publishing, Cham, pp 307–319

Chandra R, Takeuchi H, Hasegawa T (2012) Methane production from lignocellulosic agricultural crop wastes: a review in context to second generation of biofuel production. Renew Sust Energ Rev 16:1462–1476

Chhipa H (2017) Nanofertilizers and nanopesticides for agriculture. Environ Chem Lett 15:15–22

Delfani M, Firouzabadi M, Farrokhi N, Makarian H (2014) Some physiological responses of black-eyed pea to iron and magnesium nanofertilizers. Commun Soil Sci Plant Anal 45(4):530–540

Dissanayake C, Chandrajith R (2009) Phosphate mineral fertilizers, trace metals and human health. J Natl Sci Found Sri Lanka 37(3):153–165

Duhan JS, Kumar R, Kumar N, Kaur P, Nehra K, Duhan S (2017) Nanotechnology: the new perspective in precision agriculture. Biotechnol Rep 24(15):11–23

Fawzy M, Nasr M, Adel S, Helmi S (2018) Regression model, artificial neural network, and cost estimation for biosorption of Ni(II)-ions from aqueous solutions by *Potamogeton pectinatus*. Int J Phytoremediation 20(4):321–329

Ganzoury M, Allam N (2015) Impact of nanotechnology on biogas production: a mini-review. Renew Sust Energ Rev 50:1392–1404

Giannousi K, Avramidis I, Dendrinou-Samara C (2013) Synthesis, characterization and evaluation of copper based nanoparticles as agrochemicals against *Phytophthora infestans*. RSC Adv 3:21743–21752

Gopinath K, Gowri S, Karthika V, Arumugam A (2014) Green synthesis of gold nanoparticles from fruit extract of Terminalia arjuna, for the enhanced seed germination activity of *Gloriosa superba*. J Nanostruct Chem 4:115

Guha T, Ravikumar KVG, Mukherjee A, Mukherjee A, Kundu R (2018) Nanopriming with zero valent iron (nZVI) enhances germination and growth in aromatic rice cultivar (*Oryza sativa* cv. Gobindabhog L.). Plant Physiol Biochem 127:403–413

Haghighi M, Teixeira da Silva J (2014) The effect of carbon nanotubes on the seed germination and seedling growth of four vegetable species. J Crop Sci Biotechnol 17(4):201–208

Ho VA, Le PT, Nguyen TP, Nguyen CK, Nguyen VT, Tran NQ (2015) Silver core-shell nanoclusters exhibiting strong growth inhibition of plant-pathogenic fungi. J Nanomater 2015:7

Husen A, Siddiqi K (2014) Phytosynthesis of nanoparticles: concept, controversy and application. Nanoscale Res Lett 9:229

Hussein M, Yahaya A, Zainal Z, Kian L (2005) Nanocomposite-based controlled release formulation of an herbicide, 2,4-dichlorophenoxyacetate incapsulated in zinc–aluminium-layered double hydroxide. Sci Technol Adv Mater 6(8):956–962

Kah M, Hofmann T (2014) Nanopesticide research: current trends and future priorities. Environ Int 63:224–235

Kalteh M, Alipour ZT, Ashraf S, Aliabadi MM, Nosratabadi AF (2014) Effect of silica nanoparticles on Basil (*Ocimum basilicum*) under salinity stress. J Chem Health Risks (JCHR) 4(3):49–55

Lahiani MH, Chen J, Irin F, Puretzky AA, Green MJ, Khodakovskaya MV (2015) Interaction of carbon nanohorns with plants: uptake and biological effects. Carbon 81:607–619

Liu R, Lal R (2015) Potentials of engineered nanoparticles as fertilizers for increasing agronomic productions. Sci Total Environ 514:131–139

Lyu S, Wei X, Chen J, Wang C, Wang X, Pan D (2017) Titanium as a beneficial element for crop production. Front Plant Sci 8:597

Mahakham W, Theerakulpisut P, Maensiri S, Phumying S, Sarmah AK (2016) Environmentally benign synthesis of phytochemicals-capped gold nanoparticles as nanopriming agent for promoting maize seed germination. Sci Total Environ 573:1089–1102

Mahakham W, Sarmah A, Maensiri S, Theerakulpisut P (2017) Nanopriming technology for enhancing germination and starch metabolism of aged rice seeds using phytosynthesized silver nanoparticles. Sci Rep 7:8263

Marchiol L (2018) Nanotechnology in agriculture: new opportunities and perspectives. https://doi.org/10.5772/intechopen.74425

Mattos B, Magalhaes W (2016) Biogenic nanosilica blended by nanofibrillated cellulose as support for slow-release of tebuconazole. J Nanopart Res 18:274

Mittal A, Chisti Y, Banerjee U (2013) Synthesis of metallic nanoparticles using plant extracts. Biotechnol Adv 31:346–356

Nasr M, Tawfik A, Ookawara S, Suzuki M, Kumari S, Bux F (2015) Continuous biohydrogen production from starch wastewater via sequential dark-photo fermentation with emphasize on maghemite nanoparticles. J Ind Eng Chem 21:500–506

Noonari A, Mahar R, Sahito A, Brohi K (2019) Anaerobic co-digestion of canola straw and banana plant wastes with buffalo dung: effect of Fe_3O_4 nanoparticles on methane yield. Renew Energy 133:1046–1054

Ocsoy I, Paret ML, Ocsoy MA, Kunwar S, Chen T, You M, Tan W (2013) Nanotechnology in Plant Disease Management: DNA-Directed Silver Nanoparticles on Graphene Oxide as an Antibacterial against Xanthomonas perforans. ACS Nano 7(10):8972–8980

Paret ML, Vallad GE, Averett DR, Jones JB, Olson SM (2013) Photocatalysis: effect of light-activated nanoscale formulations of TiO(2) on *Xanthomonas perforans* and control of bacterial spot of tomato. Phytopathology 103(3):228–236

Parveen A, Mazhari B, Rao S (2016) Impact of bio-nanogold on seed germination and seedling growth in *Pennisetum glaucum*. Enzym Microb Technol 95:107–111

Prasad R (2014) Synthesis of silver nanoparticles in photosynthetic plants. J Nanoparticles:963961. https://doi.org/10.1155/2014/963961

Prasad R, Kumar V, Prasad K (2014) Nanotechnology in sustainable agriculture: present concerns and future aspects. Afr J Biotechnol 13(6):705–713

Prasad R (2014) Synthesis of silver nanoparticles in photosynthetic plants. Journal of Nanoparticles, Article ID 963961, http://dx.doi.org/10.1155/2014/963961

Prasad R, Bhattacharyya A, Nguyen QD (2017) Nanotechnology in sustainable agriculture: Recent developments, challenges, and perspectives. Front Microbiol 8:1014. https://doi.org/10.3389/fmicb.2017.01014

Prasad R, Jha A and Prasad K (2018) Exploring the Realms of Nature for Nanosynthesis. Springer International Publishing (ISBN 978-3-319-99570-0) https://www.springer.com/978-3-319-99570-0

Qian K, Shi T, Tang T, Zhang S, Liu X, Cao Y (2011) Preparation and characterization of nano-sized calcium carbonate as controlled release pesticide carrier for validamycin against Rhizoctonia solani. Microchim Acta 173(1–2):51–57

Rai M, Yadav A, Gade A (2008) CRC 675—current trends in phytosynthesis of metal nanoparticles. Crit Rev Biotechnol 28(4):277–284

Rai PK, Kumar V, Lee S, Raza N, Kim K-H, Ok YS, Tsang DCW (2018) Nanoparticle-plant interaction: implications in energy, environment, and agriculture. Environ Int 119:1–19

Raja K, Sowmya R, Sudhagar R, Moorthy PS, Govindaraju K, Subramanian KS (2019) Biogenic ZnO and Cu nanoparticles to improve seed germination quality in blackgram (*Vigna mungo*). Mater Lett 235:164–167

Rajesh S, Raja D, Rathi J, Sahayaraj K (2012) Biosynthesis of silver nanoparticles using *Ulva fasciata* (Delile) ethyl acetate extract and its activity against *Xanthomonas campestris* pv. malvacearum. J Biopest 5:119–128

Raut W, Lakkakula R, Kolekar S, Mendhulkar D, Kashid B (2009) Phytosynthesis of silver nanoparticle using *Gliricidia sepium* (Jacq.). Curr Nanosci 5(1):117–122

Rico CM, Lee SC, Rubenecia R, Mukherjee A, Hong J, Peralta-Videa JR, Gardea-Torresdey JL (2014) Cerium oxide nanoparticles impact yield and modify nutritional parameters in wheat (*Triticum aestivum* L.). J Agric Food Chem 62:9669–9675

Saharan V, Mehrotra A, Khatik R, Rawal P, Sharma SS, Pal A (2013) Synthesis of chitosan based nanoparticles and their in vitro evaluation against phytopathogenic fungi. Int J Biol Macromol 62:677–683

Sangeetha J, Thangadurai D, Hospet R, Purushotham P, Manowade KR, Mujeeb MA, Mundaragi AC, Jogaiah S, David M, Thimmappa SC, Prasad R, Harish ER (2017) Production of bionanomaterials from agricultural wastes. In: Prasad R, Kumar M, Kumar V (eds) Nanotechnology. Springer Nature Singapore Pte Ltd., Singapore, pp 33–58

Sayedena SV, Pilehvar B, Abrari-Vajari K, Zarafshar M, Eisvand HR (2018) Effects of seed nano-priming with multiwall carbon nanotubes (MWCNT) on seed germination and seedlings growth parameters of mountain ash (*Sorbus luristanica Bornm.*). Iran J For Poplar Res (IJFPR) 26(2):pe202–pe213

Shanmugam C, Sivasubramanian G, Parthasarathi B, Baskaran K, Balachander R, Parameswaran VR (2016) Antimicrobial, free radical scavenging activities and catalytic oxidation of benzyl alcohol by nano-silver synthesized from the leaf extract of *Aristolochia indica* L.: a promenade towards sustainability. Appl Nanosci 6(5):711–723

Taran NY, Gonchar OM, Lopatko KG, Batsmanova LM, Patyka MV, Volkogon MV (2014) The effect of colloidal solution of molybdenum nanoparticles on the microbial composition in rhizosphere of *Cicer arietinum* L. Nanoscale Res Lett 9(1):289

Vithanage M, Seneviratne M, Ahmad M, Sarkar B, Ok YS (2017) Contrasting effects of engineered carbon nanotubes on plants: a review. Environ Geochem Health 39(6):1421–1439

Xiumei L, Fudao Z, Shuqing Z, Xusheng H, Rufang W, Zhaobin F, Yujun W (2005) Responses of peanut to nano-calcium carbonate. Plant Nutr Fertil Sci 11(3):385–389

Yatim N, Shaaban A, Dimin M, Yusof F (2015) Statistical evaluation of the production of urea fertilizer-multiwalled carbon nanotubes using plackett burman experimental design. Procedia Soc Behav Sci 195:315–323

Yearla S, Padmasree K (2016) Exploitation of subabul stem lignin as a matrix in controlled release agrochemical nanoformulations: a case study with herbicide diuron. Environ Sci Pollut Res 23(18):18085–18098

Chapter 14
Advances in Bio-coaters for Nanoparticles and Biodegradable Delivery Systems in Agriculture and Food Industry: Toward a Safer and Eco-friendly Nanotechnology

Melissa Marlene Rodríguez-Delgado, Cesar Martinez-Ledezma, and Juan Francisco Villarreal-Chiu

Contents

14.1 Introduction

The protection and maintenance of agriculture fields have become an important international issue since any potential loss of food will directly affect the welfare of a global population that is constantly growing. Therefore, the agricultural industry faces today, a critical challenge to satisfy the current demands in food production, while avoiding adverse impacts on the environment and human health. Facing this challenge will not be easy since over the years, there has been a steady loss of arable land and water resources due to poor resource management or the continued erosion of soils. Furthermore, the inefficient fertilization of crops by conventional fertilizers based on inorganic forms of nitrogen and phosphorus tends to pollute urban and rural water resources by causing eutrophication (Ranjan et al. 2017). Therefore, the FAO has included as one of its main objectives for its 2030 agenda the

M. M. Rodríguez-Delgado · C. Martinez-Ledezma · J. F. Villarreal-Chiu (✉)
Universidad Autónoma de Nuevo León, Facultad de Ciencias Químicas, San Nicolás de los Garza, NL, México
e-mail: juan.villarrealch@uanl.edu.mx

R. Prasad et al. (eds.), *Nanobiotechnology in Bioformulations*, Nanotechnology in the Life Sciences, https://doi.org/10.1007/978-3-030-17061-5_14

implementation of sustainable agriculture practices that assures the protection of the ecosystems and eradicate hunger (Merino et al. 2018).

The use of nanotechnology by agri-food industries is relatively new in comparison with the development of pharmaceuticals and their drug delivery systems. However, it has demonstrated a great potential in the enhancement of crop productivity, the monitoring and detection of plant diseases, and enhancing food quality (Duhan et al. 2017; Prasad et al. 2014, 2017a, b). Its use in agriculture and food industries was initially reported by the US Department of Agriculture (USDA) in September 2003 (Joseph and Morrison 2006). Since then, food industries have been focused on the unique physical, chemical, biological, and mechanical properties of nanoparticles (NPs, materials in the scale of 1–99 nm) for the development of smart packaging technologies. These enable the food-package interaction, preventing the growth of microorganisms in post-processed food (Ranjitha et al. 2017). In this sense, the main developments in this field have been related to the preparation of bio-nanocapsules or nanocarriers containing elements that enhance the shelf-life extension, such as edible coatings from natural polymers that create a semi-permeable barrier that allows gas exchange and inhibits oxidative reactions (Mastromatteo et al. 2015).

Nevertheless, the indiscriminate use of products containing NPs has increased their presence in the environment over the years, provoking adverse effects on plant morphology, physiology, and genetics. It has also been demonstrated the capacity of NPs to penetrate different plant tissues, suggesting the potential introduction and bioaccumulation of these materials into the food chain (Kumar et al. 2017). Therefore, the development of biocompatible and biodegradable NPs needs further exploration to eliminate the potential risk of the use of these technologies. Hence, this chapter emphasizes the advances in novel bio-coatings used in metallic and nonmetallic NPs to enhance their biocompatibility for a safer application in agriculture and food industry, toward an environmentally friendly nanotechnology.

14.2 Functionalization of Nanoparticles

Nanoparticles have unique properties derived from their size and morphology that have revolutionized the applications of well-known elements and compounds. However, NPs can be submitted to further chemical modifications in a process known as functionalization, through the addition of functional groups derived from organic molecules such as thiols, disulfides, amines, nitriles, carboxylic acids, phosphines, and other biomolecules. These changes in the surface of the NPs, which can be performed through the formation of covalent bonds, cross-linking, or encapsulation by hydrophobic interactions, provide them with chemical properties that can be used for specific applications (Subbiah et al. 2010). Therefore, it is essential to evaluate the resulting size, distribution, surface charge, and nature of the functionalized NPs to identify their potential bioactivity or in vivo function and applications.

The functionalization of NPs can be accomplished by (1) direct functionalization, (2) post-functionalization, or (3) encapsulation (Fig. 14.1). Direct functional-

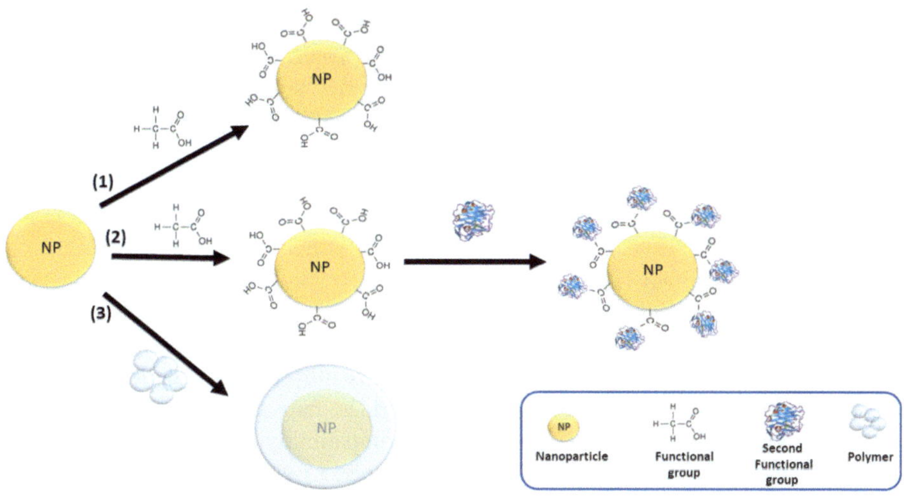

Fig. 14.1 Schematic representation of functionalization methods: (1) direct functionalization, where a conjugating agent is used to directly bind a functional group; (2) post-functionalization, which uses a first binding group to attach to the NP a secondary functional group; and (3) encapsulation through a polymer that covers the NP

ization occurs when one of the reactive groups of a bifunctional compound, such as thiols, phosphine oxides, phosphonates, or carboxylates groups, attaches to the NP surface, while the second group provides the actual functionality. For example, Li et al. (2007) reported the successful immobilization of trypsin through amine groups onto magnetic NPs. Despite being a one-step process, direct functionalization may present some drawbacks, such as steric interference or interaction between modified and pristine NPs (Kim and Bawendi 2003).

Post-functionalization describes the process in which a bifunctional molecule is anchored to the surface of the NP by one functional group, while the second group reacts through a chemical reaction (commonly hydrolysis/condensation reactions) with a second molecule that will act as the coupling agent (silanes compounds, thiols, and amines), completing the final form of the functionalized NP (nanoparticle-chelating/coupling agent-functional group) (Subbiah et al. 2010). Koh et al. modified γ-Fe_2O_3 NPs with aminopropyltriethoxy silane for the attachment of IgG antibodies (Koh et al. 2006). Meanwhile, Shen et al. (2008) achieved the binding of triphenylphosphine to gold nanoparticles through thiol groups. This strategy offers a strong binding of the functional groups with large surface coverage. However, it requires several steps and the prevention of isolated clusters of functional groups onto the NPs.

In the functionalization by encapsulation, the NPs are confined in the core of micron-sized spheres made from a semipermeable material, commonly polymers. In this method, the retention of the NPs is strongly related to the physical properties of the polymer, which is unstable under certain conditions of pH and temperature (Fernández-Fernández et al. 2013). Current methodologies for the functionalization of NPs by encapsulation take advantage of the recent development of smart polymers. For example, Reyes-Ortega et al. (2018) reported a protocol to achieve the coating of magnetite nanoparticles with a temperature-responsive polymer, by a hydrothermal method combined with an oil/water interface designed for therapeutic purposes.

Functionalized metallic NPs have gained considerable popularity in many scientific fields, particularly in electronics, biosensors, cosmetic industries, and biomedicine. However, the same properties that give NPs their application, such as their size, composition, and a wide range of chemical and physical properties, are known to generate adverse effects on human health. Since NPs can interact with some components of the immune system, they can activate stress-related genes, or cytokines, which are molecules involved in the pro-inflammatory process, or even cause membrane disruption. In this context, the scientific community has recently moved their attention to the use of biomolecules (DNA, amino acids, polysaccharides, fatty acids) for the functionalization of NPs to enhance and control their biocompatibility and stability in aqueous solutions (Pandey and Prajapati 2018). For example, Li et al. (2004) used lipids to functionalize nanoparticles for the attachment of an integrin antagonist (anti-FLK-1 antibody) for the development of an antiangiogenesis therapy.

14.2.1 Bio-functionalization

Bio-functionalization refers to the process in which biomolecules, such as DNA, proteins, antibodies, or enzymes, are anchored to NPs, resulting in hybrid biocompatible materials. This has been taken even a step further by employing methods of biosynthesis or green synthesis, where NPs are synthesized on "bio" conditions, ranging from biomolecule precursors to the use of extracts from plants or microorganisms as reaction matrix (Prasad et al. 2018). Under "bio" conditions, biomolecules act as the reducing agent of metal ions and as a capping agent, which is responsible for the biocompatibility of the functionalized NP (Ocsoy et al. 2018). This has been achieved by authors like Wirges et al. who proposed the synthesis of silver nanocrystals bio-functionalized with dialdehyde groups from modified DNA (Wirges et al. 2009) and Leng et al. who reported the green synthesis of gold nanoparticles using proteins like collagen, lysozyme, bovine serum albumin, hemoglobin, pepsin, and trypsin to reduce gold-protein complexes (Leng et al. 2016). On the other hand, Agarwal et al. reported the green synthesis of ZnO NPs using plant extracts from *Aloe vera*, *Cocus nucifera*, *Gossypium*, *Moringa oleífera*, *Azadirachta indica*, *Parthenium hysterophorus*, *Plectranthus amboinicus*, and *Vitex negundo* (Agarwal et al. 2017), while Kuppusamy et al. (2016) were able to synthesize Ag NPs using extract from *Alternanthera sessilis*, *Andrographis paniculata*, *A. mexicana*, *Artemisia nilagirica*, *Boswellia serrata*, *Carica papaya*, *Cassia fistula*, *Cinnamon zeylanicum*, *Citrullus colocynthis*, *Citrus sinensis*, *Dillenia indica*, and *Dioscorea bulbifera*.

On the other hand, it has been reported that some microorganisms, ranging from bacteria, yeast, and fungi, can process minerals directly from the environment producing a wide range of NPs (Prasad et al. 2016; Ganesh et al. 2018). For example, the biosynthesis of Ag NPs has been reported on bacterial strains of *Corynebacterium* sp., *Escherichia coli*, *Morganella* sp., *Bacillus cereus*, *Bacillus licheniformis*, and *Corynebacterium glutamicum*, while Au NPs biosynthesis has been reported on *Pseudomonas aeruginosa*, *Rhodococcus* sp., and *Ureibacillus thermosphaericus*, same for Zn and Fe NPs, where bacterial species of *Desulfobacteriaceae* and *Aquaspirillum magnetotacticum* have been identified as responsible for their biosynthesis. Cd NPs have been identified on the cytoplasm of *Lactobacillcus* sp. and *Rhodobacter sphaeroides.* Some fungi, such as *Fusarium solani*, *Coriolus versicolor*, and *Aspergillus niger*, are also known to synthesize Ag NPs (Prasad 2016, 2017; Ganesh et al. 2018). More recently, cultures of the yeast *Saccharomyces cerevisae* were used in the synthesis of biocompatible cadmium telluride quantum dots (Kowshik et al. 2002).

14.3 Application of Bio-coated Nanoparticles in Agriculture

In recent years, nanotechnology has played an essential role in agriculture, enabling sustainable practices that allow the enhancement of crops yield while avoiding adverse effects in the environment. This has been known as precision farming. The use of biofunctionalized nanoparticles and the nanoencapsulation of fertilizers and pesticides have allowed the development of smart delivery systems. The resulting nanofertilizers, nanopesticides, and even nanosensors (represented in Fig. 14.2) have been used to reduce the evaporation and leaching of agrochemicals applied in crop fields (Prasad et al. 2014, 2017a; Chhipa 2017).

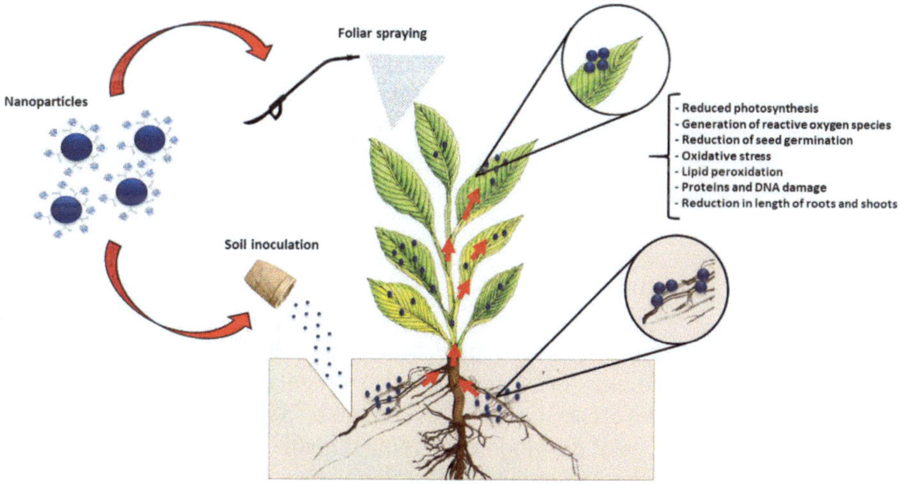

Fig. 14.2 Schematic representation of delivery of pesticides/nutrients from nanoparticles, absorption, and translocation

14.3.1 *Bio-coated Nanofertilizers*

14.3.1.1 Macronutrients

Nanofertilizers are considered as nutrient carriers of nanometric dimension. Their development has been focused on the delivery systems for the main macronutrients required by crops, such as nitrogen, potassium, and phosphorous. These macronutrients are biofunctionalized into NPs, where the material's surface facilitates the retention of these compounds and regulates their release into the crops. This process enables a continuous supply of nutrients to plants, improving their production yield (Chhipa 2017). In this way, nanofertilizers have been able to reduce the amount of agriproducts introduced to crop fields, reducing the runoff of unused conventional fertilizers, such as ammonium salts, urea, nitrates, or phosphates, which results in environmental pollution (Duhan et al. 2017). This is environmentally important as it has been estimated that around 30–60% of N, 30–50% of K, and 80–90% of P from fertilizers are lost to the environment (Duhan et al. 2017). Successful biomaterials based on zeolite, halloysite, montmorillonite, and bentonite have been reported to produce nanoclays coated with ammonia for the controlled release of nitrogen in Indian crops (Sharmila 2010; Subramanian and Tarafdar 2011).

In recent years, novel strategies have been adopted to develop innovative eco-friendly nanomaterials for agriculture (Prasad et al. 2017b). For instance, biodegradable polymeric chitosan (Corradini et al. 2010) and kaolin (Wilson et al. 2008) NPs have been used for to production of NPK-rich fertilizers of controlled release. Chitosan NPs, on the other hand, have been bio-coated with polymethacrylic acid (PMAA) to enhance the loading capacity of NPK-rich fertilizers (Hasaneen et al. 2014). Similarly, Ni et al. (2011) reported the use of natural attapulgite clay bio-coated with ethylcellulose film and sodium carboxymethyl cellulose/hydroxyl ethylcellulose hydrogel for the release of urea, ammonium sulfate, and ammonium chloride. Zeolites, minerals commonly found in nature, have also being used as a base material due to its honeycomb-like layered crystal structure which acts as a reservoir for fertilizers (Subramanian et al. 2015). The rich nanoporous surface of silica enables this material to be bio-functionalized with urease for the development of a delivery system for nitrogen (Hossain et al. 2008). A detailed list of bio-coated nanofertilizers is summarized in Table 14.1.

14.3.1.2 Micronutrients

Chemical elements like manganese, copper, boron, iron, molybdenum, and zinc are considered essential nutrients for the healthy development of plants. However, these are required at trace levels as higher concentrations may be harmful to them. Therefore, the technological development of functionalized NPs has been focused on the encapsulation of these microelements into microspheres of polymers, such as gelafundine, which enables them to dissolve in soil or be sprayed directly to leaves. These microcapsules have been designed to react with stress enzymes, organic

Table 14.1 Bio-coated nanoparticles employed as nanofertilizers

Bio-coating	NPs nature	Method of functionalization	Architecture	Size	Nutrients	References
Urea-aminopropyltrimethoxysilane	Zeolite	Chemical modification with HF	Nanoparticles	–	Nitrogen	Hidayat et al. (2015)
NPK fertilizer	Chitosan	Polymerizing methacrylic acid	Nanoparticles	17–25 nm	Nitrogen (N), phosphorus (P), and potassium (K)	Hasaneen et al. (2014)
Glutaraldehyde	Chitosan	Cross-linked	Microspheres		Indole-3-acetic acid (IAA) and naphthalene-1-acetic acid (NAA), also known as auxins (plant hormones)	Fan et al. (2012)
Genipin	Chitosan	Cross-linked	Microspheres		Urea	Hussain et al. (2012)
Soft wood of *Gliricidia sepium*	Urea-modified hydroxyapatite	Encapsulation	Nanoparticles	19–25 nm	Nitrogen	Kottegoda et al. (2011)
Ethylcellulose film and sodium carboxymethylcellulose (CMC) and hydroxyethylcellulose (HEC) hydrogel	Attapulgite	Cross-linked	–	–	Urea, ammonium sulfate, and ammonium chloride	Ni et al. (2011)
Organic polymer	Kaolinite	Adsorption	Clays	–	Nitrogen	Subramanuan and Tarafdar (2011)
NPK fertilizer		Polymerizing methacrylic acid	Nanoparticles	78 nm	Nitrogen (N), phosphorus (P), and potassium (K)	Corradini et al. (2010)
Sodium tripolyphosphate	Chitosan	Cross-linked	Microspheres	79–1470 μm	Brassinosteroids (BA) and diosgenin	Quiñones et al. (2010)
Urease	Silica	Adsorption	Gel	–	Nitrogen	Hossain et al. (2008)
Wax	Urea	Spraying molten sulphur	Granulates	–	Sulfur	Shaviv (2000)
Alginate	Chitosan	Ionotropic gelation	Nanoparticles	450 nm	Hormone gibberellic acid	Espirito et al. (2017)
Tripolyphosphate				195 nm		

acids, or phenolic substances of the plants, enabling the combined degradation of the polymer and the controlled release of the micronutrients (Subramanian et al. 2015; Corredor et al. 2009). In this sense, Tao et al. (2012) reported the synthesis of a formulation based on chitosan modified with 1-naphthyl acetic acid, a plant growth hormone. Their results showed that the delivery of the hormone was dependent on pH, with release periods up to 55 days. Meanwhile, Fan et al. encapsulated indole-3-acetic acid and naphthalene-1-acetic acid on chitosan microspheres, demonstrating a maximum release of 60% of the hormones after 120 h of exposition (Fan et al. 2012).

14.3.2 Bio-coated and Natural Carriers as Nanopesticides

The productivity of crops depends on many factors. However, the adverse effects of plant pathogens are responsible for a significant loss of the global production (Khater et al. 2017). FAO estimates that between 20 and 40 percent of the annual global crop production is lost due to these pests, with substantial costs to the global economy of US$220 billion (FAO 2017). The primary pathogenic agents that infect plants comprise fungi, bacteria, viruses and insects (Neethirajan et al. 2018), and the current technology for the protection of crops is based on the delivery of high volume of pesticides that controls them. However, the stability and effect of these chemicals are dependent on environmental factors, such as soil properties and climatic conditions (Mattos et al. 2017b). The indiscriminate use of biocides has damaged the ecosystems and increased pest resistance over the years while affecting beneficial soil microbial communities (Duhan et al. 2017). In this context, there is an increasing concern to replace or minimize the conventional use of highly toxic agrochemicals, by developing efficient delivery systems of biocides. Therefore, the use of natural coatings for NPs, such as biopolymers, has gained relevance since these materials are ecological and low cost and their biodegradation products act as natural compost to the soil (Merino et al. 2018). Table 14.2 summarizes all biodegradable carriers employed as pesticides delivery systems.

14.3.2.1 Chitosan

Chitosan is a polysaccharide commonly found in crustaceans and cell walls of fungi and Gram-positive bacteria. Micro- and nanoparticles of this natural polymer have been widely studied for the delivery of agrochemicals since it is quickly absorbed by plants through the cell membrane. This strategy improves the bioavailability of chemicals coated into the NPs (Kashyap et al. 2017). In this sense, several biocides have been tested to date. For instance, Paula et al. reported the use of microspheres of chitosan and cashew tree gum coated with a larvicide extracted from *Lippia sidoides* (Paula et al. 2011). Meanwhile, Guan et al. encapsulated the insecticide imidacloprid within a copolymer of chitosan and sodium alginate (Guan et al. 2008).

Table 14.2 Biodegradable carriers employed as pesticides delivery systems

Bio-carrier	Coating	Method of functionalization	Architecture	Biocide	Pest	Crop/plant	Effect	References
Chitosan/lactide copolymer	Nano-precipitation			Pyraclostrobin	*Colletotrichum gossypii*	–	Foliar fungicide	Xu et al. (2014)
Carboxymethyl chitosan	Azidobenzaldehyde	Photo-cross-linking	Shell cross-linked nanocapsules	Methomyl	Armyworm larvae		Insecticide	Sun et al. (2014)
Chitosan/sodium tripolyphosphate	–	–	Nanoparticle	Paraquat		Maize (*Zea mays*); mustard (*Brassica* sp.)	Herbicide	Grillo et al. (2014)
Chitosan	Saponin	Cross-linking	Nanoparticle		Phytopathogenic fungi *Alternaria alternata*, *Macrophomina phaseolina*, and *Rhizoctonia solani*	–	Fungicide	Saharan et al. (2013)
	Cu	Entrapment						
Carboxymethyl chitosan	Ricinoleic acid	–	Nanoparticle	Azadirachtin	–	–	Insecticide	Feng and Peng (2012)
Chitosan	Cashew gum	Encapsulation	Beads	Essential oil of *Lippia sidoides*	*A. aegypti* larval	NA	Larvicide	Paula et al. (2011)
Alginate/chitosan	NA	NA	Nanoparticle	Paraquat	–		Herbicide	dos Santos Silva et al. (2011)
Chitosan and sodium alginate	–	–	Nanocapsules	Imidacloprid	*Martianus dermestoides* (beetle)	–	Insecticide	Guan et al. (2008)

(continued)

Table 14.2 (continued)

Bio-carrier	Coating	Method of functionalization	Architecture	Biocide	Pest	Crop/plant	Effect	References
Chitosan	Octadecanol glycidyl ether as hydrophobic groups and sulfate	Entrapment	Nanomicelles	Rotenone	Beetle	Legumes	Insecticide	Lao et al. (2010)
Lignin	Ethylcellulose and dibutylsebacate	Encapsulation	Nanogranules	Chloridazon	Weed		Herbicide	Fernández-Pérez et al. (2011)
	–	Encapsulation	Nanoparticles	Diuron	–	Canola (*Brassica rapa*)	Herbicide/alguicide	Yearla and Padmasree (2016)
Cellulose–lignin	–	Cross-linked	Hydrogel	Polyphenols	–	–	–	Ciolacu et al. (2012)
Cellulose/silica	–	Encapsulation	Nanocomposite	Tebuconazole		–	Fungicide	Mattos and Magalhães (2016)
Lipids	–	–	Lipid nanoparticles	Atrazine and simazine	*Raphanus raphanistrum*	–	Herbicide	de Oliveira et al. (2015)
Liposomes	Chitosan	Encapsulation	Nanoliposomes	Etofenprox or alpha-cypermethrin	–	–	Insecticide	Bang et al. (2011)
Sodium alginate	–	Cross-linking	Nanoemulsion	Imidacloprid	Jassids/leafhoppers	Okra (bhindi) crop	Insecticide	Kumar et al. (2014)
Sodium alginate	–	–	Nanocapsules	Pyridalyl	Larvae of *Helicoverpa armigera*		Larvicide	Saini et al. (2014)

Oil/water micelle	–		Microemulsion	Nanopermethrin	*Culex quinquefasciatus*	–	Larvicide	Anjali et al. (2010)
Poly(ethylene glycol) (PEG)	–	–	Nanomicelles	Thiram	–	–	Fungicide	Kaushik et al. (2013)
				Imidacloprid			Insecticide	Adak et al. (2012)
				Beta-cyfluthrin				Loha et al. (2011)
				Thiamethoxam				Sarkar et al. (2012)
				Carbofuran				
Copolymers: poly(citric acid)-poly(ethylene glycol)-poly(citric acid) (PCA-PEG-PCA) ABA	–	–	Nanoparticles	Imidacloprid	Glyphodes pyloalis, moth	–	Insecticide	Memarizadeh et al. (2014a)
	–			Indoxacarb	–	–	Larvicide	Memarizadeh et al. (2014b)
Polyvinylpyridine-co-styrene	–	–	Nanoparticles	Tebuconazole and chlorothalonil	Fungus *Gloeophyllum trabeum*	Southern pine sapwood	Fungicide	Liu et al. (2001)
Starch hydrogel	-	-	Hydrogel	Carbendazim	–	–	Fungicide	Bai et al. (2015)
Zein	–	–	Nanoparticles	Geraniol and R-citronellal	*Tetranychus urticae* Koch mite	–	Insecticide	Oliveira et al. (2018)
Tobacco mild green mosaic virus	–	–	Virus	Crystal violet	*Caenorhabditis elegans* nematodes	–	Anthelmintic	Chariou and Steinmetz (2017)

In addition to its native form, several amphiphilic derivatives of chitosan have been employed as a carrier of pesticides. For example, Sun et al. reported the synthesis of amphiphilic nanocapsules of carboxymethyl chitosan with an aqueous core of azidobenzaldehyde to develop a controlled release of methomyl, which generated 100% mortality of armyworm larvae over 7 days (Sun et al. 2014). Amphiphilic derivates such as alginate/chitosan (dos Santos Silva et al. 2011) and chitosan/tripolyphosphate (Grillo et al. 2014) have been used to encapsulate the herbicide paraquat. Amphiphilic chitosan NPs activated with saponin and Cu were tested against the phytopathogenic fungi *Alternaria alternata*, *Macrophomina phaseolina*, and *Rhizoctonia solani*. Results indicated that Cu-chitosan NPs had a better performance (87.4% inhibition rate) than those of chitosan alone (Saharan et al. 2013). This has also been demonstrated for pesticides rotenone (Lao et al. 2010) and azadirachtin (Feng and Peng 2012).

14.3.2.2 Lignin

Lignin is a phenolic polymer produced in the tissues of higher plants. It has been related to important biological activities, such as antiviral, antibacterial, and antiparasitic (Ciolacu et al. 2012). Several studies have already reported the utilization of lignin as a matrix for controlled release of agrochemicals, including imidacloprid (Fernandez-Perez et al. 1998), phenylurea (Cotterill and Wilkins 1996), fluometuron (Zhao and Wilkins 2003), and diuron (Yearla and Padmasree 2016). However, more recent studies have reported important improvements in the properties of lignin NPs when coated with cationic polyelectrolytes, such as poly(diallyldimethylammonium chloride). This coating allows an increase in the stability toward strongly basic pH (Richter et al. 2016). On the other hand, the use of different contents of ethylcellulose coating onto lignin granules increases the efficiency of chloridazon, which is released gradually after 800 h (Fernández-Pérez et al. 2011).

14.3.2.3 Cellulose

Cellulose is the most abundant renewable polymer, which possesses mechanical strength, biocompatibility, hydrophilicity, and high sorption capacity, making it a great nanomaterial in delivery systems (Qiu and Hu 2013). For instance, a nanocomposite composed by biogenic nanosilica and nanofibrillated cellulose was used in the controlled release of tebuconazole. The loaded nanoparticles were blended with cellulose nanofibrils. However, the release rate was slower in cellulose nanofibrils in comparison with nanocomposite (Mattos and Magalhães 2016). Carboxymethylcellulose has also employed in the nanoencapsulation of herbicide clodinafop-propyrgyl in polymeric core-shell nanoparticles (Kumar et al. 2015).

14.3.2.4 Others

Some other biomaterials/biopolymers have been used as a matrix for the delivery of pesticides. For instance, Anjali et al. (2010) developed a polymer-free biocide based on nanopermethrin as larvicidal. Nanopermethrin was prepared by evaporation of oil in water microemulsion. The particle size of nanodispersion in water was 151. The studies were carried out against *Culex quinquefasciatus* showing a 100% of mortality within 6 h. The amphiphilic polymer poly(ethylene glycol) (PEG) was employed to encapsulate several pesticides such as thiram (Kaushik et al. 2013), imidacloprid (Adak et al. 2012), beta-cyfluthrin (Loha et al. 2011), carbofuran (Pankaj et al. 2012), and thiamethoxam (Sarkar et al. 2012). Furthermore, the polymers act as barriers to moisture, diminishing the rate of deterioration in seeds (Kaushik et al. 2013). On the other hand, Memarizadeh et al. (2014b) reported the encapsulation of imidacloprid and indoxacarb (Memarizadeh et al. 2014a) by direct linear triblock of a copolymer composed of poly(citric acid)-poly(ethylene glycol)-poly(citric acid).

The activity of fungicides tebuconazole and chlorothalonil were also tested by their incorporation into a polymeric nanoparticle composed by polyvinylpyridine-co-styrene (100–250 nm), showing a decrease of the activity as the content of styrene. The nanoparticles suspended in water were introduced into solid southern pine sapwood and then exposed to the fungus *Gloeophyllum trabeum*, resulting in the fungal resistance of the wood (Liu et al. 2001).

The use of lipids has also been employed as delivery systems of pesticides. For instance, de Oliveira et al. (2015) developed solid lipid nanoparticles containing the herbicides atrazine and simazine. The formulations were useful in pre- and post-treatment of *Raphanus raphanistrum*. Chitosan-coated nanoliposomes containing etofenprox or alpha-cypermethrin have also been reported (Bang et al. 2011).

Sodium alginate is another nanomaterial tested as nanopesticide. Kumar et al. (2014) developed sodium alginate nanoparticles by water-in-oil emulsion for imidacloprid for the treatment of okra (bhindi) crop. Meanwhile, Saini et al. (2014) evaluated the larvicide activity of pyridalyl, against of *Helicoverpa armigera*, in a nanosuspension of sodium alginate in comparison with bulk material and commercial formulation.

Zein is a protein from corn which is biocompatible, biodegradable and has low environmental toxicity. This protein was proposed by Oliveira et al. (2018) as a carrier for repellents geraniol and R-citronellal, against *Tetranychus urticae*. Bail et al. (2015) propose starch-based hydrogels as carriers of carbendazim. Meanwhile, Chariou and Steinmetz (2017) established the use of tobacco mild green mosaic virus as a carrier of the anthelmintic drug crystal violet for *Caenorhabditis elegans* nematodes.

14.3.2.5 Bio-coated Metallic and Nonmetallic Nanoparticles

Some nanoparticles such as Ag, Cu, Au, and silica work as fungicide and bactericide carriers; however information regarding the effect of these nanomaterials on plants indicates that its exposure generates suppression of growth (impact on germination and elongation of the seed) and reduces content of chlorophyll and carotenoids, absorption of nutrients, and generation of ROS (Feregrino-perez et al. 2018). Thus, some investigations have reported the use of bio-coatings/functionalization to diminish the aforementioned adverse effects. The coating of nanoparticles can be performed during/after synthesis. In general, the coating provides new physicochemical properties (size, charge, morphology) that determine their fate in an organism (Dias et al. 2011). A list of bio-coated nanopesticides is summarized in Table 14.3.

For instance, Park et al. (2006) reported the use of nanoparticles of silver combined with silica molecules and water-soluble polymer against fungi *R. solani*, *B. cinerea*, *M. grisea*, *Pythium ultimum*, and *Colletotrichum gloeosporioides*. On the other hand, Kanhed et al. (2014) established the synthesis of CuNPs coated with cetyltrimethylammonium bromide (CTAB) against fungi *C. lunata*, *A. alternata*, *F. oxysporum*, and *Phoma destructive*.

The bio-coating achieved by the polymerization of citric acid onto the surface of oxidized multi-walled carbon nanotubes (MWCNT-*g*-PCA) enabled the nanomaterial to be soluble in water and to trap water-soluble chemical species and metal ions. Pesticides such as zineb and mancozeb were encapsulated, showing a better performance against *Alternaria alternata* fungi in comparison with bulk pesticide (Sarlak et al. 2014).

Furthermore, biogenic silica derived from *Equisetum arvense* was synthesized and employed as a carrier a neem bark extract cross-linked with polycarboxylic acids. In comparison with the unloaded biogenic silica nanoparticles, an enhancement in the thermal stability was observed as well a successful development against *Acromyrmex crassispinus* ant (Mattos et al. 2017a).

14.3.3 Bio-coated and Natural Nanoparticles as Smart Packages

Food packaging represents one of the main concerns in food industries since it plays a crucial role in food quality (Carbone et al. 2016). In this sense, nanotechnology has provided a series of nanomaterials that enhance the protection of the product and prolong their shelf life. These materials coupled with the use of biodegradable polymers in food packaging have attracted much interest due to their biocompatibility and environmentally friendly properties (Elsabee and Abdou 2013; Dhall 2013; Prasad et al. 2017c). For example, Ag NPs have been incorporated into cellulose for the protection of beef meat (Fernandez et al. 2010a) and fresh-cut melon (Fernandez et al. 2010b). AgNPs have also been incorporated into sodium alginate for the protection of pears and carrots against *E. coli* and *Staphylococcus aureus* (Fayaz et al. 2009), while their incorporation into pullulan has produced an edible coating for meat that protects it against *Listeria monocytogenes* and *S. aureus* (Morsy et al. 2014; Khalaf et al. 2013). Not only other natural biopolymers, such as hydroxypropyl methylcellulose (de Moura et al. 2012) and starch (Abreu et al. 2015), have been used successfully but also alternative forms of silver have demonstrated their

Table 14.3 Bio-coated nanoparticles employed as delivery systems of nanopesticides

Bio-coating	NP	Method of functionalization	Architecture	Pest	Biocide	References
Neem	Silica	Cross-linked with polycarboxylic acids (1,2,3,4-butanetetracarboxylic acid	Nanoparticle	Biocide/ant specie, *Acromyrmex crassispinus*	Neem bark extract	Mattos et al. (2017a)
Citric acid	Multi-walled carbon nanotubes	Polymerization	nanotubes	*Alternaria alternata* fungi	Zineb and mancozeb	Sarlak et al. (2014)
CTAB	Cu	Polymerization	Nanoparticles	*C. lunata*, *A. alternata*, *F. oxysporum*, and *Phoma destructive*	Cu	Kanhed et al. (2014)
Silica	Ag	Polymerization	Nanoparticles	*R. solani*, *B. cinerea*, *M. grisea*, *Pythium ultimum*, *and Colletotrichum gloeosporioides*	Ag	Park et al. (2006)

application in food industries. For instance, silver montmorillonite NPs were used to evaluate the prolongation of shelf life in fresh-cut carrots (Costa et al. 2012), fresh fruit salad (Costa et al. 2011), and Fior di Latte cheese (Gammariello et al. 2011; Mastromatteo et al. 2015). However, despite the notorious extension in the shelf life of these products, sensorial parameters were crucial to determine the product acceptability.

Additional materials have been used for smart packages, such as ZnO NPs, which have also been employed as an antimicrobial agent in food packages for the preservation of fresh-cut apples (Li et al. 2011), kiwifruit (Meng et al. 2014), and meat (Khalaf et al. 2013; Panea et al. 2014).

Despite the full range of benefits from the use of NPs on food packaging, an increasing number of studies have exposed the migration of NPs into food. The rate of release of NPs into food has been demonstrated to be directly dependent on the type of food and its temperature of storage. While this migration has been related to sensory changes, it may as well present potential toxicity to human health (Echegoyen and Nerín 2013). To avoid these drawbacks, new strategies and nanomaterials are being studied, for instance, the development of novel nanoemulsions used toward the encapsulation of functional compounds (Silva et al. 2012) or the development of composite co-coatings of hydrocolloids and lipids, combining the structural properties of hydrocolloids and the water barrier of lipids (Yousuf et al. 2018). Finally, there is also the coupling of bacteriocins into NPs, providing them with the capacity to inhibit the growth of microorganisms. Some bacteriocins used to date are nisin (extracted from *Lactococcus lactis*) and carnocyclin A (extracted from *Carnobacterium maltaromaticum*) (O'Connor et al. 2015).

14.4 Conclusions and Future Work

Certainly, nanotechnology has played a tremendous role in agriculture and food industries during the last decade. The wide range of nanomaterials and their applications have increased the production yield of crops, the quality of food, and their shelf life. While the development of novel strategies on food packaging based on nanotechnology is still in progress, the use of NPs in combination with biopolymers, bio-nanocomposites, and nanoemulsions has demonstrated to be very efficient to inhibit the growth of harmful microorganisms. The encapsulation of bacteriocins should be highlighted in the future.

Despite the evident benefit of NPs in agriculture and food industries, the information about their risk and potentially harmful effects on human health is still limited. Therefore, more studies are required on the mobility of NPs during food packaging, their migration through the food chain, as well as the potential uptake of nanoformulations by farmers during field applications. Thus, the development of biocompatible and biodegradable carriers, such as polysaccharides and proteins, should be furtherly studied to create new responsive (pH, salinity, temperature) delivery systems or smart packaging.

Finally, further and extensive studies addressing cost-effectiveness and risk assessment need to be performed to create better legislation for the imminent use of nanotechnology in our crop fields, food products, and daily life.

References

Abreu AS, Oliveira M, de Sá A et al (2015) Antimicrobial nanostructured starch based films for packaging. Carbohydr Polym 129:127–134

Adak T, Shakil NA, Walia S (2012) Development of controlled release formulations of imidacloprid employing novel nano-ranged amphiphilic polymers. Environ Sci Health B 47:217–225

Agarwal H, Kumar SV, Rajeshkumar S (2017) A review on green synthesis of zinc oxide nanoparticles-an eco-friendly approach. Resour Technol 3:406–413

Anjali CH, Khan SS, Margulis-Goshen K et al (2010) Formulation of water-dispersible nanopermethrin for larvicidal applications. Ecotoxicol Environ Saf 73:1932–1936

Bai C, Zhang S, Huang L et al (2015) Starch-based hydrogel loading with carbendazim for controlled-release and water absorption. Carbohydr Polym 125:376–383

Bang SH, Hwang IC, Yu YM et al (2011) Influence of chitosan coating on the liposomal surface on physicochemical properties and the release profile of nanocarrier systems. J Microencapsul 28:595–604

Carbone M, Tommasa D, Sabbatella G (2016) Silver nanoparticles in polymeric matrices for fresh food packaging. J King Saud Univ Sci 28:273–279. https://doi.org/10.1016/j.jksus.2016.05.004

Chariou PL, Steinmetz NF (2017) Delivery of pesticides to plant parasitic nematodes using tobacco mild green mosaic. ACS Nano 11:4719–4473. https://doi.org/10.1021/acsnano.7b00823

Chhipa H (2017) Nanofertilizers and nanopesticides for agriculture. Environ Chem Lett 15:15–22. https://doi.org/10.1007/s10311-016-0600-4

Ciolacu D, Oprea AM, Anghel N et al (2012) New cellulose-lignin hydrogels and their application in controlled release of polyphenols. Mater Sci Eng C 32:452–463

Corradini E, de Moura MR, Mattoso LHC (2010) A preliminary study of the incorporation of NPK fertilizer into chitosan nanoparticles. Express Polym Lett 4:509–515

Corredor E, Testillano P, Coronado M et al (2009) Nanoparticle penetration and transport in living pumpkin plants: in situ subcellular identification. Plant Biol 9:45

Costa C, Conte A, Buonocore GG, Del Nobile MA (2011) Antimicrobial silver-montmorillonite nanoparticles to prolong the shelf life of fresh fruit salad. Int J Food Microbiol 148:164–167. https://doi.org/10.1016/j.ijfoodmicro.2011.05.018

Costa C, Conte A, Buonocore GG et al (2012) Calcium-alginate coating loaded with silver-montmorillonite nanoparticles to prolong the shelf-life of fresh-cut carrots. Food Res Int 48:164–169. https://doi.org/10.1016/j.foodres.2012.03.001

Cotterill JV, Wilkins RM (1996) Controlled release of phenylurea herbicides from a lignin matrix: release kinetics and modification with urea. J Agric Food Chem 44:2908–2912

de Moura MR, Mattoso LHC, Zucolotto V (2012) Development of cellulose-based bactericidal nanocomposites containing silver nanoparticles and their use as active food packaging. J Food Eng 109:520–524

de Oliveira JL, Campos EVR, da Silva CMG et al (2015) Solid lipid nanoparticles co-loaded with simazine and atrazine: preparation, characterization, and evaluation of herbicidal activity. J Agric Food Chem 63:422–432

De Oliveira JL, Campos VR, Pereira AES et al (2018) Zein nanoparticles as eco-friendly carrier systems for botanical repellents aiming sustainable agriculture. J Agric Food Chem 66:1330–1340. https://doi.org/10.1021/acs.jafc.7b05552

Dhall RK (2013) Advances in edible coatings for fresh fruits and vegetables: a review. Crit Rev Food Sci Nutr 53:435–450. https://doi.org/10.1080/10408398.2010.541568

Dias AM, Hussain A, Marcos AS, Roque AC (2011) A biotechnological perspective on the application of iron oxide magnetic colloids modified with polysaccharides. Biotechnol Adv 29:142–155

dos Santos Silva, Cocenza DS M, Grillo R, De NFS M et al (2011) Paraquat-loaded alginate/chitosan nanoparticles: preparation, characterization and soil sorption studies. J Hazard Mater 190:366–374

Duhan JS, Kumar R, Kumar N et al (2017) Nanotechnology: the new perspective in precision agriculture. Biotechnol Rep 15:11–23. https://doi.org/10.1016/j.btre.2017.03.002

Echegoyen Y, Nerín C (2013) Nanoparticle release from nano-silver antimicrobial food containers. Food Chem Toxicol 62:16–22

Elsabee MZ, Abdou ES (2013) Chitosan based edible films and coatings: a review. Mater Sci Eng C 33:1819–1841. https://doi.org/10.1016/j.msec.2013.01.010

Espirito A, Pereira S, Mayara P et al (2017) Chitosan nanoparticles as carrier systems for the plant growth hormone gibberellic acid. Colloids Surf B Biointerfaces 150:141–152. https://doi.org/10.1016/j.colsurfb.2016.11.027

Fan L, Jin R, Le X et al (2012) Chitosan microspheres for controlled delivery of auxins as agrochemicals. Microchim Acta 176:381–387

FAO (2017) The future of food and agriculture – trends and challenges, Rome

Fayaz AM, Balaji K, Girilal M, Kalaichelvan PT, Venkatesan R (2009) Mycobased synthesis of silver nanoparticles and their incorporation into sodium alginate films for vegetable and fruit preservation. J Agric Food Chem 57:6246–6252

Feng BH, Peng LF (2012) Synthesis and characterization of carboxymethyl chitosan carrying ricinoleic functions as an emulsifier for azadirachtin. Carbohydr Polym 88:576–582

Feregrino-Perez AA, Magaña-López E, Guzmán C, Esquivel K (2018) A general overview of the benefits and possible negative effects of the nanotechnology in horticulture. Sci Hortic (Amsterdam) 238:126–137. https://doi.org/10.1016/j.scienta.2018.03.060

Fernandez A, Picouet P, Lloret E (2010a) Cellulose-silver nanoparticle hybrid materials to control spoilage-related microflora in absorbent pads located in trays of fresh-cut melon. Int J Food Microbiol 142:222–228

Fernandez A, Picouet P, Lloret E (2010b) Reduction of the spoilage-related microflora in absorbent pads by silver nanotechnology during modified atmosphere packaging of beef meat. J Food Prot 73:2263–2269

Fernández-Fernández M, Sanromán MÁ, Moldes D (2013) Recent developments and applications of immobilized laccase. Biotechnol Adv 31:1808–1825. https://doi.org/10.1016/j.biotechadv.2012.02.013

Fernandez-Perez MF, Gonzalez-Pradas E, Urena-Amate M (1998) Controlled release of imidacloprid from a lignin matrix: water release kinetics and soil mobility study. J Agric Food Chem 46:3828–3834

Fernández-Pérez M, Villafranca-Sánchez M, Flores-Céspedes F, Daza-Fernández I (2011) Ethylcellulose and lignin as bearer polymers in controlled release formulations of chloridazon. Carbohydr Polym 83:1672–1679

Gammariello D, Conte A, Buonocore GG, Del Nobile MA (2011) Bio-based nanocomposite coating to preserve quality of Fior di latte cheese. J Dairy Sci 94:5298–5304

Ganesh R, Karuppusamy I, Dattatraya G et al (2018) A comprehensive review on green nanomaterials using biological systems: recent perception and their future applications. Colloids Surf B Biointerfaces 170:20–35. https://doi.org/10.1016/j.colsurfb.2018.05.045

Grillo R, Pereira AE, Nishisaka CS et al (2014) Chitosan/tripolyphosphate nanoparticles loaded with paraquat herbicide: an environmentally safer alternative for weed control. J Hazard Mater 278:163–171

Guan H, Chi D, Yu J, Li X (2008) A novel photodegradable insecticide: preparation, characterization and properties evaluation of nano-Imidacloprid. Pestic Biochem Physiol 92:83–91

Hasaneen MNA, Abdel-Aziz HMM, El-Bialy D, Omer AM (2014) Preparation of chitosan nanoparticles for loading with NPK fertilizer. Afr J Biotechnol 13:3158–3164. https://doi.org/10.5897/AJB2014.13699

Hidayat R, Fadillah G, Chasanah U et al (2015) Effectiveness of urea nanofertilizer based aminopropyltrimethoxysilane (APTMS)-zeolite as slow release fertilizer system. Afr J Agric Res 10:1785–1788
Hossain K, Monreal M, Sayari K (2008) Adsorption of urease on PE-MCM-41 and its catalytic effect on hydrolysis of urea. Colloid Surf B 62:42–50
Hussain MR, Devi RR, Maji TK (2012) Controlled release of urea from chitosan microspheres prepared by emulsification and cross-linking method. Iran Polym J 21:473–479
Joseph T, Morrison M (2006) Nanotechnology in agriculture food: a nanoforum report. www.nanoforum.org
Kanhed P, Birla S, Gaikwas S et al (2014) In vitro antifungal efficacy of copper nanoparticles against selected crop pathogenic fungi. Mater Lett 115:13–17
Kashyap PL, Kumar S, Srivastava AK (2017) Nanodiagnostics for plant pathogens. Environ Chem Lett 15:7–13. https://doi.org/10.1007/s10311-016-0580-4
Kaushik P, Shakil NA, Kumar J et al (2013) Development of controlled release formulations of thiram employing amphiphilic polymers and their bioefficacy evaluation in seed quality enhancement studies. J Environ Sci Health B 48:677–685
Khalaf HH, Sharoba AM, El-Tanahi HH, Morsy MK (2013) Stability of antimicrobial activity of pullulan edible films incorporated with nanoparticles and essential oils and their impact on Turkey deli meat quality. J Food Dairy Sci 4:557–573
Khater M, De la Escosura-Muñiz A, Merkoçi A (2017) Biosensors for plant pathogen detection. Biosens Bioelectron 93:72–86. https://doi.org/10.1016/j.bios.2016.09.091
Kim SJ, Bawendi MG (2003) Oligomeric ligands for luminescent and stable nanocrystal quantum Dots. J Am Chem Soc 125:14652–14653
Koh I, Wang X, Varughese B et al (2006) Magnetic iron oxide nanoparticles for biorecognition: evaluation of surface coverage and activity. J Phys Chem B 110:1553–1558
Kottegoda N, Munaweera I, Madusanka N, Karunaratne V (2011) A green, slow-release fertilizer composition based on urea-modified hydroxyapatite nanoparticles encapsulated wood. Curr Sci 101:1–7
Kowshik M, Deshmukh N, Vogel W et al (2002) Microbial synthesis of semiconductor CdS nanoparticles, their characterization, and their use in the fabrication of an ideal diode. Biotechnol Bioeng 78:583–588
Kumar S, Bhanjana G, Sharma A et al (2014) Synthesis, characterization and on field evaluation of pesticide loaded sodium alginate nanoparticles. Carbohydr Polym 101:1061–1067
Kumar S, Bhanjana G, Sharma A et al (2015) Herbicide loaded carboxymethyl cellulose nanocapsules as potential carrier in agrinanotechnology. Sci Adv Mater 7:1143–1148
Kumar D, Singh S, Singh S et al (2017) An overview on manufactured nanoparticles in plants: uptake, translocation, accumulation and phytotoxicity. Plant Physiol Biochem 110:2–12. https://doi.org/10.1016/j.plaphy.2016.07.030
Kuppusamy P, Yusoff MM, Maniam GP, Govindan N (2016) Biosynthesis of metallic nanoparticles using plant derivatives and their new avenues in pharmacological applications–an updated report. Saudi Pharm J 24:473–484
Lao SB, Zhang ZX, Xu HH, Jiang GB (2010) Novel amphiphilic chitosan derivatives: synthesis, characterization and micellar solubilization of rotenone. Carbohydr Polym 82:1136–1142
Leng Y, Fu L, Ye L et al (2016) Protein-directed synthesis of highly monodispersed, spherical gold nanoparticles and their applications in multidimensional sensing. Sci Rep 6:28900
Li L, Wartchow CA, Danthi SN et al (2004) A novel antiangiogenesis therapy using an integrin antagonist or anti-Flk-1 antibody coated 90Y-labeled nanoparticles. Int J Radiat Oncol Biol Phys 58:1215–1227
Li Y, Xu X, Deng C et al (2007) Immobilization of trypsin on superparamagnetic nanoparticles for rapid and effective proteolysis. J Proteome Res 6:3849–3855
Li X, Li W, Jiang Y et al (2011) Effect of nano-ZnO coated active packaging on quality of fresh-cut 'Fuji' apple. Int J Food Sci Technol 46:1947–1955
Liu Y, Yan L, Heiden P, Laks P (2001) Use of nanoparticles for controlled release of biocides in solid wood. J Appl Polym Sci 79:458–465

Loha KM, Shakil NA, Kumar J et al (2011) Release kinetics of beta-cyfluthrin from its encapsulated formulations in water. J Environ Sci Health B 46:201–206
Mastromatteo M, Conte A, Lucera A et al (2015) Packaging solutions to prolong the shelf life of Fiordilatte cheese: bio-based nanocomposite coating and modified atmosphere packaging. LWT Food Sci Technol 60:230–237. https://doi.org/10.1016/j.lwt.2014.08.013
Mattos BD, Magalhães WLE (2016) Biogenic nanosilica blended by nanofibrillated cellulose as support for slow-release of tebuconazole. J Nanopart Res 18:1–10
Mattos BD, Rojas OJ, Magalh WLE (2017a) Biogenic silica nanoparticles loaded with neem bark extract as green, slow-release biocide. J Clean Prod 142:4206–4213. https://doi.org/10.1016/j.jclepro.2016.11.183
Mattos BD, Tardy BL, Magalhães WLE, Rojas OJ (2017b) Controlled release for crop and wood protection: recent progress toward sustainable and safe nanostructured biocidal systems. J Control Release 262:139–150. https://doi.org/10.1016/j.jconrel.2017.07.025
Memarizadeh N, Ghadamyari M, Adeli M, Talebi K (2014a) Preparation, characterization and efficiency of nanoencapsulated imidacloprid under laboratory conditions. Ecotoxicol Environ Saf 107:77–83
Memarizadeh N, Ghadamyari M, Adeli M, Talebi K (2014b) Linear-dendritic copolymers/indoxacarb supramolecular systems: biodegradable and efficient nano-pesticides. Environ Sci Process Impacts 16:2380–2389
Meng X, Zhang M, Adhikari B (2014) The effects of ultrasound treatment and nano-zinc oxide coating on the physiological activities of fresh-cut kiwifruit. Food Bioprocess Technol 7:126–132
Merino D, Casalongué C, Alvarez VA (2018) Polysaccharides as eco-nanomaterials for agricultural applications polysaccharides as eco-nanomaterials for agricultural applications. In: Martínez LMT et al (eds) Handbook of ecomaterials. Springer Nature Switzerland AG
Morsy MK, Khalaf HH, Sharoba AM et al (2014) Incorporation of essential oils and nanoparticles in pullulan films to control foodborne pathogens on meat and poultry products. J Food Sci 79:M675–M682
Neethirajan S, Ragavan KV, Weng X (2018) Agro-defense: biosensors for food from healthy crops and animals. Trends Food Sci Technol 73:25–44. https://doi.org/10.1016/j.tifs.2017.12.005
Ni B, Liu M, Lu S et al (2011) Environmentally friendly slow release nitrogen fertilizer. J Agric Food Chem 59:10169–10175
O'Connor PM, Ross RP, Hill C, Cotter PD (2015) Antimicrobial antagonists against food pathogens: a bacteriocin perspective. Curr Opin Food Sci 2:51–57
Ocsoy I, Tasdemir D, Mazicioglu S et al (2018) Biomolecules incorporated metallic nanoparticles synthesis and their biomedical applications. Mater Lett 212:45–50. https://doi.org/10.1016/j.matlet.2017.10.068
Pandey RK, Prajapati VK (2018) Molecular and immunological toxic effects of nanoparticles. Int J Biol Macromol 107:1278–1293. https://doi.org/10.1016/j.ijbiomac.2017.09.110
Panea B, Ripoll G, González J et al (2014) Effect of nanocomposite packaging containing different proportions of ZnO and Ag on chicken breast meat quality. J Food Eng 123:104–112. https://doi.org/10.1016/j.jfoodeng.2013.09.029
Pankaj, Shakil NA, Kumar J et al (2012) Bioefficacy evaluation of controlled release formulations based on amphiphilic nano-polymer of carbofuran against *Meloidogyne incognita* infecting tomato. J Environ Sci Health B 47:520–528
Park HJ, Kim SH, Kim HJ, Choi SH (2006) A new composition of nanosized silica-silver for control of various plant diseases. Plant Pathol J 22:295–302
Paula HC, Sombra FM, Cavalcante d F, Abreu FO R, de Paula RC (2011) Preparation and characterization of chitosan/cashew gum beads loaded with *Lippia sidoides* essential oil. Mater Sci Eng C 31:173–178
Prasad R, Kumar V, Prasad KS (2014) Nanotechnology in sustainable agriculture: present concerns and future aspects. Afr J Biotechnol 13(6):705–713
Prasad R (2016) Advances and Applications through Fungal Nanobiotechnology. Springer, International Publishing Switzerland (ISBN: 978-3-319-42989-2)

Prasad R (2017) Fungal Nanotechnology: Applications in Agriculture, Industry, and Medicine. Springer Nature Singapore Pte Ltd. (ISBN 978-3-319-68423-9)
Prasad R, Bhattacharyya A, Nguyen QD (2017a) Nanotechnology in sustainable agriculture: Recent developments, challenges, and perspectives. Front Microbiol 8:1014. https://doi.org/10.3389/fmicb.2017.01014
Prasad R, Kumar M, Kumar V (2017b) Nanotechnology: An Agriculture paradigm. Springer Nature Singapore Pte Ltd. (ISBN: 978-981-10-4573-8)
Prasad R, Kumar V and Kumar M (2017c) Nanotechnology: Food and Environmental Paradigm. Springer Nature Singapore Pte Ltd. (ISBN 978-981-10-4678-0)
Prasad R, Pandey R, Barman I (2016) Engineering tailored nanoparticles with microbes: quo vadis. WIREs Nanomed Nanobiotechnol 8:316–330. doi: 10.1002/wnan.1363
Prasad R, Jha A and Prasad K (2018) Exploring the Realms of Nature for Nanosynthesis. Springer International Publishing (ISBN 978-3-319-99570-0) https://www.springer.com/978-3-319-99570-0
Qiu X, Hu S (2013) "Smart" materials based on cellulose: a review of the preparations, properties, and applications. Materials (Basel) 6:738–781. https://doi.org/10.3390/ma6030738
Quiñones JP, García YC, Curiel H, Covas CP (2010) Microspheres of chitosan for controlled delivery of brassinosteroids with biological activity as agrochemicals. Carbohydr Polym 80:915–921
Ranjan S, Dasgupta N, Lichtfouse E (eds) (2017) Nanoscience in food and agriculture 5. Springer Nature Switzerland AG
Ranjitha K, Rao DVS, Shivashankara KS et al (2017) Shelf-life extension and quality retention in fresh-cut carrots coated with pectin. Innov Food Sci Emerg Technol 42:91–100. https://doi.org/10.1016/j.ifset.2017.05.013
Reyes-Ortega F, Delgado A, Schneider E et al (2018) Magnetic nanoparticles coated with a thermosensitive polymer with hyperthermia properties. Polymers (Basel) 10:2–15. https://doi.org/10.3390/polym10010010
Richter AP, Bharti B, Armstrong HB et al (2016) Synthesis and characterization of biodegradable lignin nanoparticles with tunable surface properties. Langmuir 32:6468–6477
Saharan V, Mehrotra A, Khatik R et al (2013) Synthesis of chitosan based nanoparticles and their in vitro evaluation against phytopathogenic fungi. Int J Biol Macromol 62:677–683
Saini P, Gopal M, Kumar R, Srivastava C (2014) Development of pyridalyl nanocapsule suspension for efficient management of tomato fruit and shoot borer (Helicoverpa armigera). J Environ Sci Health B 49:344–351
Sarkar DJ, Kumar J, Shakil NA, Walia S (2012) Release kinetics of controlled release formulations of thiamethoxam employing nano-ranged amphiphilic PEG and diacid based block polymers in soil. J Environ Sci Health A Tox Hazard Subst Environ Eng 47:1701–1712
Sarlak N, Taherifar A, Salehi F (2014) Synthesis of nanopesticides by encapsulating pesticide nanoparticles using functionalized carbon nanotubes and application of new nanocomposite for plant disease treatment. J Agric Food Chem 62:4833–4838
Sharmila RC (2010) Nutrient release pattern of nano-fertilizer formulations (Ph.D.Thesis). Tamil Nadu Agricultural University, Coimbatore, Tamil Nadu, India,
Shaviv A (2000) Advances in controlled-release fertilizers. Adv Agron 71:1–49. https://doi.org/10.1016/S0065-2113(01)71011-5
Shen Q, Wang X, Fu D (2008) The amplification effect of functionalized gold nanoparticles on the binding of anticancer drug dacarbazine to DNA and DNA bases. Appl Surf Sci 255:577–580
Silva HD, Cerqueira MA, Vicente AA (2012) Nanoemulsions for food applications: development and characterization. Food Bioprocess Technol 5:854–867
Subbiah R, Veerapandian M, Yun KS (2010) Nanoparticles: functionalization and multifunctional applications in biomedical sciences. Curr Med Chem 17:4559–4577
Subramanian K, Manikandan A, Thirunavukkarasu M, Rahale-Sharmila C (2015) Nanofertilizers for balanced crop nutrition. In: Rai M et al (eds) Nanotechnologies in food and agriculture. Springer, Cham, pp 69–80
Subramanuan K, Tarafdar J (2011) Prospects of nanotechnology in Indian farming. Indian J Agric Sci 81:887–893

Sun C, Shu K, Wang W et al (2014) Encapsulation and controlled release of hydrophilic pesticide in shell cross-linked nanocapsules containing aqueous core. Int J Pharm 463:108–114

Tao S, Pang R, Chen C et al (2012) Synthesis, characterization and slow release properties of O-naphthylacetyl chitosan. Carbohydr Polym 88:1189–1194

Wilson MA, Tran NH, Milev AS et al (2008) Nanomaterials in soils. Geoderma 146:291–302. https://doi.org/10.1016/j.geoderma.2008.06.004

Wirges CT, Timper J, Fischler M et al (2009) Controlled nucleation of DNA metallization. Angew Chem Int Ed 48:219–223

Xu L, Cao LD, Li FM et al (2014) Utilization of chitosanlactide copolymer nanoparticles as controlled release pesticide carrier for pyraclostrobin against *Colletotrichum gossypii* Southw. J Dispers Sci Technol 35:544–550

Yearla SR, Padmasree K (2016) Exploitation of subabul stem lignin as a matrix in controlled release agrochemical nanoformulations: a case study with herbicide diuron. Environ Sci Pollut Res 23:18085–18098

Yousuf B, Sha O, Kumar A (2018) Recent developments in shelf-life extension of fresh-cut fruits and vegetables by application of di ff erent edible coatings: a review. LWT Food Sci Technol 89:198–209. https://doi.org/10.1016/j.lwt.2017.10.051

Zhao J, Wilkins RM (2003) Controlled release of the herbicide, fluometuron, from matrix granules based on fractionated organosolv lignins. J Agric Food Chem 51:4023–4028

Chapter 15
Green Synthesis Approaches of Nanoagroparticles

Lilian Rodrigues Rosa Souza, Argus Cezar da Rocha Neto, César Rodrigues da Silva, Leonardo Pereira Franchi, and Tiago Alves Jorge de Souza

Contents

L. R. R. Souza
Department of Chemistry, FFCLRP-USP, University of São Paulo-USP, Ribeirão Preto, SP, Brazil

A. C. da Rocha Neto · C. R. da Silva
Department of Agronomic Engineering, Adventist University of São Paulo-UNASP, Engenheiro Coelho, SP, Brazil

L. P. Franchi
Department of Genetics, FMRP-USP, São Paulo University-USP, Ribeirão Preto, SP, Brazil

T. A. J. de Souza (✉)
Department of Agronomic Engineering, Adventist University of São Paulo-UNASP, Engenheiro Coelho, SP, Brazil

Department of Genetics, FMRP-USP, São Paulo University-USP, Ribeirão Preto, SP, Brazil

R. Prasad et al. (eds.), *Nanobiotechnology in Bioformulations*, Nanotechnology in the Life Sciences, https://doi.org/10.1007/978-3-030-17061-5_15

15.1 Introduction

The world population grew from 2.5 billion in 1950 to 7.6 billion in 2017. By the year 2050, the world population would have an estimate of 9 billion (United Nation 2017). An adult person needs on average 2900 kcal per day. In theory, the current world food production can only provide in average 2800 kcal per day per capita, nearly sufficient to feed the world population. However, it is important to state that this division is far from equal in the different countries. Whereas in developed countries the daily average consumption of food provides 3500 kcal per capita, in poor countries, people may not obtain even 2000 kcal/day (Carvalho 2006).

In this sense, the increase of the global food supplies requires the modernization and intensification of agricultural techniques since the farmlands for expanding the agriculture are not as available as decades ago. Intensifying agriculture involves the use of several knowledge fields, such as biotechnology, botany, chemistry, genetics, and microbiology, among others, in order to improve crop varieties and tolerance to environmental factors and also microorganisms (Schreinemachers and Tipraqsa 2012).

Currently, field crops and their products are routinely treated with synthetic products (Frankova et al. 2016) such as insecticides, fungicides, bactericides, herbicides, and others in order to help diminish the losses caused by animal pests (e.g., insects, mites, nematodes, and rodents) and plant pathogens (e.g., bacteria, fungi and viruses, or weeds). Although these modern technologies aimed to decrease field losses and increase the shelf life of many agricultural products, their systematical usage is known to increase the risk of poisoning non-target organisms (Eddleston and Bateman 2012), where farmers and consumers in developing countries are much more vulnerable to the risks arising from them, e.g., cancer and other diseases, than those from developed countries (WHO 2017). Even after the publication of *The International Code of Conduct on the Distribution and Use of Pesticides* by the Food and Agriculture Organization of the United Nations (FAO 2006), which supports the increase of food security while at the same time protects human health and environment, the usage of pesticides increased year after year (Dich et al. 1997; Eddleston et al. 2002; Schreinemachers and Tipraqsa 2012; Mew et al. 2017).

Carvalho (2006) stated that both developed and developing countries are moving in opposite directions in terms of synthetic pesticide use, i.e., while developed countries tend to use fewer pesticides than before, developing countries are using more. This corroborates the Schreinemachers and Tipraqsa (2012) study where it was suggested that a 1% increase in crop output per hectare was associated with a 1.8% increase in pesticide use per hectare.

15.1.1 Nanoagroparticles

The search for methods to decrease or even retard field losses and increase the shelf life of agricultural products is of great interest to all sectors involved in their production and preservation, and thus, many different solutions have been proposed.

In this sense, the use of nanotechnology in the agricultural field has the desired potential to make a positive impact by alleviating the problems related to the synthetic pesticides as aforementioned.

Nanoagroparticles can be defined as nano-sized particles that can act efficiently as fungicides, insecticides, herbicides, pesticides, and also plant growth-promoting factors (Baker et al. 2017; Sangeetha et al. 2017). One of the major concerns related to the production of these new nanoparticles (NPs) resides in the green synthetic approaches, i.e., creating them without any use of toxic elements during its synthesis. Once most of the methods often bound with different implications, e.g., high heat, toxic elements, or sophisticated facilities (Kavitha et al. 2013) the scientific communities are engaged in employing biogenic sources for its synthesis, e.g., plants, microbes, or their products, which can act as reductive agents and mediate the synthesis or even stabilize the nanoagroparticles (Baker et al. 2017; Prasad 2014, Prasad et al. 2016, 2018).

15.1.1.1 Nanoagroparticles Types

Nanoagroparticles have different sizes and functions depending on the methods involved in their synthesis. Therefore, the choice of the correct method becomes essential to obtain nanoagroparticles (e.g., fungicides, bactericides, insecticides, herbicides, carriers and sensors) with different characteristics, mechanisms of actions and environments, increasing their potential and diminishing their concentration (Baker et al. 2017) (Table 15.1). In this manner, nanoagroparticles can be considered as a single product or the mixture of at least two different products which will enhance the bioavailability of the active ingredients while avoiding the use of several adjuvants which may be toxic for non-target organisms (Anjali et al. 2010; Iavicoli et al. 2017; Suresh et al. 2013).

Several methods can be used to obtain nanoagroparticles with different purposes, e.g., nanomycosynthesis, nanoplantsynthesis, polymeric nanospheres, nanocapsules, nanogels, and nanofibers can be used as formulations in order to slow and control the release of the desired active ingredients, serving as protective reservoirs and carriers (Almenar et al. 2007; Kayaci et al. 2014; Balaguer et al. 2014; Alghuthaymi et al. 2015; Luiz et al. 2017; da Rocha Neto et al. 2018). Other complex nanoformulations for pesticide delivery including solid lipid nanoparticles, coated liposomes, or other formulations involving the use of inorganic nanoparticles associated with organic ingredients (i.e., calcium carbonate or mesoporous silica) have been also investigated due their low toxicity and residues (Iavicoli et al. 2017).

Although many benefits of nanoagroparticles have been suggested, concerns regarding their application on agriculture have emerged, mainly on the biosafety of these new products and their long-term effects on humans and environment, since their improved solubility and bioavailability may affect the complex interplay between the environmental conditions and also on the different pathways related to the human health (Sangeetha et al. 2017b). Thus, several toxicological research should be performed, e.g., toxicokinetic and dynamic behavior of the adsorbed products (Krug 2014; Shatkin and Kim 2015; Iavicoli et al. 2017).

Table 15.1 Types, synthesis methods, and application of nanoagroparticles in agriculture

Type of nanoparticle	Synthesis	Applications	References
Crop protection			
Copper	Chemical	Fungicidal	Kanhed et al. (2014)
Copper and Silver	Chemical	Fungicidal	Ouda (2014)
Silver	Biological	Fungicidal	Khadri et al. (2013)
Silver	Biological	Bactericidal	El-Rahman and Mohammad (2013)
Silver	Biological	Pesticide/insecticide	Ali et al. (2015)
Zinc oxide	Chemical	Bactericidal	Hafez et al. (2014)
Seed treatment			
Silver	Biological	Seed dressing	Anand and Kulothungan (2014)
Silver	Biological	Surface sterilize of seed crops	Morsy et al. (2014)
Formulation of herbicides			
Carboxymethyl chitosan	Chemical	Herbicidal	Yu et al. (2015)
Chitosan	Chemical	Herbicidal	Grillo et al. (2014)
Polyepsilon-caprolactone	Chemical	Herbicidal	Pereira et al. (2014)
Biosensing			
Carbon	Chemical	Herbicide detection	Luo et al. (2014)
Graphene	Chemical	Herbicide detection	Zhao et al. (2014)
Gold	Chemical	Herbicide detection	Boro et al. (2011)
Gold	Chemical	Organophosphates detection	Kang et al. (2010)
Silver	Chemical	Herbicide detection	Dubas and Pimpan (2008)
Plant growth promotion			
Iron, cobalt, copper	Chemical	Increase germination rate	Ngo et al. (2014)
Silver	Chemical	Plant growth	Syu et al. (2014)
Zinc oxide	Chemical	Flowering and seed productivity	Laware and Raskar (2014)

Adapted from Baker et al. (2017)

15.2 Nanoagroparticles Applications

15.2.1 Nanoagroparticles as Nanocarriers

Nanocarriers are fundamental components in novel nanoagroparticle formulations, acting as vehicles to different molecules, increasing their bioavailability, or even protecting and stabilizing sensitive molecules, minimizing side effects as the environment pollution or human/animals' intoxication by providing means for active targeting. The formulations of nanoagroparticles are typically 1–300 nm in size and loaded with the desirable active ingredient, as stated before. Several different types of molecules can be used as nanocarriers, e.g., polymeric or solid lipid nanoparticles, lipid or albumin nanocapsules, liposomes and micelles, nanovesicles, dendrimers, nanogels, nanoemulsions (NE), and nanosuspensions (Karande et al. 2015; Vyas et al. 2004). In addition, several materials, e.g., polymers (such as cyclodextrins and chitosan), lipids (such as NPLSs), ceramics (such as nanoclay), and carbon nanotubes, have been explored as nanocarriers for nanoagroparticles synthesis. However, their use will be restricted to their biocompatibility with the desirable molecules. In this regard, several examples are given below.

15.2.1.1 Nanoagroparticles as Fungicides

Plants can be infected by different kinds of microbial pathogens that are present in the soil, water, and air, or even in other propagative materials, e.g., seeds or tubers, corms suckers, and setts. Whatever may be the source of the initial inoculum, if the pathogen is able to find favorable conditions for further development, systemic symptoms are induced in tissues or organs of the plant (Narayanasamy 2011). Fungal diseases are responsible for more than 70% of the diseases in the different crops, leading to reduction in crop yield and economic losses (Agrios 2005). Thus, the use of synthetic fungicides continues to be the main strategy against these pathogens worldwide. However, fungicides are reported to present deleterious effects on all different kinds of life, including non-target organisms such as humans, plants, and other animals (da Rocha Neto et al. 2015, 2016; Luiz et al. 2015; Darolt et al. 2016; Felipini et al. 2016).

In this regard, nanoagroparticles can be used as fungicides individually or through the complex of one or more active(s) ingredient(s) with another molecule of interest. Recent studies demonstrated the use of these nanoagroparticles as an effective alternative against several fungal pathogens. For instance, Oh et al. (2006) demonstrated the strong antifungal effect of Ag-SiO_2 nanoparticles against *Botrytis cinerea*, while Medda et al. (2014) reported that silver nanoparticles synthetized by *Aloe vera* leaf exhibited antifungal activity against *Rhizopus stolonifer* and *Aspergillus* sp., inhibiting hyphae growth, conidial germination, and normal budding process. Silver nanoparticles were also found to exhibit antifungal activity against *Fusarium oxysporum* (Kasprowicz et al. 2010), *Colletotrichum gloeosporioides* (Aguilar-Mendez et al. 2011), *Bipolaris sorokiniana*, and *Magnaporthe grisea* (Jo et al. 2009).

Copper nanoparticles were found to suppress the development of *B. cinerea* and *Alternaria alternata* (Ouda 2014). Similarly, nanoparticles of zinc oxide and magnesium oxide exhibited antifungal activity against *A. alternata*, *F. oxysporum*, *R. stolonifer*, and *Mucor plumbeus* inhibiting their spore germination (Wani and Shah 2012). Zinc oxide nanoparticles (ZnONPs) were also found to inhibit *Penicillium expansum* mycelial growth and patulin production (Ramy and Ahmed 2013; He et al. 2010), while $ZnTiO_3$ nanopowder showed a higher efficiency to impair the growth of *Aspergillus niger* than ZnO (Ruffolo et al. 2010). Rao and Paria (2013) reported the fungicidal efficacy of sulfur nanoparticles against *Fusarium solani* and *Venturia inaequalis*, disturbing the cell wall which in turn disturbed the intracellular content.

Different strategies and materials can be used to create other types of nanoagroparticles, e.g., liposomes, proteins, and carbohydrates, among others. In terms of carbohydrates, the natural cationic linear polymer chitosan is mainly used in the nanoencapsulation of active ingredients, exhibiting antimicrobial and antioxidant activities against a wide range of pathogenic fungus by itself (Martínez-Camacho et al. 2010; Fathi et al. 2014; Shao et al. 2015; Ribeiro-Santos et al. 2017). Moreover, cyclodextrins which are well known due to their truncated cone-shaped oligosaccharides with a lipophilic central cavity and a hydrophilic outside surface are also commonly used as a host to lipophilic active ingredients, being used in order to allow a controlled release of the ingredients against the pathogens (Almenar et al. 2007, 2009).

15.2.1.2 Nanoagroparticles as Bactericides

Several scientific studies reported the use of nanoparticles as potential bactericides against a wide range of both gram-positive and gram-negative pathogenic bacteria which seemed to be an alternative of some antibacterial agents, overcoming the bacterial resistance against the commonly used active ingredients (Baker et al. 2015, 2017; Zhang et al. 2017). According to Paulkumar et al. (2014), silver nanoparticles synthetized from *Piper nigrum* leaf and stem extracts were capable of inhibiting the development of the two Gram-negative bacteria *Citrobacter freundii* and *Erwinia cacticida*. Moreover, Kim et al. (2007) conduced a detailed study, reporting that Ag nanoparticles were able to completely inhibit the growth of *Escherichia coli* and *Staphylococcus aureus*, even in low concentrations. Moreover, the anti-biofilm potential of Ag nanoparticles against *Pseudomonas aeruginosa* and *Staphylococcus epidermidis* was also assessed by Kalishwaralal et al. (2009).

The antibacterial activity of Ag nanoparticles synthetized from *Helianthus tuberosus* against *Ralstonia solanacearum* and *Xanthomonas axonopodis* was reported by Aravinthan et al. (2015). According to their study, the synthetized nanoparticles were able to significantly reduce the bacterial growth in low concentrations. Such bactericidal effect was already observed by Bhor et al. (2014) using the Ag nanoparticles synthetized from *Nephrolepis exaltata* against *X. axonopodis*. Bactericides can also be encapsulated into hosts in order to form nanoagroparticles. Wen et al. (2016) used ß-cyclodextrin to nanoencapsulate into polyvinyl alcohol (PVA) nanofibers the essential oils of cinnamon, clove, artemisia, and eucalyptus and evaluated the bactericide effects of the electrospun nanofibrous against *E. coli* and *S. aureus*. The nanofibers containing the essential oils were more effective to control both bacterial development than the chemical control used, reaching an inhibition zone of 30 mm.

Similarly, Chen and Liu (2016) used cellulose sulfate-based films with slow release with mustard essential oil entrapped into ß-cyclodextrins to control *E. coli* and *S. aureus*. The results showed a great antimicrobial activity of the films, completely inhibiting *E. coli* development at 0.02% of essential oil concentration, while for *S. aureus* the complete inhibition occurred with a 0.18% concentration of the essential oil.

15.2.1.3 Nanoagroparticles as Insecticides

Insects have several natural enemies that play a key role in reducing their population levels below to those causing economic losses. Many biological agents have been used for this purpose, e.g., bacteria, fungus, virus, and insects, among others, where *Bacillus thuringiensis* is the most widely applied species of bacteria used for biological control (Dimetry and Hussein 2016).

In this sense, several studies reported the effectivity of nanoparticles to control insects and pests (Bhattacharyya et al. 2016a, b; Buhroo et al. 2017). For example, nano-silica, a type of nanomaterial prepared from silica, can be used to control insects through its absorption into the cuticular lipids by physisorption, leading to death by physical means, once it prevents the cuticular lipids of the insects to act as a water barrier, allowing their death by desiccation (Ulrichs et al. 2005). Silica can also be used as porous hollow nanoparticles (PHSNs) and loaded with different kinds of insecticide. The PHSNs capability and efficiency to deliver water-soluble molecules such as validamycin can be a promising strategy when a prolonged release is needed for plants (Liu et al. 2006). Similar strategies can also be used for oil-in-water formulation, such as nanoemulsions or oil-loaded solid lipid nanoparticles (Liu et al. 2006). Yang et al. (2009) reported the insecticidal effect of polyethylene glycol-coated nanoparticles loaded with garlic essential oil against an adult of the red flour beetle (*Tribolium castaneum*), controlling up to 80% of the beetle incidence.

The applications of other types of nanoagroparticles, e.g., silver, aluminum oxide, zinc oxide, and titanium oxide, were also reported. Goswami et al. (2010) observed that the aforementioned nanoparticles were able to control the rice weevil and grasserie disease in silk-worm (*Bombyx mori*) up to 95%, in both solid and liquid formulations. Similar results were found by Stadler et al. (2010) who reported the insecticidal effects of nanostructured alumina against rice weevil (*Sitophilus oryzae*) and the beetle *Rhyzopertha dominica* known to cause great damage in stored food supplies worldwide. The nanostructure alumina decreased the development of the pests by increasing their mortality after 3 days, being cheaper and more reliable than the usual treatments made with pesticides.

15.2.1.4 Nanoagroparticles as Herbicides

Currently, one of the major problems during farming is related to undesirable weeds, which can compete with cultivated crops for nutrients, light, water, growth space, etc. (Baker et al. 2017). In this scenario, it was estimated that 0.79 million tons of herbicides were used in different crops (Kiely et al. 2004), such as soya (*Glycine max*); sugar cane (*Saccharum officinarum*) and corn (*Zea mays*) (Azania et al. 2010).

One of the most popular herbicides used by farmers is atrazine (1-chloro-3-ethylamino-5-isopropylamino-2,4,6-triazine) due to its potential to prevent pre- and postemergence of broadleaf weeds by impairing the photosynthesis and ATP synthesis of weeds (Rodrigues and Almeida 2011; Hess 2000). In addition, non-selective herbicides like paraquat and glyphosate are commonly used to aid harvesting time, allowing an increased production in function of time, reducing the interference caused by weeds (Alcantara and Wyse 1988; Jeffery et al. 1998).

However, there is a great debate regarding the toxicology of this product to the environment, animals, humans, and microorganisms due its persistence, mobility, and chemical profile, where such substances may cause genotoxic effects to different organisms exposed to its application, where a long-term exposure may cause cancer, mutations, or even death (Queiroz et al. 2011).

In order to avoid such situations, nanoherbicides can be used as safe tools, since they are reported to bear chemical stability, solubility, bioavailability, photodecomposition, and soil sorption (Baker et al. 2017). In a recent study, Yu et al. (2015) created nanoherbicides by cross-linking the disulfide bonds of the herbicide diuron with chitosan, being able to release the herbicide based on the glutathione concentration, promoting plant growth and reducing its toxicity.

Similarly, Grillo et al. (2014) already reported the use of atrazine as herbicide for the nanoformulation with chitosan and sodium triphosphate, reducing its toxicity compared to its sole active component application and reducing the risks to the environment. Other studies demonstrated an increase of atrazine activity when a formulation based on nanoparticles of epsilon-caprolactone was used (Pereira et al. 2014). Finally, nanoherbicides can also be developed using molecules as nanoclay, which will stabilize the herbicide. For example, a study performed by Katz and Mishael (2013) demonstrated that molecules of montmorillonite can act as surfactants, controlling the release of the herbicide and thus extending its action.

15.2.1.5 Nanoagroparticles as Nanosensors

One of the critical issues in agriculture which can lead to a major destruction of crops on the field is the diseases caused by fungi, bacteria, viruses, and nematodes. Thus, an early and on-site detection of these pathogens can avoid such losses. In this sense, one of the major problems encountered in was related to on-site detection systems. These problems could be easily overcome by the use of modern nanosensors, which can monitor the physicochemical properties of several analytes, e.g., pesticides, toxins, pathogens, and other pollutants, in places difficult to reach, since they are highly sensitive and can detect up to ppb level without the aim of sophisticated laboratory infrastructure or even trained professionals (Baker et al. 2017; Shaoqin et al. 2008). Moreover, nanosensors may also offer the opportunity to assess crop growth and field conditions, e.g., moisture levels, pH, fertility, temperature, nutrient status, and concentrations (Iavicoli et al. 2017).

Nanoagroparticles based on gold were successfully used by Baker et al. as nanosensors to detect the levels of the pesticide dichlorodiphenyltrichloroethane (DDT) in the environment (2013). Similarly, carbon nanotubes were used by Zhang et al. (2014) to monitor environmental samples such as methyl parathion and paraoxon (insecticide) and their metabolites. Several nanosensors were developed, in order to achieve the minimum necessary data to the development nanosensors. The elaboration of these systems is challenging as they require a complete investigation of nanomaterials sensitivity to common residues, pathogens, and environmental parameters. Only then the fabrication and validation of a suitable detection instrument will be available to a complex environment such as the agricultural production system (Iavicoli et al. 2017).

Nowadays, field crops and their products are routinely treated with synthetic products (Frankova et al. 2016) such as insecticides, fungicides, bactericides, herbicides, and others in order to help diminish the losses caused by animal pests (e.g., insects, mites, nematodes, and rodents) and plant pathogens (e.g., bacteria, fungi and viruses, or weeds). Although these modern technologies aimed to decrease field losses and increase shelf life of many agricultural products, their systematical usage is known to increase the risk of poisoning non-target organisms (Eddleston and Bateman 2012), where farmers and consumers in developing countries are much more vulnerable to the risks arising from cancer and other diseases, than those from developed countries (WHO 2017).

Considering the diversity of nanoagroparticles applications (Fig. 15.1) and the limitations related with the conventional synthesis methods of these particles, alternative green synthesis methods should be developed in order to guarantee, at the

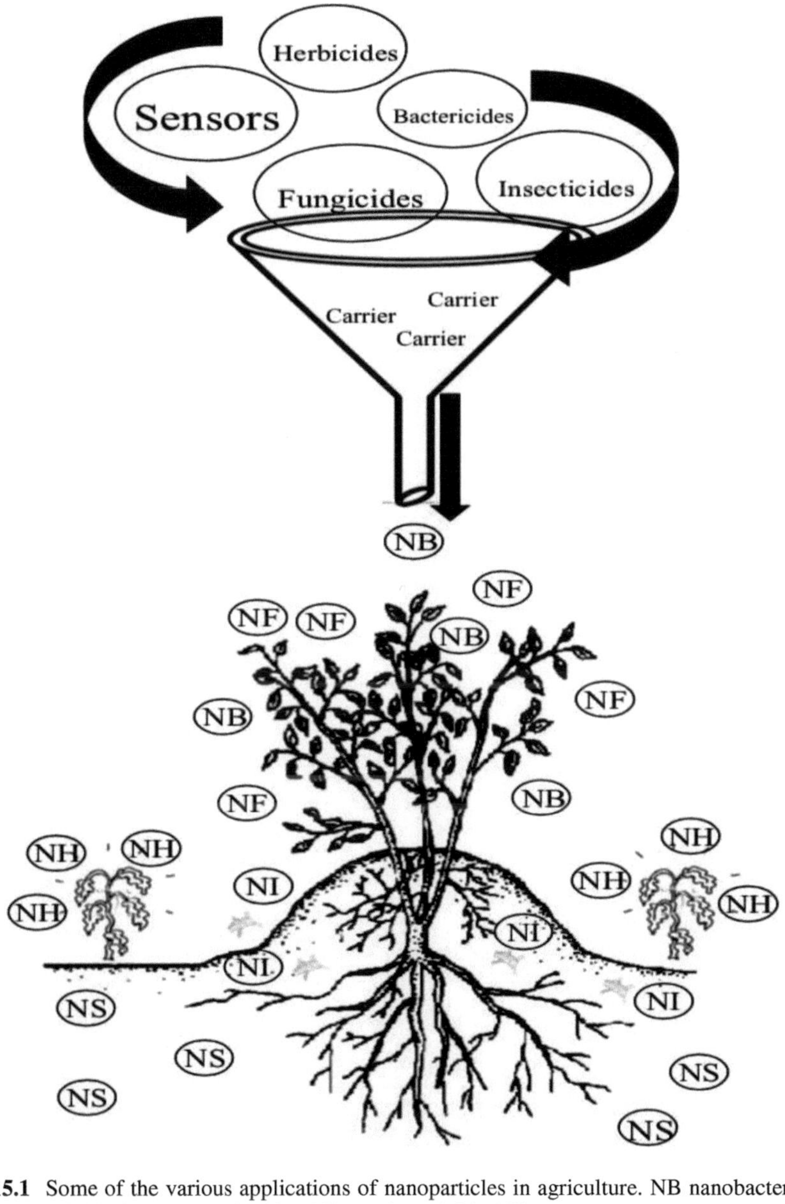

Fig. 15.1 Some of the various applications of nanoparticles in agriculture. NB nanobactericide, NF nanofungicide, NH nanoherbicides, NI nanoinsecticide, NS nanosensor

same time, the efficiency of these particles and the absence of environmental impact caused by their employment.

15.3 Methods of Nanoagroparticles Synthesis

The principles of nanoagroparticles formulations are related with the (i) increase of solubility in the poorly soluble compound and the (ii) compound release in a targeted manner protecting it against premature degradation (Kah et al. 2013). Furthermore, it is important to think about low toxicity to non-target species and low residues of the compounds. In view of these facts, these compounds should be packaged into a protective matrix, and a range of formulation have been suggested, nanoemulsion, nanocapsules (use of polymers), nanospheres, porous silica nanoparticles, and inorganic NPs (metals, nanoclays, and metal oxides), as illustrated in Fig. 15.2 (Liu et al. 2014).

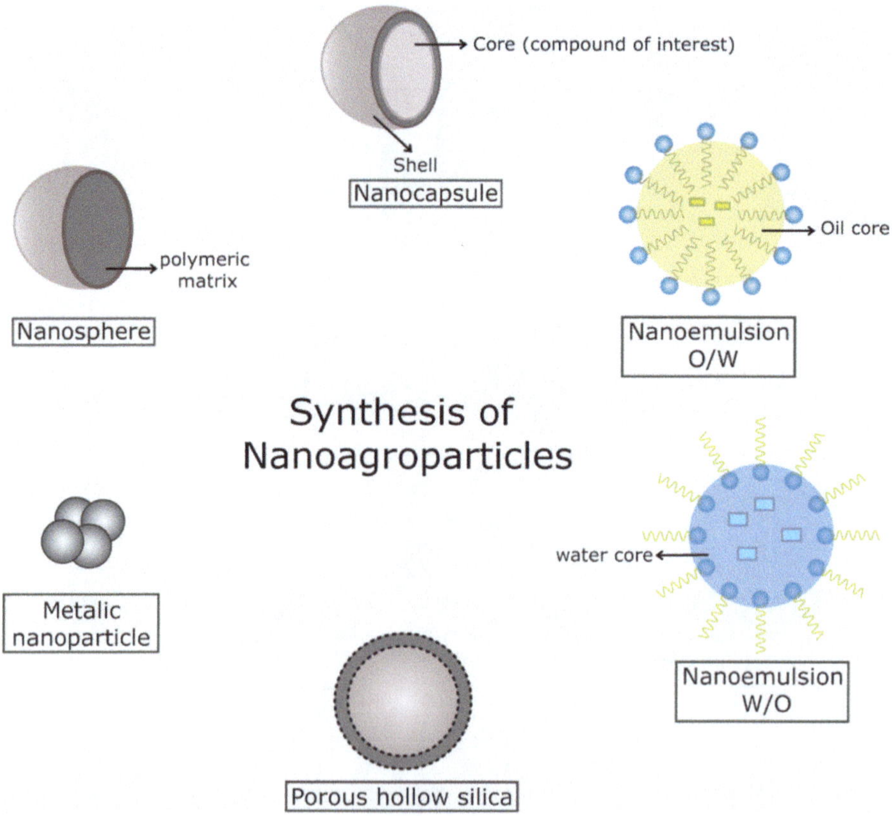

Fig. 15.2 Different strategies for the synthesis of nanoagroparticles

Nanoemulsions (NE) are biphasic dispersion, with an average diameter of <500 nm, of two immiscible liquids which can be water in oil (W/O) and oil-in-water (O/W) or multiple emulsions such as oil-in-water-in-oil (O/W/O) and water-in-oil-in-water (W/O/W) which the dispersed particles are themselves emulsions and are stabilized by an surfactant with amphiphilic properties. Although the NE have a tendency to separate due to the Ostwald ripening effect, they have high kinetic stability (Kah et al. 2013; Singh et al. 2017).

The synthesis of NE can be categorized into high-energy, low-energy or a combination of both. The methods of high-energy depend on mechanical equipment to create powerful disruptive forces for size reduction which includes high-pressure homogenizers, high-shear stirring, and ultrasound generator. The disadvantage of these techniques is the high energy cost. The methods with low energy can be employed by spontaneous emulsification and phase inversion temperature methods. These methods employ the energy stored in the system to produce ultrafine particles (Kah et al. 2013; Singh et al. 2017).

The fungicide Tebuconazole NE was synthesized using low-energy method by Díaz-Blancas and colleagues (Díaz-Blancas et al. 2016). According to phase diagrams, the zone O/W was employed in NE formulation, and although this formulation showed to be promisor, its efficiency against fungi in crops was not tested (Díaz-Blancas et al. 2016). Zhao and coworkers investigated the synthesis of Fenpropathrin and Cyflumetofen NE and their application. The best ratio for these active components was 39/1 of fenpropathrin/cyflumetofen, and the NE effect showed to be efficient due to its high absorption of plant leaves (Zhao et al. 2016).

Nanocapsules exhibit a core-shell structure and can act as a reservoir for the compound, while nanospheres have a polymeric matrix. Although nanocapsules may present advantages such as reduced phytotoxicity, higher spraying surface, and higher uptake, their synthesis can be challenging due to the difficulty to keep an amount of compound sufficiently high relative to the amount of polymer (Kah et al. 2013).

Nanocapsules can be synthesized during the polymerization of monomers or from preformed polymers (Cătălin et al. 2017). Jia and colleagues (2014), for example, prolonged the foliar pesticide retention by using the method of preformed polymers to synthesize the polydopamine coated avermectin (Av@PDA) (Jia et al. 2014). The method of preformed polymers was used by Loha et al. in the synthesis of nanocapsules of poly(ethylene glycol) with the insecticide β-Cyfluthrin (Loha et al. 2011).

Porous hollow silica nanoparticles (PHSNPs) have versatility in their applications because the compound can be located in different parts of the particles: external, in pore channels, and in the internal core, which can exhibit different profiles of release of the compound (Kah et al. 2013). The PHSNPs can be synthesized by preloading and post-loading method. The preloading method consists of the solubilization of the compound of interest in the template material prior to the formation of silica shells. The post-loading method is based on the diffusion of the compound of interest into the interior by high-pressure supercritical fluid or immersion. The preloading method has the advantage of reducing the possibility of attachment of the compound of interest at the external surface of the shells, which increases the protection of the compound (Jampílek and Kráľová 2017).

Liu et al. (2006) synthesized PHSNPs with the pesticide validamycin by post-loading method, and they observed that the adsorption of validamycin occurred at the pore channels on the shell, external surface, or internal core of PHSNPs, which leads to different rates of release. Furthermore, after the initial release, the release mechanism continued for a long time indicating the efficiency of the treatment (Liu et al. 2006). Another application of PHSNPs was investigated by Wen et al. (2005). They synthesized PHSNPs measuring 100 nm and with a pore size of 4.5 nm by sol-gel process and observed that avermectin may be loaded on the external surface, in the core of PHSNPs, and in the pore channels which can lead to a multistage release. Furthermore, they observed that the increase in pH and temperature enhances the release of avermectin (Wen et al. 2005).

Metal NPs can also be used in agriculture due to their biocidal properties, and different metal nanoparticles can be employed: AgNPs, copper nanoparticles (CuNPs), and titanium dioxide nanoparticles (TiO_2NPs). There are numerous different methodologies for synthesis of metal NPs according to the size, structure, and shape among which can be cited: the classic citrate reduction of silver nitrate for AgNPs synthesis (Qin et al. 2010), the borohydride reduction of copper nitrate salt for CuNPs synthesis (Abdulla-Al-Mamun et al. 2009), and hydrothermal synthesis of TiO_2 (Wu et al. 2002).

They act in three different ways: (i) uptake of metallic ions causing harmful effects for DNA replication, (ii) formation of superoxide radicals causing the death of the target organism, and (iii) accumulation in the cell membrane, leading to membrane damage (Hayles et al. 2017). AgNPs have been widely employed due to their bactericidal and virucidal efficiency and adaptability to different substrates, which can facilitate their utilization (Hayles et al. 2017). Kim and coworkers (2012) tested the efficiency of AgNPs against 18 different plant pathogenic fungi and observed that these nanoparticles are efficient in the treatment; however, this efficiency depends on the concentration and the type of AgNPs (Kim et al. 2012). The antifungal activity of CuNPs was also studied by Kanhed and colleagues. They reported that CuNPs in a range of 5–10 nm were efficient against all the following plant pathogenic fungi: *Phoma destructiva*, *Curvularia lunata*, *Alternaria alternate*, and *Fusarium oxysporum* (Kanhed et al. 2014). The bactericidal effect of TiO_2NPs was investigated against *Xanthomonas* sp., which causes a disease in roses affecting their production. The TiO_2NPs with Zn in a concentration range of 500–800 ppm were applied in field and significantly reduced bacterial spot when compared to the control sample (Paret et al. 2013).

15.3.1 Problems Related to Conventional Methodologies

Conventional methodologies of nanoagroparticles are related to various implications such as sophisticated facility, non-biodegradable compounds, toxicity to non-target organisms, and high demand for energy. The zinc oxide nanoparticles (ZnONPs) can also be used in agriculture due to its potential to boost the yield and growth of crops. However, some chemical synthesis of ZnONPs such as pyrolysis, precipitation, and hydrothermal methods can present some toxic chemicals adsorbed on these NPs (Sabir et al. 2014).

Some nanocapsules synthesis methods (Campos et al. 2015), during the emulsion step, use organic solvents such as chloroform and acetone which generate a non-biodegradable residue. Toxicity of nanocapsules loaded with atrazine and the nanoformulation of solid lipid with atrazine presented toxicity to the soil organism *Caenorhabditis elegans*. The lethality of these nanoformulations increases in a dose-dependent manner (Jacques et al. 2017).

The method of high energy for nanoemulsion synthesis is quite straightforward as the higher the energy input, the smaller the size of the particles. However, the energy required to obtain particles in the nanometer scale is cost-inefficient because only 0.1% of the energy produced is used for emulsification (Solans and Solé 2012). Due to the tendency to use methodologies according to the green chemistry, a high input of energy in a method is not desirable, and in this contest, the high-energy method for synthesis of nanoemulsions can be replaced by the low-energy method.

15.3.2 Green Synthesis of Nanoagroparticles

In view of the problems that conventional synthesis of nanoagroparticles can cause, the green synthesis can be an alternative to develop less harmful products by using biodegradable materials besides the reduction of the environmental impact during

Table 15.2 Green methodologies and products employed in the synthesis of nanoagroparticles

Nanoagroparticles	Green component	Active compound	Methodology	References
Nanocapsule	Alginate-chitosan	Acetamiprid	Ionic pregelation and polyelectrolyte complexation	Kumar et al. (2015)
	Poly(epsilon-caprolactone)	Atrazine	Oil-in-water method	Pereira et al. (2014)
Nanospheres	Chitosan	Chitosan	Biological Synthesis	Sathiyabama and Parthasarathy (2016)
	Poly(lactic acid)	Metazachlor		Stloukal et al. (2012)
Nanoemulsion	Essential oil from *Satureja hortensis*	Essential oil from *Satureja hortensis*	Low energy	Hazrati et al. (2017)
	Essential oil *from Ageratum conyzoides*, *Achillea fragrantissima*, and *Tagetes minuta*	Essential oil *from Ageratum conyzoides*, *Achillea fragrantissima*, and *Tagetes minuta*	High-pressure homogenization	Nenaah et al. (2015)

the synthesis (Kah and Hofmann 2014). The polymers considered for the synthesis of nanoagroparticles are similar to those used in pharmaceutical products and the most used for this purpose are chitosan, pectin, alginates, polyethylene glycol (PEG), and lecithin (Kah and Hofmann 2014). Table 15.2 lists the different methodologies and green substances of nanoagroparticles.

The green synthesis of nanocapsules can use a variety of natural polymers and can be combined to achieve better release conditions. Kumar et al. (2015), for example, synthesized alginate-chitosan nanocapsules for controlled release of the insecticide acetamiprid by ionic pregelation and polyelectrolyte complexation method. This green synthesis showed to be effective due to the superior controlled releasing compared to the commercial formulation besides to be effective in different ranges of pH in soil (Kumar et al. 2015). Chitosan nanospheres were studied by Baker et al. (2017) against *Fusarium graminearum*, the fungus that causes the fusarium head blight in wheat. Significant inhibition of mycelial growth and number of colonies was observed, and maximum inhibition was observed using concentrations in a range of 1000 and 5000 ppm of chitosan nanospheres (Kheiri et al. 2016).

Glyphosate isopropylamine (IPA), an eco-friendly herbicide with rapid degradation and deactivation in the soil, was used in the synthesis of NE with less than 200 nm for narrow-leaved weed *Eleusine indica* and showed an ED_{50} of 0.40 kg a.e./ha (Chaw et al. 2012). The use of garden savory (*Satureja hortensis*) essential oil was also investigated in the NE synthesis for herbicidal application against *Amaranthus retroflexus* and *Chenopodium album*. The higher toxic effect was observed at 800 $\mu L \cdot L^{-1}$, and comparing the roots and shoots length, it was observed that the root length was more inhibited (Hazrati et al. 2017).

15.4 Environmental Impact

15.4.1 Nanoagroparticle Release Mechanisms

The controlled release of active compounds in nanoagroparticles is one of the advantages of their utilization and has been encouraged due to the biodegradability, target delivery, and low costs.

In this process occurs the regulated transfer of the active compound from the matrix to the target in order to maintain a certain concentration level during a specific period of time. The difference between a conventional system of agrochemical application and the controlled release offered by nanoagroparticles is that the first has a high initial dose which rapidly decreases to a level lower than the effective, while the second can maintain an effective level during a specific time as shown in Fig. 15.3 (Roy et al. 2014).

When these nanoagroparticles are used, their release is primarily controlled by diffusion processes in which they absorb water, and the encapsulated compound diffuses out of the matrix. In this release process, the diffusion of the active com-

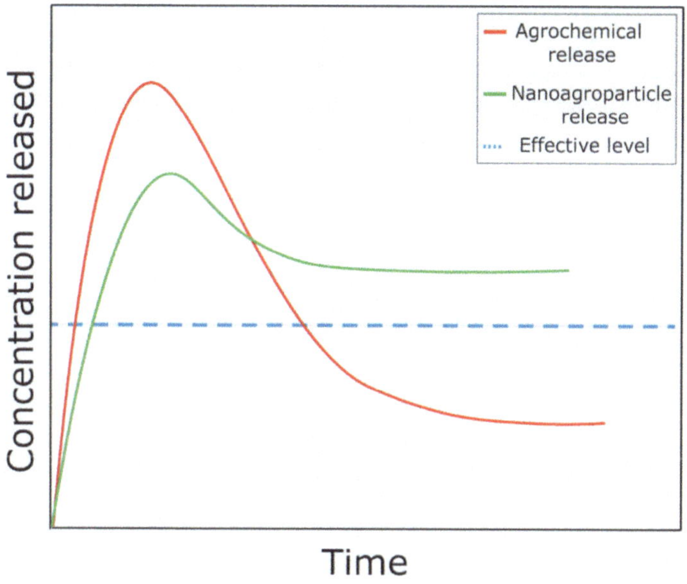

Fig. 15.3 Illustration of different releases between common application of agrochemicals and controlled release from nanoagroparticles

pound depends on the matrix properties of the nanoparticle/nanocapsule, environmental conditions, and the size of the nanoagroparticle (Pereira et al. 2015).

The affinity between the polymeric matrix (in nanoagroparticles made of polymers) and the active compound is an important factor to be investigated. When the polymeric matrix has hydrophilic nature and the active compound has low water solubility, the result is a poor chemical affinity between these components, and this may impair the release of the active compound, since the stronger is the interaction between them, the slower is the release rate (Pereira et al. 2015; Jampílek and Kráľová 2017).

When bonds between polymer and active compounds occur, a hydrolysis reaction is needed for the release of the active compound, and their solubility in water increases resulting in the reaction acceleration (Jampílek and Kráľová 2017). Adak and colleagues developed a nanoagroparticle synthesized with poly(ethylene glycols) for encapsulation of imidacloprid. They compared the release of imidacloprid in water and in the nanoformulation with the controlled release and observed that the diffusion of imidacloprid in water ranged from 0.22 to 0.37 days, while in nanoformulation, these values ranged from 2.32 to 9.31 days which evidences the efficiency of nanoformulation with controlled release (Adak et al. 2012).

Recently, Xiang et al. (2018) investigated the mechanism and the release efficiency of the herbicide prometryn (PMT) from porous calcium carbonate matrix. According to mathematical models, the release mechanism of PMT alone is the dissolution (it fitted better in the linear model), and the PMT encapsulated mechanism was the diffusion through the pores of calcium carbonate matrix (it fitted better in parabolic diffusion model). They also observed that while the PMT release alone finished in 2 h, the PMT encapsulated reached 86.7% of the release after 12 h, which indicates its higher efficiency (Xiang et al. 2018).

In the case of nanofertilizers, it is also possible that nanoagroparticles enter into plant cells if their size were smaller than the sizes of cell wall pores (5–20 nm). The primary mechanism is the dissolution of these nanofertilizers in water/soil solution (as described in Eqs. (15.1) and (15.2) by Liu and Lal (2015)); in this way, plants absorb the soluble ions indiscriminately such as conventional fertilizers (Liu and Lal 2015).

$$Ca_5\left(PO_4\right)_3OH\left(NPs\right) \rightleftharpoons 5Ca^{2+} + 3PO_4^{3-} + OH^- \quad (15.1)$$

$$ZnO\left(NPs\right) + H_2O \rightleftharpoons Zn^{2+} + 2OH^- \quad (15.2)$$

Yuvaraj and Subramanian (2014) reported that the zinc encapsulated by a manganese hollow core-shell can be employed as nanofertilizer for rice. The release of nanofertilizer was much higher than $ZnSO_4$ after 696 and 408 h, respectively (Yuvaraj and Subramanian 2014). The release of N was investigated by Kottegoda and colleagues by urea-hydroxyapatite (urea-HA) naoagroparticles. They reported that the release of N in urea-HA follows the diffusion mechanism according to Fickian kinetics. Furthermore, the rate release of N for urea-HA is 12 times slower compared to the pure urea, and it took 3820 s to release 86% of N for urea-HA, while 99% of N from pure urea were released in the first 320 s of exposure to water (Kottegoda et al. 2017).

In green nanoagroparticles made of biodegradable polymers, the mechanism of the release follows the same mechanism of other kinds of nanoagroparticles such as hydrolysis and surface degradation. The surface degradation depends on the geometry and surface area which means that it is controlled by the radius to thickness ratio of the particles (Roy et al. 2014).

15.4.2 Bioavailability

The nanoformulations have proved their higher efficiency, compared to the pure active compound due to their higher bioavailability and increased uptake. While an increased uptake is desirable, it is important to evaluate the risks to the non-target organisms. The bioavailability depends on different factors such as the target organism, matrix of nanoagroparticles properties, and the location of the active compound in the matrix, which can affect, for example, the access of microorganism to the active compound since it will be more available at the surface than in the core (Kah and Hofmann 2014).

The bioavailability also depends on the size of the nanoagroparticles: smaller particles enhance the spreading of nanoagroparticle on the treated surface and enhance the penetration in the organism (Hayles et al. 2017). This effect was observed by Xu et al. (2010) by comparing the effect of neem oil nanoemulsion size to the median lethal time value (LT_{50}) in *Sarcoptes scabie* var. *cuniculi* larvae. The nanoemulsions, the emulsion in water, and liquid paraffin neem presented an LT_{50} of 81.7, 95.5, and 156.6 min, respectively, indicating that the size of nanoemulsion is an important parameter in the bioavailability of the nanoagroparticle (Xu et al. 2010). Several nanoformulations showed the increase of uptake by the target organism, and fortunately, this occurs without increasing the risk to non-target organisms (as discussed in the topic 15.4.3) (Hayles et al. 2017).

15.4.3 Non-target Effects of the Green Nanoagroparticles

When green nanoagroparticles are used in the environment, is important to assess their risks to non-target organisms in order to attest their efficiency in not harming the environment. Some studies revealed that several green nanoagroparticles failed to show toxicity against non-target organisms. Pereira and colleagues (2014) investigated the efficiency of poly(epsilon-caprolactone) nanocapsules containing atrazine herbicide against *Brassica* sp. and the toxic effects to the non-target organism *Zea mays*. The germination of *Zea mays* was not affected by the atrazine nanoformulation and achieved a germination index of 90% indicating that the development of the plants was not affected. Furthermore, the nanocapsules showed to be effective against *Brassica* sp. (Pereira et al. 2014).

Poly(epsilon-caprolactone) nanocapsules containing triazine herbicide was also studied by Clemente and coworkers (2014). They showed the toxicity of these particles to non-target organisms (alga *Pseudokirchneriella subcapitata* and the microcrustacean *Daphnia similis)*. Although this formulation was toxic to *Daphnia similis,* it was observed with a lower toxicity to the alga compared to the pure herbicide (Clemente et al. 2014).

Rawani and coworkers (2013) green-synthesized mosquitocidal AgNPs with high efficiency against mosquitoes and without any toxic effects to their predators such as *Diplonychus annulatum* (Rawani et al. 2013). The organisms of soil can also be affected by nanoagroparticles, and because of this, the toxicity of the green nanocapsule chitosan/tripolyphosphate (CS/TPP) with the herbicide paraquat was tested in *Caenorhabditis elegans* by Jacques et al. (2017). They observed that CS/TPP did not affect the reproduction and body length even at higher concentrations (Jacques et al. 2017). In general, limited evidence of green nanoagroparticles toxicity was reported (as observed by Clemente and coworkers (Clemente et al. 2014)), and most of the LC_{50} calculated for non-target organisms have been found higher compared to the LC_{50} of the target species (Benelli et al. 2018).

15.4.4 Degradation Kinetics of Nanoagroparticles

The investigation of the sorption and degradation processes of nanoagroparticles is very important for assessing environmental exposure of nanoagroparticles (Kah et al. 2018a). The persistence ($t_{1/2}$ or DT_{50}) is related to the degradation of the active compound and can be influenced by its release from the matrix. In order to investigate this parameter, incubation experiments to determine the total concentrations of the nanoformulation and pure active compound are desirable. In view of this fact, two types of processes are possible: (i) when both pure active compound and nanoformulation of active compound are equally accessible for degradation and (ii) when the active compound in nanoformulation is not degraded at all (Kookana et al. 2014).

One of the processes of pesticide degradation in the abiotic hydrolysis, which consists of a reaction with OH^- and H_3O^+ species, converts it into molecules with reduced biological activity. When the pesticide is a nanoformulation, such as nanoemulsion O/W, this formulation protects it against the reaction of degradation by hydrolysis due to the different environment surrounding the active compound (Hayles et al. 2017). The nanoformulations can protect the active compound from degradation. Kah and colleagues (2018b) reported the photodegradation of clothianidin nanoformulation. The photodegradation half-lives (DT_{50}) for nanoformulations are lower than the pure compound in presence of fertilizers probably due to complexation with NH_4^+ or Fe present in the fertilizers, quenching, or light absorption of the colored fertilizers solution (Kah et al. 2018b). The effects of the nanoformulation of the pesticide bifenthrin were evaluated in two different soils (loam and sand) and compared to the application of pure bifenthrin in the soil. It was observed that the nanoformulation had prolonged persistence in both soils. The reasons for these results can be explained by the delaying of the compound degradation and by its toxicity for microorganisms which can degrade it (Kah et al. 2016).

15.5 Discussion

15.5.1 Recent Advances in the Nanoagroparticles Synthesis

The concern about the reduction of the harmful impacts caused by nanoagroparticles is increasing, and the green methods which are non-toxic or have lower toxicity are desirable (as discussed in the topic 15.4.3). In this scenario, a synthesis method which uses microorganisms is starting to gain attention. An example is the nanobiofungicides, which denote the class of fungicides derived from bacteria, animal, fungi, or plant products. A combination of bacteria and fungi to colonize and defeat plant pathogens can be very effective; furthermore, since these microorganisms are found in the soil, they offer an eco-friendly strategy to combat pests (Bhattacharyya et al. 2016).

The use of microorganisms with nanoparticles can be effective as nanopesticide. An example is the use of *Bacillus thuringiensis* bacteria, an eco-friendly biopesticide, against *Spodoptera littoralis.* The action of this pesticide, which is slow, can be enhanced by synthesizing a composite of sodium titanate nanotubes with *Bacillus thuringiensis.* Moreover, this nanopesticide showed to be effective as an eco-friendly nanopesticide by affecting the larval and pupal duration of *Spodoptera littoralis* (Zaki et al. 2017). Bionano-hybrid agroparticles are a complex nanosystem based on bioconjugation chemistry (in which biomolecules and nanoparticles are conjugated) and are promisors in combating the drug-resistant pathogens. The procedure is highly dependent on physicochemical and biochemical properties of nanomaterial and bioactive compound, and the interaction between them is based on electrostatic forces and functional parts (which can form covalent bonds and hydrophobic interactions) forming functionalized nanoparticles in a reversible manner (Baker et al. 2017).

The use of polymeric nanoparticles for coating biofertilizers is also a recent advancement in nanoagroparticle synthesis. Microorganisms such as fungal mycorrhizae, *Rhizobium*, *Azotobacter*, *Azospirillum*, and blue-green algae can be used as biofertilizers because they convert organic matter into essential elements for plant nutrition. The W/O nanoemulsions are one of the techniques to distribute microorganisms through liquid formulations, although sedimentation can occur and decrease their efficiency. To overcome this problem, hydrophobic silica nanoparticles can be used in improving the cell viability (Duhan et al. 2017).

15.5.2 The Benefits of Nanoagroparticles in Agriculture

When common pesticides are used, large proportions of them are required and sometimes do not reach their target (these proportions can vary from 10% to 75% depending on the method of application) and can result in environmental contamination. In this context, the use of nanoagroparticles can reduce the amount of pesticide needed for crop protection due to the controlled release, target delivery and enhanced bioavailability (Kah et al. 2018a).

Besides the use of nanoagroparticles in crop protection (such as nanopesticide), it is possible to employ nanotechnology to enhance the production by using nanofertilizers. About 50–70% of N applied in soil is not used by plants due to the incorporation in soil organic matter and leaching in water; the nanofertilizer can reduce this N (and other elements) loss. The nanofertilizers can be divided into three categories: (i) nanoagroparticles made of macronutrients, (ii) nanoagroparticles made of micronutrients, and (iii) nanoagroparticles as carriers for macronutrients. Nanoagroparticles made of macronutrients are chemically comprised of one or more macronutrients such as N, P, Ca, and K, which can supply these essential nutrients to plants. Nanoagroparticles with micronutrients such as Fe, Mn, Zn, Cu, and Mo contain these elements in trace levels for the healthy growth of plants (Hong et al. 2013; Liu and Lal 2015; Kah et al. 2018a).

The use of urea-coated hydroxyapatite as N nanofertilizer and also as a source of P was investigated by Kottegoda and coworkers (2017) which added this nanofertilizer in a rice farm field. The use of this nanofertilizer resulted in better yield using a 50% lower concentration of N, indicating that besides its efficiency, this nanofertilizer reduces costs (Kottegoda et al. 2017). Carbon nanotubes (both single and multi-walled) can also be added to enhance plants growth. Joshi (2014) studied the influence of multi-walled carbon nanotubes (MWCNTs) in germination and growth of bread wheat (*Triticum aestivum* L.) by considering the field conditions. They observed that compared to the control samples, samples with MWCNTs addition presented denser roots which facilitate the uptake of P and K and the grain yield increased 63%. Furthermore, xylem and phloem were dilated by approximately 80% which enhances the transport of water and nutrients, and no DNA damage was observed (Joshi et al. 2018). According to Tiwari and coworkers (2014), the addition of low concentrations of MWCNTs (20 mg L^{-1}) to maize (*Zea mays*) increased the water absorption and, subsequently, increased the concentration of Ca and Fe which are essential nutrients. Furthermore, these researches observed an increase of maize biomass (Tiwari et al. 2014).

15.5.3 Future Perspectives

Recently, the nanoagroparticles have been extensively studied due to the concern about the high toxicity of the current agrochemicals, which affect the environment and human health. The use of nanoagroparticles such as nanopesticides, nanoherbicides, and nanofertilizers showed to be efficient against target organisms. This efficiency allows the use of a small concentration of active compounds decreasing the residues and reduces the costs and avoiding or decreasing harmful effects on non-target organisms. Furthermore, the use of these nanoformulations aims to maximize the crop yield and at the same time decreases the input (pesticides, herbicides, fertilizers) (Grillo et al. 2014).

However, there are some challenges that should be faced (Fig. 15.4). Unfortunately, considering the necessity to prevent nanoagroparticles harmful effects, there are only a few studies evaluating the bioavailability of these particles in the environment. It is also necessary for the particles to be tested in large crops in order to measure their real impact in the agricultural production and attest their

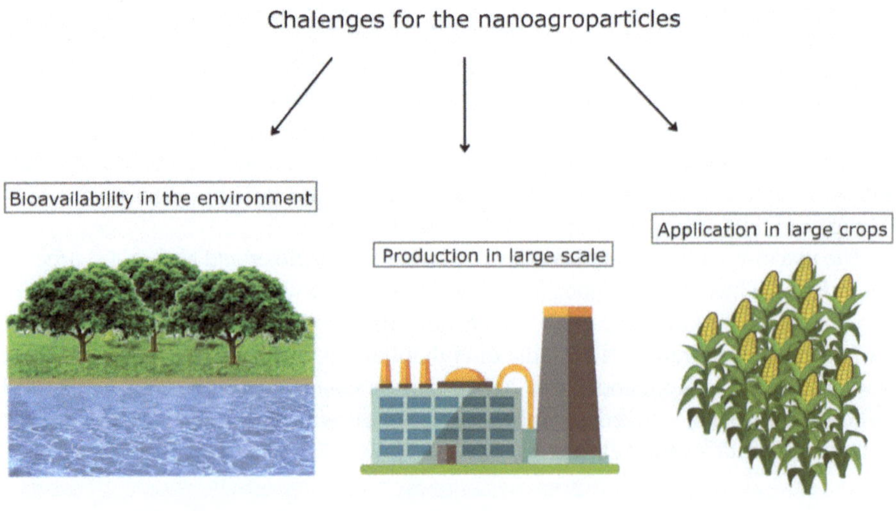

Fig. 15.4 The futures perspectives: the challenges for the nanoagroparticles

efficiency. Moreover, the cost of nanoagroparticles production in large scale needs to be reduced and optimized in order to replace the production of common agrochemicals. Nevertheless, it is important to note that the challenges described above can be overcome. The nanoagroparticles are promisors, have been efficient for agricultural application, and have potential to replace the common agrochemicals.

15.5.4 Conclusions

The use of nanoagroparticles can be promising due to their controlled release and the possible lower toxic effects to the environment. Great advances have been observed in the research, and new legislations have been created (e.g., the Regulation (EC) No. 1107/2009, by European Commission) in order to fix the basis of a responsible strategy for the development of these nanoformulations (Villaverde et al. 2017). This strategy is important to guaranty the security related to employment of nanoagroparticles in agriculture. In the future, it is expected that the nanotechnology will be included in all the sectors of science and technology including agriculture, which will increase the quantity and the quality of the crops yield without affecting the environment and human health.

References

Abdulla-Al-Mamun M, Kusumoto Y, Muruganandham M (2009) Simple new synthesis of copper nanoparticles in water/acetonitrile mixed solvent and their characterization. Mater Lett 63:2007–2009

Adak T, Kumar J, Shakil NA, Walia S (2012) Development of controlled release formulations of imidacloprid employing novel nano-ranged amphiphilic polymers. J Environ Sci Heal Part B Pestic Food Contam Agric Wastes 47:217–225

Agrios GN (2005) Plant pathology. Elsevier, San Diego, p 952

Aguilar-Mendez MA, Martin-Martínez ES, Ortega-Arroyo L, Portillo GC, Sanchez-Espindola E (2011) Synthesis and characterization of silver nanoparticles: effect on phytopathogen Colletotrichum gloeosporioides. J Nanopart Res 13:2525–2532

Alcantara EN, Wyse DL (1988) Glyphosate as harvest aid for corn (*Zea mays*). Weed Technol 2:410–413

Alghuthaymi MA, Almoammar H, Rai M et al (2015) Myconanoparticles: synthesis and their role in phytopathogens management. Biotechnol Biotechnol Equip 29:221–226

Ali SM, Yousef NMH, Nafady NA (2015) Application of biosynthesized silver nanoparticles for the control of land snail *Eobania vermiculata* and some plant pathogenic fungi. J Nanomater 2015:1–10

Almenar E, Auras R, Rubino M, Harte B (2007) A new technique to prevent the main post harvest diseases in berries during storage: inclusion complexes β-cyclodextrin-hexanal. Int J Food Microbiol 118:164–172

Almenar E, Catala R, Hernandez-Muñoz P, Gavara R (2009) Optimization of an active package for wild strawberries based on the release of 2-nonanone. Food Sci Technol 42:587–593

Anand R, Kulothungan S (2014) Silver mediated bacterial nanoparticles as seed dressing against crown rot pathogen of groundnut. Arch Appl Sci Res 6:109–113

Anjali CH, Sudheer KS, Margulis-Goshen K, Magdassi S, Mukherjee A, Chandrasekaran N (2010) Formulation of water-dispersible nanopermethrin for larvicidal applications. Ecotoxicol Environ Saf 73:1932–1936

Aravinthan A, Govarthanan M, Selvam K, Praburaman L, Selvankumar T, Balamurugan R, Kamala-Kannan S, Kim JH (2015) Sunroot mediated synthesis and characterization of silver nanoparticles and evaluation of its antibacterial and rat splenocyte cytotoxic effects. Int J Nanomedicine 10:1977–1983

Azania CAM, Azania AAPM, Schiaveto AR, Pizzo IV, Marcari M, Panin IEL, Oliveira C (2010) Eficácia de herbicidas no controle de espécies de corda-de-viola em cana-de-açucar. STAB 29:41–45

Baker S, Rakshith D, Kavitha KS et al (2013) Plants: emerging as nanofactories towards facile route in synthesis of nanoparticles. Bioimpacts 3(3):111–117

Baker S, Satish S (2015) Biosynthesis of gold nanoparticles by Pseudomonas veronii AS41G inhabiting Annona squamosa. Spectrochim Acta A Mol Biomol Spectrosc 150:691–695

Baker S, Volova T, Prudnikova SV et al (2017) Nanoagroparticles emerging trends and future prospect in modern agriculture system. Environ Toxicol Pharmacol 53:10–17

Balaguer MP, Fajardo P, Gartner H et al (2014) Functional properties and antifungal activity of films based on gliadins containing cinnamaldehyde and natamycin. Int J Food Microbiol 173:62–71

Benelli G, Maggi F, Pavela R et al (2018) Mosquito control with green nanopesticides: towards the One Health approach? A review of non-target effects. Environ Sci Pollut Res 25:10184–10206

Bhattacharyya A, Duraisamy P, Govindarajan M et al (2016) Nano-biofungicides: emerging trend in insect pest control. In: Prasad R (ed) Advances and applications through fungal nanobiotechnology. Springer, Cham, pp 307–3019

Bhattacharyya A, Duraisamy P, Govindarajan M, Buhroo AA, Prasad R (2016a) Nano-biofungicides: Emerging trend in insect pest control. In: Prasad R (ed) Advances and applications through fungal nanobiotechnology. Springer International Publishing Switzerland, pp 307–319

Bhattacharyya A, Prasad R, Buhroo AA, Duraisamy P, Yousuf I, Umadevi M, Bindhu MR, Govindarajan M, Khanday AL (2016b) One-pot fabrication and characterization of silver nanoparticles using *Solanum lycopersicum*: An eco-friendly and potent control tool against Rose Aphid, *Macrosiphum rosae*. J Nanoscience, Article ID 4679410, 7 pages, 2016. https://doi.org/10.1155/2016/4679410

Buhroo AA, Nisa G, Asrafuzzaman S, Prasad R, Rasheed R, Bhattacharyya A (2017) Biogenic silver nanoparticles from *Trichodesma indicum* aqueous leaf extract against *Mythimna separata* and evaluation of its larvicidal efficacy. J Plant Protect Res 57(2):194–200, DOI:10.1515/jppr-2017-0026

Bhor G, Maskare S, Hinge S, Singh L, Nalwade A (2014) Synthesis of silver nanoparticles by using leaflet extract of *Nephrolepis exaltata* L. and evaluation of antibacterial activity against human and plant pathogenic bacteria. Asian J Pharm Technol Innov 02(07):2014

Boro RC, Kaushal J, Nangia Y, Wangoo N, Bhasinc A, Suri CR (2011) Gold nanoparticles catalyzed chemiluminescence immunoassay for detection of herbicide 2,4-dichlorophenoxyacetic acid. Analyst 136:2125–2130

Campos EVR, De Oliveira JL, Da Silva CMG et al (2015) Polymeric and solid lipid nanoparticles for sustained release of carbendazim and tebuconazole in agricultural applications. Sci Rep 5:1–14

Carvalho FP (2006) Agriculture, pesticides, food security and food safety. Environ Sci Pol 9:685–692

Cătălin BP, Gudovan D, Gudovan I (2017) Nanopesticides: a new paradigm in crop protection. In: Grumezescu AM (ed) New pesticides and soil sensors. Elsevier, London, pp 129–192

Chaw JL, Basri M, Omar D et al (2012) Green nano-emulsion intervention for water-soluble glyphosate isopropylamine (IPA) formulations in controlling *Eleusine indica* (*E. indica*). Pestic Biochem Physiol 102:19–29

Chen G, Liu B (2016) Cellulose sulfate based film with slow-release antimicrobial properties prepared by incorporation of mustard essential oil and ß-cyclodextrin. Food Hydrocoll 55:100–107

Clemente Z, Grillo R, Jonsson M et al (2014) Ecotoxicological evaluation of poly(ε-caprolactone) nanocapsules containing triazine herbicides. J Nanosci Nanotechnol 14:4911–4917

da Rocha Neto AC, Maraschin M, Di Piero RM (2015) Antifungal activity of salicylic acid against Penicillium expansum and its possible mechanisms of action. Int J Food Microbiol 215:64–70
da Rocha Neto AC, Luiz C, Maraschin M, Di Piero RM (2016) Efficacy of salicylic acid to reduce Penicillium expansum inoculum and preserve apple fruits. Int J Food Microbiol 221:54–60
da Rocha Neto AC, de Oliveira da Rocha AB, Maraschin M et al (2018) Factors affecting the entrapment efficiency of β-cyclodextrins and their effects on the formation of inclusion complexes containing essential oils. Food Hydrocoll 77:509–523
Darolt JC, da Rocha Neto AC, Di Piero RM (2016) Effects of the protective, curative, and eradicative applications of chitosan against Penicillium expansum in apples. Braz J Microbiol 47:1014–1019
Díaz-Blancas V, Medina DI, Padilla-Ortega E et al (2016) Nanoemulsion formulations of fungicide tebuconazole for agricultural applications. Molecules 21:1–12
Dich J, Zahm SH, Hanberg A, Adami HO (1997) Pesticides and cancer. Cancer Causes Control 8:420–443
Dimetry NZ, Hussein HM (2016) Role of nanotechnology in agriculture with special reference to pest control. Int J PharmTech Res 9:121–144
Dubas ST, Pimpan V (2008) Humic acid assisted synthesis of silver nanoparticles and its application to herbicide detection. Mater Lett 62:2661–2663
Duhan JS, Kumar R, Kumar N et al (2017) Nanotechnology: the new perspective in precision agriculture. Biotechnol Rep 15:11–23
Eddleston M, Bateman DN (2012) Pesticides. Medicine (Baltimore) 40:147–150
Eddleston M, Karalliedde L, Buckley N et al (2002) Pesticide poisoning in the developing world - a minimum pesticides list. Lancet 360:1163–1167
El-Rahman AFA, Mohammad TGM (2013) Green synthesis of silver nanoparticle using *Eucalyptus globulus* leaf extract and its antibacterial activity. J Appl Sci Res 9(10):6437–6440
Fathi M, Martin A, McClements DJ (2014) Nanoencapsulation of food ingredients using carbohydrate based delivery systems. Trends Food Sci Technol 39:18–39
Felipini RB, Boneti JI, Katsurayama Y et al (2016) Apple scab control and activation of plant defence responses using potassium phosphite and chitosan. Eur J Plant Pathol 145:929–939
Food and Agriculture Organization of the United Nations (2006) International code of conduct on the distribution and use of pesticides. Adopted by the hundred and twenty-third session of the FAO Council in November 2002
Frankova A, Smid J, Bernardos A et al (2016) The antifungal activity of essential oils in combination with warm air flow against postharvest phytopathogenic fungi in apples. Food Control 68:62–68
Goswami A, Roy I, Sengupta S, Debnath N (2010) Novel applications of solid and liquid formulations of nanoparticles against insect pests and pathogens. Thin Solid Films 519:1252–1257
Grillo R, Pereira AES, Nishisaka CS, Lima RD, Oehlke K, Greiner R, Leonardo F, Fraceto LF (2014) Chitosan/tripolyphosphate nanoparticles loaded with paraquat herbicide, an environmentally safer alternative for weed control. J Hazard Mater 278:163–171
Hafez EE, Hassan HS, Elkady M, Salama E (2014) Assessment of antibacterial activity for synthesized zinc oxide nanorods against plant pathogenic strains. Int J Sci Technol Res 3:318–324
Hayles J, Johnson L, Worthley C, Losic D (2017) Nanopesticides: a review of current research and perspectives. In: Grumezescu AM (ed) New pesticides and soil sensors. Elsevier Inc., Amsterdam, pp 193–225
Hazrati H, Saharkhiz MJ, Niakousari M, Moein M (2017) Natural herbicide activity of *Satureja hortensis* L. essential oil nanoemulsion on the seed germination and morphophysiological features of two important weed species. Ecotoxicol Environ Saf 142:423–430
He L, Liu Y, Mustapha A, Lin M (2010) Antifungal activity of zinc oxide nanoparticles against Botrytis cinerea and Penicillium expansum. Microbiol Res 166:207–215
Hess FD (2000) Light-dependent herbicides: an overview. Weed Sci 48:160–170
Hong J, Peralta-Videa JR, Gardea-Torresdey JL (2013) Nanomaterials in agricultural production: benefits and possible threats? In: Shamim N, Sharma VK (eds) Sustainable nanotechnology

and the environment: advances and achievements. American Chemical Society, Washington, D.C., pp 73–90

Iavicoli I, Leso V, Beezhold DH, Shvedova AA (2017) Nanotechnology in agriculture: opportunities, toxicological implications, and occupational risks. Toxicol Appl Pharmacol 329:96–111

Jacques MT, Oliveira JL, Campos EVR et al (2017) Safety assessment of nanopesticides using the roundworm *Caenorhabditis elegans*. Ecotoxicol Environ Saf 139:245–253

Jampílek J, Kráľová K (2017) Nanopesticides: preparation, targeting, and controlled release. In: New pesticides and soil sensors. Elsevier, London, pp 81–127

Jeffery EM, Shaw DR, Barrentine WL (1998) Herbicide combinations for preharvest weed desiccation in early maturing soybean (*Glycine max*). Weed Technol 12:157–165

Jia X, Sheng WB, Li W et al (2014) Adhesive polydopamine coated avermectin microcapsules for prolonging foliar pesticide retention. ACS Appl Mater Interfaces 6:19552–19558

Jo YK, Kim BH, Jung G (2009) Antifungal activity of silver ions and nanoparticles on phytopathogenic fungi. Plant Dis 93:1037–1043

Joshi A (2014) Utility of Carbon Nanotubes for Enhancement of Crop (Wheat) in Agriculture. Nanoscitech Volume: Tata McGraw Hill, USA :554–555

Joshi A, Kaur S, Dharamvir K et al (2018) Multi-walled carbon nanotubes applied through seed-priming treatment influence multiple mechanisms to stimulate growth and yield of bread wheat. J Sci Food Agric 98:3148–3160

Kah M, Hofmann T (2014) Nanopesticide research: current trends and future priorities. Environ Int 63:224–235

Kah M, Beulke S, Tiede K, Hofmann T (2013) Nanopesticides: state of knowledge, environmental fate, and exposure modeling. Crit Rev Environ Sci Technol 43:1823–1867

Kah M, Weniger AK, Hofmann T (2016) Impacts of (Nano)formulations on the fate of an insecticide in soil and consequences for environmental exposure assessment. Environ Sci Technol 50:10960–10967

Kah M, Singh KR, Gogos A, Bucheli T (2018a) A critical evaluation of nanopesticides and nanofertilizers against their conventional analogues. Nat Nanotechnol 13(8):677–684

Kah M, Walch H, Hofmann T (2018b) Environmental fate of nanopesticides: durability, sorption and photodegradation of nanoformulated clothianidin. Environ Sci Nano 5:882–889

Kalishwaralal K, Banumathi E, Ram KPS, Deepak V, Muniyandi J, Eom SH, Gurunathan S (2009) Silver nanoparticles inhibit VEGF induced cell proliferation and migration in bovine retinal endothelial cells. Colloids Surf B Biointerfaces 73:51–57

Kang T, Wang F, Lu L, Zhang Y, Liu T (2010) Methyl parathion sensors based on gold nanoparticles and Nafion film modified glassy carbon electrodes. Sensors Actuators B Chem 145:104–109

Kanhed P, Birla S, Gaikwad S, Gade A, Seabra AB, Rubilar O, Duran N, Rai M (2014) In vitro antifungal efficacy of copper nanoparticles against selected crop pathogenic fungi. Mater Lett 115:13–17

Karande P, Trasatti JP, Chandra D (2015) Novel approaches for the delivery of biologics to the central nervous system. In: Novel approaches and strategies for biologics, vaccines and cancer therapies. Elsevier, New York, pp 59–88

Kasprowicz MJ, Kozio M, Gorczyca A (2010) The effect of sil- ver nanoparticles on phytopathogenic spores of Fusarium culmorum. Can J Microbiol 56:247–253

Katz H, Mishael YG (2013) Reduced herbicide leaching by in situ adsorption of herbicide-micelle formulations to soils. J Agric Food Chem 62:50–57

Kavitha KS, Baker S, Rakshith D, Kavitha HU, Yashwantha RHC, Harini BP, Satish S (2013) Plants as green source towards synthesis of nanoparticles. Int Res J Biol Sci 2:66–76

Kayaci F, Sen HS, Durgun E, Uyar T (2014) Functional electrospun polymeric nanofibers incorporating geraniol-cyclodextrin inclusion complexes: high thermal stability and enhanced durability of geraniol. Food Res Int 62:424–431

Khadri H, Alzohairy M, Janardhan A, Kumar AP, Narasimha G (2013) Green synthesis of silver nanoparticles with high fungicidal activity from olive seed extract. Adv Nanopart 2:241–246

Kheiri A, Moosawi Jorf SA, Mallihipour A et al (2016) Application of chitosan and chitosan nanoparticles for the control of Fusarium head blight of wheat (*Fusarium graminearum*) in vitro and greenhouse. Int J Biol Macromol 93:1261–1272

Kiely T, Donaldson D, Grube A (2004) Pesticide industry sales and usage. 2000 and 2001 market estimates available at: http://wwwepagov/pesticides/pestsales/01pestsales/market_estimates2001pdf. Accessed 25 Oct 2012

Kim JS, Kuk E, Yu KN, Kim JH, Park SJ, Lee HJ, Kim SH, Park YK, Park YH, Hwang CY, Kim YK, Lee YS, Jeong DH, Cho MH (2007) Antimicrobial effects of silver nanoparticles. Nanomedicine 3:95–101

Kim SW, Jung JH, Lamsal K et al (2012) Antifungal effects of silver nanoparticles (AgNPs) against various plant pathogenic fungi. Mycobiology 40:53–58

Kookana RS, Boxall ABA, Reeves PT et al (2014) Nanopesticides: guiding principles for regulatory evaluation of environmental risks. J Agric Food Chem 62:4227–4240

Kottegoda N, Sandaruwan C, Priyadarshana G et al (2017) Urea-hydroxyapatite nanohybrids for slow release of nitrogen. ACS Nano 11:1214–1221

Krug HF (2014) Nanosafety research – are we on the right track? Angew Chem Int Ed 3: 12304–12319

Kumar S, Chauhan N, Gopal M et al (2015) Development and evaluation of alginate-chitosan nanocapsules for controlled release of acetamiprid. Int J Biol Macromol 81:631–637

Laware S, Raskar S (2014) Influence of zinc oxide nanoparticles on growth, flowering and seed productivity in onion. Int J Curr Microbiol Appl 3:874–881

Liu R, Lal R (2015) Potentials of engineered nanoparticles as fertilizers for increasing agronomic productions. Sci Total Environ 514:131–139

Liu F, Wen LX, Li ZZ et al (2006) Porous hollow silica nanoparticles as controlled delivery system for water-soluble pesticide. Mater Res Bull 41:2268–2275

Liu W, Yao J, Cai M et al (2014) Synthesis of a novel nanopesticide and its potential toxic effect on soil microbial activity. J Nanopart Res 16:2677

Loha KM, Shakil NA, Kumar J et al (2011) Release kinetics of β-cyfluthrin from its encapsulated formulations in water. J Environ Sci Health B Pestic Food Contam Agric Wastes 46:201–206

Luiz C, Rocha Neto AC, Di Piero RM (2015) Resistance to xanthomonas gardneri in tomato leaves induced by polysaccharides from plant or microbial origin. J Plant Pathol 97:119–127

Luiz C, Rocha Neto AC, Franco PO, Robson Marcelo DP (2017) Nanoemulsions of essential oils and aloe polysaccharides: antimicrobial activity and resistance inducer potential against Xanthomonas fragariae. Trop Plant Pathol 42:370–381

Luo M, Liu D, Zhao L, Han J, Liang Y, Wang P, Zhou Z (2014) A novel magnetic ionic liquid modified carbon nanotube for the simultaneous determination of aryloxyphenoxy-propionate herbicides and their metabolites in water. Anal Chim Acta 852:88–96

Martínez-Camacho AP, Cortez-Rocha MO, Ezquerra-Brauer JM et al (2010) Chitosan composite films: thermal, structural, mechanical and antifungal properties. Carbohydr Polym 82:305–315

Medda S, Hajra A, Dey U, Bose P, Mondal NK (2014) Biosynthesis of silver nanoparticles from *Aloe vera* leaf extract and antifungal activity against Rhizopus sp. and Aspergillus sp. Appl Nanosci 5:875

Mew EJ, Padmanathan P, Konradsen F et al (2017) The global burden of fatal self-poisoning with pesticides 2006–15: systematic review. J Affect Disord 219:93–104

Morsy MK, Khalaf HH, Sharoba AM, El-Tanahi HH, Cutter CN (2014) Incorporation of essential oils and nanoparticles in pullulan films to control foodborne pathogens on meat and poultry products. J Food Sci 79:M675–M682

Narayanasamy P (2011) Detection of fungal pathogens in plants. In: Microbial plant-pathogens detection and disease diagnosis. Springer, Dordrecht, p 291

Nenaah GE, Ibrahim SIA, Al-Assiuty BA (2015) Chemical composition, insecticidal activity and persistence of three Asteraceae essential oils and their nanoemulsions against *Callosobruchus maculatus* (F.). J Stored Prod Res 61:9–16

Ngo QB, Dao TH, Nguyen HC, Tran XT, Nguten TV, Khuu TD, Huynh TH (2014) Effects of nanocrystalline powders (Fe, Co and Cu) on the germination, growth, crop yield and product quality of soybean (Vietnamese species DT-51). Adv Nat Sci Nanosci Nanotechnol 5:1–7

Oh SD, Lee S, Choi SH, Lee IS, Lee YM, Chun JH, Park HJ (2006) Synthesis of Ag and Ag-SiO2 nanoparticles by Y-irradiation and their antibacterial and antifungal efficiency against *Salmonella enterica* serovar typhimurium and Botrytis cinerea. Colloids Surf A Physicochem Eng Asp 275:228–233

Ouda SM (2014) Antifungal activity of silver and copper nanoparticles on two plant pathogens, *Alternaria alternata* and Botrytis cinerea. Res J Microbiol 9:34–42

Paret ML, Palmateer AJ, Knox GW (2013) Evaluation of a light-activated nanoparticle formulation of titanium dioxide with zinc for management of bacterial leaf spot on Rosa "Noare". Hortscience 48:189–192

Paulkumar K, Gnanajobitha G, Vanaja M, Rajeshkumar S, Malarkodi C, Pandian K, Annadurai G (2014) *Piper nigrum* leaf and stem assisted green synthesis of silver nanoparticles and evaluation of its antibacterial activity against agricultural plant pathogens. Sci World J 2014:829894

Pereira AES, Grillo R, Mello NFS et al (2014) Application of poly(epsilon-caprolactone) nanoparticles containing atrazine herbicide as an alternative technique to control weeds and reduce damage to the environment. J Hazard Mater 268:207–215

Pereira EI, Giroto AS, Bortolin A et al (2015) Perspectives in nanocomposites for the slow and controlled release of agrochemicals: fertilizers and pesticides Elaine. In: Nanotechnologies in food and agriculture. Springer, Cham, pp 241–265

Prasad R (2014) Synthesis of silver nanoparticles in photosynthetic plants. Journal of Nanoparticles, Article ID 963961, http://dx.doi.org/10.1155/2014/963961

Prasad R, Pandey R, Barman I (2016) Engineering tailored nanoparticles with microbes: quo vadis. WIREs Nanomed Nanobiotechnol 8:316–330. doi: 10.1002/wnan.1363

Prasad R, Jha A and Prasad K (2018) Exploring the Realms of Nature for Nanosynthesis. Springer International Publishing (ISBN 978-3-319-99570-0) https://www.springer.com/978-3-319-99570-0

Qin Y, Ji X, Jing J et al (2010) Size control over spherical silver nanoparticles by ascorbic acid reduction. Colloids Surf A Physicochem Eng Asp 372:172–176

Queiroz GMP, Silva MR, Bianco RJF, Pinheiro A, Kaufmann V (2011) Transporte de glifosato pelo escoamento superficial e por lixiviação em um solo agrícola. Quim Nova 34:190–195

Ramy SY, Ahmed OF (2013) In vitro study of the antifungal efficacy of zinc oxide nanoparticles against Fusarium oxysporum and Penicillium expansum. Afr J Microbiol Res 7:1917–1923

Rao KJ, Paria S (2013) Use of sulfur nanoparticles as a green pesticide on *Fusarium solani* and Venturia inaequalis phytopathogens. RSC Adv 3:10471–10478

Rawani A, Ghosh A, Chandra G (2013) Mosquito larvicidal and antimicrobial activity of synthesized nano-crystalline silver particles using leaves and green berry extract of *Solanum nigrum* L. (Solanaceae: Solanales). Acta Trop 128:613–622

Ribeiro-Santos R, Andrade M, de Melo NR, Sanches-Silva A (2017) Use of essential oils in active food packaging: recent advances and future trends. Trends Food Sci Technol 61:132–140

Rodrigues BN, Almeida FS (2011) Guia de herbicidas. Grafmarke, Londrina, p 639

Roy A, Singh SK, Bajpai J, Bajpai AK (2014) Controlled pesticide release from biodegradable polymers. Cent Eur J Chem 12:453–469

Ruffolo SA, La Russa MF, Malagodi M, Oliviero Rossi C, Palermo AM, Crisci GM (2010) ZnO and ZnTiO3 nanopowders for antimicrobial stone coating. Appl Phys A Mater Sci Process 100:829–834

Sabir S, Arshad M, Chaudhari SK (2014) Zinc oxide nanoparticles for revolutionizing agriculture: synthesis and applications. Sci World J 2014:1–8

Sangeetha J, Thangadurai D, Hospet R, Harish ER, Purushotham P, Mujeeb MA, Shrinivas J, David M, Mundaragi AC, Thimmappa AC, Arakera SB, Prasad R (2017) Nanoagrotechnology for soil quality, crop performance and environmental management. In: Prasad R, Kumar M, Kumar V (eds) Nanotechnology. Springer Nature Singapore Pte Ltd., Singapore, pp 73–97

Sangeetha J, Thangadurai D, Hospet R, Purushotham P, Karekalammanavar G, Mundaragi AC, David M, Shinge MR, Thimmappa SC, Prasad R, Harish ER (2017b) Agricultural nanotechnology: Concepts, benefits, and risks. In: Prasad R, Kumar M, Kumar V (eds) Nanotechnology. Springer Nature Singapore Pte Ltd., Singapore, pp 1–77

Sathiyabama M, Parthasarathy R (2016) Biological preparation of chitosan nanoparticles and its in vitro antifungal efficacy against some phytopathogenic fungi. Carbohydr Polym 151:321–325

Schreinemachers P, Tipraqsa P (2012) Agricultural pesticides and land use intensification in high, middle and low income countries. Food Policy 37:616–626

Shao X, Cao B, Xu F et al (2015) Effect of postharvest application of chitosan combined with clove oil against citrus green mold. Postharvest Biol Technol 99:37–43

Shaoqin L, Lang Y, Xiuli Y, Zhaozhu Z, Zhiyong T (2008) Recent Advances in Nanosensors for Organophosphate Pesticide Detection. Adv Powder Technol 19 (5): 419–441

Shatkin JA, Kim B (2015) Cellulose nanomaterials: life cycle risk assessment and environmental health and safety roadmap. Environ Sci Nano 2:497–499

Singh Y, Meher JG, Raval K et al (2017) Nanoemulsion: concepts, development and applications in drug delivery. J Control Release 252:28–49

Solans C, Solé I (2012) Nano-emulsions: formation by low-energy methods. Curr Opin Colloid Interface Sci 17:246–254

Stadler T, Buteler M, Weaver DK (2010) Novel use of nanostructured alumina as an insecticide. Pest Manag Sci 66(6):577–579

Stloukal P, Kucharczyk P, Sedlarik V et al (2012) Low molecular weight poly(lactic acid) microparticles for controlled release of the herbicide metazachlor: preparation, morphology, and release kinetics. J Agric Food Chem 60:4111–4119

Suresh KRS, Shiny PJ, Anjali CH, Jerobin J, Goshen KM, Magdassi S, Mukherjee A, Chandrasekaran N (2013) Distinctive effects of nano-sized permethrin in the environment. Environ Sci Pollut Res Int 20:2593–2602

Syu YY, Hung JH, Chen JC, Chuang HW (2014) Impacts of size and shape of silver nanoparticles on Arabidopsis plant growth and gene expression. Plant Physiol Biochem 83:57–64

Tiwari DK, Dasgupta-Schubert N, Villaseñor Cendejas LM et al (2014) Interfacing carbon nanotubes (CNT) with plants: enhancement of growth, water and ionic nutrient uptake in maize (*Zea mays*) and implications for nanoagriculture. Appl Nanosci 4:577–591

Ulrichs C, Mewis I, Goswami A (2005) Crop diversification aiming nutritional security in West Bengal: biotechnology of stinging capsules in nature's water-blooms. Ann Tech Issue State Agric Technol Service Assoc:1–18

United Nation (2017). Available at: https://www.un.org/development/desa/en/news/population/world-population-prospects-2017.html

Villaverde JJ, Sevilla-Morán B, López-Goti C et al (2017) An overview of nanopesticides in the framework of European legislation. In: New pesticides and soil sensors. Elsevier Inc., pp 227–271. Academic Press, Cambridge, Massachusetts, USA

Vyas SP, Kannan ME, Jain S, Mishra V, Singh P (2004) Design of liposomal aerosols for improved delivery of rifampicin to alveolar macrophages. Int J Pharm 269:37–49

Wani AH, Shah MA (2012) A unique and profound effect of MgO and ZnO nanoparticles on some plant pathogenic fungi. J Appl Pharm Sci 2:40–44

Wen L-X, Li Z-Z, Zou H-K et al (2005) Controlled release of avermectin from porous hollow silica nanoparticles. Pest Manag Sci 61:583–590

Wen P, Zhu D-H, Wu H et al (2016) Encapsulation of cinnamon essential oil in electrospun nanofibrous film for active food packaging. Food Control 59:366–376

World Health Organization (2017) Cancer. Available at: http://www.who.int/cancer/en/

Wu M, Lin G, Chen D et al (2002) Sol-hydrothermal synthesis and hydrothermally structural evolution of nanocrystal titanium dioxide. Chem Mater 14:1974–1980

Xiang Y, Han J, Zhang G et al (2018) Efficient synthesis of starch-regulated porous calcium carbonate microspheres as a carrier for slow-release herbicide. ACS Sustain Chem Eng 6:3649–3658

Xu J, Fan QJ, Yin ZQ et al (2010) The preparation of neem oil microemulsion (*Azadirachta indica*) and the comparison of acaricidal time between neem oil microemulsion and other formulations in vitro. Vet Parasitol 169:399–403

Yang FL, Li XG, Zhu F, Lei CL (2009) Structural characterization of nanoparticles loaded with garlic essential oil and their insecticidal activity against *Tribolium castaneum* (Herbst) (Coleoptera: Tenebrionidae). J Agric Food Chem 57(21):10156–10162

Yu Z, Sun X, Song H, Wang W, Ye Z, Shi L, Ding K (2015) Glutathione-responsive carboxymethyl chitosan nanoparticles for controlled release of herbicides. Mater Sci Appl 6:591–604

Yuvaraj M, Subramanian KS (2014) Controlled-release fertilizer of zinc encapsulated by a manganese hollow core shell. Soil Sci Plant Nutr 61:319–326

Zaki AM, Zaki AH, Farghali AA, Abdel-Rahim EF (2017) Sodium titanate -Bacillus as a new nanopesticide for cotton leaf-worm. J Pure Appl Microbiol 11:725–732

Zhang W, Asiri AM, Liu D, Du D, Lin Y (2014) Nanomaterial-based biosensors for environmental and biological monitoring of organophosphorus pesticides and nerve agents. Trends Anal Chem 54:1–10

Zhang X, Liu Z, Shen W, Gurunathan S (2017) Silver nanoparticles: synthesis, characterization, properties, applications, and therapeutic approaches. Int J Mol Sci 17:1534

Zhao L, Seth A, Wibowo N, Zhao CX, Mitter C, Yu Middelberg APJ (2014) Nanoparticle vaccine. Vaccine 32:327–337

Zhao H, Lan Y, Nan C et al (2016) Preparation of 8% fenpropathrin·cyflumetofen nano-emulsion and its performance. Sci Agric Sin 49:2700–2710

Chapter 16
Development of Nano-Bioformulations of Nutrients for Sustainable Agriculture

Tanveer Bilal Pirzadah, Bisma Malik, Tariq Maqbool, and Reiaz Ul Rehman

Contents

16.1 Introduction

The global agricultural production faces great challenges like stagnation in crop yields, low nutrient use potency, declining soil fertility, multi-nutrient deficiencies, shrinking of agricultural land, water accessibility, and a dearth of labor due to evacuation of individuals from farming (Godfray et al. 2010; FAO 2017). FAO reports that due to the industrialization and urbanization, the quality of the soil has been degraded and thus poses a great threat to the agricultural sector to meet the global food production and to cater the food crises of the world's ever-increasing

T. B. Pirzadah · B. Malik · R. U. Rehman (✉)
Department of Bioresources, University of Kashmir, Srinagar, India

T. Maqbool
Department of Nanotechnology, University of Kashmir, Srinagar, India

R. Prasad et al. (eds.), *Nanobiotechnology in Bioformulations*, Nanotechnology in the Life Sciences, https://doi.org/10.1007/978-3-030-17061-5_16

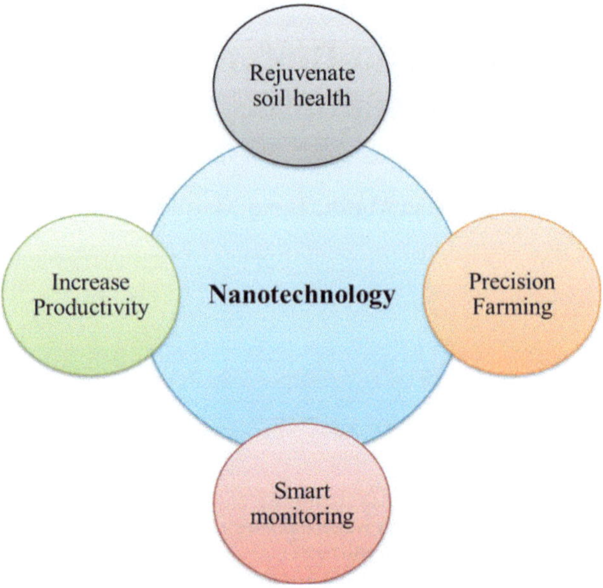

Fig. 16.1 Various domains of nanotechnology in the agricultural sector

population (FAO 2017). Excessive use of synthetic fertilizers and pesticides to enhance the crop yield is definitely not an appropriate choice for long term in the light of the fact that these synthetic fertilizers are considered as double-edged sword, which, from one viewpoint, enhance the crop yield, but at the same time they possess deleterious effect on the soil microflora and thus reduce its fertility. Besides, it causes irreparable damage to the soil structure and disturbs the equilibrium in the food chain across ecosystem, which might in turn leads to genetic mutations in future generations of consumers. Thus, the increased reliance on synthetic agricultural additives during and post green revolution has created severe issues pertaining to sustainability, environmental impact, and health hazards. Therefore, nanotechnology has emerged as a promising area to tackle the above problems, especially in the agriculture sector. Nano-agribusiness is an emerged field to enhance crop yield, rejuvenate soil health, provide precision farming, and stimulate plant growth (Prasad et al. 2014, 2017a, b) (Fig. 16.1).

16.2 Integration of Biofertilizers with Nanotechnology

An innovative approach of using green fertilizers, known as biofertilizers, has recently been used as a promising alternative to chemical fertilizers for the sustainable agriculture (Mishra et al. 2015). However, this approach of using biofertilizers confronted certain issues like poor shelf-life, stability, efficiency, and performance

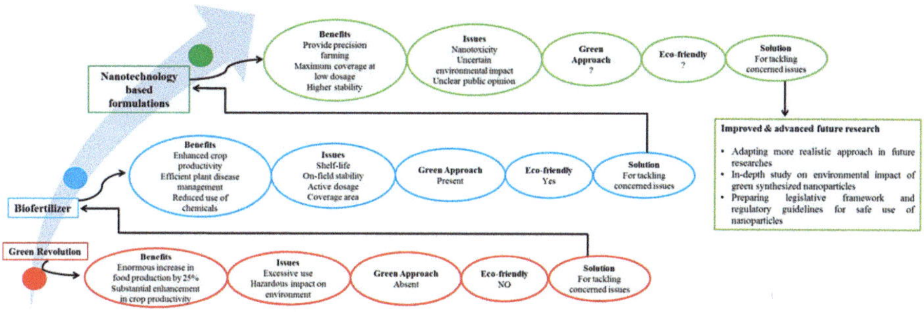

Fig. 16.2 Schematic representation of different revolutions in the field of agriculture with their respective benefits, issues, and solutions

under fluctuating climatic conditions, and the most significant factor is high dosage requirements for maximal coverage area. In recent years, nanobiotechnology has emerged as a tool in the agriculture sector to promote growth and productivity by designing and developing the ultra-small particles having some unique features like surface area to volume ratio and physicochemical properties as compared to conventional approaches. Nanotechnology has gained a rapid momentum in the modern agriculture system and is expected to become a main thrust in near future by offering potential applications. Nanotechnology involves the multidisciplinary approach and is considered as the sixth most revolutionary technology in the current era (Knell 2010). The overall prominence and development of nanotechnology industry could be foreseen by the way that its fairly estimated worth will reach to US$ 75.8 billion by 2020 due to its noteworthy development at worldwide level (Research and Markets 2015). No doubt, the nanotechnology has speedily added to worldwide development by conveying solid applications in numerous previously mentioned mechanical segments, but at the same time, nanotechnology can possibly profit society by altering the agrarian segment. Essentially, this innovation approach has sponsored the farming-based business segment with yearly development rate of 25% (US$ 1.08 billion). Besides, it is evaluated that joining of cutting edge nanotechnology in agribusiness would push the worldwide monetary development to approximately US$ 3.4 trillion by 2020 (Sabourin and Ayande 2015). This clearly indicates how agri-nanobiotechnology plays a pivotal role in the agricultural sector, without any negative impact on the environment and other regulatory issues of biosafety. Agri-nanobiotechnology is an innovative green technology which provides the solution to global food security, sustainability, and climate change (Mishra et al. 2014; Sangeetha et al. 2017a, b) (Fig. 16.2). In the current era, nano-fertilizers are available in the market, and some of the nano-fertilizers commercialized in the market are listed in Table 16.1.

Table 16.1 Nano-fertilizers available in the market (Prasad et al. 2017a)

Commercial product	Content	Company
Nano-Gro™	Plant growth regulator and immunity enhancer	Agro Nanotechnology Corp., FL, United States
Nano Green	Extracts of corn, grain, soybeans, potatoes, coconut, and palm	Nano Green Sciences, Inc., India
Nano-Ag Answer®	Microorganism, sea kelp, and mineral electrolyte	Urth Agriculture, CA, United States
Biozar Nano-Fertilizer	Combination of organic materials, micronutrients, and macromolecules	Fanavar Nano-Pazhoohesh, Markazi Company, Iran
Nano Max NPK Fertilizer	Multiple organic acids chelated with major nutrients, amino acids, organic carbon, organic micro nutrients/trace elements, vitamins, and probiotic	JU Agri Sciences Pvt. Ltd., Janakpuri, New Delhi, India
Master Nano Chitosan Organic Fertilizer	Water-soluble liquid chitosan, organic acid and salicylic acids, phenolic compounds	Pannaraj Intertrade, Thailand
TAG NANO (NPK, PhoS, Zinc, Cal, etc.) fertilizers	Proteino-lacto-gluconate chelated with micronutrients, vitamins, probiotics, seaweed extracts, humic acid	Tropical Agro-system India (P) Ltd., India

16.3 Advantages of Using Nano-biofertilizers

Chemical fertilizers are applied on the crops through soil either by spraying or surface broadcasting. But the primary drawback that determines the mode of action involves the final concentration of the fertilizers available to the plant. It has been observed that only a small amount meets the target site because the bulk of the fertilizers are lost due to surface runoff, leaching, and evaporation of some volatile chemicals and microbial or photo-degradation of chemicals, thus posing a great threat to the environment (Tilman et al. 2002). Reports revealed that most of these fertilizers (nitrogen, 40–70%; phosphorus, 80–90%; and potassium, 50–90%) are lost in the ecosystem and thus are not available to the plant which in turn causes a great economic loss (Ombodi and Saigusa 2000). Therefore, it is indispensable to promote green approach that leads to more nutrition and improvement by reducing the utilization of synthetic fertilizers. In comparison to synthetic fertilizers that are used in bulk quantity (80–140 kg/hectare), nanotechnology targets precision farming by involving small amount of these fertilizers. Besides, the problems that are associated with bio-formulations are overcome by nano-biofertilizers (Auffan et al. 2009). The nano-fertilizers possess the efficiency to reduce nutrient loss via leaching and prevent brisk modifications in their chemical nature which in turn increases the nutrient use efficiency and thus addressing fertilizer-related environmental concerns. Moreover, the application of nanotechnology in agriculture counteracts the problems such as crop yield, food security, climate change, and sustainability (Mishra et al. 2014). Nano-fertilizers are a nano-structured formulation that delivers nutrients to the plants, allowing dynamic uptake or gradual discharge of active ingredients. Nanoparticles are regarded as efficient vehicles to carry nutrients to the target site by encapsulation or in the form of emulsion of nanoscale dimension. However, the surface coating of nanomaterials on fertilizer particles holds the material more firmly due to higher surface tension than the conventional surfaces and thus aids in controlled release (Brady and Weil 1999; DeRosa et al. 2010; Bhattacharyya et al. 2016). Cui et al. (2010) reported that nano-fertilizers possess a great advantage over conventional fertilizers as they exhibit specificity, reduced toxicity, and gradual release of nutrients.

16.4 Synthesis of Nano-biofertilizers Using Microbes

It has been felt that nanoparticles can be synthesized by novel methods so that the particles are environment-friendly, i.e., no toxic chemicals are used during synthesis. Currently, the use of biological entities has emerged as a novel method for the synthesis of nanoparticles. The biological systems for synthesis of nanoparticles utilize mostly microbes since they naturally produce inorganic materials either intracellularly or extracellularly, for example, magnetostatic bacteria used for magnetite, diatoms used for silicious material, and S-layer bacteria used for gypsum and calcium carbonate (Sastry et al. 2003). Nano-biotechnological way for the synthesis of nanoparticles possesses many advantages, such as use of known microbial technologies and processes for scaling up of biomass (Prasad et al. 2016). This leads to economic viability, possibility of readily covering large surface areas by suitable growth of the microbes, which is a major advantage in the field of agriculture for easier production of biofertilizers. The green synthesis of nano-biofertilizer involves the use of microbial enzymes for breakdown of the respective salts into nanoparticles (Duhan et al. 2017). Further, the organic polymers can play important role in ecosystems by accumulating biologically important elements and also by retaining soil moisture after aggregating soil particles (Ding et al. 2015). Extracellular polymeric substances (EPS) play an important role in cell aggregation, cell adhesion, and biofilm formation that subsequently protect cells from a hostile environment (Ding et al. 2015). Furthermore, certain polysaccharides from microbial sources are surface active, and thus attempts have been made to use them as metal chelaters (Sutherland 1998) emulsifiers (Cirigliano and Carman 1984) and flocculants (Choi et al. 1998) in industrial and environmental fields/domain. Such use of microbial polysaccharides has infused renewed interest in its production and characteristics (Raliya et al. 2013). Worldwide efforts are being done in this direction to make the nano-biofertilizer technology cost-effective. The formulation of any nano-fertilizer should be in such a way that they possess all desired properties such as high solubility, stability, effectiveness, time-controlled release, enhanced targeted activity with effective concentration, and less eco-toxicity with safe, easy mode of delivery and disposal (Torney et al. 2007; Prasad et al. 2014, 2017a). Recently, myconanotechnology has emerged as an attractive field where fungi can be used to synthesize the nanoproducts which possess great application in agriculture sector. Fungi play a lead role in the biosynthesis of nanoparticles because of the potent efficiency in extracellular as well as intracellular enzyme production compared to other microorganisms like bacteria and actinomycetes (Rai et al. 2009; Narayanan and Sakthivel

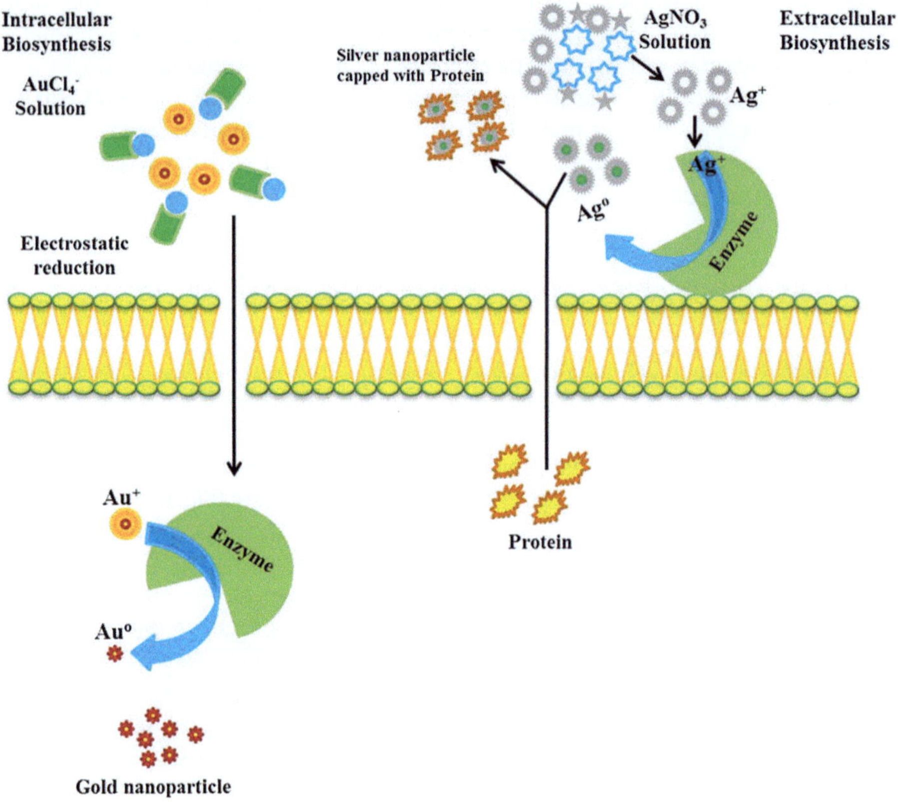

Fig. 16.3 Mycogenic synthesis of nanoparticles using intracellular and extracellular mechanisms

2010; Prasad 2016, 2017; Prasad et al. 2018a). Some microorganisms possess innate ability to survive in extremophilic conditions such as high metal concentrations, and this is due to some important mechanisms like efflux systems, oozing out some organic acids that cause precipitation of metals through redox reactions or chelate formations. The mode of nanomaterial fabrication using mycogenic approach is represented in Fig. 16.3. Table 16.2 summarizes the list of various species of fungi used in the biosynthesis of nanomaterials.

16.5 Targeting of Nanoparticles

Nanoparticles have great potential to deliver nutrients to specific target sites in living systems. The loading of nutrients on the nanoparticles is usually done by means of following approaches like absorption on nanoparticles, attachment on nanoparticles

Table 16.2 List of fungi used in the biosynthesis of nanoparticles

Fungal species	Nanoparticle	Mode of biosynthesis	Size (nm)	References
Penicillium sp.	Ag	Extracellular	16–40	Sadowski et al. (2008)
Alternaria alternate	Ag	Extracellular	20–60	Gajbhiye et al. (2009)
A. clavatus	Ag	Extracellular	10–25	Verma et al. (2010)
A. alternate	Au	Extracellular	12 ± 5	Sarkar et al. (2011)
A. clavatus	Au	Intracellular	20–35	Verma et al. (2011)
A. flavus	Ag	Extracellular	7	Moharrer et al. (2012)
Trichoderma reesei	Ag	Extracellular	5–50	Vahabi et al. (2011)
T. viride	Ag	Extracellular	5–40	Fayaz et al. (2010)
T. harzianum	Ag	Extracellular	30–50	Singh and Balaji (2011)
Verticillium sp.	Magnetite	Extracellular	100–400	Bharde et al. (2006)
Neurospora crassa	Ag	Intra-/extracellular	11	Castro-Longoria et al. (2011)
Pleurotus sajor caju	Ag	Extracellular	5–50	Nithya and Ragunathan (2009)
Verticillium luteoalbum	Au	Intracellular	≥10	Gericke and Pinches (2006)
Fusarium semitectum	Au, Au–Ag	Extracellular	18–80	Sawle et al. (2008)
Aspergillus niger	Ag	Extracellular	15–20	Gade et al. (2008)
Cochliobolus lunatus	Ag	Extracellular	3–21	Salunkhe et al. (2011)

mediated by ligands, encapsulation in nano-particulate polymeric shell, entrapment of polymeric nanoparticles, and synthesis of nanoparticles composed of the nutrient itself. Corradini et al. (2010) evaluated the interaction and stability of chitosan nanoparticle suspensions containing N, P, and K fertilizers which can be useful for agricultural applications. Kottegoda et al. (2011) synthesized urea-modified hydroxyapatite (HA) nanoparticles for gradual release of nitrogen with the crop growth. These nano-fertilizers showed initially burst and subsequently slow release of nitrogen up to 60 days of plant growth compared to commercial fertilizer which shows release only up to 30 days. The large surface area of HA facilitates the large amount of urea attachment on the HA surface. Strong interaction between HA nanoparticles and urea contributes to the slow and controlled release of urea. Similarly, polymer-based mesoporous nanoparticles can also provide efficient carrier system to agrochemical compounds which improves the efficiency and economical utilization. Mesoporous silica nanoparticles (150 nm) have been reported to entrap urea. It has been observed that 15.5% of urea was loaded inside the nanoparticles pores and demonstrated a controlled urea release profile in soil and water. The study revealed at least fivefold improvement in release period (Wanyika et al. 2012). Zinc solubility and dissolution kinetics of ZnO nanoparticles and bulk ZnO particles coated on macronutrient fertilizers (urea and monoammonium phosphate) have been compared by Milani et al. (2012). They reported that coated monoammonium phosphate granules show faster dissolution rate.

16.6 Upgradation of Biofertilizers to Smart Fertilizer

The mode of fertilizer application influences their efficiency and impact on plant systems. Nowadays attention to biofertilizer has increased due to the advancement in countries' research and development, prices of chemical fertilizers, and attention to sustainable agricultural systems (Yosefi et al. 2011). There are some evidences in support of biofertilizers, including that plant growth and yield increase may be stimulated by plant growth promoting bacteria due to their ability of N_2-fixing, phosphate solubility, and production of plant growth hormones (Sahin et al. 2004; Prasad et al. 2015, 2018b). Biofertilizer with 50% of chemical fertilizers (N and P) led to an increase in plant growth, plant height, branch numbers, and fresh and dry weight of safflower in comparison with chemical fertilizers application alone (Ojaghloo et al. 2007). Also the utilization of *Azotobacter* biofertilizer, bio-phosphate fertilizer, and organic fertilizers, with half rate of chemical fertilizers increased the grain yield of safflower (Ojaghloo et al. 2007). Mirzaei et al. (2010) applied *Azotobacter* and *Azospirillum* bacteria in different levels of nitrogen for safflower plant. Their results showed that combined application of these two types of bacteria increased plant growth characteristics and reduced nitrogen fertilizer application by 50%. Cereal yield responses to inoculation may also depend on plant genotype, bacterial strains, and soil type as well as environmental conditions (Salantur et al. 2005). Biofertilizers are able to fix atmospheric nitrogen in the available form for plant (Chen 2006) and have beneficial effects upon plant growth by production of antibiotic (Zahir et al. 2004). *Azotobacter* is used as biofertilizer in the cultivation of most crops (Yasari and Patwardhan 2007). Nowadays, nanotechnology has progressively moved away from the experimental to the practical areas (Baruah and Dutta 2009). For example, the development of slow/controlled release fertilizers and conditional release of pesticides and herbicides, on the basis of nanotechnology, has become critically important for promoting the development of environment-friendly and sustainable agriculture. Indeed, nanotechnology has provided the feasibility of exploiting nanoscale or nanostructured materials as fertilizer carriers or controlled-release vectors for building of so-called "smart fertilizer" as new alternatives to enhance nutrient use efficiency and reduce the costs of environmental protection (Cui et al. 2010; Chinnamuthu and Boopathi 2009).

16.7 Nanotechnology: An Important Tool in Precision Farming

Nanotechnology emerges as an essential tool in precision farming or smart agriculture by maximizing the output such as crop yield and reducing inputs like fertilizer and pesticide usage. It involves the integration of information technology, nano-biosensors, global positioning system (GPS), and remote sensing technology to monitor climatic conditions and agronomic problems and helps to determine maximum crop efficiency. Nano-biosensors act as smart diagnostic tools to enhance crop yield by providing precise information to farmers that helps them to make better decisions and thus makes agriculture more sustainable (McBratney et al. 2005; Prasad et al. 2014, 2017a, b).

16.8 Agribusiness and Nanotechnology

It is reported that global agribusiness market during the financial year 2010 ranges from US$ 20.7 billion to US$ 0.98 trillion. As per the recent reports, it is predicted that the market will enhance to more than US$ 3.4 trillion by 2020. Currently the largest market for nano-agrochemicals is USA with about US$ 3.7 billion followed

Table 16.3 Nanotech products and their agronomic application

Year	Nano-products	Institute	Applications	References
2015	Nano-sized nutrients (ZnO and TiO_2 nanoparticles)	Washington University in St. Louis	Boost in growth and antioxidants in tomatoes	Raliya et al. (2015)
2015	Acetamprid-loaded alginate-chitosan nanocapsules	GJUS & T Hisar, India	Improved delivery of agrochemicals in the field, better efficacy, better control of application/dose.	Kumar et al. (2015)
2012	Macronutrient fertilizers coated with zinc oxide nanoparticles	University of Adelaide, AU, CSIRO Land and Water, AU, Kansas State University, US	Enhancement of nutrients absorption by plants and the delivery of nutrients to specific sites	Milani et al. (2015)
2012	Zeolites and Nano-clays	Geohumus-Frankfurt, DE	Water retention and slow release of agrochemicals for proper absorption by the plants	http://www.geohumus.com/us/products.html
2012	Nanoemulsion	VIT University, INDIA	Neem oil (*Azadirachta indica*) nanoemulsion as larvicidal agent	Anjali et al. (2012)
2011	Primo MAXX	Syngenta, Greensboro, NC, USA	Grass growth regulatory	Prasad et al. (2014)
2008	Hydrolyzed collagen/sodium alginate nanocomposite	Sichuan University, Chengdu, Sichuan, China	Preservation of loquat and cherry	Jia et al. (2008)
2007	Nanosensors	(University of Crete, GR)	Pesticide detection with a liposome-based nano-biosensor	Vamvakaki and Chaniotakis (2007)
2002	Biodegradable thermoplastic starch (TPS)	Pusan National University, Korea	Good tensile strength and lowered water permeability	Park et al. (2002)

by Japan (US$ 750 million) and European Union (US$ 1.2 billion). Besides, 400 industries are actively involved in nanotechnology-based research worldwide, and this number is going to increase in near future. The various nano-tech products used in agriculture generated by different companies are depicted in Table 16.3.

16.9 Ethical Issues in Using Nano-fertilizers

Although nanotechnology has revolutionized the world and gained a rapid momentum in every sector, at the same time there are certain ethical issues and safety measures which are taken into consideration. People all over the world are very conscious regarding the ill effects of the nano-fertilizer because it is directly related to humans. This led to the origin of nano-toxicology which is responsible for assessing toxicological potential of these nano-products but at the same time promotes safety design and use of nanoparticles. This approach needs a comprehensive experimental analysis of the nano-tech products to ensure the safety of the consumer. Currently, there is a common assumption that due to the small size of these nanoparticles they are able to enter into the food chain and ultimately into cells and tissues where they may cause a severe damage to cells (Xia et al. 2009). Therefore, it is mandatory to devise the quality control check points to assess the toxic effects of the nano-products so as to ensure its eco-friendly nature and safety to human consumption.

16.10 Conclusion and Future Recommendations

In the current scenario, the excessive use of fertilizers in the agriculture sector to boost the production not only affects the soil quality but also has deleterious effects on the environment. Enhancing agricultural production to cater the demand of the people is essential, but keeping in mind the negative impact to the environment, novel approaches need to be considered. Nanotechnology is a promising green approach that has revolutionized the agriculture sector. The use of microbial enzymes for the biosynthesis of nano-fertilizers gains a rapid momentum in the nano-bioformulations due to its excellent efficiency and cost-effective nature. Due to the small size of the nano-fertilizers (synthesized chemically), there is a need for risk assessment of our environment, particularly in terms of consumption in form of food or feed. However, nano-biofertilizers would seem to be more environmental friendly as they are synthesized from the biological form, but it does not mean that their risk assessment is not required. Furthermore, there is a need to develop more technologies for the synthesis of nanoparticles with microorganisms well suited and adapted within the rhizosphere of a particular plant. This could help in developing the crop-/plant-specific nano-biofertilizers (biomimetic approach) and thus help in improving the yield. This technology would not only help us currently but also the coming future generations and thus play an active role in food security of the world.

References

Anjali CH, Sharma Y, Mukherjee A, Chandrasekaran N (2012) Neem oil (*Azadirachta indica*) nanoemulsion-a potent larvicidal agent against *Culex quinquefasciatus*. Pest Manag Sci 68:158–163

Auffan M, Rose J, Bottero JY, Lowry GV, Jolivet JP, Wiesner MR (2009) Towards a definition of inorganic nanoparticles from an environmental, health and safety perspective. Nat Nanotechnol 4:634–641. https://doi.org/10.1038/nnano.2009.242

Baruah S, Dutta J (2009) Nanotechnology applications in sensing and pollution degradation in agriculture. Environ Chem Lett J 7:191–204

Bharde A, Rautray D, Bansal V, Ahmad A, Sarkar I, Yusuf SM, Sanyal M, Sastry M (2006) Extracellular biosynthesis of magnetite using fungi. Small 2(1):135–141

Bhattacharyya A, Duraisamy P, Govindarajan M, Buhroo AA, Prasad R (2016) Nano-biofungicides: emerging trend in insect pest control. In: Prasad R (ed) Advances and applications through fungal nanobiotechnology. Springer International Publishing, Cham, pp 307–319

Brady NR, Weil RR (1999) In: Brady NR, Weil RR (eds) The nature and properties of soils. Prentice Hall, New Jersey, pp 415–473

Castro-Longoria E, Vilchis-Nestor AR, Avalos-Borja M (2011) Biosynthesis of silver, gold and bimetallic nanoparticles using the filamentous fungus *Neurospora Crassa*. Colloids Surf B Biointerfaces 83:42–48

Chen J (2006) The combined use of chemical and organic fertilizers for crop growth and soil fertility. Int. workshop on sustained management of the soil rhizosphere system for efficient crop production and fertilizer use. 16–20 October, Thailand

Chinnamuthu CR, Boopathi PM (2009) Nanotechnology and agroecosystem. Madras Agric J 96:17–31

Choi CW, Yoo SA, Oh IH, Park SH (1998) Characterization of an extracellular flocculating substance produced by a planktonic cyanobacterium, Anabaena sp. Biotechnol Lett 20:643–646

Cirigliano MC, Carman GM (1984) Isolation of a bioemulsifier from Candida lipolytica. Appl Environ Microbiol 48:747–750

Corradini E, De Moura MR, Mattoso LHC (2010) A preliminary study of the incorporation of NPK fertilizer into chitosan nanoparticles. Express Polym Lett 4:509–515

Cui HX, Sun CJ, Liu Q, Jiang J, Gu W (2010) Applications of nanotechnology in agrochemical formulation, perspectives, challenges and strategies. In: International conference on Nanoagri, Sao Pedro, Brazil, pp 28–33

DeRosa MC, Monreal C, Schnitzer M, Walsh R, Sultan Y (2010) Nanotechnology in fertilizers. Nat Nanotechnol 5:91

Ding Z, Bourven I, Guibaud G, van Hullebusch ED, Panico A, Pirozzi F, Esposito G (2015) Role of extracellular polymeric substances (EPS) production in bioaggregation: application to wastewater treatment. Appl Microb Biotechnol 99(23):9883–9905

Duhan JS, Kumar R, Kumar N, Kaur P, Nehra K, Duhan S (2017) Nanotechnology: the new perspective in precision agriculture. Biotechnol Rep 15:11–23

FAO (2017) The future of food and agriculture – trends and challenges. FAQ, Rome

Fayaz M, Balaji K, Girilal M, Yadav R, Kalaichelvan PT, Venketesan R (2010) Biogenic synthesis of silver nanoparticles and their synergistic effect with antibiotics: a study against gram-positive and gram-negative bacteria. Nanomed Nanotechnol Biol Med 6:103–109

Gade A, Bonde PP, Ingle AP, Marcato P, Duran N, Rai MK (2008) Exploitation of *Aspergillus niger* for synthesis of silver nanoparticles. J Biobaased Mater Bioenergy 2(3):1–5

Gajbhiye M, Kesharwani J, Ingle A, Gade A, Rai M (2009) Fungus mediated synthesis of silver nanoparticles and their activity against pathogenic fungi in combination with fluconazole. Nanomed Nanotechnol Biol Med 5:382–386

Gericke M, Pinches A (2006) Biological synthesis of metal nanoparticles. Hydrometallurgy 83:132–140

Godfray CHJ, Crute IR, Haddad L, Lawrence D, Muir JF, Nisbett N, Pretty J, Robinson S, Toulmin C, Whiteley R (2010) The future of the global food system. Philos Trans R Soc B 365:2769–2777. http://www.geohumus.com/us/products.html. (Accessed 20 Oct 2016)
Jia LR, Xia J, Zhou N, Chen WY (2008) Preservation of fruits by hydrolyzed collagen/sodium alginate nanoparticles latex. Food Mach 1:46–50
Knell M (2010) Nanotechnology and the sixth technological revolution. In: Cozzens SE, Wetmore JM (eds) Nanotechnology and the challenges of equity, equality and development. Springer, Dordrecht, pp 127–143. https://doi.org/10.1007/978-90-481-9615-9_8
Kottegoda N, Munaweera I, Madusanka N, Karunaratne V (2011) A green slow-release fertilizer composition based on urea-modified hydroxyapatite nanoparticles encapsulated wood. Curr Sci 101:73–78
Kumar S, Chauhan N, Gopal M, Kumar R, Dilbaghi N (2015) Development and evaluation of alginate-chitosan nanocapsules for controlled release of acetamiprid. Int J Biol Macromol 81:631–637
McBratney A, Whelan B, Ancev T, Bouma J (2005) Future directions of precision agriculture. Precis Agric 6:7–23
Milani N, McLaughlin MJ, Stacey SP, Kirkby JK, Hettiarachchi GM, Beak DG, Cornelis G (2012) Dissolution kinetics of macronutrient fertilizers coated with manufactured zinc oxide nanoparticles. J Agric Food Chem 60:3991–3998
Milani N, Hettiarachchi GM, Kirby JK, Beak DG, Stacey SP, McLaughlin MJ (2015) Fate of zinc oxide nanoparticles coated onto macronutrient fertilizers in an alkaline calcareous soil. PLoS One:10. https://doi.org/10.1371/journal.pone.0126275
Mirzaei A, Vazan S, Naseri R (2010) Response of yield components of safflower (*Carthamus tinctorius* L.) to seed inoculation with Azotobacter and Azosprilium and different nitrogen levels under dry land condition. World Appl Sci J 11(10):1287–1291
Mishra S, Singh A, Keswani C, Singh HB (2014) Nanotechnology: exploring potential application in agriculture and its opportunities and constraints. Biotech Today 4:9–14. https://doi.org/10.5958/2322-0996.2014.00011.8
Mishra S, Singh A, Keswani C, Saxena A, Sarma BK, Singh HB (2015) Harnessing plant-microbe interactions for enhanced protection against phytopathogens. In: Arora NK (ed) Plant microbes symbiosis: applied facets. Springer, New Delhi, pp 111–125
Moharrer S, Mohammad B, Gharamohammad RA, Yargol M (2012) Biological synthesis of silver nanoparticles by Aspergillus flavus, isolated from soil of Ahar copper mine. Indian J Sci Technol 5:2443–24447
Narayanan KB, Sakthivel N (2010) Facile green synthesis of gold nanostructures by NADPH-dependent enzyme from the extract of Sclerotium rolfsii. Colloids Surf A Physicochem Eng Asp 380:156–161
Nithya R, Ragunathan R (2009) Synthesis of silver nanoparticle using pleurotus *Pleurotus sajor caju* and its antimicrobial study. Dig J Nanomater Bios 4:623–629
Ojaghloo F, Farahvash F, Hassanzadeh A, Pouryusef M (2007) Effect of inoculation with azotobacter and phosphate biofertilizers on yield of safflower (*Carthamus tinctorius* L.). J Agric Sci, Islamic Azad University, Tabriz Branch 3:25–30
Ombodi A, Saigusa M (2000) Broadcast application versus band application of polyolefin coated fertilizer on green peppers grown on anisole. J Plant Nutr 23:1485–1493
Park HM, Li X, Jin CZ, Park CY, Cho WJ, Ha CS (2002) Preparation and properties of biodegradable thermoplastic starch/clay hybrids. Macromol Mater Eng 287:553–558
Prasad R, Kumar V, Prasad KS (2014) Nanotechnology in sustainable agriculture: present concerns and future aspects. Afr J Biotechnol 13:705–713
Prasad R, Kumar M, Varma A (2015) Role of PGPR in soil fertility and plant health. In: Egamberdieva D, Shrivastava S and Varma A (eds.), Plant Growth-Promoting Rhizobacteria (PGPR) and Medicinal Plants Springer International Publishing Switzerland, pp. 247–260
Prasad R (2016) Advances and applications through fungal nanobiotechnology. Springer, International Publishing, Cham. (978-3-319-42989-2)

Prasad R, Pandey R, Barman I (2016) Engineering tailored nanoparticles with microbes: quo vadis. WIREs Nanomed Nanobiotechnol 8:316–330. https://doi.org/10.1002/wnan.1363
Prasad R, Bhattacharyya A, Nguyen QD (2017a) Nanotechnology in sustainable agriculture: Recent developments, challenges, and perspectives. Front Microbiol 8:1014. https://doi.org/10.3389/fmicb.2017.01014
Prasad R, Kumar M, Kumar V (2017b) Nanotechnology: An Agriculture paradigm. Springer Nature Singapore Pte Ltd. (ISBN: 978-981-10-4573-8)
Prasad R (2017) Fungal nanotechnology: applications in agriculture, industry, and medicine. Springer Nature Singapore Pte Ltd., Singapore. (ISBN 978-3-319-68423-9)
Prasad R, Bhattacharyya A, Nguyen QD (2017) Nanotechnology in sustainable agriculture: recent developments, challenges and perspectives. Front Microbiol 8:1014. https://doi.org/10.3389/fmicb.2017.01014
Prasad R, Kumar V, Kumar M, Wang S (2018a) Fungal Nanobionics: Principles and Applications. Springer Nature Singapore Pte Ltd. (ISBN 978-981-10-8666-3) https://www.springer.com/gb/book/9789811086656
Prasad R, Gill SS, Tuteja N (2018b) Crop Improvement Through Microbial Biotechnology. Elsevier
Rai M, Yadav A, Bridge P, Gade A (2009) Myconanotechnology: a new and emerging science. In: Rai MK, Bridge PD (eds) Applied mycology. CAB International, New York, pp 258–267
Raliya R, Tarafdar JC, Gulecha K, Choudhary K, Ram R, Mal P, Saran RP (2013) Review article; Scope of nanoscience and nanotechnology in agriculture. J Appl Biol Biotech 1(03):041–044
Raliya R, Nair R, Chavalmane S, Wang WN, Biswas P (2015) Mechanistic evaluation of translocation and physiological impact of titanium dioxide and zinc oxide nanoparticles on the tomato (*Solanum lycopersicum* L.) plant. Metallomics 12:1584–1594
Research and Markets (2015) Global nanotechnology market outlook 2015–2020. Available at: http://www.prnewswire.com/news-releases/globalnanotechnology-market-outlook-2015-2020%2D%2D-industry-will-grow-to-reachus-758-billion-507155671.html
Sabourin V, Ayande A (2015) Commercial opportunities and market demand for nanotechnologies in agribusiness sector. J Technol Manag Innov 10:40–51
Sangeetha J, Thangadurai D, Hospet R, Purushotham P, Karekalammanavar G, Mundaragi AC, David M, Shinge MR, Thimmappa SC, Prasad R, Harish ER (2017a) Agricultural nanotechnology: Concepts, benefits, and risks. In: Prasad R, Kumar M, Kumar V (eds.). Nanotechnology, Springer Nature Singapore Pte Ltd, pp. 1–17
Sangeetha J, Thangadurai D, Hospet R, Harish ER, Purushotham P, Mujeeb MA, Shrinivas J, David M, Mundaragi AC, Thimmappa AC, Arakera SB, Prasad R (2017b) Nanoagrotechnology for soil quality, crop performance and environmental management. In: Prasad R, Kumar M, Kumar V (eds.). Nanotechnology, Springer Nature Singapore Pte Ltd, pp. 73–97
Sadowski Z, Maliszewska GB, Polowczyk I, Kozlecki T (2008) Synthesis of silver nanoparticles using microorganisms. Mater Sci Pol 26:419–425
Sahin F, Cakmakci R, Kantar F (2004) Sugar beet and barley yields relation to inoculation with N2-fixing and phosphate solubilizing bacteria. Plant Soil 265:123–129
Salantur A, Ozturk A, Akten S, Sahin F, Donmez F (2005) Effect of inoculation with nonindigenous and indigenous rhizobacteria of Erzurum (Turkey) origin on growth and yield of spring barley. Plant Soil 275:147–156
Salunkhe RB, Patil SV, Salunke BK, Patil CD, Sonawane AM (2011) Studies on silver accumulation and nanoparticle synthesis by *Cochliobolus lunatus*. Appl Biochem Biotechnol 165:221–234
Sarkar J, Chattopadhyay D, Patra S, Deo SS, Sinha S, Ghosh M, Mukherjee A, Acharya K (2011) *Alternaria alternata* mediated synthesis of protein capped silver nanoparticles and their genotoxic activity. Dig J Nanomater Biostruct 6:563–573
Sastry M, Ahmad A, Khan MI, Kumar R (2003) Biosynthesis of metal nanoparticles using fungi and actinomycete. Curr Sci 85(2):162–170

Sawle BD, Salimath B, Deshpande R, Bedre MD, Prabhakar KB, Venkataraman A (2008) Biosynthesis and stabilization of Au and Au–Ag alloy nanoparticles by fungus, *Fusarium semitectum*. Sci Technol Adv Mater 9:9035012

Singh P, Balaji R (2011) Biological synthesis and characterization of silver nanoparticles using the fungus *Trichoderma harzianum*. Asian J Exp Biol Sci 2:600–605

Sutherland IW (1998) Novel and established applications of microbial polysaccharides. Trends Biotechnol 16(1):41–46

Tilman D, Cassman KG, Matson PA, Naylor R, Polasky S (2002) Agricultural sustainability and intensive production practices. Nature 418:671–677

Torney F, Trewyn BG, Lin VS-Y, Wang K (2007) Mesoporous silica nanoparticles deliver DNA and chemicals into plants. Nat Nanotechnol 2:295–300

Vahabi K, Ali Mansoori G, Karimi S (2011) Biosynthesis of silver nanoparticles by fungus *Trichoderma reesei* (a route for largescale production of AgNPs). Insci J 1:65–79

Vamvakaki V, Chaniotakis NA (2007) Pesticide detection with a liposome-based nano-biosensor. Biosens Bioelectron 22:2848–2853

Verma VC, Kharwar RN, Gange AC (2010) Biosynthesis of antimicrobial silver nanoparticles by the endophytic fungus *Aspergillus clavatus*. Nanomedicine 5:33–40

Verma VC, Singh SK, Solanki R, Prakash, S (2011) Biofabrication of anisotropic gold nanotriangles using extract of endophytic Aspergillus clavatus as a dual functional reductant and stabilizer. Nanoscale Res Lett 6:16–22

Wanyika H, Gatebe E, Kioni P, Tang Z, Gao Y (2012) Mesoporous silica nanoparticles carrier for urea: potential applications in agrochemical delivery systems. J Nanosci Nanotechnol 12:2221–2228

Xia T, Li N, Nel AE (2009) Potential health impact of nanoparticles. Annu Rev Public Health 30:137–150

Yasari E, Patwardhan AM (2007) Effects of (*Azotobacter* and *Azospirillum*) inoculants and chemical fertilizers on growth and productivity of canola (*Brassica napus* L.).Asian J. Plant Sci 6(1):77–82

Yosefi K, Galavi M, Ramrodi M, Mousavi SR (2011) Effect of bio-phosphate and chemical phosphorus fertilizer accompanied with micronutrient foliar application on growth, yield and yield components of maize (Single Cross 704). Aust J Crop Sci 5(2):175–180

Zahir A, Arshad ZM, Frankenberger WF (2004) Plant growth promoting rhizobacteria: applications and perspectives in agriculture. Adv Agron 81:97–168

Chapter 17
Applications of Nanoparticles in Wastewater Treatment

Simranjeet Singh, Vijay Kumar, Romina Romero, Kankan Sharma, and Joginder Singh

Contents

Simranjeet Singh and Vijay Kumar contributed equally.

S. Singh
Department of Biotechnology, Lovely Professional University, Phagwara, Punjab, India

Punjab Biotechnology Incubators, Mohali, Punjab, India

Regional Advanced Water Testing Laboratory, Mohali, Punjab, India

V. Kumar
Regional Ayurveda Research Institute for Drug Development, Gwalior, Madhya Pradesh, India

R. Romero
Technological Development Unit (UDT), Universidad de Concepcion, Coronel, Chile

K. Sharma · J. Singh (✉)
Department of Biotechnology, Lovely Professional University, Phagwara, Punjab, India
e-mail: joginder.15005@lpu.co.in

R. Prasad et al. (eds.), *Nanobiotechnology in Bioformulations*, Nanotechnology in the Life Sciences, https://doi.org/10.1007/978-3-030-17061-5_17

17.1 Introduction

Emerging pollutants in wastewater streams are mostly chemical substances that are nonbiodegradable and persist in the environment, bioaccumulate through the food web, and pose a risk of causing adverse effects not only to human health but also to the environment, microflora, etc. (Larramendy and Soloneski 2015; Kumar et al. 2015a, b; Sidhu et al. 2019). Disposal of wastewater directly into streams, rivers, etc., without a proper management system, is one of the major global challenges and has a negative effect on the water quality in surface water bodies (Naidoo and Olaniran 2013). A total of 38,354 million liters per day (MLD) of sewage and wastewater is generated from the major cities of India, which have a sewage treatment capacity of only 11,786 MLD. Likewise, 13,468 MLD of industrial wastewater is generated across India, but the sewage treatment capacity is only 8080.8 MLD (Kaur et al. 2012). In Punjab, an estimated 1528.26 MLD of wastewater (sewage class generation I) is generated from 19 major cities, but they have a sewage treatment capacity of only 411 MLD (CPCB Report 2009–2010). Wastewater contains various pathogenic organisms such as bacteria, viruses, protozoans, and helminths, which are directly associated with severe diseases such as hepatitis A, campylobacteriosis, and leptospirosis (Lin and Ganesh 2013; Kumar et al. 2014a, b; Singh et al. 2016, 2017a, b; Kaur et al. 2018). Various efforts have been made to treat wastewater produced by different anthropogenic activities (Kumar et al. 2013; Sahu 2014; Ranade and Bhandari 2014). The treatment processes involve coagulation–flocculation, activated carbon adsorption, ozonation and advanced oxidation processes, membrane processes, membrane bioreaction, attached growth treatment processes, etc. (Altmann et al. 2014; Liang et al. 2014; Jegatheesan et al. 2016; Bollmann et al. 2016; Bonvin et al. 2016). The aforementioned methods are competent in treating the particular particulates to an extent but are incapable of removing other pollutants or contaminants present in the wastewater, especially endocrine-disruptor chemicals (Gosavi and Sharma 2014; Arbabi et al. 2015). Irrespective of the benefit of treating specific substances, until now there has been no effective method that could eliminate the pollutants present in the water (Luo et al. 2014). Thus, there is a need for optimization of treatment processes. Moreover, a collaborative approach is needed to develop an eco-friendly method that can prevent further discharge of various pollutants into the water and overcome all of the limitations of wastewater management systems.

Nanomaterials are materials measuring between 1 and 100 nm in at least one dimension (Stark et al. 2015). The properties of nanoparticles—such as their magnetic, optical, and electrical properties—are significantly dissimilar from those of conventional materials. Nanomaterials are known to have the characteristics of high adsorption, catalytic activities, and high reactivity (Khan et al. 2017). In the past few decades, nanoparticles have attracted active research and development, and have been efficaciously applied in various fields such as biology, sensing, medicine, and catalytic chemistry (Biju 2014; Dauthal and Mukhopadhyay 2016; Kamaly et al. 2016). In addition, use of nanoparticles in wastewater treatment has drawn widespread attention. Because nanoparticles combine small size with a large surface area, they have strong adsorption reactivity and capacity (Mauter et al. 2018). Various contaminants such as inorganic anions, organic pollutants, emerging pollutants, and bacteria have been reported worldwide to be magnificently disintegrated by numerous kinds of nanomaterials (Varjani et al. 2017; Méndez et al. 2017). Nanoparticles are promising tools for applicability in various wastewater ecosystems, including zerovalent nanoparticles, carbon nanotubes (CNTs), nanocomposites, and metal oxide nanoparticles (MNPs) (Usmania et al. 2017; Prasad and Thirugnanasanbandham 2019).

17.2 Potential Applications of Nanomaterials in Water and Wastewater Treatment

17.2.1 Zerovalent Metal Nanoparticles

The term "zerovalent metal nanoparticles" usually refers to elemental metallic iron with zero charge-bearing Fe atoms. It is usually used to convert oxidized materials or as a bulk reducing agent, which may sometimes be hazardous (Tsarev et al. 2016). Zerovalent metal nanoparticles are known for their distinctive optical, mechanical, magnetic, electronic, and catalytic properties, and are used for remediation of contaminated soils, sediments, water bodies, and wastewater (Ju-Nam and Lead 2016; Prasad et al. 2014, 2017). Zinc, silver, and Fe are commonly known zerovalent metal nanoparticles (Sharma et al. 2015). Different types of zerovalent nanoparticles are illustrated in Fig. 17.1.

17.2.1.1 Zinc Nanoparticles

Zinc nanoparticles (Zn NPs) are leached indirectly into ecosystems as a result of their progressive use in sunscreen formulations, packaging, paints, plastics, food supplements, and cosmetic products (Nowack and Bucheli 2007). The annual production of Zn NPs in the European Union has increased to 1.6 metric tonnes (Sun et al. 2014). Zn NPs are considered an alternative to iron (Fe) nanoparticles, as zinc is a stronger reductant than Fe, has rich negative reduction potential, and has been extensively used in wastewater treatment (Yan et al. 2013; Fu et al. 2014). The rate of degradation of contaminants in wastewater treated with zinc particles is much higher than that of wastewater treated with zerovalent iron (ZVI) nanoparticles. Most studies on treatment of wastewater by using Zn NPs have been based on a dehalogenation reaction (Mahgoub and Samaras 2014; Ghosh et al. 2017). Various studies have been conducted to assess the potent role of Zn NPs in conventional activated sludge (CAS) processes for wastewater treatment (Puay et al. 2015; Zheng et al. 2011). It has also been reported that in a sequencing batch reactor (SBR), Zn NPs at a low concentration (5.0 mg/L) hindered the growth of nitrifying bacteria, resulting in a decline in NH_4^+-N removal (Hou et al. 2013). Another study confirmed that Zn NPs showed a 50% inhibitory concentration (IC_{50}) of 13.1 mg/L for

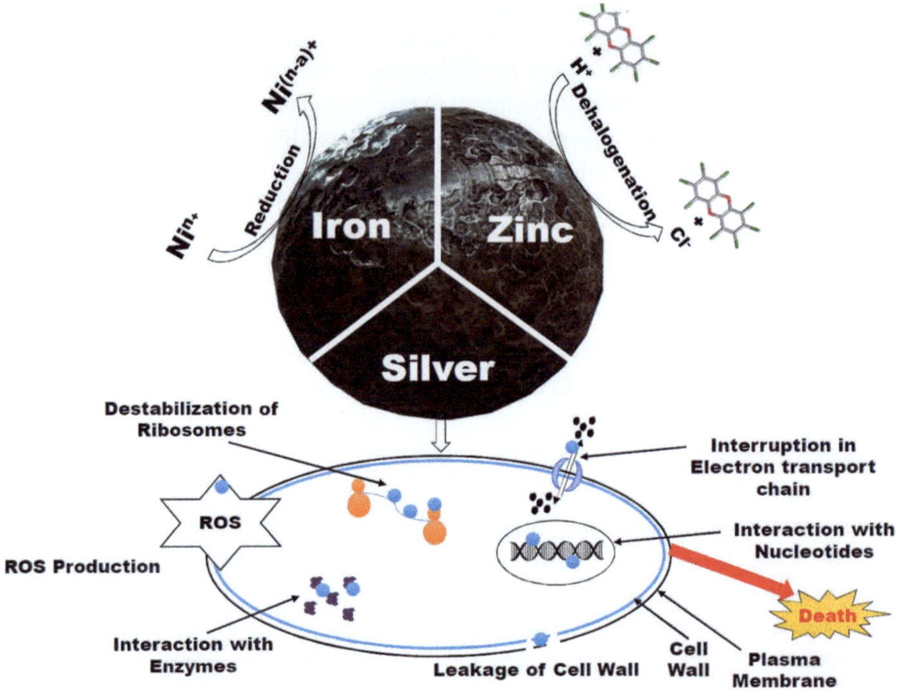

Fig. 17.1 Zerovalent nanoparticles used in wastewater management. ROS reactive oxygen species

ammonium-oxidizing bacteria (Liu et al. 2011a, b). Zn NPs have also been reported to achieve efficient degradation of octachlorodibenzo-p-dioxin into its substituent chlorinated congeners under ambient conditions (Bokare et al. 2013).

Although numerous studies have demonstrated that pollutant reduction by Zn NPs could be successful, it has been limited to degradation of halogenated and chlorinated organic compounds. Treatment of other types of contaminants by Zn NPs has rarely been reported. Therefore, advanced full-scale or pilot-scale studies of Zn NPs at polluted field sites have not yet been accomplished.

17.2.1.2 Iron Nanoparticles

ZVI nanoparticles have achieved promising results in the development of remediation technologies for on-site wastewater treatment, as they have been experimentally shown to detoxify hexachlorocyclohexane (lindane), carbon tetrachloride, vinyl chloride, chlorinated organic compounds (COCs), and trichloroethane (Crane and Scott 2012). They serve as potent and cost-effective electron donors, and most studies on use of Fe nanoparticles in wastewater have been based on a reductive dehalogenation reaction (Anjum et al. 2016). In anaerobic conditions, they are oxidized by molecules of water or protons, resulting in generation of Fe(II) and H_2, which act as potent reducing agents for various contaminants present in wastewater. They react with contaminants, resulting in oxidation of Fe(II) to Fe(III) to form $Fe(OH)_3$. $Fe(OH)_3$ acts as an effective flocculant, which facilitates removal of organic and inorganic contaminants from various ecosystems (Klačanová et al. 2013). ZVI nanoparticles oxidize and degrade a diverse range of organic contaminants, as they transfer two electrons to an oxygen molecule to form hydrogen peroxide, which is further reduced to a water molecule (Lee 2015). Furthermore, the reaction between hydrogen peroxide and Fe(II) generates hydroxyl radicals with a strong oxidizing ability against a number of organic compounds (Hao et al. 2015). Because of their properties of oxidation, precipitation, reduction, and adsorption, nanomaterials are applied worldwide for removal of contaminants such as nitroaromatic compounds, inorganic anions, phosphates, radio elements, nitrates, phenols, organic dyes, and chlorinated and halogenated organic compounds (Lu et al. 2016).

Although ZVI nanoparticles have many advantages, they also have some disadvantages such as separation difficulty, aggregation, and oxidation from the contaminant degradation system (Guan et al. 2015). Various new approaches—such as surface coating, emulsification, doping with essential metal ions, or conjugation with supports—are used to improve the efficiency of ZVI nanoparticles (Dutta et al. 2015a, b). These new approaches are supposed to augment the reactivity, aggregation, and dispersibility of ZVI and prevent its separation from the degraded system.

17.2.1.3 Silver Nanoparticles

As silver is a good antimicrobial agent, silver nanoparticles (Ag NPs) have been extensively used for disinfection of water against a wide range of pathogens, viruses, fungi, and bacteria. They reportedly attach to the bacterial cell membrane, thus increasing its permeability by generating free radicals (Le et al. 2012). They damage the cell membrane of the cell, finally resulting in apoptosis. Ag NPs act on the phosphorus and sulfur elements of DNA and destroy them. Sometimes, dissolution of Ag NPs releases (antimicrobial) Ag^+ ions, which come into contact with the thiol group of essential enzymes, incapacitate them, and interrupt normal life functions (Rudakiya and Pawar 2017; Aziz et al. 2014, 2015, 2016, 2019). However, their tendency to accumulate in liquid media reduces their effectiveness for long-term use. In in situ remediation experiments, immobilization of Ag NPs on cellulose fibers has been reported, in which the sheets showed antibacterial properties against suspensions of *Enterococcus faecalis*, *Escherichia coli*, and inactivated bacteria in a filtration system (Chitra and Annadurai 2014; Peiris et al. 2017). Moreover, Ag NPs manufactured by various chemical reductions have been incorporated into polyethersulfone (PES) microfiltration (MF) membranes (Ferreira et al. 2015). The action of microorganisms on the membranes was observed to be remarkably suppressed. The PES–Ag NP membranes demonstrated strong antimicrobial properties and held prodigious potential for application in wastewater treatment. Ag NPs kill pathogenic bacteria by inducing physical perturbation with oxidative stress, through disruption of a specific microbial process via oxidization or disturbance of a vital cellular component or structure of the cell membrane (Seo et al. 2014; Prasad 2014; Prasad et al. 2018).

In the last two decades, research on immobilization of Ag NPs on membranes and ceramic materials has drawn considerable attention because of their potential for use in biofouling reduction and disinfection in sewage and wastewater treatment (Amin et al. 2014). To improve their efficiency, high-porosity filters (built with sawdust and clay) have been used now-a-days for removal of *E. coli* and other pathogens from wastewater (Moosa and Muhsen 2017). With the introduction of colloidal chemistry, filters with colloidal nanoparticles improve the filter performance and remove the pathogenic bacteria *E. coli* by up to 97.8–100% (Abebe et al. 2014).

17.2.2 Metal Oxide Nanoparticles

Metal oxide nanoparticles (MNPs) have shown prodigious potential as an environmentally friendly, low-cost, and sustainable wastewater treatment technology. Various applications of nanoparticles are listed in Table 17.1. These nanoparticles possess a short intraparticle diffusion distance and a high specific surface area, are compressible without a significant surface area reduction, have more adsorption sites, and are easy to reuse. Moreover, some of them are superparamagnetic, which results in adsorption performance superior to that of activated carbon (Corsi et al. 2018). The adsorption is mainly achieved by complexation between dissolved metals and molecules of oxygen in metal oxide groups. It is well known that efficient adsorption of heavy metals on the outer surface on MNPs is triggered along the walls by rate-limiting intraparticle diffusion (Ilankoon 2014). Because of their large number of surface reaction sites (edges, vacancies, and corners), high specific surface area, and short intraparticle diffusion distance, these nanoparticles have a high adsorption capacity and faster kinetics (Sa and Premalatha 2016).

A reduction in the particle size of nanomagnetite increases the adsorption capacity by up to 100 times, and this is attributed to the increase in the specific surface area, suggesting a nanoscale effect (Nassar 2012). This effect has been ascribed to the modification of nanoparticles on the magnetite surface, with development of new adsorption sites. Metal oxides of titanium, zinc, aluminum, and iron are effective, low-cost adsorbents for heavy metals and radionuclides (Dave and Chopda 2014).

Metal oxide–based nanoparticles have been well documented in previous studies to remove multiple heavy metals such as chromium, nickel, cadmium, copper, mercury, arsenic, and lead (Gunatilake 2015). They can be easily reused and regenerated by changing the pH of the solution, which maintains the adsorption capacities of the MNPs (Kunduru et al. 2017). Their ease of separation, low cost, ease of regeneration, and high adsorption capacity make MNPs economically and technological advantageous (Fig. 17.2).

17.2.2.1 TiO_2 Nanoparticles

TiO_2 nanoparticles are the most extensively used photocatalyst nanoparticles, particularly in wastewater treatment, because of their high chemical stability, cost effectiveness, photocatalytic generation of reactive oxygen species (ROS), and low

Table 17.1 Comparison of different nanomaterials for removal of various contaminants (dyes, heavy metals, etc.), with their adsorption capacities and rate constants

Category	Contaminant	Adsorbent material	Nanomaterial used	Adsorption capacity (mg/g)	Rate constant (k_1, h^{-1})	Reference
Dyes	Reactive red M-2BE	Activated carbon	Carbon-based	260.7	1.503	Machado et al. (2011)
	Reactive blue 29 (RB29)	Single-walled carbon nanotubes	Carbon-based	496	–	Nadafi et al. (2011)
	Acid red 18	Single-walled carbon nanotubes	Carbon-based	166.67	21.12	Shirmardi et al. (2013)
	Reactive red 120	Single-walled carbon nanotubes	Carbon-based	426.49	–	Bazrafshan et al. (2013)
	Reactive red M-2BE	Multiwalled carbon nanotubes	Carbon-based	335.7	2.860	Machado et al. (2011)
	Maxilon blue	Multiwalled carbon nanotubes	Carbon-based	260.7	–	Alkaim et al. (2015)
	Bromothymol blue	Oxidized multiwalled carbon nanotubes	Carbon-based	55	0.042	Ghaedi et al. (2012)
	Reactive blue 19	MgO	Metal oxide–based	166.7	4.2	Moussavi and Mahmoudi (2009)
	Reactive blue 198	MgO	Metal oxide–based	123.5	5.4	Moussavi and Mahmoudi (2009)
	Neutral red (NR)	Graphene oxide/Fe_3O_4 hybrid composite	Metal oxide–based	171.3	–	Xie et al. (2012)
	Methylene blue (MB)	Graphene oxide/Fe_3O_4 hybrid composite	Carbon-based	167.2	–	Xie et al. (2012)
	Methylene blue	Reduced graphene oxide/TiO_2	Carbon and metal oxide hybrid	467.6	3.1278	Wang et al. (2016a)
	Methylene blue	Polyvinyl alcohol	Polymer-based	123.3	4.854	Agarwal et al. (2016)

(continued)

Table 17.1 (continued)

Category	Contaminant	Adsorbent material	Nanomaterial used	Adsorption capacity (mg/g)	Rate constant (k_1, h^{-1})	Reference
	Bromothymol blue (BTB)	Polyvinyl alcohol	Polymer-based	276.2	4.266	Agarwal et al. (2016)
	Malachite green (MG)	Polyacrylamide/$Ni_{0.02}Zn_{0.98}O$ nanocomposite	Polymer-based	–	6.12	Kumar et al. (2014)
	Rhodamine B	Polyacrylamide/$Ni_{0.02}Zn_{0.98}O$ nanocomposite	Polymer-based	–	8.88	Kumar et al. (2014)
	Congo red	Multiwalled carbon nanotubes	Carbon-based	352.1	3.18	Zare et al. (2015)
	Maxilon blue	Multiwalled carbon nanotubes	Carbon-based	260.7	–	Alkaim et al. (2015)
Heavy metals	Cr(VI)	Single-walled carbon nanotubes	Carbon-based	1.26	–	Dehghani et al. (2015)
	Cr(VI)	Multiwalled carbon nanotubes	Carbon-based	2.35	0.42	Dehghani et al. (2015)
	Pb(II)	Diethylenetriamine-modified multiwalled carbon nanotubes	Carbon-based	58.26	–	Vuković et al. (2011)
	Cd(II)	Diethylenetriamine- modified multiwalled carbon nanotubes	Carbon-based	31.45	–	Vuković et al. (2011)
	Cd(II)	Graphene	Carbon-based	106.3	–	Zhao et al. (2011a)
	Co(II)	Graphene	Carbon-based	68.2	–	Zhao et al. (2011a)
	Pb(II)	Graphene oxide nanosheets	Carbon-based	842	–	Zhao et al. (2011b)
	Ni(II)	Graphene nanosheets	Carbon-based	3.00	0.0576	Ren et al. (2011b)
	Cd(II)	TiO_2	Metal oxide–based	16.69	2.244	Visa et al. (2009)
	Cu(II)	TiO_2	Metal oxide–based	5.18	0.927	Visa et al. (2009)
	Cu(II)	TiO_2 nanotubes/carbon nanotubes	Metal oxide–based	83–124	–	Doong and Chiang (2008)

	Pb(II)	TiO_2 nanotubes/carbon nanotubes	Metal oxide–based	192–588	–	Doong and Chiang (2008)
	Cu(II)	Fe_3O_4 magnetic nanoparticles	Metal oxide–based	61.07	–	Ozmen et al. (2010)
	Hg(II)	Modified Fe_3O_4 microspheres	Metal oxide–based	37.4 μmol g^{-1}	–	Wang et al. (2013)
	Cr(VI)	Ethylenediamine-functionalized nano-Fe_3O_4	Metal oxide–based	136.98	47.172	Zhao et al. (2010)
	Ni(II)	δ-MnO_2	Metal oxide–based	30.63	0.108	Ren et al. (2011a, b)
	Ni(II)	Graphene nanosheet/δ-MnO_2 composite	Carbon and metal oxide hybrid	46.55	0.0432	Ren et al. (2011a, b)
	Cd(II)	Al_2O_3/multiwalled carbon nanotubes	Carbon and metal oxide hybrid	27.21	5.7644	Liang et al. (2015)
	Pb(II)	MnO_2/carbon nanotubes	Carbon and metal oxide hybrid	78.74	0.816	Wang et al. (2007)
	Pb(II)	Ni@C composite nanostructures	Carbon and metal oxide hybrid	21.45	–	Ni et al. (2010)
	Cu(II)	Ni@C composite nanostructures	Carbon and metal oxide hybrid	14.3	–	Ni et al. (2010)
	Cd(II)	Ni@C composite nanostructures	Carbon and metal oxide hybrid	6.43	–	Ni et al. (2010)
	Zn(II)	Polyaniline/rice husk nanocomposite	Polymer-based	24.3	–	Ghorbani et al. (2012)
	Cr(VI)	Polypyrrole/Fe_3O_4 magnetic nanocomposite	Polymer-based	169.4	11.28	Bhaumik et al. (2011)
Pesticides	Chlorpyrifos	Reduced graphene oxide nanosheets	Carbon-based	1200	–	Maliyekkal et al. (2013)
	Endosulfan	Reduced graphene oxide nanosheets	Carbon-based	1100	–	Maliyekkal et al. (2013)

(continued)

Table 17.1 (continued)

Category	Contaminant	Adsorbent material	Nanomaterial used	Adsorption capacity (mg/g)	Rate constant (k_1, h^{-1})	Reference
	Malathion	Reduced graphene oxide nanosheets	Carbon-based	800	–	Maliyekkal et al. (2013)
Others	Nitrofurazone	Powder-activated carbon	Carbon-based	50.8	0.1129	Wu and Xiong (2016)
	Nitrofurazone	Multiwalled carbon nanotubes	Carbon-based	59.9	0.2082	Wu and Xiong (2016)
	Trichloroethylene	Al_2O_3/multiwalled carbon nanotubes	Carbon and metal oxide hybrid	19.84	1.1048	Liang et al. (2015)
	BTB	Polyvinyl alcohol	Polymer-based	276.2	4.266	Agarwal et al. (2016)

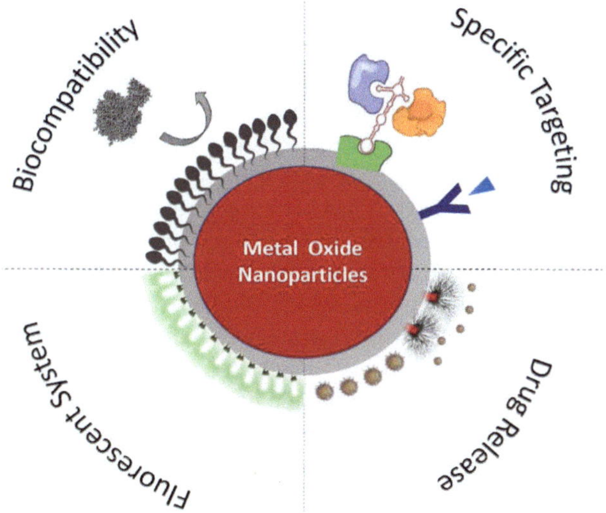

Fig. 17.2 Metal oxide nanoparticles and their applications

toxicity to humans (Qu et al. 2013). They have been used and studied as catalysts in the ozonation process, which provides firm and complete mineralization of organic contaminants (Panahi et al. 2018). In the last two decades, usage of TiO_2 nanoparticles has drawn the attention of many researchers because of their ability to kill a wide range of viruses, including *Herpes simplex* virus (Hajkova et al. 2007), MS2 bacteriophages (Cho et al. 2011), hepatitis B virus (Zan et al. 2007), and poliovirus 1 (Liga et al. 2011). The minimum concentration of TiO_2 used to kill pathogens in wastewater ranges from 0.1 to 1 g in 1 L and totally depends on the intensity of the wavelength (light) and the particle size (Ibhadon and Fitzpatrick 2013). The antagonistic effects of TiO_2 are associated with generation of ROS (H_2O_2 and OH^- free radicals), which are formed during ultraviolet (UV)-A irradiation by reductive and oxidative pathways (Li et al. 2014). It has also been documented that plastic containers coated with TiO_2 inactivate fecal coliforms at a concentration of 3000 colony-forming units (cfu) per 100 mL within a short span of time (Gelover et al. 2006). To improve photocatalytic inactivation of viruses and bacteria, doping is usually done with silver molecules, which facilitate removal of pathogens from wastewater. Development of solar disinfection (SoDis) units is also facilitated by nano-TiO_2, which provides pure and safe drinking water in some developing countries.

17.2.2.2 Zinc Oxide Nanoparticles

Zinc oxide nanoparticles (ZnO NPs) have emerged as a promising technology for photocatalytic degradation, with high transparency to visible light and UV absorption efficiency. They are efficient candidates for treatment of wastewater because of their distinctive properties such as a strong oxidation ability, good photocatalytic properties, and a wide bandgap in the UV spectral region (Rana et al. 2018). They are released directly into the environment from a wide number of application in formulations, paints, packaging, cosmetic products, etc. (Wang et al. 2017). It has also been reported that ZnO NPs induce a significant decrease in the treatment of biological nitrogen from wastewater (Zheng et al. 2011). They exhibit good antibacterial activities against a wide range of pathogenic bacteria, but the mechanisms of these are still uncertain, although it has been suggested that photocatalytic generation of ROS molecules is the primary mechanism. Furthermore, disorganization of the cellular membrane and penetration of the cellular envelope by ZnO NPs also hinder bacterial growth (Beyth et al. 2015; Bhuyan et al. 2015). To improve the effectiveness of ZnO NPs, a common strategy is metal doping (Badreddine et al. 2018). Several types of metal dopants have been tested, including codopants, rare-earth dopants, cationic dopants, and anionic dopants (Lee et al. 2016). Numerous studies have confirmed that coupling with other semiconductors—such as reduced graphene oxide (RGO), graphene oxide (GO), TiO_2, SnO_2, CeO_2, and CdO—is a practicable approach to enhance the photodegradation properties of ZnO NPs (Azarang et al. 2015). In addition to the dearth of data concerning the transformation of ZnO NPs throughout the effluent treatment process, there is a scarcity of data regarding the effects of product formulation chemistry on the fate of ZnO NPs throughout the waste product treatment process (Mostafaii et al. 2017). This is a vital information gap, as ZnO NPs are often dispersed in special formulations to facilitate their stability and inclusion in products such as sunscreens and cosmetics (Lohani et al. 2014).

17.2.2.3 Iron Oxide Nanoparticles

The utilization and synthesis of iron oxide nanoparticles with novel functions and properties have been extensively studied because of their high surface volume, super paramagnetism, and nanoscale size (Ali et al. 2016). In particular, the ease of their synthesis, modification, or coating, and the potential to control or alter matter on the atomic scale could facilitate unparalleled adaptability. Moreover, iron oxide nanoparticles—with chemical inertness, biocompatibility, and low toxicity—show a marvelous potential in amalgamation with biotechnology (Dinali et al. 2017).

The potential of iron oxide nanoparticles to remove various contaminants has been demonstrated in both laboratory and on-site field tests (Gutierrez et al. 2017). Current applications of iron oxide nanoparticles in treatment of wastewater are based on adsorptive technologies (which use iron oxide nanoparticles as a kind of immobilization carrier or a nanosorbent to enhance efficient removal of contaminants) or photocatalytic technologies (which use iron oxide nanoparticles as semiconductor photocatalysts to converts contaminants into less toxic compounds) (Saharan et al. 2014). The mechanisms underlying adsorption from contaminated sites or wastewater include electrostatic interaction, magnetic selective adsorption, ligand combination, and surface binding (Yang and Yin 2017). The adsorption of both inorganic and organic contaminants occurs via a surface binding mechanism; subsequently, contaminants either diffuse into the adsorbent or are adsorbed via electrostatic interactions for additional interactions with functional groups. To extend their applicability, chemical treatment or modification of iron oxide nanoparticles are crucial to improve their adsorption capability (Das et al. 2017).

17.2.3 Other Nanoparticles

17.2.3.1 Carbon Nanotubes

Carbon nanotubes (CNTs) have scaffold functions in wastewater purification systems and have been extensively used as adsorbents for inorganic, organic, and biological wastewater pollutants (Das et al. 2014a, b). In addition to that, they are also utilized as hybrid catalysts in nanocarriers for enzyme immobilization, electrocatalysis, and photocatalysis systems (Zhou et al. 2018a, b). CNTs are a member of the carbon family and are subclassified into single-walled CNTs (SWCNTs) and multiwalled CNTs (MWCNTs), which are differentiated by the layering system of the nanotubes (Sears et al. 2010). CNTs have drawn the attention of many researchers in the field of nanotechnology because of their excellent electronic properties (Guzmán-Verri and Voon 2007), mechanical properties (Esawi et al. 2010), potential applications (Liu et al. 2011a, b), high chemical stability (Okpalugo et al. 2005), and high thermal stability (Kundu et al. 2008). Because of their high specific adsorption surface area (Peigney et al. 2001), layered structure (Pande et al. 2009), and hollowness (Zhang and Zhu 2006), they make a perfect adsorbent. They are commercially used to accelerate catalysis rates (Volder et al. 2013), to desalinate brackish water and seawater (Das et al. 2014a, b), mitigation of various pathogens, contaminants and complexity of various wastewater matrices (Qu et al. 2013).

The external surface, peripheral groove, inner site, and interstitial channel constitute the four possible sites for adsorption on CNTs (Ren et al. 2011a, b). Most contaminants adsorb on the inner and external surfaces of CNTs. CNTs are prone to developing bundles along the length of the tube axis, resulting in formation of peripheral grooves and interstitial channels. Functionalization of CNTs increases their chemical reactivity, enhances their colloidal activity, and mediates pollutant adsorption (Apul and Karanfil 2015). CNT functionalization is done by endohedral filling, treatment with covalent chemical agents, and noncovalent wrapping (Spitalsky et al. 2010). The interaction forces between contaminants and CNTs are mesopore filling, p–p electron coupling, hydrophobic interactions, ion exchange, electrostatic interactions, hydrogen bonding, and covalent bonding. Tremendous adsorption by CNTs in removal of organic 1,2-dichlorobenzene (DCB), cadmium, and lead from water has also been reported (Scida et al. 2011).

The major drawback of this technology is to transfer this to large-scale application systems or pilot studies. Further development of CNT-based water purification technologies should be encouraged to improve other characteristics such as desalination capacity, compatibility with industrial settings, scalability, material costs, energy savings, antifouling, high water permeability, and robustness (Sadhu et al. 2018).

17.2.3.2 Nanocomposites

Various research studies have been conducted to focus on creating multifunction or synergism by addition of nanomaterials to inorganic or polymeric membranes. Nanoparticles used for these applications include photocatalytic nanomaterials (e.g., TiO_2 and bimetallic nanoparticles), antimicrobial nanoparticles (e.g., CNTs and nano-Ag), and MNPs (e.g., zeolite Al_2O_3 and TiO_2) (Thoniyot et al. 2015). They have broad applications in various fields such as drug delivery systems, soil quality improvement, wastewater treatment, and biological sciences (Bogue 2011). The adsorption of various contaminants (such as dyes, heavy metal ions, and pesticides) from contaminated wastewater with nanocomposites has gained significant consideration in the past few years (Rasalingam et al. 2014). Nanocomposites have large surface areas and various other characteristics that make them efficient tools for removal of pollutants (Amin et al. 2014). Addition of various photoactive or antimicrobial nanomaterials to nanocomposite membranes makes them more reactive to minimize fouling (Huh and Kwon 2011; Rodrigues et al. 2017). Ceramic and polymeric membranes containing TiO_2 have been found to efficiently remove pathogens and contaminants under UV-A irradiation. The degradation of organic contaminants and inactivation of pathogenic bacteria by the photocatalayst TiO_2 makes nanocomposite biomembranes less vulnerable to biological and organic fouling (Qu et al. 2013; Berekaa 2016). Embedment of a chitosan layer on the nanocomposite membrane polyethylene glycol diacrylate or a polyacrylic acid layer has been found to potently inhibit Gram-negative and Gram-positive bacteria (Zhao et al. 2015).

In the same fashion, embedment or incorporation of nanocomposite membranes with various functional nanoparticles (e.g., antimicrobial, photocatalytic, and catalytic nanoparticles) into wastewater treatment can also be developed (Jardón-Maximino et al. 2018; Guerra et al. 2018). Photocatalytic nanoparticles embedded into nanocomposite materials result in the introduction of UV light into the system. An outside-in submerged ultrafiltration (UF) membrane or microfiltration (MF) reactor configuration can be employed to permit ease of access to photocatalytic light, using UV sources. The possible use of various optical fibers in submerged UF membrane or MF reactors is also promising (Zheng et al. 2017; Oller et al. 2018; Rahimi and Mosleh 2018).

17.2.3.3 Quantum Dots

Quantum dot (QD) nanoparticles can be used for detection of water pollutants and for demonstration of connections between specific pollutants in wastewater through changes in their optical emission signals (Zhang et al. 2017). They function better than dye-based sensors because of their longer emissions and compatibility with molecularly imprinted polymers (MIPs) (Foguel et al. 2017). The imprinted fluorescence transduction platform binds to the target pollutant in the matrix and emits fluorescence in a ratio equal to that of the amount of the target pollutant (Wang et al. 2016a, b). Over time, there has been extensive contamination of surface water and groundwater, consisting of heavy metals, organic–inorganic pollutants, polycyclic aromatic hydrocarbons, etc. (Dutta et al. 2015a, b). Although many techniques have emerged that can contribute to the removal of pollutants, in a few cases there is the presence of micropollutants in low concentrations that are nearly impossible to remove (Kumar et al. 2016, 2017; Elmizadeh et al. 2018). A sensitive nanosensor based on synthetic ligand–coated CdTe QDs has been developed for rapid detection of Cr(III) ions in water and wastewater samples (Elmizadeh et al. 2018). Therefore, QDs, in association with MIPs, are used for their unique quality of selective molecular binding (Sun et al. 2018). For a decade, MIP-QD-based fluorescence has been synthesized by different groups, with the advantages of both MIPs and magnetic nanoparticles or QDs. One study resulted in preparation of a highly sensitive molecularly imprinted fluorescence sensor with the help of CdSe QDs as a signal transducer and mesoporous silica nanoparticles as an imprinting material for fluorescent sensing of bisphenol A (BPA). Mn^{2+}-doped ZnS QDs, CdSe QDs, and CdTe QDs are some groups that have been fabricated on molecularly imprinted film (Zhou et al. 2014). Mn-doped ZnS QDs are anchored on an MIP for chemiluminescence detection of 4-nitrophenol in tap water. MIP-QD nanospheres have shown high recognition selectivity in aqueous media and have fast adsorption and desorption kinetics (Yu et al. 2017). A template molecule grafted on the surface of Mn-doped ZnS QDs has been used to develop a sensitive sensor for easy and fast determination of tetrabromobisphenol A in water, using Mn-doped ZnS QDs (Feng et al. 2019). MIP-coated graphene QDs (GQDs) have been used for determination of paranitrophenol in water (Zhou et al. 2014). QD composites are able to select and target a pollutant to remove it from sewage water. The combination of QDs and MIPs has shown efficient results in removal of pollutants from wastewater, and in the future this technology will revolutionize environment management.

17.3 Conclusions and Future Prospects

Wastewater treatment and reuse help to maintain and manage the water balance, but also raise questions. Various studies conducted to date have been aimed at improving the safety and health of humans and environmental fauna. Nanomaterials equipped with unique physical and chemical properties have great potential to remove contaminants. The idea of nanomaterial development has been presented to prioritize prospects for its application.

Metal oxide–containing nanomaterials are effective photocatalysts with superior performance under the influence of visible light. Being nanosorbent, metal oxides have shown effective results in removal of heavy metals and organic pollutants.

Metal oxide–containing nanomaterials are favored for absorbing heavy metals and organic pollutants, as they show promise for use in various applications. They are considered to be the immobilization carriers, which are rarely addressed, and they can also be used as support carriers in biosensors and biosorbents. Their success has been credited to their physical and chemical properties, but their applications in wastewater treatment are still limited.

This review has focused on the range of nanoparticle-based technologies that have either been proposed or are on the verge of active development. Many techniques are still at the pilot or experimental phase, and various difficulties have been encountered in research on metal oxide–containing nanomaterials for in vivo and in vitro applications. The extensive diversity of iron oxides includes various physical and chemical forms with great potential for use in pioneering applications in the environmental sector.

In conclusion, engineered nanoparticles are already very much in use and provide a noninvasive tool for wastewater treatment. However, because of uncertainty regarding their potential effects on human health and the environment, the fate of nanomaterials in the environment must be assessed before they can be used in various applications. A dramatic increase in research on nanomaterials has evidenced that their discharge into the environment could have huge impacts, and such discharges into the environment are expected to increase with the growth of their industrial applications. Coordination of research is needed to close the knowledge gaps in the area of risk management and risk assessment of nanomaterial technology. Future research addressing the safety of these systems and their economics and scalability is expected to help overcome current limitations and allow the use of nanomaterials to revolutionize the treatment of wastewater.

References

Abebe LS, Smith JA, Narkiewicz S, Oyanedel-Craver V, Conaway M, Singo A, Amidou S, Mojapelo P, Brant J, Dillingham R (2014) Ceramic water filters impregnated with silver nanoparticles as a point-of-use water-treatment intervention for HIV-positive individuals in Limpopo Province, South Africa: a pilot study of technological performance and human health benefits. J Water Health 12(2):288–300

Agarwal S, Sadegh H, Monajjemi M, Hamdy AS, Ali GA, Memar AO, Shahryari-Ghoshekandi R, Tyagi I, Gupta VK (2016) Efficient removal of toxic bromothymol blue and methylene blue from wastewater by polyvinyl alcohol. J Mol Liq 218:191–197

Ali A, Hira Zafar MZ, ul Haq I, Phull AR, Ali JS, Hussain A (2016) Synthesis, characterization, applications, and challenges of iron oxide nanoparticles. Nanotechnol Sci Appl 9:49

Alkaim AF, Sadik Z, Mahdi DK, Alshrefi SM, Al-Sammarraie AM, Alamgir FM, Aljeboree AM (2015) Preparation, structure and adsorption properties of synthesized multiwall carbon nanotubes for highly effective removal of maxilon blue dye. Korean J Chem Eng, 32(12):2456–2462

Altmann J, Ruhl AS, Zietzschmann F, Jekel M (2014) Direct comparison of ozonation and adsorption onto powdered activated carbon for micropollutant removal in advanced wastewater treatment. Water Res 55:185–193

Amin MT, Alazba AA, Manzoor U (2014) A review of removal of pollutants from water/wastewater using different types of nanomaterials. Adv Mater Sci Eng 2014:1–24

Anjum M, Miandad R, Waqas M, Gehany F, Barakat MA (2016) Remediation of wastewater using various nano-materials. Arab J Chem. https://doi.org/10.1016/j.arabjc.2016.10.004

Apul OG, Karanfil T (2015) Adsorption of synthetic organic contaminants by carbon nanotubes: a critical review. Water Res 68:34–55

Arbabi M, Hemati S, Amiri M (2015) Removal of lead ions from industrial wastewater: a review of removal methods. Int J Epidemiol Res 2:105–109

Azarang M, Shuhaimi A, Yousefi R, Jahromi SP (2015) One-pot sol–gel synthesis of reduced graphene oxide uniformly decorated zinc oxide nanoparticles in starch environment for highly efficient photodegradation of methylene blue. RSC Adv 5:21888–21896

Aziz N, Fatma T, Varma A, Prasad R (2014) Biogenic synthesis of silver nanoparticles using Scenedesmus abundans and evaluation of their antibacterial activity. J Nanoparticles, Article ID 689419, http://dx.doi.org/10.1155/2014/689419

Aziz N, Faraz M, Pandey R, Sakir M, Fatma T, Varma A, Barman I, Prasad R (2015) Facile algae-derived route to biogenic silver nanoparticles: Synthesis, antibacterial and photocatalytic properties. Langmuir 31:11605–11612 http://dx.doi.org/10.1021/acs.langmuir.5b03081

Aziz N, Pandey R, Barman I, Prasad R (2016) Leveraging the attributes of Mucor hiemalis-derived silver nanoparticles for a synergistic broad-spectrum antimicrobial platform. Front Microbiol 7:1984. http://dx.doi.org/10.3389/fmicb.2016.01984

Aziz N, Faraz M, Sherwani MA, Fatma T, Prasad R (2019) Illuminating the anticancerous efficacy of a new fungal chassis for silver nanoparticle synthesis. Front Chem 7:65. http://dx.doi.org/10.3389/fchem.2019.00065

Badreddine K, Kazah I, Rekaby M, Awad R (2018) Structural, morphological, optical, and room temperature magnetic characterization on pure and Sm-doped ZnO nanoparticles. J Nanomater 2018:1–11

Berekaa MM (2016) Nanotechnology in wastewater treatment; influence of nanomaterials on microbial systems. Int J Curr Microbiol App Sci 5:713–726

Beyth N, Houri-Haddad Y, Domb A, Khan W, Hazan R (2015) Alternative antimicrobial approach: nano-antimicrobial materials. Evid Based Complement Altern Med 2015:1–16

Bhaumik M, Maity A, Srinivasu VV, Onyango MS (2011) Enhanced removal of Cr(VI) from aqueous solution using polypyrrole/Fe_3O_4 magnetic nanocomposite. J Hazard Mater 190:381–390

Bhuyan T, Mishra K, Khanuja M, Prasad R, Varma A (2015) Biosynthesis of zinc oxide nanoparticles from *Azadirachta indica* for antibacterial and photocatalytic applications. Mater Sci Semicond Process 32:55–61

Biju V (2014) Chemical modifications and bioconjugate reactions of nanomaterials for sensing, imaging, drug delivery and therapy. Chem Soc Rev 43:744–764

Bogue R (2011) Nanocomposites: a review of technology and applications. Assembly Autom 31:106–112

Bokare V, Jung JL, Chang YY, Chang YS (2013) Reductive dechlorination of octachlorodibenzo-p-dioxin by nanosized zero-valent zinc: modeling of rate kinetics and congener profile. J Hazard Mater 250:397–402

Bollmann AF, Seitz W, Prasse C, Lucke T, Schulz W, Ternes T (2016) Occurrence and fate of amisulpride, sulpiride, and lamotrigine in municipal wastewater treatment plants with biological treatment and ozonation. J Hazard Mater 320:204–215

Bonvin F, Jost L, Randin L, Bonvin E, Kohn T (2016) Super-fine powdered activated carbon (SPAC) for efficient removal of micropollutants from wastewater treatment plant effluent. Water Res 90:90–99

Bazrafshan E, Ahmadabadi M, Mahvi A.H (2013) Reactive Red-120 removal by activated carbon obtained from cumin herb wastes. Fres Environ Bull 22(2a):584–590

Central Pollution Control Board (2010) Status of water quality in India 2009. Central Pollution Control Board, Ministry of Environment and Forests, Government of India, New Delhi

Chitra K, Annadurai G (2014) Antibacterial activity of pH-dependent biosynthesized silver nanoparticles against clinical pathogen. Biomed Res Int 2014:725165

Cho M, Cates EL, Kim JH (2011) Inactivation and surface interactions of MS-2 bacteriophage in a TiO_2 photoelectrocatalytic reactor. Water Res 45:2104–2110

Corsi I, Winther-Nielse M, Sethi R, Punta C, Della Torre C, Libralato G, Cinuzzi F (2018) Ecofriendly nanotechnologies and nanomaterials for environmental applications: key issue and consensus recommendations for sustainable and ecosafe nanoremediation. Ecotoxicol Environ Saf 154:237–244
Crane RA, Scott TB (2012) Nanoscale zero-valent iron: future prospects for an emerging water treatment technology. J Hazard Mater 211:112–125
Das R, Ali ME, Hamid SBA, Ramakrishna S, Chowdhury ZZ (2014a) Carbon nanotube membranes for water purification: a bright future in water desalination. Desalination 336:97–109
Das R, Hamid SBA, Ali ME, Ismail AF, Annuar MSM, Ramakrishna S (2014b) Multifunctional carbon nanotubes in water treatment: the present, past and future. Desalination 354:160–179
Das R, Vecitis CD, Schulze A, Cao B, Ismail AF, Lu X, Chen J, Ramakrishna S (2017) Recent advances in nanomaterials for water protection and monitoring. Chem Soc Rev 46:6946–7020
Dauthal P, Mukhopadhyay M (2016) Noble metal nanoparticles: plant-mediated synthesis, mechanistic aspects of synthesis, and applications. Ind Eng Chem Res 55:9557–9577
Dave PN, Chopda LV (2014) Application of iron oxide nanomaterials for the removal of heavy metals. J Nanotechnol 2014:1–14
Dehghani MH, Taher MM, Bajpai AK, Heibati B, Tyagi I, Asif M, Agarwal S, Gupta VK (2015) Removal of noxious Cr(VI) ions using single-walled carbon nanotubes and multi-walled carbon nanotubes. Chem Eng J 279:344–352
Dinali R, Ebrahiminezhad A, Manley-Harris M, Ghasemi Y, Berenjian A (2017) Iron oxide nanoparticles in modern microbiology and biotechnology. Crit Rev Microbiol 43:493–507
Doong RA, Chiang LF (2008) Coupled removal of organic compounds and heavy metals by titanate/carbon nanotube composites. Wat Sci Tech 58:1985–1992
Dutta D, Thakur D, Bahadur D (2015a) SnO_2 quantum dots decorated silica nanoparticles for fast removal of cationic dye (methylene blue) from wastewater. Chem Eng J 281:482–490
Dutta DK, Borah BJ, Sarmah PP (2015b) Recent advances in metal nanoparticles stabilization into nanopores of montmorillonite and their catalytic applications for fine chemicals synthesis. Cat Rev Sci Eng 57:257–305
Elmizadeh H, Soleimani M, Faridbod F, Bardajee GR (2018) A sensitive nano-sensor based on synthetic ligand-coated CdTe quantum dots for rapid detection of Cr(III) ions in water and wastewater samples. Colloid Polym Sci 296:1581–1590
Esawi AMK, Morsi K, Sayed A, Taher M, Lanka S (2010) Effect of carbon nanotube (CNT) content on the mechanical properties of CNT-reinforced aluminium composites. Compos Sci Technol 70:2237–2241
Feng J, Tao Y, Shen X, Jin H, Zhou T, Zhou Y, Lee YI (2019) Highly sensitive and selective fluorescent sensor for tetrabromobisphenol-A in electronic waste samples using molecularly imprinted polymer coated quantum dots. Microchem J 144:93–101
Ferreira AM, Roque ÉB, Fonseca FVD, Borges CP (2015) High flux microfiltration membranes with silver nanoparticles for water disinfection. Desalin Water Treat 56:3590–3598
Foguel MV, Pedro NTB, Zanoni MVB, Sotomayor MDPT (2017) Molecularly imprinted polymer (MIP): a promising recognition system for development of optical sensor for textile dyes. Procedia Tech 27:299–300
Fu F, Dionysiou DD, Liu H (2014) The use of zero-valent iron for groundwater remediation and wastewater treatment: a review. J Hazard Mater 267:194–205
Gelover S, Gómez LA, Reyes K, Leal MT (2006) A practical demonstration of water disinfection using TiO_2 films and sunlight. Water Res 40:3274–3280
Ghaedi M, Khajehsharifi H, Yadkuri AH, Roosta M, Asghari A (2012) Oxidized multiwalled carbon nanotubes as efficient adsorbent for bromothymol blue. Toxicol Environ Chem 94:873–883
Ghorbani M, Eisazadeh H, Ghoreyshi AA (2012) Removal of zinc ions from aqueous solution using polyaniline nanocomposite coated on rice husk. Iran J Energ Environ 3:83–88
Ghosh A, Nayak AK, Pal A (2017) Nano-particle-mediated wastewater treatment: a review. Curr Pollut Rep 3:17–30
Gosavi VD, Sharma S (2014) A general review on various treatment methods for textile wastewater. J Environ Sci Comput Sci Eng Technol 3:29–39

Guan X, Sun Y, Qin H, Li J, Lo IM, He D, Dong H (2015) The limitations of applying zero-valent iron technology in contaminants sequestration and the corresponding countermeasures: the development in zero-valent iron technology in the last two decades (1994–2014). Water Res 75:224–248

Guerra F, Attia M, Whitehead D, Alexis F (2018) Nanotechnology for environmental remediation: materials and applications. Molecules 23:1760

Gunatilake SK (2015) Methods of removing heavy metals from industrial wastewater. Methods 1:12–18

Gutierrez AM, Dziubla TD, Hilt JZ (2017) Recent advances on iron oxide magnetic nanoparticles as sorbents of organic pollutants in water and wastewater treatment. Rev Environ Health 32:111–117

Guzmán-Verri GG, Voon LLY (2007) Electronic structure of silicon-based nanostructures. Phys Rev B 76:075131

Hajkova P, Spatenka P, Horsky J, Horska I, Kolouch A (2007) Photocatalytic effect of TiO_2 films on viruses and bacteria. Plasma Process Polym 4:S397–S401

Hao C, Feng F, Wang X, Zhou M, Zhao Y, Ge C, Wang K (2015) The preparation of Fe_2O_3 nanoparticles by liquid phase–based ultrasonic-assisted method and its application as enzyme-free sensor for the detection of H_2O_2. RSC Adv 5(27):21161–21169

Hou L, Xia J, Li K, Chen J, Wu X, Li X (2013) Removal of ZnO nanoparticles in simulated wastewater treatment processes and its effects on COD and NH_4^+-N reduction. Water Sci Technol 67:254–260

Huh AJ, Kwon YJ (2011) “Nanoantibiotics”: a new paradigm for treating infectious diseases using nanomaterials in the antibiotics resistant era. J Control Release 156:128–145

Ibhadon AO, Fitzpatrick P (2013) Heterogeneous photocatalysis: recent advances and applications. Catalysts 3:189–218

Ilankoon N (2014) Use of iron oxide magnetic nanosorbents for Cr(VI) removal from aqueous solutions: a review. J Eng Res Appl 4:55–63

Jardón-Maximino N, Pérez-Alvarez M, Sierra-Ávila R, Ávila-Orta CA, Jiménez-Regalado E, Bello AM, González-Morones P, Cadenas-Pliego G (2018) Oxidation of copper nanoparticles protected with different coatings and stored under ambient conditions. J Nanomater 2018:1–8

Jegatheesan V, Pramanik BK, Chen J, Navaratna D, Chang CY, Shu L (2016) Treatment of textile wastewater with membrane bioreactor: a critical review. Bioresour Technol 204:202–212

Ju-Nam Y, Lead J (2016) Properties, sources, pathways, and fate of nanoparticles in the environment. Engineered Nanoparticles and the Environment: Biophysicochemical Processes and Toxicity 4:95–117

Kamaly N, Yameen B, Wu J, Farokhzad OC (2016) Degradable controlled-release polymers and polymeric nanoparticles: mechanisms of controlling drug release. Chem Rev 116:2602–2663

Kaur R, Wani SP, Singh AK, Lal K (2012) Wastewater production, treatment and use in India. In: National report presented at the 2nd Regional Workshop on Safe Use of Wastewater in Agriculture, New Delhi, 16–18 May 2012, pp 1–13

Kaur P, Singh S, Kumar V, Singh N, Singh J (2018) Effect of rhizobacteria on arsenic uptake by macrophyte *Eichhornia crassipes* (Mart.) Solms. Int J Phytoremediation 20(2):114–120

Khan I, Saeed K, Khan I (2017) Nanoparticles: properties, applications and toxicities. Arab J Chem. https://doi.org/10.1016/j.arabjc.2017.05.011

Klačanová K, Fodran P, Šimon P, Rapta P, Boča R, Jorik V, Miglierini M, Kolek E, Čaplovič L (2013) Formation of Fe(0)-nanoparticles via reduction of Fe(II) compounds by amino acids and their subsequent oxidation to iron oxides. J Chem 2013:961629. https://doi.org/10.1155/2013/961629

Kumar V, Upadhyay N, Singh S, Singh J, Kaur P (2013) Thin-layer chromatography: comparative estimation of soil's atrazine. Curr World Environ 8(3):469–472

Kumar A, Sharma G, Naushad M, Singh P, Kalia S (2014) Polyacrylamide/$Ni_{0.02}Zn_{0.98}O$ nanocomposite with high solar light photocatalytic activity and efficient adsorption capacity for toxic dye removal. Ind Eng Chem Res 53:15549–15560

Kumar V, Upadhyay N, Kumar V, Kaur S, Singh J, Singh S, Datta S (2014a) Environmental exposure and health risks of the insecticide monocrotophos—a review. J Biodivers Environ Sci 5:111–120
Kumar V, Singh S, Manhas A, Singh J, Singla S, Kaur P (2014b) Bioremediation of petroleum hydrocarbon by using *Pseudomonas* species isolated from petroleum contaminated soil. Orient J Chem 30(4):1771–1776
Kumar V, Singh S, Kashyap N, Singla S, Bhadrecha P, Kaur P (2015a) Bioremediation of heavy metals by employing resistant microbial isolates from agricultural soil irrigated with industrial waste water. Orient J Chem 31(1):357–361
Kumar V, Singh S, Singh J, Upadhyay N (2015b) Potential of plant growth promoting traits by bacteria isolated from heavy metal contaminated soils. Bull Environ Contam Toxicol 94:807–815
Kumar V, Kaur S, Singh S, Upadhyay N (2016) Unexpected formation of N′-phenyl-thiophosphorohydrazidic acid O,S-dimethyl ester from acephate: chemical, biotechnical and computational study. 3 Biotech 6(1):1
Kumar V, Singh S, Singh R, Upadhyay N, Singh J (2017) Design, synthesis, and characterization of 2,2-bis(2,4-dinitrophenyl)-2-(phosphonatomethylamino) acetate as a herbicidal and biological active agent. J Chem Biol 10(4):179–190
Kundu S, Wang Y, Xia W, Muhler M (2008) Thermal stability and reducibility of oxygen-containing functional groups on multiwalled carbon nanotube surfaces: a quantitative high-resolution XPS and TPD/TPR study. J Phys Chem C 112:16869–16878
Kunduru KR, Nazarkovsky M, Farah S, Pawar RP, Basu A, Domb AJ (2017) Nanotechnology for water purification: applications of nanotechnology methods in wastewater treatment. In: Grumezescu AM (ed) Water purification. Academic, London, pp 33–74
Larramendy ML, Soloneski S (2015) Emerging pollutants in the environment: current and further implications. InTechOpen. https://doi.org/10.5772/59332
Le AT, Le TT, Tran HH, Dang DA, Tran QH, Vu DL (2012) Powerful colloidal silver nanoparticles for the prevention of gastrointestinal bacterial infections. Adv Nat Sci Nanosci Nanotechnol 3:045007
Lee C (2015) Oxidation of organic contaminants in water by iron-induced oxygen activation: a short review. Environ Eng Res 20:205–211
Lee KM, Lai CW, Ngai KS, Juan JC (2016) Recent developments of zinc oxide based photocatalyst in water treatment technology: a review. Water Res 88:428–448
Li M, Yin JJ, Wamer WG, Lo YM (2014) Mechanistic characterization of titanium dioxide nanoparticle-induced toxicity using electron spin resonance. J Food Drug Anal 22:76–85
Liang CZ, Sun SP, Li FY, Ong YK, Chung TS (2014) Treatment of highly concentrated wastewater containing multiple synthetic dyes by a combined process of coagulation/flocculation and nanofiltration. J Membr Sci 469:306–315
Liang J, Liu J, Yuan X, Dong H, Zeng G, Wu H, Wang H, Liu J, Hua S, Zhang S, Yu Z (2015) Facile synthesis of alumina-decorated multi-walled carbon nanotubes for simultaneous adsorption of cadmium ion and trichloroethylene. Chem Eng J 273:101–110
Liga MV, Bryant EL, Colvin VL, Li Q (2011) Virus inactivation by silver doped titanium dioxide nanoparticles for drinking water treatment. Water Res 45:535–544
Lin J, Ganesh A (2013) Water quality indicators: bacteria, coliphages, enteric viruses. Int J Environ Health Res 23:484–506
Liu G, Wang D, Wang J, Mendoza C (2011a) Effect of ZnO particles on activated sludge: role of particle dissolution. Sci Total Environ 409:2852–2857
Liu L, Ma W, Zhang Z (2011b) Macroscopic carbon nanotube assemblies: preparation, properties, and potential applications. Small 7:1504–1520
Lohani A, Verma A, Joshi H, Yadav N, Karki N (2014) Nanotechnology-based cosmeceuticals. ISRN Dermatol 2014:843687
Lu H, Wang J, Stoller M, Wang T, Bao Y, Hao H (2016) An overview of nanomaterials for water and wastewater treatment. Adv Mater Sci Eng 2016:1–10

Luo Y, Guo W, Ngo HH, Nghiem LD, Hai FI, Zhang J, Liang S, Wang XC (2014) A review on the occurrence of micropollutants in the aquatic environment and their fate and removal during wastewater treatment. Sci Total Environ 473:619–641

Machado FM, Bergmann CP, Fernandes TH, Lima EC, Royer B, Calvete T, Fagan SB (2011) Adsorption of Reactive Red M-2BE dye from water solutions by multi-walled carbon nanotubes and activated carbon. J Hazard Mater 192:1122–1131

Mahgoub S, Samaras P (2014) Nanoparticles from biowastes and microbes: focus on role in water purification and food preservation. In: Proceedings of the 2nd international conference on sustainable solid waste management, Athens, 12–14 June 2014, pp 1–39

Maliyekkal SM, Sreeprasad TS, Krishnan D, Kouser S, Mishra AK, Waghmare UV, Pradeep T (2013) Graphene: a reusable substrate for unprecedented adsorption of pesticides. Small 9:273–283

Mauter MS, Zucker I, Perreault F, Werber JR, Kim JH, Elimelech M (2018) The role of nanotechnology in tackling global water challenges. Nat Sustain 1:166

Méndez E, González-Fuentes MA, Rebollar-Perez G, Méndez-Albores A, Torres E (2017) Emerging pollutant treatments in wastewater: cases of antibiotics and hormones. J Environ Sci Health A 52:235–253

Moosa AA, Muhsen MF (2017) Ceramic filters impregnated with silver nanoparticles for household drinking water treatment. Am J Mater Sci 7:232–239

Mostafaii G, Chimehi E, Gilasi H, Iranshahi L (2017) Investigation of zinc oxide nanoparticles effects on removal of total coliform bacteria in activated sludge process effluent of municipal wastewater. J Environ Sci Technol 10:49–55

Moussavi G, Mahmoudi M (2009) Removal of azo and anthraquinone reactive dyes from industrial wastewaters using MgO nanoparticles. J Hazard Mater 168:806–812

Nadafi K, Mesdaghinia A, Nabizadeh R, Younesian M, Rad MJ (2011) The combination and optimization study on RB29 dye removal from water by peroxy acid and single-wall carbon nanotubes. Desalin Water Treat 27:237–242

Naidoo S, Olaniran AO (2013) Treated wastewater effluent as a source of microbial pollution of surface water resources. Int J Environ Res Public Health 11:249–270

Nassar NN (2012) Iron oxide nanoadsorbents for removal of various pollutants from wastewater: an overview. In: Bhatnagar A (ed) Application of adsorbents for water pollution control. Bentham Science, Sharjah, pp 81–118

Ni Y, Jin L, Zhang L, Hong J (2010) Honeycomb-like Ni@C composite nanostructures: synthesis, properties and applications in the detection of glucose and the removal of heavy-metal ions. J Mater Chem 20:6430–6436

Nowack B, Bucheli TD (2007) Occurrence, behavior and effects of nanoparticles in the environment. Environ Pollut 150:5–22

Okpalugo TIT, Papakonstantinou P, Murphy H, McLaughlin J, Brown NMD (2005) High resolution XPS characterization of chemical functionalised MWCNTs and SWCNTs. Carbon 43:153–161

Oller I, Miralles-Cuevas S, Aguera A, Malato S (2018) Monitoring and removal of organic micro-contaminants by combining membrane technologies with advanced oxidation processes. Curr Org Chem 22:1103–1119

Ozmen M, Can K, Arslan G, Tor A, Cengeloglu Y, Ersoz M (2010) Adsorption of Cu(II) from aqueous solution by using modified Fe_3O_4 magnetic nanoparticles. Desalination 254:162–169

Panahi Y, Mellatyar H, Farshbaf M, Sabet Z, Fattahi T, Akbarzadehe A (2018) Biotechnological applications of nanomaterials for air pollution and water/wastewater treatment. Mater Today Proc 5:15550–15558

Pande S, Singh BP, Mathur RB, Dhami TL, Saini P, Dhawan SK (2009) Improved electromagnetic interference shielding properties of MWCNT–PMMA composites using layered structures. Nanoscale Res Lett 4:327

Peigney A, Laurent C, Flahaut E, Bacsa RR, Rousset A (2001) Specific surface area of carbon nanotubes and bundles of carbon nanotubes. Carbon 39:507–514

Peiris MK, Gunasekara CP, Jayaweera PM, Arachchi ND, Fernando N (2017) Biosynthesized silver nanoparticles: are they effective antimicrobials? Mem Inst Oswaldo Cruz 112:537–543

Prasad R, Bhattacharyya A, Nguyen QD (2017) Nanotechnology in sustainable agriculture: Recent developments, challenges, and perspectives. Front Microbiol 8:1014. doi: 10.3389/fmicb.2017.01014

Prasad R (2014) Synthesis of silver nanoparticles in photosynthetic plants. Journal of Nanoparticles, Article ID 963961, http://dx.doi.org/10.1155/2014/963961

Prasad R, Kumar V, Prasad KS (2014) Nanotechnology in sustainable agriculture: present concerns and future aspects. Afr J Biotechnol 13(6):705–713

Prasad R, Thirugnanasanbandham K (2019) Advances Research on Nanotechnology for Water Technology. Springer International Publishing https://www.springer.com/us/book/9783030023805

Prasad R, Jha A and Prasad K (2018) Exploring the Realms of Nature for Nanosynthesis. Springer International Publishing (ISBN 978-3-319-99570-0) https://www.springer.com/978-3-319-99570-0

Puay NQ, Qiu G, Ting YP (2015) Effect of zinc oxide nanoparticles on biological wastewater treatment in a sequencing batch reactor. J Clean Prod 88:139–145

Qu X, Alvarez PJ, Li Q (2013) Applications of nanotechnology in water and wastewater treatment. Water Res 47:3931–3946

Rahimi MR, Mosleh S (2018) Intensification of textile wastewater treatment processes. In: ul-Islam S, Butola BS (eds) Advanced textile engineering materials. Scrivener, Salem, pp 329–387. https://doi.org/10.1002/9781119488101.ch9

Rana N, Ghosh KS, Chand S, Gathania AK (2018) Investigation of ZnO nanoparticles for their applications in wastewater treatment and antimicrobial activity. Indian J Pure Appl Phys 56(1):19–25

Ranade VV, Bhandari VM (2014) Industrial wastewater treatment, recycling and reuse. Butterworth-Heinemann, Oxford

Rasalingam S, Peng R, Koodali RT (2014) Removal of hazardous pollutants from wastewaters: applications of TiO_2–SiO_2 mixed oxide materials. J Nanomater 2014:617405. https://doi.org/10.1155/2014/617405

Ren X, Chen C, Nagatsu M, Wang X (2011a) Carbon nanotubes as adsorbents in environmental pollution management: a review. Chem Eng J 170:395–410

Ren Y, Yan N, Wen Q, Fan Z, Wei T, Zhang M, Ma J (2011b) Graphene/δ-MnO_2 composite as adsorbent for the removal of nickel ions from wastewater. Chem Eng J 175:1–7

Rodrigues SM, Demokritou P, Dokoozlian N, Hendren CO, Karn B, Mauter MS, Sadik OA, Safarpour M, Unrine JM, Viers J, Welle P (2017) Nanotechnology for sustainable food production: promising opportunities and scientific challenges. Environ Sci Nano 4:767–781

Rudakiya DM, Pawar K (2017) Bactericidal potential of silver nanoparticles synthesized using cell-free extract of *Comamonas acidovorans*: in vitro and in silico approaches. 3 Biotech 7:92

Sa RM, Premalatha M (2016) Applications of nanotechnology in waste water treatment: a review. Imp J Interdiscip Res 2(11):1500–1511

Sadhu SD, Garg M, Meena PL (2018) Nanotechnology based separation systems for sustainable water resources. In: Mishra AK, Hussain CM (eds) Nanotechnology for sustainable water resources. Scrivener, Beverly, pp 523–558

Saharan P, Chaudhary GR, Mehta SK, Umar A (2014) Removal of water contaminants by iron oxide nanomaterials. J Nanosci Nanotechnol 14:627–643

Sahu O (2014) Reduction of organic and inorganic pollutant from waste water by algae. Int Lett Nat Sci 8:1–8

Scida K, Stege PW, Haby G, Messina GA, García CD (2011) Recent applications of carbon-based nanomaterials in analytical chemistry: critical review. Anal Chim Acta 691:6–17

Sears K, Dumée L, Schütz J, She M, Huynh C, Hawkins S, Duke M, Gray S (2010) Recent developments in carbon nanotube membranes for water purification and gas separation. Materials 3:127–149

Seo Y, Hwang J, Kim J, Jeong Y, Hwang MP, Choi J (2014) Antibacterial activity and cytotoxicity of multi-walled carbon nanotubes decorated with silver nanoparticles. Int J Nanomedicine 9:4621

Sharma VK, Filip J, Zboril R, Varma RS (2015) Natural inorganic nanoparticles—formation, fate, and toxicity in the environment. Chem Soc Rev 44:8410–8423

Shirmardi M, Mahvi AH, Mesdaghinia A, Nasseri S, Nabizadeh R (2013) Adsorption of acid red18 dye from aqueous solution using single-wall carbon nanotubes: kinetic and equilibrium. Desalin Water Treat 51:6507–6516

Sidhu GK, Singh S, Kumar V, Datta S, Singh D, Singh J (2019) Toxicity, monitoring and biodegradation of organophosphate pesticides: a review. Crit Rev Environ Sci Technol. https://doi.org/10.1007/s00128990044

Singh S, Singh N, Kumar V, Datta S, Wani AB, Singh D, Singh J (2016) Toxicity, monitoring and biodegradation of the fungicide carbendazim. Environ Chem Lett 14:317–329

Singh S, Kumar V, Chauhan A, Datta S, Wani AB, Singh N, Singh J (2017a) Toxicity, degradation and analysis of the herbicide atrazine. Environ Chem Lett 16(1):211–237. https://doi.org/10.1007/s10311-017-0665-8

Singh S, Kumar V, Upadhyay N, Singh J, Singla S, Datta S (2017b) Efficient biodegradation of acephate by *Pseudomonas pseudoalcaligenes* PS-5 in the presence and absence of heavy metal ions [Cu(II) and Fe(III)], and humic acid. 3 Biotech 7(4):262. https://doi.org/10.1007/s13205-017-0900-9

Spitalsky Z, Tasis D, Papagelis K, Galiotis C (2010) Carbon nanotube–polymer composites: chemistry, processing, mechanical and electrical properties. Prog Polym Sci 35:357–401

Stark WJ, Stoessel PR, Wohlleben W, Hafner A (2015) Industrial applications of nanoparticles. Chem Soc Rev 44:5793–5805

Sun T, Gottschalk F, Hungerbuhler K, Nowack B (2014) Comprehensive probabilistic modelling of environmental emissions of engineered nanomaterials. Environ Pollut 185:69–76

Sun A, Chai J, Xiao T, Shi X, Li X, Zhao Q, Li D, Chen J (2018) Development of a selective fluorescence nanosensor based on molecularly imprinted–quantum dot optosensing materials for saxitoxin detection in shellfish samples. Sensors Actuators B Chem 258:408–414

Thoniyot P, Tan MJ, Karim AA, Young DJ, Loh XJ (2015) Nanoparticle–hydrogel composites: concept, design, and applications of these promising, multi-functional materials. Adv Sci 2:1400010

Tsarev S, Collins RN, Fahy A, Waite TD (2016) Reduced uranium phases produced from anaerobic reaction with nanoscale zerovalent iron. Environ Sci Technol 50:2595–2601

Usmania MA, Khan I, Bhatd AH, Pillaie RS, Ahmadf N, Haafizg MM, Ovesh M (2017) Current trend in the application of nanoparticles for waste water treatment and purification: a review. Curr Org Synth 14:1–21

Varjani SJ, Gnansounou E, Pandey A (2017) Comprehensive review on toxicity of persistent organic pollutants from petroleum refinery waste and their degradation by microorganisms. Chemosphere 188:280–291

Visa M, Carcel RA, Andronic L, Duta A (2009) Advanced treatment of wastewater with methyl orange and heavy metals on TiO_2, fly ash and their mixtures. Catal Today 144:137–142

Volder DMF, Tawfick SH, Baughman RH, Hart JA (2013) Carbon nanotubes: present and future commercial applications. Science 339:535–539

Vuković GD, Marinković AD, Škapin SD, Ristić MĐ, Aleksić R, Perić-Grujić AA, Uskoković PS (2011) Removal of lead from water by amino modified multi-walled carbon nanotubes. Chem Eng J 173:855–865

Wang SG, Gong WX, Liu XW, Yao YW, Gao BY, Yue QY (2007) Removal of lead(II) from aqueous solution by adsorption onto manganese oxide-coated carbon nanotubes. Sep Purif Technol 58:17–23

Wang Z, Wu D, Wu G, Yang N, Wu A (2013) Modifying Fe_3O_4 microspheres with rhodamine hydrazide for selective detection and removal of Hg^{2+} ion in water. J Hazard Mater 244:621–627

Wang H, Gao H, Chen M, Xu X, Wang X, Pan C, Gao J (2016a) Microwave-assisted synthesis of reduced graphene oxide/titania nanocomposites as an adsorbent for methylene blue adsorption. Appl Surf Sci 360:840–848
Wang Z, Shan Y, Xu L, Wu G, Lu X (2016b) Determination of the azo dye, sunset yellow, using carbon paste electrode modified with molecularly imprinted polymer. Indian J Chem Sect A Inorg Phys Theor Anal 55(12):1458–1464
Wang MM, Wang J, Cao R, Wang SY, Du H (2017) Natural transformation of zinc oxide nanoparticles and their cytotoxicity and mutagenicity. J Nanomater 2017:1–12
Wu Y-Y, Xiong Z-H (2016) Multi-walled carbon nanotubes and powder-activated carbon adsorbents for the removal of nitrofurazone from aqueous solution. J Dispers Sci Technol 37:613–624
Xie G, Xi P, Liu H, Chen F, Huang L, Shi Y, Hou F, Zeng Z, Shao C, Wang J (2012) A facile chemical method to produce superparamagnetic graphene oxide–Fe_3O_4 hybrid composite and its application in the removal of dyes from aqueous solution. J Mater Chem 22:1033–1039
Yan W, Lien HL, Koel BE, Zhang WX (2013) Iron nanoparticles for environmental clean-up: recent developments and future outlook. Environ Sci Process Impacts 15:63–77
Yang JC, Yin XB (2017) $CoFe_2O_4$@MIL-100 (Fe) hybrid magnetic nanoparticles exhibit fast and selective adsorption of arsenic with high adsorption capacity. Sci Rep 7:40955
Yu J, Wang X, Kang Q, Li J, Shen D, Chen L (2017) One-pot synthesis of a quantum dot-based molecular imprinting nanosensor for highly selective and sensitive fluorescence detection of 4-nitrophenol in environmental waters. Environ Sci Nano 4:493–502
Zan L, Fa W, Peng T, Gong ZK (2007) Photocatalysis effect of nanometer TiO_2 and TiO_2-coated ceramic plate on Hepatitis B virus. J Photochem Photobiol B 86:165–169
Zare K, Sadegh H, Shahryari-Ghoshekandi R, Maazinejad B, Ali V, Tyagi I, Agarwal S, Gupta VK (2015) Enhanced removal of toxic Congo red dye using multi walled carbon nanotubes: kinetic, equilibrium studies and its comparison with other adsorbents. J Mol Liq 212:266–271
Zhang XX, Zhu CC (2006) Field-emission lighting tube with CNT film cathode. Microelectron J 37:1358–1360
Zhang J, Wang H, Xiao Y, Tang J, Liang C, Li F, Dong H, Xu W (2017) A simple approach for synthesizing of fluorescent carbon quantum dots from tofu wastewater. Nanoscale Res Lett 12:611
Zhao YG, Shen HY, Pan SD, Hu MQ, Xia QH (2010) Preparation and characterization of amino-functionalized nano-Fe_3O_4 magnetic polymer adsorbents for removal of chromium(VI) ions. J Mater Sci 45:5291–5301
Zhao G, Li J, Ren X, Chen C, Wang X (2011a) Few-layered graphene oxide nanosheets as superior sorbents for heavy metal ion pollution management. Environ Sci Technol 45:10454–10462
Zhao G, Ren X, Gao X, Tan X, Li J, Chen C, Huang Y, Wang X (2011b) Removal of Pb(II) ions from aqueous solutions on few-layered graphene oxide nanosheets. Dalton Trans 40:10945–10952
Zhao F, Yao D, Guo R, Deng L, Dong A, Zhang J (2015) Composites of polymer hydrogels and nanoparticulate systems for biomedical and pharmaceutical applications. Nano 5:2054–2130
Zheng X, Wu R, Chen Y (2011) Effects of ZnO nanoparticles on wastewater biological nitrogen and phosphorus removal. Environ Sci Technol 45:2826–2832
Zheng X, Shen ZP, Shi L, Cheng R, Yuan DH (2017) Photocatalytic membrane reactors (PMRs) in water treatment: configurations and influencing factors. Catalysts, 7(8):224.
Zhou Y, Qu ZB, Zeng Y, Zhou T, Shi G (2014) A novel composite of graphene quantum dots and molecularly imprinted polymer for fluorescent detection of paranitrophenol. Biosens Bioelectron 52:317–323
Zhou S, Huo D, Goines S, Yang TH, Lyu Z, Zhao M, Gilroy KD, Wu Y, Hood ZD, Xie M, Xia Y (2018a) Enabling complete ligand exchange on the surface of gold nanocrystals through the deposition and then etching of silver. J Am Chem Soc 140(38):11898–11901. https://doi.org/10.1021/jacs.8b06464
Zhou Z, Zhang Y, Shen Y, Liu S, Zhang Y (2018b) Molecular engineering of polymeric carbon nitride: advancing applications from photocatalysis to biosensing and more. Chem Soc Rev 47:2298–2321

Chapter 18
Emerging Trends in Nanobiosensor

Vinita Kumari, Sarushi Rastogi, and Vasudha Sharma

Contents

V. Kumari (✉)
School of Engineering Sciences and Technology, Jamia Hamdard, New Delhi, India

S. Rastogi · V. Sharma
Department of Food Technology, School of Interdisciplinary Sciences, Jamia Hamdard, New Delhi, India

R. Prasad et al. (eds.), *Nanobiotechnology in Bioformulations*, Nanotechnology in the Life Sciences, https://doi.org/10.1007/978-3-030-17061-5_18

18.1 Introduction to Nanobiosensors

Nanobiosensors are emerging as a powerful tool in area of sensing and diagnostics. These sensors are developed due to integration of nanotechnology with biosensors. These nanomaterial-based biosensors are much more sophisticated, reliable, and sensitive in comparison to traditional biosensors. A wide range of nanobiosensors is available nowadays in order to address various diagnostic problems. These highly sensitive nanobiosensors are particularly attractive for providing detection of extremely low concentration of analyte. They have applications in clinical diagnosis, environmental monitoring, food analysis, forensic sciences, and drug delivery systems. The clinical application includes detection of presence of pathogens and biomarkers, analysis of drugs in body fluids, monitoring of oxidative stress, and sensing of nucleic acids. In the near future, this technology is going to revolutionize personalized diagnostics based on point-of-care (POC) system. Nanobiosensors can also be used for food analysis to detect contaminants including pathogenic microbes and pesticides. Environmental applications include detection of pollutants, pesticides, toxins, and heavy metals.

Most common nanomaterials used to develop biosensors are carbon nanotubes, graphene, nanowires, quantum dots, and nanoparticles which have the capability of biological and chemical functionalization. With rapid progress in the area of nanobiosensor and its integration with microfluidics, electronics, and mechanical devices, the new generation of applications like lab on a chip devices (LOC), microarray, biochips, and drug delivery systems are going to be advanced. In this chapter, various nanomaterials, their application in nanobiosensing, and recent advancement in nanobiosensor applications with future perspectives are discussed.

18.2 Nanomaterials for Biosensing Applications

It is highly desirable to enhance feature of biosensors like sensitivity and rapidity to detect analyte at very low concentration for the development of LOC and point-of-care (POC) applications in diagnostics, environmental monitoring, and food analysis. Use of carbon nanotubes, graphene and its derivatives, nanowires, quantum dots, and metal nanoparticles (Fig. 18.1) have resulted in rapid nanobiosensors with much enhanced sensitivity due to catalytic activity, electrical feature, optical properties, and very high surface area-to-volume ratio of these nanomaterials. Minute size and disposability of nanobiosensors make the nanomaterials attractive for biosensing applications. Table 18.1 represents an overview of many nanomaterials used in developing nanobiosensors with enhanced features.

In this section, nanomaterial like carbon nanotube, nanowires, quantum dots, graphene and its derivatives and metal nanoparticles (gold, silver, and magnetic) are discussed. These materials are categorized as one dimensional (1D), two dimensional (2D), and three dimension (3D) on the basis of quantum confinement with novel magnetic, catalytic, and optical properties.

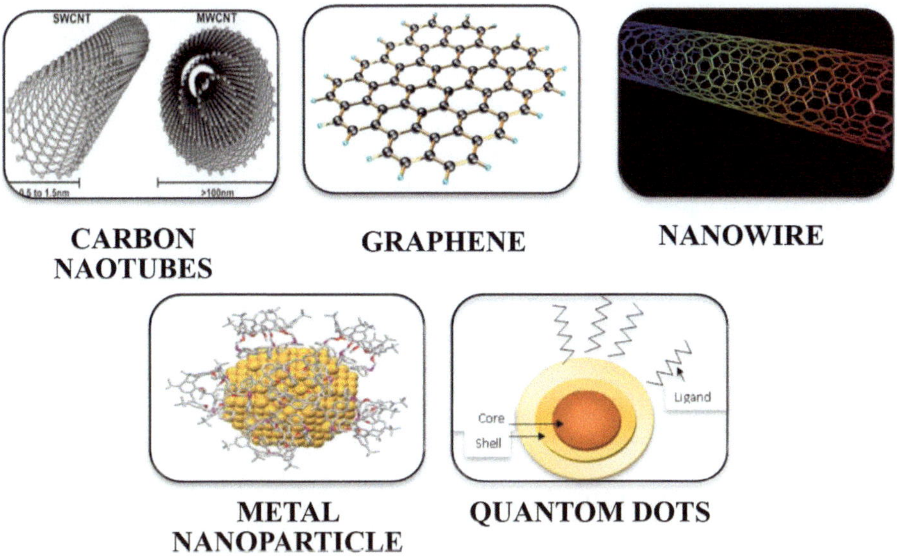

Fig. 18.1 Types of nanomaterials used in biosensor

Table 18.1 Nanomaterials used in developing Nanobiosensors

Nanomaterial used	Method of synthesis	Properties	References
Carbon nanotube	Carbon vapor discharge (CVD), carbon arc discharge, laser ablation method and n-hexane pyrolysis	Good mechanical strength, high thermal stability	Ibrahim and Saeed (2013), Salvetat et al. (1999)
Nanowires	Vapor-liquid-solid method, solution liquid-solid, electrochemical deposition, and vapor phase epitaxy	Large surface area to volume ratio, high flexibility, and low thermal conductivity	Dasgupta et al. (2014)
Quantum dots	Physical methods used are molecular beam epitaxy (MBE), ion implantation, e-beam lithography, and X-ray lithography and chemical methods include colloidal synthesis	High molar extinction coefficients and high photostability	Altintas et al. (2017)
Metal Nanoparticles	1. MNP: Chemical method involves co-precipitation, thermal decomposition, hydrothermal synthesis, and micro-emulsion physical methods involve laser evaporation and milling, hydrothermal synthesis 2. Silver NP: Conventional physical methods including spark discharging evaporation-condensation, pyrolysis, and laser ablation 3. GNP: Turkevich and brust-schiffrin method, microwave, biological methods	High surface-to-volume ratio, high mechanical strength, good optical properties and good catalytic activity, good thermal conductivity	Khan et al. (2017)
Graphene	Scotch tape method, exfoliation, Chemical vapor deposition, intercalation method	Good thermal conductivity, stiffness, impermeability, and electrical conductivity	Randviir et al. (2014), Zheng et al. (2015)

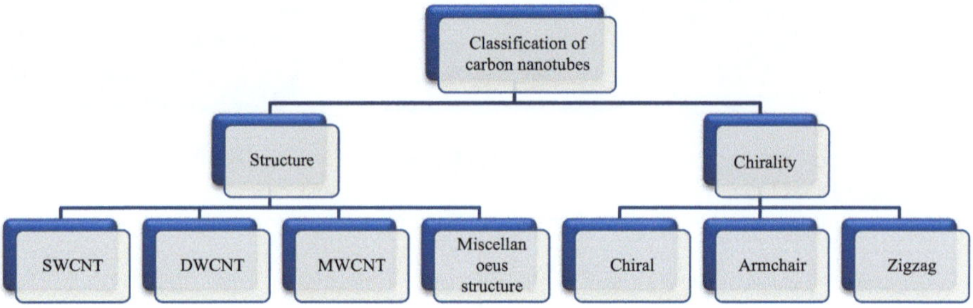

Fig. 18.2 Classification of carbon nanotube based on structure

18.2.1 Carbon Nanotubes

Carbon nanotubes (CNTs) are tubular carbon-based fiber structure with diameter in <1 nm to 50 nm range. These structures were first observed by Iijima in 1991 and made huge impact on nanoscience and nanotechnology. The carbon nanotubes are considered to be rolled form of graphene sheets in which electrons in sp^2 hybridized carbon atoms get delocalized. The electrical properties of carbon nanotube are exhibited due to delocalization of free valence electrons and making it suitable for use in transistors and emitters. CNTs are highly narrow, tubular, lightweight, and highly flexible material with high strength (Salvetat et al. 1999).

Carbon nanotube is categorized into single-walled carbon nanotubes (SWCNTs), double-walled carbon nanotubes (DWCNTs), multiwalled carbon nanotube (MWCNT), and several other structures (Varshney 2014) as in Fig. 18.2.

The various synthesis methods of carbon nanotube are derived from basic idea:

$$\text{Carbon source} \xrightarrow[\text{catalyst}]{\varnothing} \text{Carbon nanotube}$$

The four major methods for synthesis of carbon nanotube are carbon vapor discharge (CVD), carbon arc discharge, laser ablation method, and n-hexane pyrolysis (Ibrahim and Saeed 2013).

18.2.2 Nanowires

Nanowires have two-dimensional quantum confinement but one unconfined direction available for electrical conduction. Synthesis of nanowire follows top-down and bottom-up approaches, which include various methods like vapor-liquid-solid method, solution liquid-solid method, electrochemical deposition, and vapor phase epitaxy (Dasgupta et al. 2014). Nanowires' properties are solely dependent on their diameter, composition, and catalyst used. Some of the properties include large surface area-to-volume ratio, high flexibility, and low thermal conductivity.

18.2.3 Graphene

Graphene (GN) is a two-dimensional nanomaterial with planar monolayer sheet. Here, carbon atoms are sp^2 hybridized and densely packed in honeycomb crystal lattice. Graphene was discovered by A. K. Geim and K. S. Novoselov in 2007. It is gaining worldwide attention because of its chemical inertness, high electron mobility, high flexibility, and optical transmittance (Randviir et al. 2014). Overlapping of delocalized pi electron system allows free movement of electrons without scattering and contributes to electronic property of graphene (Zheng et al. 2015).

18.2.4 Nanoparticles

Nanoparticles are nanoscale particles with size range less than 100 nm. Some of the features of nanoparticles are high surface-to-volume ratio, high mechanical strength, good optical properties, and good catalytic activity. High surface area-to-volume ratio is related to high sensitivity towards target analyte. These nanoparticles (Nikam et al. 2014; Khan et al. 2017; Prasad et al. 2016) have enhanced optical and thermal properties depending on size, shape, and interparticle distance.

Silver, gold, and magnetic nanoparticles are most commonly used for biosensing applications. Silver and gold nanoparticles are generally synthesized by chemical methods. Magnetic nanoparticles (MNP) are synthesized by chemical and physical methods using materials with magnetic properties at room temperature such as metals like iron, cobalt, nickel, alloys (CoPt, FePt, FeNi), and metallic oxides.

18.2.5 Quantum Dots

Brus in 1983 first characterized quantum dots (QDs) which behave like artificial atoms due to quantum confinement effects. Quantum dots are fluorescent semiconductor crystals of size ranging 2–10 nm, consisting of a semiconductor core coated by a shell and exhibiting discrete energy levels. Core is usually made of elements from groups II or VI or groups III or V. Colloidal quantum dot can be synthesized through chemical method like colloidal synthesis and physical methods like molecular beam epitaxy (MBE). The properties of quantum dots vary with its composition and size. QDs are capable of absorbing white light and re-emitting different colors. Florescence of quantum dots are determined by their band gaps. With the increment in size of quantum dot, distance between conduction band and valence band increases and thus affecting its fluorescence. At size range 2–4 nm, they emit fluorescence in higher frequency range giving blue color, while at size more than 6 nm, they exhibit red color. Building a shell of larger band gap semiconductor material around QDs can improve fluorescence quantum yield (Altintas et al. 2017).

18.3 Application of Nanomaterials in Biosensing

Highly sensitive nanobiosensors commonly comprise of a nanomaterial along with biological recognition molecule. These recognition biomolecules are immobilized with nanomaterials like CNT, graphene, metal nanoparticles, nanowires, and quantum dots to develop nanobiosensors. The nanomaterials are known for their unique properties which are quite different from bulk.

18.3.1 Carbon Nanotube (CNT)-Based Nanobiosensors

Features of CNT like chemical species adsorption, better electron transfer rate, and modification of the resistivity as well as band gap make it an ideal material for sensing applications. Fusion of CNT with biosensing devices has facilitated the development of highly sensitive electrochemical nanobiosensors for various applications like detection of viruses, cancer, and chemical pollutants.

Bhattacharya et al. (2011) has developed a CNT-based highly sensitive biosensor for virus detection (avian metapneumovirus (aMPV)) utilizing chemical functionalization for carbon nanotubes to tailor the interactions with viruses and respective antiviral antibodies. In this nanobiosensor, formation of antigen-antibody complex is reflected by variation in conductance.

Lung cancer can be diagnosed by detecting volatile organic compounds (VOCs) 1, 2, 4-trimethybenzene (TMB) and decane in breath. Liu et al. (2010) has reported an organic material-functionalized single-walled carbon nanotube (SWNT) biosensor to detect these VOCs. This SWNT-based sensor is reported to be useful for noninvasive lung cancer screening depicted in Fig. 18.3.

CA-125 antigen is a biomarker for ovarian cancer which is difficult to detect at early stages. Mandal et al. (2017) have developed an interdigitated nanobiosensor for detection of this antigen by coating electrode surface with CA-125 antibody-functionalized CNTs (Fig. 18.4). This sensor is reported to be capable of detecting the ovarian cancer antigen from a micro volume of sample. Enhanced sensitivity of this nanobiosensor is the result of higher capacitance values, high surface-to-volume ratio, and better electron transfer rate of CNT.

Baldo et al. (2016) have reported cost-effective and fast MWCNT-based amperometric immunosensor for detection of arginase (ARG-1). Human arginase-1 is reported to be an important biomarker in several pathological conditions like autoimmune inflammation in the central nervous system, acute liver injury, different types of cancers, and obstructive nephropathy. To develop nanobiosensor for ARG-1, anti-ARG1 antibody has been immobilized on MWCNT-NHS-sensitive layer to work as resistor. This device is capable to detect ARG-1in the range 30–100 ng/mL which is comparable to ELISA.

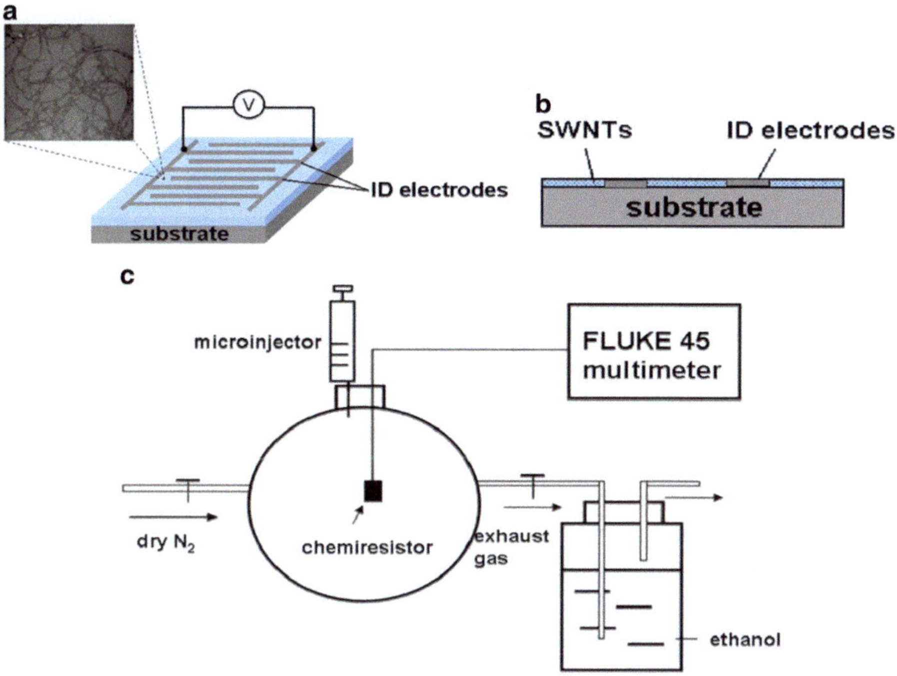

Fig. 18.3 Schematic of the test device. (**a**) the interdigitated electrode coated with SWNTs, (**b**) cross-section view of the electrode, and (**c**) the ichnography of test device. (Reprinted from Physica E: Low-dimensional Systems and Nanostructures, 44, Liu et al. (2010), Single-walled carbon nanotube-based biosensors for the detection of volatile organic compounds of lung cancer, 367–372, Copyright (2018), with permission from Elsevier)

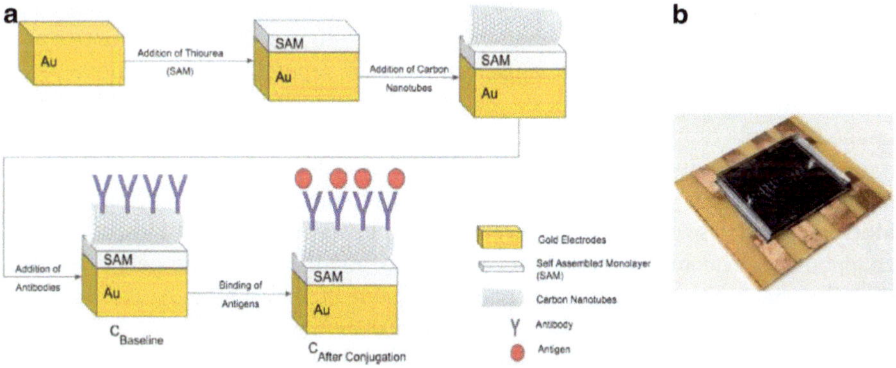

Fig. 18.4 (**a**) Flow diagram of various layer formations on top of the sensor (**b**) real image of the biosensor. (Reprinted from Mandal et al. (2017), Copyright (2018), with permission from Elsevier)

A disposable CNT-based field effect transistor (FET) biosensor for the detection of domoic acid (DA) has been developed by Marques et al. (2017). Domoic acid is a neurotoxin associated with shellfish poisoning in seawater. To develop nano-FET for DA, CNT dispersion is deposited on FET surface and anti-DA immobilized on this FET. As sample containing DA is tested on the nanobiosensor, drain current (I_d) is found to be reduced. With increase in concentration of DA, I_d reduced further. This CNT-based nano-FET has been capable of detecting domoic acid in the range of 10–500 ng/L.

For detection of urea in biological samples like blood or urine, a highly sensitive CNT-FET biosensor is developed by Melzer et al. (2016) using the principle of enzyme-substrate interaction. To develop this nanobiosensor, MWCNT solution drop is casted on gate electrode. The selective detection is reported to be achieved by immobilizing urease enzyme on the biosensor surface. This nano-FET is further modified with polymeric ion-selective membranes and pH-sensitive layers. In this work, concentration of the urea-urease interaction product is detected by the change in respective pH value of electrolyte, which is further reflected by modification in drain current (I_d).

18.3.2 Nanowire-Based Biosensors

Efficient transport of electrons and optical excitation are the unique factors which enable nanowire to be used in biosensors. Most commonly used nanowires in biosensors involve metallic, semiconducting, insulating, and molecular nanowires, which act as promising tool for field use applications such as label-free DNA sensing, virus detection, and environmental monitoring (Patolsky et al. 2006).

Organophosphorus pesticides (OP) are widely used in agriculture. Long-term exposure to low concentration of these OP compounds can damage immune and nervous system (Ghorab and Khalili 2015). An ultrasensitive palladium-copper nanowire-based electrochemical biosensor has been reported for quantitative determination of organophosphate pesticide (malathion) in fruits and vegetables based on the principle of inhibition of acetylcholinesterase (AChE) activity. On exposure to malathion, activity of AChE inhibited resulting in decrease in current. The bimetallic nanomaterials combination used in this nanobiosensor showed chemical stability as well as produced synergistic effect between catalytic activity and conductivity. The result was achieved due to good catalytic property, electron mobility, and high surface area of nanowires (Song et al. 2017).

Food-borne pathogens cause disease outbreaks every year. *E. coli, Salmonella, Listeria monocytogenes, Shigella,* and *Staphylococcus aureus* found in milk, meat, fruits, and vegetables release deadly toxins which can be fatal for humans. Ali et al. (2018) devised an impedance-based electronic biosensor consisting of interdigital silver electrodes with silver nanowires uniformly decorated on the facilitating electrical connection between electrodes. This biosensor measured variation in impedance

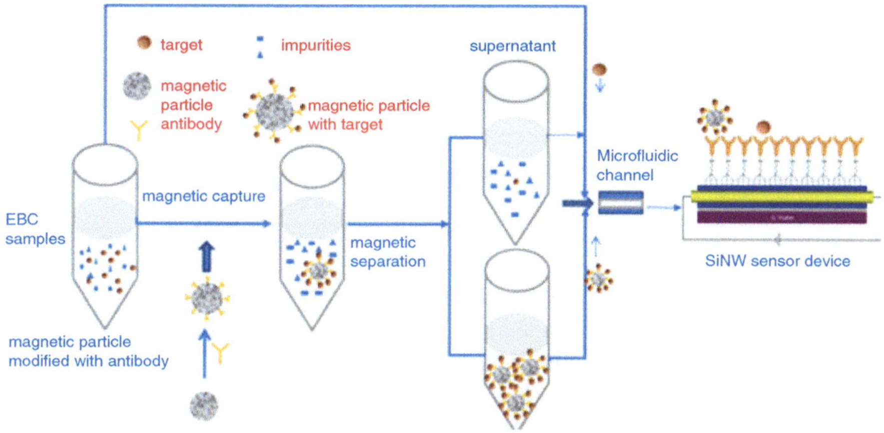

Fig. 18.5 Experimental procedure of virus and 8 iso PGF 2a detection in EBC samples using the SiNW sensor device with and without the magnetic concentrating; EBC samples collected were diluted 100-fold and transported to the sensor device at a flow rate of 170 μL/min. (Reprinted (adapted) with permission from Shen et al. (2012), Copyright (2012) American Chemical Society)

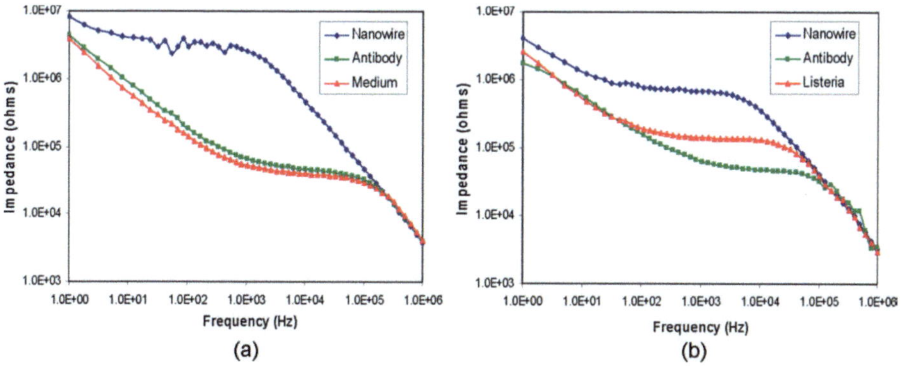

Fig. 18.6 Typical impedance spectra of a TiO_2 nanowire bundle microelectrode-based impedance immunosensor for antibody immobilization and sample detection: (**a**) detection of growth medium without bacteria (control) and (**b**) detection of *Listeria monocytogenes* at a concentration of 4.65 × 103 cfu/mL. (Reprinted (adapted) with permission from Wang et al. (2008). Copyright (2018) American Chemical Society)

depending on the concentration of three different bacteria, namely, *Escherichia coli* strains JM 109 and DH5-α and *Salmonella typhimurium*.

A nano-FET device based on silicon nanowire (SiNW) has been developed by Shen et al. (2012) for selective detection of influenza virus (H3N2) from exhaled breath condensate samples collected from the flu patients. In this device, SiNW functionalized with H3N2/H1N1 antibodies has been used for selectively detecting the virus antigen through change in conductance due to antigen-antibody interaction (Fig. 18.5).

TiO_2 nanowire bundle has been employed in an impedance immunosensor for *Listeria monocytogenes* detection. The nanostructured TiO_2 properties such as large surface area, good biocompatibility, ease of fabrication, and good chemical and photochemical stabilities have been found suitable for nanobiosensor development. TiO_2 nanowire is immobilized with monoclonal antibodies to specifically capture *L. monocytogenes*. This nanowire-antibody-bacteria complex caused change in impedance correlated to bacterial number (Fig. 18.6) (Wang et al. 2008).

Nanowires have also been used for detection of DNA for application in agriculture, forensics, paternity, and medicine. Rahman et al. (2016) has developed SiNW-based electrical nanobiosensor for early-stage diagnosis of dengue virus. In this sensor, single-stranded DNA-functionalized SiNW has been used between source and drain of FET. Alteration in current level between source and drain terminal after hybridization of target DNA with immobilized probe DNA is used as the main working principle in this sensor. The nanobiosensor's sensitivity and stability is enhanced using oxygen plasma technology for efficient use in point-of-care devices.

A glucose sensor has been reported by Liu et al. (2017) based on 3D copper foam (CF)-supported copper oxides nanowire arrays and nanoflowers. These nanostructures exhibited enhanced electrochemical catalytic activity toward glucose oxidation through the enormously increased surface area. The resulting nanosensor for glucose detection reported to provide high sensitivity, selectivity, reproducibility, and stability.

18.3.3 Graphene-Based Nanobiosensors

Graphene (GN) and its water-soluble derivative graphene oxide (GO) are, nowadays, very popular material to develop nanobiosensors for various applications like DNA, glucose, cholesterol, and dopamine sensing. Graphene is capable of improving overall performance of biosensor due to its large surface area, excellent electrical conductivity, and great electron mobility, resulting in enhancement of its selectivity and sensitivity (Suvarnaphaet and Pechprasarn 2017).

Graphene oxide (GO) is found to be useful to develop nanobiosensors due to its capability of excellent biomolecular adsorption and mediating capacity for chemical reactions. It is also reported that GO shows electrocatalytic activities toward small molecules like dopamine, hydrogen peroxide, and NADH, and there is direct electrochemistry of enzymes with graphene oxide. So enzyme biosensors have been developed based on graphene oxide. Several DNA sensors have also been developed on the basis of adsorption of single-stranded DNA (ss-DNA) on GO. As both ss-DNA and GO are negatively charged, high salt concentration is used to screen the electrostatic repulsion between them. The attractive forces between DNA and graphene oxide include hydrophobic interaction, π-π stacking, van der Waals forces, and hydrogen bonding. While probing the interaction between DNA and GO, Liu et al. (2013) explained the mechanisms of DNA sensing on the surface of graphene oxide. GO possesses strong fluorescence quenching ability, and it has strong affinity toward single-stranded DNA (ss-DNA) as well. DNA and aptamers labelled with fluorophore have been used extensively with GO to develop nanobiosensors for detecting nucleic acids, small molecules, and proteins.

Zhang et al. (2015) have reported the development of enzyme-modified graphene solution-gated transistor sensor for glucose sensing. Regular monitoring of glucose level in body is required in disease like diabetes mellitus. This nanosensor has been developed by surface modification of the graphene gate electrode using enzyme glucose oxidase (GOx). Catalysis of glucose oxidation with GOx resulted in generation of H_2O_2 and potential drop on gate electrode, which further affected channel current. Another method reported by Zhu et al. (2016) involved a graphene-based nano-FET for affinity-based detection of glucose. In this sensor, graphene functionalized with boronic acid is used for specific binding with glucose. After interaction with glucose, changes in electrical properties of graphene and detectable signal production have been reported. Kwak et al. (2012) have developed a flexible glucose nanobiosensor using graphene-based FET. In this nano-FET, enzyme GOx is used to functionalize graphene to induce the catalytic reaction of glucose. For this purpose, PET has been used as a flexible substrate and graphene film as channel in FET. Source and drain electrodes have been developed by silver paint and epoxy resin on the PET substrate. Through the measurements of differential drain-source current and Dirac point shift, fabricated nano-FET device detected the presence of glucose.

A rapid and highly sensitive nanobiosensor for multicolor fluorescent analysis of DNA has been designed by He et al. (2010) based on very high quenching efficiency of graphene oxide. Binding of fluorescent ss-DNA probe with graphene resulted in quenching of fluorescence, while strong emission obtained after hybridization of probe with complementary target. The large surface area of GO is exploited in this nanobiosensor for simultaneous quenching of several DNA probes labelled with different dyes. This led to the development of a multicolor sensor for detection of several DNA targets in single sample.

A graphene-based electrochemical biosensor has been developed by Rasheed and Sandhyarani (2014) for low concentration detection of BRCA1 gene related to breast cancer. This biosensor used sandwich method in which there is hybridization of capture probe (DNA-c) and reporter probe (DNA-r) with target probe DNA (DNA-t) on graphene-modified glassy carbon electrode in a sandwich arrangement. The electrochemical detection performed with chronoamperometry and cyclic voltammetry as shown in Fig. 18.7.

A method of nonenzymatic and electrochemical detection of cholesterol has been developed by graphene modification with β-cyclodextrin (Grp-β-CD) (Agnihotri et al. 2015). Redox indicator methylene blue (MB) after adding to Grp-β-CD forms a complex Grp-β-CD-MB. Working mechanism of this nanosensor is based on electrochemical detection of replacement of methylene blue by cholesterol molecule using differential pulse voltammetric (DPV) technique.

Biosensor for sensitive and rapid detection of dopamine (DA) is required for clinical and other applications. DA is an electroactive neurotransmitter, so electrochemical sensing can be used to detect it. Yang et al. (2012) have used this approach

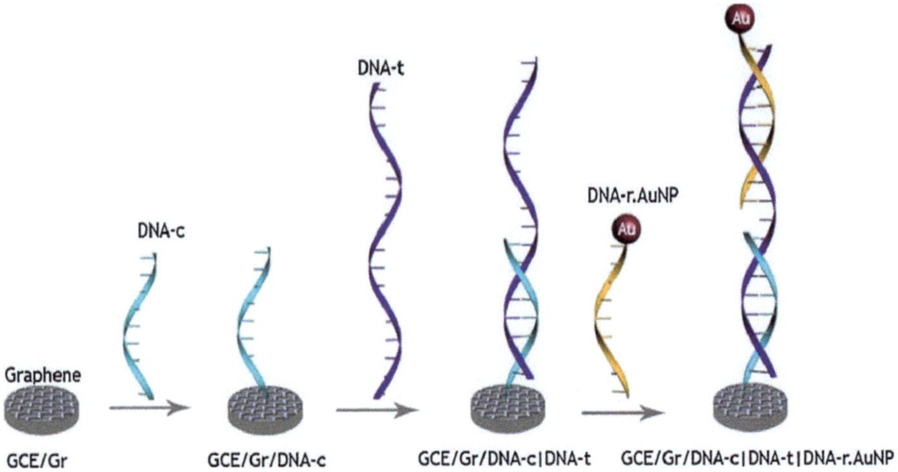

Fig. 18.7 The schematic representations of the various stages of sensor fabrication. (Reprinted from Rasheed and Sandhyarani (2014), Copyright (2018), with permission from Elsevier)

to develop a disposable working electrode based on reduced graphene oxide and gold nanoparticles on ITO-coated glass substrate for sensing presence of DA in meat samples. This sensor is found to have good selectivity for DA against other electrochemical interfering species like uric acid (UA) and ascorbic acid (AA).

18.3.4 Metal Nanoparticles-Based Nanobiosensors

Size-related properties of metal nanoparticles have been assisting in scientific researches for development of novel sensing techniques. Localized surface plasmon resonance, fluorescence enhancement/quenching, surface-enhanced Raman scattering (SERS), and electrochemical activity-related properties have been explored for metallic nanoparticles to be used in a wide range of applications such as pathogen testing, biomarker identification, and DNA testing (Doria et al. 2012). Gold, silver, and magnetic nanoparticles have been used to develop several nanobiosensors (Malik et al. 2013). Applications of metal nanoparticles in nanobiosensors are summarized in Table 18.2.

Gold nanoparticle (GNP) is known for its unique optical, electrical, and catalytic properties. Nanobiosensors based on GNP are developed for applications in diagnostics; environmental contaminants such as heavy metal ions, pathogens, and toxins; and food applications (Paul and Tiwari 2015). Gold nanoparticle-based nanosensors have also been developed for detecting different classes of enzyme activity in analytes, i.e., hydrolases, transferase, and lyase through colorimetric and fluorescence resonance energy transfer (FRET)-based assays (Hutter and Maysinger 2013).

Table 18.2 Applications of metal nanoparticles in nanobiosensors

Metal nanoparticle	Principle of detection	Applications	References
GNP	Fluorescence quenching	Detection of acryalmide in food based on fluorescent quenching	Asnaashari et al. (2018)
	Aggregation principle	Detection of malathion in food	Bolat and Abaci (2018)
	SPR	Detection of 2,4,6-trinitrotoluene (TNT)	Tan et al. (2016)
	Electrochemical	Label-free detection of DNA	Mohammad et al. (2014)
MNP	Fluorescence quenching	Detection of *E.coli* in food samples	Shelby et al. (2017)
AgNP	Colorimetric	Detection of cadmium and lead in water	Kumar and Anthony (2014)
	Electrochemical	Detection of Human chorionic gonadotropin (hCG) biomarker for prostate tumor and gestational choriocarcinoma in blood and urine; Detection of β-adrenergic agonists like ractopamine (RAC), salbutamol (SAL), and clenbuterol (CLB)	Xia et al. (2017), Wang et al. (2013)

Gold nanoparticle-based nanosensors are mainly based on its fluorescence quenching ability, surface plasmon, and aggregation mechanism. GNP is an excellent material for colorimetric biosensors development due to ease of functionalize and color display depending on state of aggregation, size, and shape.

Asnaashari et al. (2018) have reported a biosensor for acrylamide in food products based on fluorescence quenching of FAM, a fluorophore, using DNA and GNP. Acrylamide is a neurotoxic and carcinogen substance developed in some food products during high temperature processing like roasting, frying, and baking. In this nanosensor, remarkable difference in fluorescence intensity is reported in the presence of acrylamide (Fig. 18.8). Presence of the acrylamide in the sample led to acrylamide-ss-DNA adduct formation, which after binding with GNP surface resulted in quenching of fluorescence from FAM.

A highly sensitive biosensor for detection of malathion on the basis of aggregation of GNP has been developed. *Malathion* is a broad-spectrum organophosphorus pesticide. In this nanosensor, a cationic polymer poly (diallyldimethylammonium chloride) (PDDA) and aptamer are used to control aggregation of GNP. PDDA in free form interacts with GNP leading to its aggregation. In absence of malathion, apatamer binds with PDDA and inhibits aggregation of GNP. But in presence of malathion, apatamer interacts with it and free PDDA causing aggregation of GNP and change of color from red to blue (Bolat and Abaci 2018).

SPR is an optical phenomenon which can be amplified using gold nanoparticles. Biosensors for detecting different analytes are developed using SPR principle in the presence of GNP. A GNP-amplified SPR-based aptamer biosensor has been

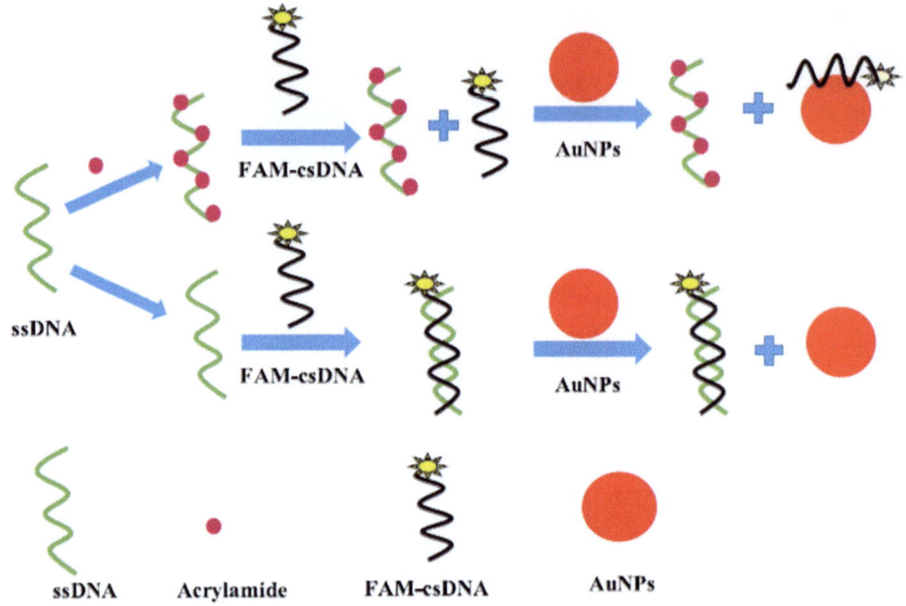

Fig. 18.8 Schematic description of acrylamide detection using fluorescent biosensor. (Reprinted from Asnaashari et al. (2018), Copyright (2018), with permission from Elsevier)

developed for detection of 2, 4, 6-trinitrotoluene (TNT), an explosive. The basic mechanism involved formation of anti-TNT peptide aptamer-GNP complexes from which TNT is capable to bind (Tan et al. 2016).

Electrochemical biosensors work on the principle of generating electrical signals as a result of biological interaction. Conductivity and catalytic property of GNP have been exploited for facilitating electron transfer between the immobilized biomolecules and surface of electrode and as catalyst in electrochemical sensors. This has led to development of GNP-based electrochemical immunosensor and DNA biosensor with enhanced performance. For label-free DNA detection, SiO_2 thin films functionalized with GNP used for attachment of DNA probes. Presence of target DNA was detected by change in capacitance of developed system after hybridization (Mohammed et al. 2014).

Similar to GNP, silver nanoparticles (AgNP) are also being used in several nanobiosensors. A multiplexed, electrochemical immunosensor for simultaneous detection of β-adrenergic agonists: ractopamine (RAC), salbutamol (SAL), and clenbuterol (CLB), has been developed. These β-adrenergic agonists are used as growth promoter in animals. For the development of sensor, silver-palladium alloy nanoparticles (AgPd NP) are used as signal label for antibodies and reduced graphene oxide (rGO) as substrate material for electrode. The immunoreactions at electrode generated strong electrochemical signal which determined the presence of β-adrenergic agonists (Wang et al. 2013).

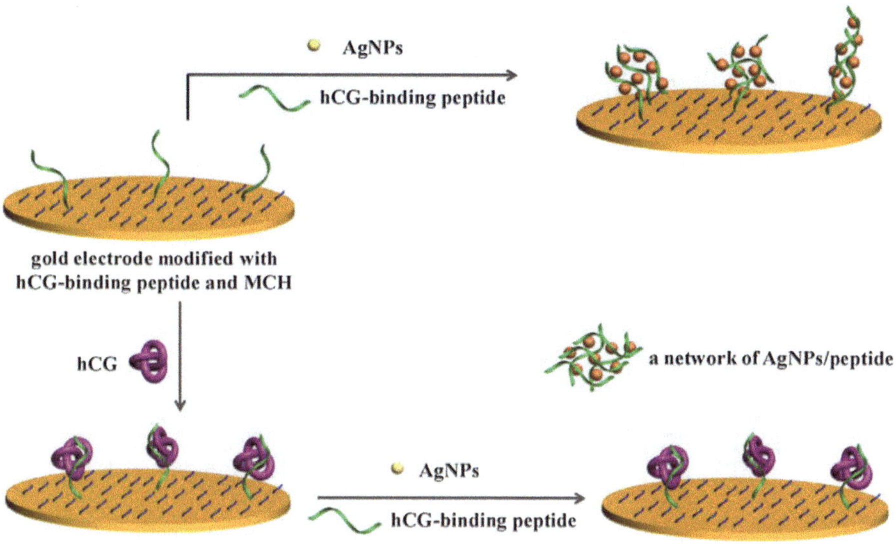

Fig. 18.9 Schematic illustration of the electrochemical method for hCG detection using AgNPs as the redox reporters and hCG-binding peptide as the receptor of hCG and the inducer of AgNPs aggregation. (Reprinted from Xia et al. (2017), Copyright (2018), with permission from Elsevier)

Another silver nanoparticles-based sensor has been reported by Xia et al. (2017) for detection of human chorionic gonadotropin (hCG) which is a biomarker for prostate tumor and gestational choriocarcinoma. The basic principle involved conversion of colorimetric assay of silver nanoparticles to electrochemical signal. The electrode surface in this electrochemical sensor is modified with hCG-specific binding peptides. This hCG-binding peptide induced aggregation of silver nanoparticles on the electrode surface as shown in Fig. 18.9. However, in the presence of hCG, aggregation is found to be inhibited and reflected by attenuation in near-sweep voltammetry (LSV) current.

Surface-modified silver nanoparticles with amino acid-based phenolic ligands have been used to detect mercury, cadmium, and lead toxicants in water samples. These heavy metal ions are known to cause central nervous defects and cancer. These ligands functionalized silver nanoparticles exhibited selective colorimetric change which further measured with UV-Vis spectroscopy in the presence of toxic Cd^{2+}, Hg^{2+}, and Pb^{2+} metal ions in aqueous solution at ppm level (Kumar and Anthony 2014).

Magnetic nanoparticles (MNP)-based nanobiosensor has been exploited for monitoring and separation of various pathogens in the processed food and raw food materials without any filtration or centrifugation procedures and making it practically effortless (Augustine et al. 2016). The basic mechanism for detection involving magnetic nanoparticles is related to formation of bacterium nanoconjugates by conjugation of MNP with appropriate bacterial ligands (Fig. 18.10). These bacterial ligands move under influence of magnetic field and move along with MNP.

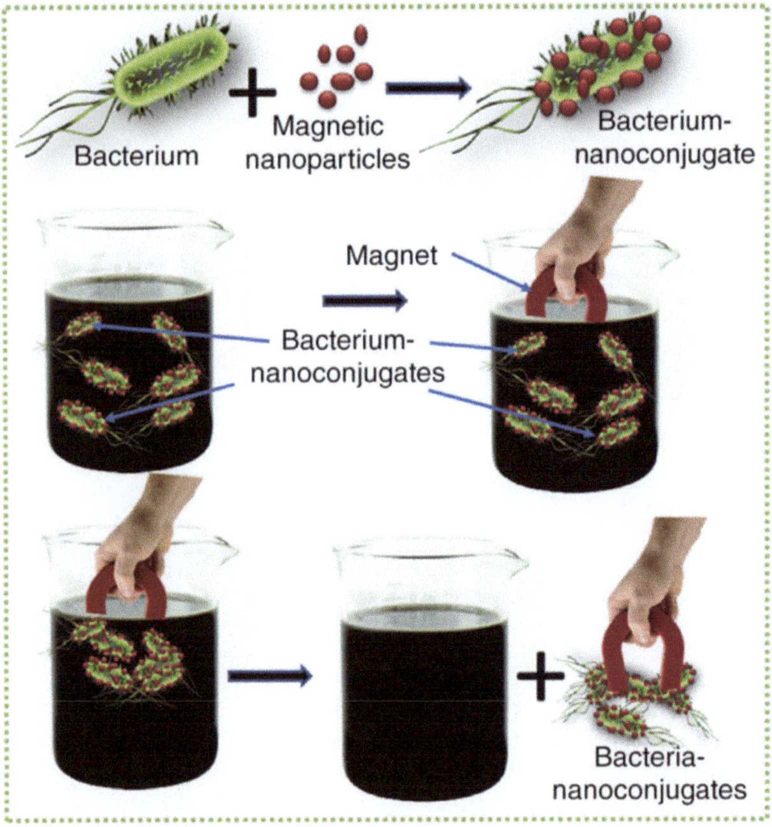

Fig. 18.10 Magnetic nanoparticle-based separation of bacteria from a contaminated solution. (Reprinted from Augustine et al. (2016), Copyright (2018), with permission from Elsevier)

The bacteria bound to MNP can be removed and decontaminated by appropriate sterilization techniques for reuse. MNPs used in electrochemical devices have shown to improve the transduction mechanism through their contact with the electrode surface, transport of a redox-active species to the electrode surface, and formation of a thin film on the electrode surface (Rocha-Santos 2014).

Aljabali et al. (2018) devised a sensitive nanosensor based on antibody conjugated iron oxide nanoparticle for detecting *Serratia marcescens* belonging to enterobacteriacia family found in starchy food causing food-borne infection. Iron oxide nanoparticle coated with *Serratia* antibody has been used to bind with cell wall of the bacteria. The magnetic properties of iron nanoparticle strengthened the detection capability.

Shelby et al. (2017) devised a magneto-fluorescent nanosensor (MFnS) for detection of *E.coli* in food samples (milk and lake water) using magnetic nanoparticles. In this nanosensor, surface of iron oxide nanoparticles conjugated with target-specific antibodies is coated with fluorescent 1,1′-Dioctadecyl-3,3,3′,3′-tetramethylindocarbocyanine perchlorate (DiI) dye. Magnetic relaxation values are reported to be changed after clustering of MFnS around the surface of bacterial contaminants. Binding of MFnS could also be detected optically due to strong signal produced by it, hence giving a dual mechanism for detection of *E. coli.*

18.3.5 Quantum Dots-Based Nanobiosensors

Quantum dots (QD) are small semiconductor crystals exhibiting unique optical and electronic properties such as broad absorption spectra, photodurability, size-tunable emissions, and large intrinsic dipole moments (Frasco and Chaniotakis 2009; Wen et al. 2017). QDs have advantages over organic fluorophores like narrow emission peaks, longer lifetime, and broad excitation ranges. Quantum dots can also be functionalized by different biological molecules.

Rennin is an enzyme secreted by kidneys for regulating blood pressure. Its evaluation is essential for assessment of rennin-related diseases such as hypertension, congestive heart failure, and cancers. Long et al. (2012) have devised a method for sensitive detection of rennin activity using quantum dots. The QD biosensor for rennin activity has advantages over other methods of rennin detection such as size-dependent emission spectra with narrow bandwidths, good resistance to chemical, and photodegradation. The sensor involved complex of streptadavin-coated QDs from which bioinylated peptide substrates are attached. These bioinylated peptide substrates labelled with Cy5 dye make QD complex to exhibit fluorescence resonance energy transfer (FRET). The QD cleaves from substrate in the presence of rennin and resulting in decreases of FRET efficiency. This results in decrease of Cy5 counts which is directly related to quantity of rennin as shown in Fig. 18.11.

Quantum dots are also used for developing a model nanosensor for protein recognition with potential to be used in future for clinical diagnostics and food analysis. For this purpose, a combination of fluorescent QDs conjugated goat anti-mouse IgG and capture antibody (rabbit anti-mouse IgG antibody) has been used. In this work, use of fluorescent QDs for protein recognition has been successfully demonstrated using UV-illuminated microscope (Xu et al. 2010).

Bhattacharyya et al. (2017) have used fluorescent CdSe QD probes with tunable excitation and emission for tuberculosis volatile organic biomarkers (TB-VOB) detection in the breath (Fig. 18.12). Emission properties of VOB-modified QDs are found to be altered exhibiting peak shift which reflected in the display of completely different colors.

Photoelectrochemical sensors based on quantum dot are gaining attraction as they are being viewed as strong alternative of biochemical and chemical molecules. The QDs are being utilized to develop electrochemical sensors in which electrode with immobilized QDs generates photocurrent upon illumination depending on type and concentration of the analyte (Yue et al. 2013).

A biosensor using fluorescent cadmium-tellurium (CdTe) quantum dots has been developed to identify human T-lymphotropic virus-1. In the developed method, two probes, acceptor and reporter probes, hybridized with target DNA, making a sandwich complex. This complex immobilized on a well having streptavidin in which

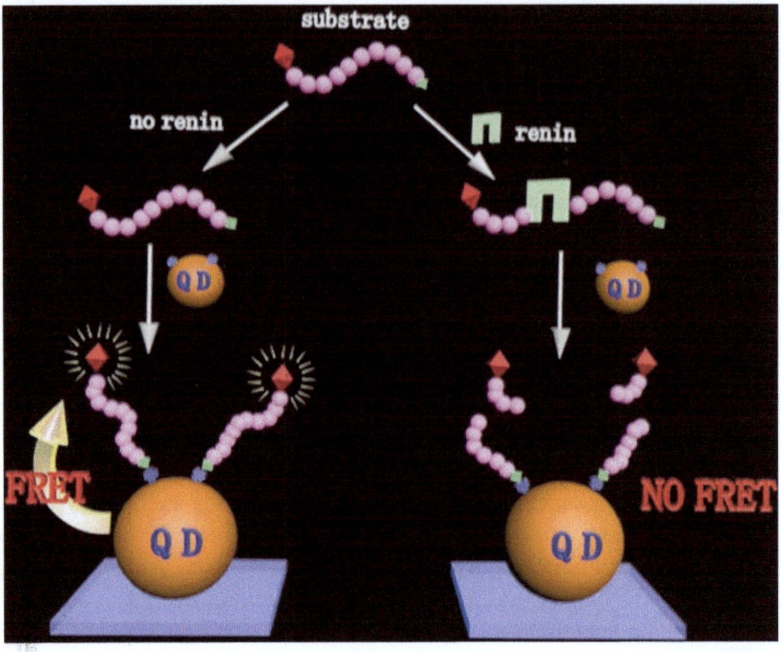

Fig. 18.11 Schematic Illustration of the Single-QD-Based Nanosensor for Renin Assay. (Reprinted (adapted) with permission from Long et al. (2012). Copyright (2009) American Chemical Society)

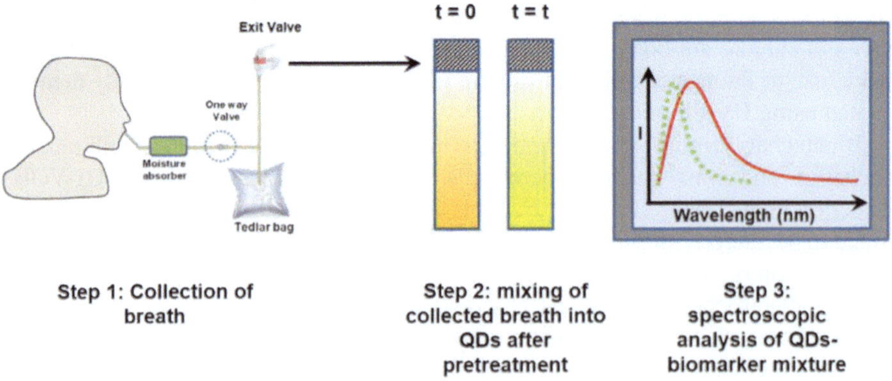

Fig. 18.12 A diagram (schematic) showing proposed method of TB biomarker detection using quantum dots. The methodology involves the collection of breath into a Tedlar bag and subsequent mixing into a QDs (or C-dot) solution. The spectral analysis performed for estimation of VOB concentration and prediction of patient health. (Reprinted from Bhattacharya et al. (2017), Copyright (2018), with permission from Elsevier)

QD solution was added. After conjugation with reporter probe, emission spectra of the quantum dots are found to be modified (Norouzi et al. 2017).

Fluorescence property of the QD has also been utilized in the detection of pesticides trichlorfon (TF). The sensor developed consists of core and multilayered shell structure of CdSe/ZnSe/ZnS QDs, streptavidin (SA), and enzyme AChE (acetylcholine esterase). QD/SA/AChE structure has been used as biosensor for TF. Hydrolytic activity of AChE and enzyme inhibition capability of trichlorfon are used in this sensor to modify the surroundings and the fluorescence of QD (Nguyen et al. 2015).

A nanobiosensor for aflatoxin has been reported using fluorescence property of QD. Fungi *Aspergillus flavus* and *Aspergillus parasiticus* produce aflatoxin B1 (AFB1), a potent carcinogen into infected agricultural products such as oilseed and cereals. Zeng et al. (2015) demonstrated an electrochemical competitive immunoassay for highly sensitive detection of aflatoxins (AFB1) using layer-by-layer (LBL) assembled PbS quantum dots conjugated with monoclonal anti-AFB1 antibody for peanut samples.

Meshik et al. (2014) have developed a QD-based aptamer nanosensor for optical detection of potassium and lead ions which are important for diagnostic purposes. To develop this sensor, thrombin-binding aptamer (TBA), InGaP QD, and GNP quencher have been used, and photoluminescence (PL) is used to measure the efficiency of quenching leading to estimation of ion concentration of analytes.

18.4 Recent Advancement in Nanobiosensor Applications

Recent advancement in nanotechnology has led to designing of innovative biosensing techniques utilizing CNT, graphene, GNP, magnetic nanoparticles, QD, and nanowires. Nanobiosensors are becoming important in the field of diagnostics, food safety, defense, and environmental monitoring. This led to integrating them as biological sensing element with applications like point-of-care (POC) devices, biochips, drug delivery, cancer biomarker detection, and lab on a chip to make them more sophisticated, rapid, cost-effective, and reliable. In this section, an overview of recent advancements in nanobiosensor applications is provided.

18.4.1 Point-of-Care Devices

Point-of-care (POC) devices have important role in diagnostics, especially for disease surveillance and clinical care. It is important to take care of affordability, portability, rapidity, and user friendliness during designing of POC devices (Gonza'lez and Merkoçi 2018; Choi 2010a). POC devices have been classified as small handheld devices and the ones which are used in laboratories. Example of the handheld point-of-care devices includes glucose meter, pregnancy test kit, and bench top devices like testing and electrolyte analysis. It has been noted that these POCs have overcome the problems faced in the conventional method of analysis and diagnosis which are time-consuming and expensive. POC devices are advantageous for rural setup also, where there is lack of facility or people can't afford the expensive diagnostic testing. This impact of POCs in diagnostic has been made possible due to conjugation with nanotechnology. The crucial function in POCs made by nanotechnology is huge improvement in transduction mechanism. The most common transduction mechanisms shown in POC's are optical and electrochemical. Integration of nanomaterial with POC provided very large surface area for biomolecule attachment, increasing specificity and helping in detecting ultra trace concentration, thus performing multiple analysis with small quantity of sample.

Many nanobiosensors developed using nanomaterials such as QDs; gold, silver, and magnetic nanoparticles; carbon nanotubes, and graphene are being used in point-of-care devices for detecting DNA, cancer, pathogenic bacteria, glucose, and hormones (Syedmoradi et al. 2017; Gonza'lez and Merkoçi 2018; Wang et al. 2016). A microfluidic POC device has been proposed for nucleic acid detection from bacteria related to tuberculosis based on a magnetic barcoding strategy. The device contained all the main units of assay loaded onto one chip for performing DNA amplification, MNPs-DNA incubation, washing, and NMR detection (Liong et al. 2013). The fluorescent nanoparticle QD and magnetic particles have been used for *Staphylococcus aureus*, methicillin-resistant *Staphylococcus aureus*, and *Klebsiella pneumoniae* detection for magnetic barcode system where the specific bacteria changed the emission spectra of fluorescent QDs (Cihalova et al. 2017). Nanowires employed on paper for recording electrocardiogram (ECG) signal by measuring changes between skin and electrode have shown a possible alternative toward wet gel electrodes (Mostafalu and Sonkusale 2015). However, use of POCs as an ideal diagnostic device holds some challenges like constraints in the development of appropriate sample preparation methods, prevention of nonspecific adsorption, standardization, automation of technology and interpretation of results need to be taken care of (Syedmoradi et al. 2017).

18.4.2 Biochips

The evolvement of biochips is finding major applications in biotechnology industry such as genomics and proteomics (Rao et al. 2012). Biochip consists of multiple tools having sensitive detection capability which has been made possible due to use of nanobiosensing principles.

A highly efficient and reusable Raman spectroscopy (SERS)-active biochip has been developed using laser scribing treatment of silver nanoparticle and graphene (AgNP-GO) composite (Fig. 18.13). AgNP on graphene surface has been used to enhance the Raman signal. By immobilizing biomolecules on graphene, controllable trapping and release of DNA could be obtained in this biochip (Han et al. 2018).

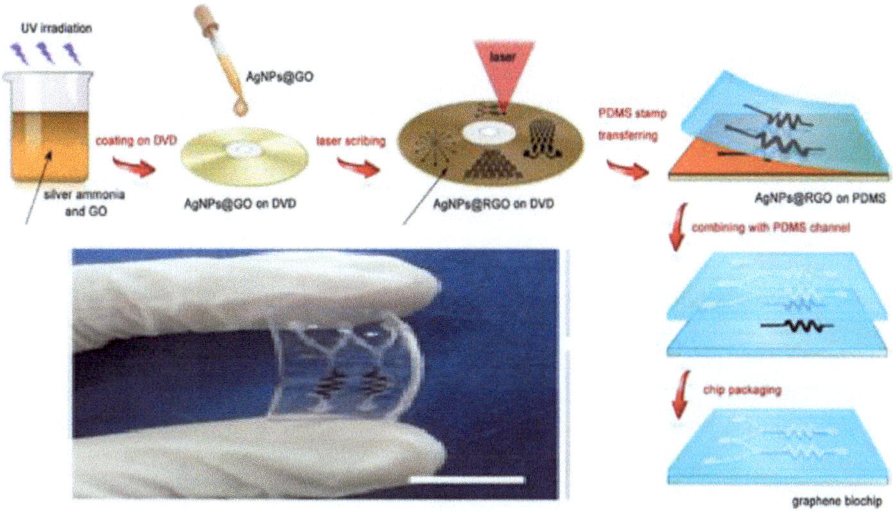

Fig. 18.13 Fabrication of AgNPs@RGO biochip. (**a**) Schematic illustration of the fabrication procedure of an AgNPs@RGO SERS biochip and Photograph of the as-prepared AgNPs@RGO SERS biochip. (Reprinted from Han et al. (2018). Copyright (2018), with permission from Elsevier)

Epidemic caused by Zika virus in 2015 has been a major issue of health concern due to its long-term effects such as severe brain defects and Guillain-Barre syndrome on fetus and adults. Afsahi et al. (2018) devised a graphene-based chip functionalized with anti-Zika NS1 providing early diagnosis for Zika virus. Capacitance of graphene chip functionalized with monoclonal antibodies gets changed after binding with varying doses of antigen (ZIKV NS1).

A biochip for prostate cancer biomarker detection has been fabricated by Lee et al. (2012) by integrating microfluidic system with silicon nanowire (SiNW) array immobilized with PSA and IL-6 antibodies. This biochip has been able to detect PSA and IL-6 in plasma for prostates cancer with fast and multiple analyte detection capability.

18.4.3 Drug Delivery

Automated drug delivery system is the area where biosensors play important role. Application of nanotechnology for biosensing in the drug delivery system is giving new route, as hugely growing population urgently needs rapid and cost-effective methods. Researchers are coming up with miniaturized systems containing nanobiosensors and control unit which can be implanted on humans for detection and monitoring of disease. The role of these biosensors is to provide feedback control by perceiving changes in their surrounding physiological or biological fluid and then taking action by controlled release of one or more drugs at desired site. The selectivity of biosensor is critical as different proteins and chemicals present in its surrounding may affect its parameters like accuracy, selectivity, and sensitivity. A variety of organic and inorganic nanomaterials and devices have been used as drug delivery carriers to enhance the therapeutic activity. The requirement for nanoparticle-based drug delivery system includes biodegradability, biocompatibility, nontoxicity, and functionality.

Noninvasive glucose sensing has been gaining significant attention nowadays. These systems modified with nanoscale materials are implanted or administered into the biological system. Nanosensors based on CNT, quantum dots, and graphene have been widely used as a replacement to current glucose sensors (Cash and Clark 2010). Early detection of cancer biomarkers or tumor cells in system may help in the survival of cancer patients. Nowadays, nanomaterial-based biosensors are being employed for cancer detection and drug delivery monitoring (Salvati et al. 2015). A study has been conducted for potential application of MWCNT-Ti sensors in orthopedics for enhancing osteoblast differentiation to promote bone growth and treat infection or inflammation next to bone implants (Sirivisoot and Pareta 2012). Such advancement in drug delivery system through nanosensors can enhance self-monitoring of disease and illness management.

18.4.4 Cancer Biomarker Detection

Uncontrolled growth of abnormal cells in body organs and system results in cancer. Different forms of cancer including lung, prostate, breast, ovarian, hematologic, skin, and colon cancer can be screened through CT-scan, X-ray, and MRI. Currently, researchers are developing methods for accurate diagnosis of cancer relying on cancer biomarkers detection. The detection of biomarker is helpful to know the progress of disease and its response to treatment. Several biosensors used for cancer screening have become more sensitive and show precise response toward biomarkers detection due to the use of nanomaterials (Choi et al. 2010b). Some of the cancer biomarkers are carcinoembryonic antigen (CEA) for cancers of the gastrointestinal tract; prostate-specific antigen (PSA), a biomarker to screen for prostate cancer; and cancer antigen 125 (CA125), an antigen in patients detected with ovarian cancer. Nanowires, quantum dots, and carbon nanotubes are some of the nanomaterials which have been utilized for detection of these cancer biomarkers (Devi et al. 2015; Jaishree and Gupta 2012).

A gold and magnetic nanoparticle-based enzyme-labelled probe for detection of carcinoembryonic antigen (CEA) has been reported. In this method, GNP coated with antibody, single-stranded DNA (ssDNA), and horseradish peroxidase (HRP) are used. A capture probe developed using MNP functionalized another antibody. The immunoreaction between target antigen and antibody had transduced into optical signals (Liu et al. 2010).

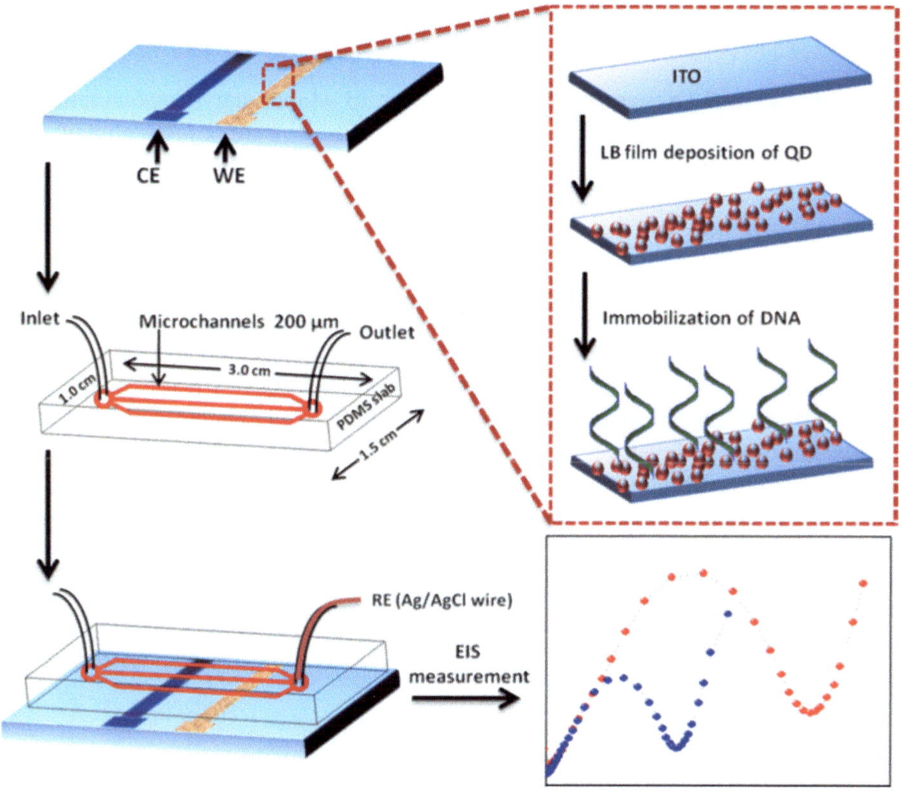

Fig. 18.14 Different steps showing the fabrication of microfluidic biochip for electrochemical detection of DNA hybridization. WE Working electrode, CE counter electrode, RE reference electrode. (Reprinted from Ghrera et al. (2015), Copyright (2018), with permission from Elsevier)

QDs-based nanobiosensor has been developed for detection of mutation in breast cancer cells, BRCA1 and BRCA2. 185delAG I deletion is breast cancer gene associated with risk of developing breast cancer. Amine functionalized DNA attached to QD and emission spectra of this system at 489 nm are found to be stronger for cDNA compared to mDNA (Eftekhari-Sis et al. 2017). Same principle has been used in cadmium selenide quantum dots used in microfluidic sensor for nucleic acid to detect chronic myelogenous leukemia (CML) (Fig. 18.14).The quantification of DNA has been done through measure in interfacial charge transfer resistance (Ghrera et al. 2015).

Carbon nanotube-based biosensors are used for detection of carbohydrate antigen 19-9 (CA19-9), a cancer biomarker for pancreatic cancer. Interdigitated gold electrodes with a thin film sensing unit containing polyethyleneimine and CNT thin film with antibodies anti-CA19-9 have been developed. Binding of CA-19-9 antigen on developed electrode is further studied using impedance spectroscopy (Thapa et al. 2017).

18.4.5 Lab on a Chip

Labs on a chip (LOC) are miniaturized device in which analytical functions are combined for biomolecule detection. The functions like sample pretreatment, recognition of analyte, transduction of the reaction, and amplification of measured signal are integrated in LOC devices. LOC have become meaningful platforms in detection of protein and DNA and diagnostics due to their numerous benefits like low volume of sample, economical, multiple sample detection, and portability (Claussen and Medintz 2012). These miniaturized LOC devices contain nanostructured sensors for detection of analytes for improved transduction mechanism. Chua and Pumera (2013) showed use of chemically modified graphene materials as electrochemical detectors in a lab-on-chip device. Nanomaterials such as quantum dots, GNP, and MNP have been used for labeling of biomarkers to be used in LOC.

For label-free detection of low-density lipoprotein (LDL) cholesterol in human serum, an impedimetric nanobiosensor based on CNT is developed for LOC device applications. In this application, CNT-NiO nanocomposite functionalized with antiapolipoprotein B, is deposited on indium-tin oxide (ITO) substrate to detect LDL by impedimetric method as illustrated in Fig. 18.15. This LOC device convertes the antigen-antibody reaction into measurable electrical signal (Ali et al. 2015).

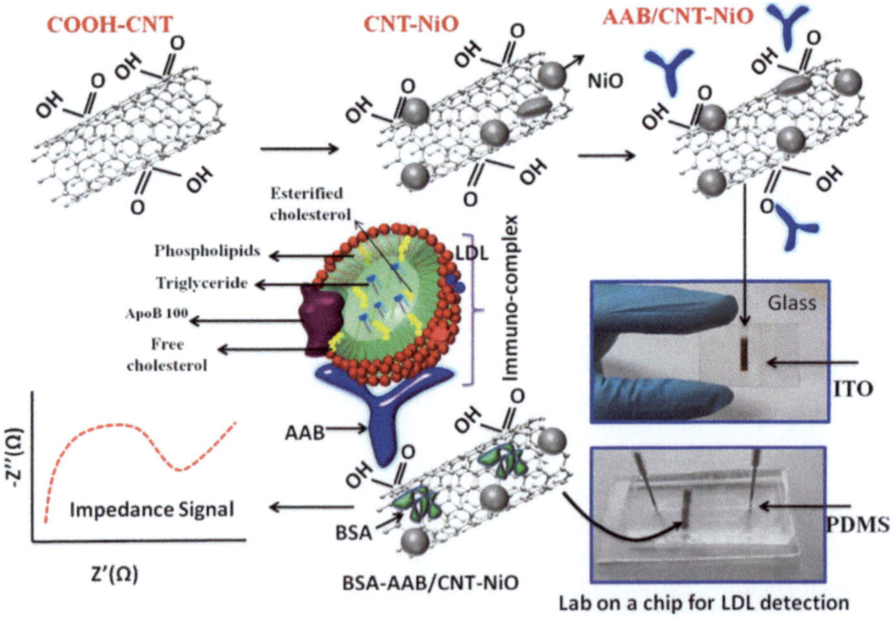

Fig. 18.15 Schematic representation of the Lab-on-a-Chip Fabrication for LDL Detection. (Reprinted (adapted) with permission from Ali et al. (2015), Copyright (2018) American Chemical Society)

An LOC device with nanobiosensing elements for food safety application has been discussed by Yang et al. (2010). In the developed device, anti-SEB antibody-CNT immobilized on polycarbonate strip for detection of staphylococcus enterotoxin (SEB) which is responsible for food-borne intoxication. This LOC device showed potential for sensitive detection of toxin with low volume and sample without power requirement.

In spite of so many advancements in the field of LOC using nanobiosensors, there are some of the current drawbacks including quantification of low concentration of sample, complexity in fabrication procedures, and difficulty in multiplex analysis.

18.5 Future Perspectives

There has been huge growth in the field of nanomaterials in recent years which have attracted attention of researchers to develop novel nanobiosensors by coupling biomolecular recognition elements with nanomaterials like CNT, graphene, metal nanoparticles, and quantum dots. Use of the nanomaterials with high surface area is important in developing rapid and highly sensitive nanobiosensors with applications in clinical monitoring, environmental applications, agriculture, bioprocess control, and food industry applications.

Recent researches are focused on development of miniaturized and portable systems based on electrical, optical, and magnetic sensing. So, the researchers are working toward the design of devices utilizing nanomaterial-based biosensors which fulfil these characteristics. Lab on a chip, point-of-care devices, automatic drug delivery systems, and biochips are examples of such devices in which nanobiosensors are used.

However, this newly growing area of nanobiosensor faces challenges regarding enhanced amplification of signals and enhancement of signal to noise ratio. Interaction mechanisms between nanomaterials and biomolecules are also not fully understood yet. Future work should be focused on understanding the interaction mechanism and using novel techniques to develop more advanced biosensors with the capability of single molecule detection of biomarkers and multiplexed assay feature. There is also requirement of cheap and disposable nanobiosensors for applications in diverse areas. Noninvasive and continuous detection of toxins, biomarkers, hormones, and drugs from body of subject is also a challenging area for nanobiosensors.

However, these nanobiosensors seem attractive as they have potential for delivery of systems like point-of-care diagnostics, automatic drug delivery, and DNA chips that can surpass conventional technologies regarding time, cost, and accuracy.

References

Afsahi S, Lerner MB, Goldstein JM, Lee J, Tang X, Bagarozzi DA Jr, Goldsmith BR (2018) Novel graphene-based biosensor for early detection of Zika virus infection. Biosens Bioelectron 100:85–88

Agnihotri N, Chowdhury AD, De A (2015) Non-enzymatic electrochemical detection of cholesterol using β-cyclodextrin functionalized graphene. Biosens Bioelectron 63:212–217

Ali MA, Solanki PR, Srivastava S, Singh S, Agrawal VV, John R, Malhotra BD (2015) Protein functionalized carbon nanotubes-based smart lab-on-a-chip. ACS Appl Mater Interfaces 7(10):5837–5846

Ali S, Hassan A, Hassan G, Eun CH, Bae J, Lee CH, Kim IJ (2018) Disposable all-printed electronic biosensor for instantaneous detection and classification of pathogens. Sci Rep 8(1):5920

Aljabali AA, Hussein E, Aljumaili O, Al Zoubi M, Altrad B, Albatayneh K, Al-razaq MAA (2018) Rapid magnetic nanobiosensor for the detection of Serratia marcescens. In IOP Conference Series: Mater Sci Eng 305(1):012005

Altintas Z, Davis F, Scheller FW (2017) Applications of quantum dots in biosensors and diagnostics. Biosens Nanotechnol Appl Health Care Diagn 3:185–199

Asnaashari M, Kenari RE, Farahmandfar R, Taghdisi SM, Abnous K (2018) Fluorescence quenching biosensor for acrylamide detection in food products based on double-stranded DNA and gold nanoparticles. Sens Actuators B Chem 265:339–345

Augustine R, Abraham AR, Kalarikkal N, Thomas S (2016) Monitoring and separation of food-borne pathogens using magnetic nanoparticles. In: Novel approaches of nanotechnology in food. Academic Press, UK, pp 271–312

Baldo S, Buccheri S, Ballo A, Camarda M, La Magna A, Castagna ME, Scalese S (2016) Carbon nanotube-based sensing devices for human Arginase-1 detection. Sens Bio-Sens Res 7:168–173

Bhattacharya M, Hong S, Lee D, Cui T, Goyal SM (2011) Carbon nanotube based sensors for the detection of viruses. Sens Actuators B Chem 155(1):67–74

Bhattacharyya D, Sarswat PK, Free ML (2017) Quantum dots and carbon dots based fluorescent sensors for TB biomarkers detection. Vacuum 146:606–613

Bolat G, Abaci S (2018) Non-enzymatic electrochemical sensing of malathion pesticide in tomato and apple samples based on gold nanoparticles-chitosan-ionic liquid hybrid nanocomposite. Sensors 18(3):773

Cash KJ, Clark HA (2010) Nanosensors and nanomaterials for monitoring glucose in diabetes. Trends Mol Med 16(12):584–593

Choi W, Lahiri I, Seelaboyina R, Kang YS (2010a) Synthesis of graphene and its applications: a review. Crit Rev Solid State Mater Sci 35(1):52–71

Choi YE, Kwak JW, Park JW (2010b) Nanotechnology for early cancer detection. Sensors 10(1): 428–455

Chua CK, Pumera M (2013) Chemically modified graphenes as detectors in lab-on-chip device. Electroanalysis 25(4):945–950

Cihalova K, Hegerova D, Jimenez AM, Milosavljevic V, Kudr J, Skalickova S, Adam V (2017) Antibody-free detection of infectious bacteria using quantum dots-based barcode assay. J Pharm Biomed Anal 134:325–332

Claussen J, Medintz IL (2012) Using nanotechnology to improve lab on a chip devices. Biochip Tissue Chip 2(4):1609–1614

Dasgupta NP, Sun J, Liu C, Brittman S, Andrews SC, Lim J, Yang P (2014) 25th anniversary article: semiconductor nanowires–synthesis, characterization, and applications. Adv Mater 26(14):2137–2184

Devi RV, Doble M, Verma RS (2015) Nanomaterials for early detection of cancer biomarker with special emphasis on gold nanoparticles in immunoassays/sensors. Biosens Bioelectron 68:688–698

Doria G, Conde J, Veigas B, Giestas L, Almeida C, Assunção M, Baptista PV (2012) Noble metal nanoparticles for biosensing applications. Sensors 12(2):1657–1687

Eftekhari-Sis B, Karaminejad S, Malekan F, Araghi HY, Akbari A (2017) CdSe quantum dots based Nano-biosensor for detection of 185delAG mutation in BRCA1 gene, responsible for breast Cancer. J Inorg Organomet Polym Mater 27(6):1911–1917

Frasco MF, Chaniotakis N (2009) Semiconductor quantum dots in chemical sensors and biosensors. Sensors 9(9):7266–7286

Geim AK, Novoselov KS (2007) The rise of graphene. Nat Mater 6(3):183–191
Ghorab MA, Khalil MS (2015) Toxicological effects of organophosphates pesticides. Int J Environ Monit Anal 3:218–220
Ghrera AS, Pandey CM, Ali MA, Malhotra BD (2015) Quantum dot-based microfluidic biosensor for cancer detection. Appl Phys Lett 106(19):193703
Han B, Zhang YL, Zhu L, Chen XH, Ma ZC, Zhang XL, Sun HB (2018) Direct laser scribing of AgNPs@ RGO biochip as a reusable SERS sensor for DNA detection. Sens Actuators B Chem 270:500–507
He S, Song B, Li D, Zhu C, Qi W, Wen Y, Fan C (2010) A graphene nanoprobe for rapid, sensitive, and multicolor fluorescent DNA analysis. Adv Funct Mater 20(3):453–459
Hutter E, Maysinger D (2013) Gold-nanoparticle-based biosensors for detection of enzyme activity. Trends Pharmacol Sci 34(9):497–507
Ibrahim and Saeed (2013) Carbon nanotubes- properties and applications: a review. Carbon Lett 14(3):131–144
Iijima S (1991) Helical microtubules of graphitic carbon. Nature 354(6348):56–58
Jaishree V, Gupta PD (2012) Nanotechnology: a revolution in cancer diagnosis. Indian J Clin Biochem 27(3):214–220
Khan I, Saeed K, Khan I (2017) Nanoparticles: properties, applications and toxicities. Arab J Chem. https://doi.org/10.1016/j.arabjc.2017.05.011, in press
Kumar VV, Anthony SP (2014) Silver nanoparticles based selective colorimetric sensor for Cd2+, Hg2+ and Pb2+ ions: tuning sensitivity and selectivity using co-stabilizing agents. Sens Actuators B Chem 191:31–36
Kwak YH, Choi DS, Kim YN, Kim H, Yoon DH, Ahn SS, Seo S (2012) Flexible glucose sensor using CVD-grown graphene-based field effect transistor. Biosens Bioelectron 37(1):82–87
Lee MH, Seong W, Lee S, Kim K (2012) Silicon nanowire based immunoassay for the detection of prostate cancer biomarkers. In Biomedical Engineering and Biotechnology (iCBEB), 2012 International Conference 1086–1088
Liong M, Hoang AN, Chung J, Gural N, Ford CB, Min C, Toner M (2013) Magnetic barcode assay for genetic detection of pathogens. Nat Commun 4:1752
Liu M, Jia C, Huang Y, Lou X, Yao S, Jin Q, Xiang J (2010) Highly sensitive protein detection using enzyme-labeled gold nanoparticle probes. Analyst 135(2):327–331
Liu B, Sun Z, Zhang X, Liu J (2013) Mechanisms of DNA sensing on graphene oxide. Anal Chem 85(16):7987–7993
Liu X, Yang W, Chen L, Jia J (2017) Three-dimensional copper foam supported CuO nanowire arrays: an efficient non-enzymatic glucose sensor. Electrochim Acta 235:519–526
Long Y, Zhang LF, Zhang Y, Zhang CY (2012) Single quantum dot based nanosensor for renin assay. Anal Chem 84(20):8846–8852
Malik P, Katyal V, Malik V, Asatkar A, Inwati G, Mukherjee TK (2013) Nanobiosensors: concepts and variations. ISRN Nanomater 2013:1–9
Mandal D, Nunna BB, Zhuang S, Rakshit S, Lee ES (2017) Carbon nanotubes based biosensor for detection of cancer antigens (CA-125) under shear flow condition. Nano-Struct Nano-Objects 15:180–185
Marques I, Pinto da Costa J, Justino C, Santos P, Duarte K, Freitas A, Rocha-Santos T (2017) Carbon nanotube field effect transistor biosensor for the detection of toxins in seawater. Int J Environ Anal Chem 97(7):597–605
Melzer K, Bhatt VD, Jaworska E, Mittermeier R, Maksymiuk K, Michalska A, Lugli P (2016) Enzyme assays using sensor arrays based on ion-selective carbon nanotube field-effect transistors. Biosens Bioelectron 84:7–14
Meshik X, Xu K, Dutta M, Stroscio MA (2014) Optical detection of lead and potassium ions using a quantum-dot-based aptamer nanosensor. IEEE Trans Nanobioscience 13(2):161–164
Mohammed AM, Rahim RA, Ibraheem IJ, Loong FK, Hisham H, Hashim U, Al-Douri Y (2014) Application of gold nanoparticles for electrochemical DNA biosensor. J Nanomater 13:1–7

Mostafalu P, Sonkusale S (2015) A high-density nanowire electrode on paper for biomedical applications. RSC Adv 5(12):8680–8687

Nguyen NH, Duong TG, Pham NT, Dao TC, Pham TN (2015) Synthesis and application of quantum dots-based biosensor. Adv Nat Sci Nanosci Nanotechnol 6(1):015015

Nikam AP, Mukesh PR, Haudhary SP (2014) Nanoparticles—an overview. J Drug Deliv Ther 3:1121–1127

Norouzi M, Zarei Ghobadi M, Golmimi M, Mozhgani SH, Ghourchian H, Rezaee SA (2017) Quantum Dot-Based Biosensor for the Detection of Human T-Lymphotropic Virus-1. Anal Lett 50(15):2402–2411

Patolsky F, Zheng G, Lieber CM (2006) Nanowire sensors for medicine and the life sciences. Future Med 1:51–65

Paul B, Tiwari A (2015) A brief review on the application of gold nanoparticles as sensors in multi dimensional aspects. J Environ Sci Toxicol Food Technol 1(4):01–07

Prasad R, Pandey R, Barman I (2016) Engineering tailored nanoparticles with microbes: quo vadis. WIREs Nanomed Nanobiotechnol 8:316–330. doi: 10.1002/wnan.1363

Quesada-González D, Merkoçi A (2018) Nanomaterial-based devices for point-of-care diagnostic applications. Chem Soc Rev 47(13):4697–4709

Rahman SFA, Yusof NA, Hashim U, Hushiarian RMN, Hamidon MN, Fathil MFM (2016) Enhanced sensing of dengue virus DNA detection using O2 plasma treated-silicon nanowire based electrical biosensor. Anal Chim Acta 942:74–85

Randviir EP, Brownson DA, Banks CE (2014) A decade of graphene research: production, applications and outlook. Mater Today 17(9):426–432

Rasheed PA, Sandhyarani N (2014) Graphene-DNA electrochemical sensor for the sensitive detection of BRCA1 gene. Sens Actuators B Chem 204:777–782

Rocha-Santos TAP (2014) Sensors and biosensors based on magnetic nanoparticles. TrAC Trends Anal Chem 62:28–36

Salvati E, Stellacci F, Krol S (2015) Nanosensors for early cancer detection and for therapeutic drug monitoring. Nanomedicine 10(23):3495–3512

Salvetat JP, Bonard JM, Thomson NH, Kulik AJ, Forro L, Benoit W, Zuppiroli L (1999) Mechanical properties of carbon nanotubes. Appl Phys A Mater Sci Process 69(3):255–260

Shelby T, Sulthana S, McAfee J, Banerjee T, Santra S (2017) Foodborne pathogen screening using magneto-fluorescent nanosensor: rapid detection of *E. Coli* O157: H7. J Vis Exp (JoVE) 127:1–7

Shen F, Wang J, Xu Z, Wu Y, Chen Q, Li X, Zhu T (2012) Rapid flu diagnosis using silicon nanowire sensor. Nano Lett 12(7):3722–3730

Sirivisoot S, Pareta RA (2012) Orthopedic carbon nanotube biosensors for controlled drug delivery. In: Nanomedicine. Woodhead Publishing, UK, pp 149–179

Song D, Li Y, Lu X, Sun M, Liu H, Yu G, Gao F (2017) Palladium-copper nanowires-based biosensor for the ultrasensitive detection of organophosphate pesticides. Anal Chim Acta 982:168–175

Suvarnaphaet P, Pechprasarn S (2017) Graphene-based materials for biosensors: a review. Sensors 17(10):2161

Syedmoradi L, Daneshpour M, Alvandipour M, Gomez FA, Hajghassem H, Omidfar K (2017) Point of care testing: the impact of nanotechnology. Biosens Bioelectron 87:373–387

Tan J, Hao B, Wang C, Ren Y, Hao H, Yang R (2016) A gold nanoparticle-enhanced surface plasmon resonance aptasensor for the detection of 2, 4, 6-trinitrotoluene. J Forensic Sci Med 2(4):195

Thapa A, Soares AC, Soares JC, Awan IT, Volpati D, Melendez ME, Oliveira ON Jr (2017) Carbon nanotube matrix for highly sensitive biosensors to detect pancreatic cancer biomarker CA19-9. ACS Appl Mater Interfaces 9(31):25878–25886

Varshney K (2014) Carbon nanotubes: a review on synthesis, properties and applications. Int J Eng Res 2(4):660–677

Rao TVN, Sukruthi SG, Raj G (2012) Biochip Technology- A gigantic Innovation. International Int J Emerg Technol Adv Eng 2:129–135

Wang R, Ruan C, Kanayeva D, Lassiter K, Li Y (2008) TiO2 nanowire bundle microelectrode based impedance immunosensor for rapid and sensitive detection of *Listeria monocytogenes*. Nano Lett 8(9):2625–2631

Wang H, Zhang Y, Li H, Du B, Ma H, Wu D, Wei Q (2013) A silver–palladium alloy nanoparticle-based electrochemical biosensor for simultaneous detection of ractopamine, clenbuterol and salbutamol. Biosens Bioelectron 49:14–19

Wang H, Sugiarto S, Li T, Ang WH, Lee C, Pastorin G (2016) Advances in nanomaterials and their applications in point of care (POC) devices for the diagnosis of infectious diseases. Biotechnol Adv 34(8):1275–1288

Wen L, Qiu L, Wu Y, Hu X, Zhang X (2017) Aptamer-modified semiconductor quantum dots for biosensing applications. Sensors 17(8):1736

Xia N, Chen Z, Liu Y, Ren H, Liu L (2017) Peptide aptamer-based biosensor for the detection of human chorionic gonadotropin by converting silver nanoparticles-based colorimetric assay into sensitive electrochemical analysis. Sens Actuators B Chem 243:784–791

Xu H, Aguilar ZP, Wang A (2010) Quantum Dot-based Sensors for Proteins. ECS Trans 25(31):1–8

Yang M, Sun S, Kostov Y, Rasooly A (2010) Lab-on-a-chip for carbon nanotubes based immunoassay detection of Staphylococcal Enterotoxin B (SEB). Lab Chip 10(8):1011–1017

Yang J, Strickler JR, Gunasekaran S (2012) Indium tin oxide-coated glass modified with reduced graphene oxide sheets and gold nanoparticles as disposable working electrodes for dopamine sensing in meat samples. Nanoscale 4(15):4594–4602

Yue Z, Lisdat F, Parak WJ, Hickey SG, Tu L, Sabir N, Bigall NC (2013) Quantum-dot-based photoelectrochemical sensors for chemical and biological detection. ACS Appl Mater Interfaces 5(8):2800–2814

Zeng X, Gao H, Pan D, Sun Y, Cao J, Wu Z, Pan Z (2015) Highly sensitive electrochemical determination of alfatoxin B1 using quantum dots-assembled amplification labels. Sensors 15(8):20648–20658

Zhang M, Liao C, Mak CH, You P, Mak CL, Yan F (2015) Highly sensitive glucose sensors based on enzyme-modified whole-graphene solution-gated transistors. Sci Rep 5:8311

Zhu Y, Hao Y, Adogla EA, Yan J, Li D, Xu K, Wang Q, Hone J, Lin Q (2016) A graphene-based affinity nanosensor for detection of low-charge and low-molecular-weight molecules. Nanoscale 8(11):5815–5819

Chapter 19
Application of Nanotechnology in Diagnosis, Drug Dissolution, Drug Discovery, and Drug Carrier

Abhishek Kumar Mishra

Contents

19.1 Introduction

In healthcare and medicine, diagnosis of a disease is a very critical step. Diagnosis is a multistep process which includes patient's history, examination, and assessment of laboratory reports. Diagnosis should be fast and accurate, having very low chances of false-positive results with a high degree of sensitivity and specificity. Fast and precise diagnosis aid earlier detection of disease and is helpful in better prognosis. Since the invention of stethoscope, scientists have endeavored to develop an advanced method of diagnostics for rapid and accurate diagnosis of diseases.

Nanotechnology (NT) in the field of molecular diagnostics extends the limits to the nanoscale and explores one more dimension of microfluidic/lab-on-a-chip technology. It becomes so fast and easier to sense the presence or activity of certain

A. K. Mishra (✉)
Himalayan School of Biosciences, Swami Rama Himalayan University, Jollygrant, Dehradun, Uttarakhand, India

R. Prasad et al. (eds.), *Nanobiotechnology in Bioformulations*, Nanotechnology in the Life Sciences, https://doi.org/10.1007/978-3-030-17061-5_19

compounds through labeling or tagging of certain nanoscale particles. Application of NT in the field of biomedical diagnostics comes under nanodiagnostics (NDs) which provides new insights for point-of-care performance in personalized medicine (Krukemeyer et al. 2015). NDs offers great advantage over the traditional methods of diagnosis such as:

(i) It provides rapid testing protocol and can be performed even in the doctor's chamber and thus provides early start of the treatment and possibly less damage to the patients.
(ii) Being highly sensitive, very small quantity of samples is required.

The very first step in NT-based testing is the detection of signals generated through the binding of labeled/tagged probes to the target biomolecules which produce characteristic signals. The probe may be quantum dots (QDs), nanoshells, and metallic nanoparticles (Challa and Kumar 2007).

Detection of biomarkers in laboratories needs sophisticated automated analyzers, too much time, and high costs which can be substituted by robust, faster, and economical devices. Nanoscale-fabricated structured devices provide diagnostic results available at the patient's bedside, i.e., point-of-care diagnosis (Mascini and Tombelli 2008).

This chapter explores the nanodiagnostics in use or in different phases of development.

19.2 Nanotechnology in Medical Diagnosis

Nanotechnology has revolutionized the era of diagnostics by overcoming the drawbacks like poor sensitivity, specificity, and reproducibility in conventional diagnostic methods like bioassays, biosensors, and imaging (Prasad et al. 2016). Nanotechnology-fabricated structured devices are elegantly small and prepared by the engineering of systems at the atomic or molecular scale. These devices can recognize very low concentrations of disease biomarkers compared with traditional tools. Application of nanotechnology in medical diagnosis includes diagnosis, prevention, and treatment (Rajasundari and Hamurugu 2011; Geho et al. 2006). Nanostructures have been successfully used in in vitro diagnostics which have made the diagnosis rapid, simpler, and more precise than earlier traditional methods. The main goal of nanotechnology is to focus on development of novel biomarkers for disease diagnosis and construction of nanobiosensors using ultrasensitive nanomaterials such as carbon nanotubes, nanoparticles, and so on. The use of nanomaterials in nanodiagnostics significantly improved the method of diagnosis in techniques such as immunohistochemistry (IHC), genotyping, cancer detection, and biomarker detection. A variety of materials are used to construct diagnostic nanodevices by manipulation of matter at the nanoscale.

19.2.1 Nanomaterials and Nanodevices for Medical Diagnosis

Nanomaterials used in medical diagnosis are very small in size and possess high surface to unit volume ratio due to which they show higher chemical reactivity, increased tensile strength, and faster electrical and magnetic responses (Jackson et al. 2017). The usefulness of nanoparticles and their unique properties are due to their size similar to that of biomolecules such as proteins and nucleic acids which makes them suitable for real-time interaction with biomolecules inside the cells. Nanoparticles can be synthesized in a number of ways. The common method of nanoparticle synthesis includes physical, chemical, and biological synthesis (Prasad 2014; Prasad et al. 2016). Generally, nanomaterials are produced by top-down and bottom-up techniques. In the top-down approach with the help of mechanical, chemical, or other form of energy, a bulk material breaks into smaller pieces. The bottom-up technique is based on chemical reactions in which materials are synthesized by allowing the precursor particles to grow in size (Chan and Kwok 2011). Biological synthesis of nanoparticles is ecofriendly and performed by manipulation of microorganisms of biomedical interests (Prasad 2017; Prasad et al. 2016, 2018; Aziz et al. 2015, 2016, 2019; Jackson et al. 2017). Nanomaterials of biomedical interests can be classified according to their structure, chemical composition, and applications. The structural basis of classification includes nanoparticles, quantum dots, nanotubes, dendrimers, and micelle formations (Nasimi and Haidari 2013; Prasad et al. 2017a). Nanomaterials can be either organic or inorganic. Inorganic structures include metal oxide nanoparticles, semimetal oxides, metal nanoparticles whereas carbon structures (nanotubes, graphene, fullerenes) and organic structures include polymer nanoparticles or dendrimers (Choi et al. 2014). The classification of nanomaterials on the basis of chemical composition is not universally acceptable as most of the nanostructures are hybrid in nature or a combination of both inorganic and organic substances as organic substances stabilize the functions of nanomaterials. Some of the nanomaterials useful in the fabrication of nanostructured devices are nanotubes, nanocrystals, nanobots, nanowires.

19.2.1.1 Nanotube

Nanotubes are cylindrical carbon molecule of 0.5–3 nm in diameter and 20–1000 nm in length and are very useful in biomedical science because of their unique properties such as extraordinary strength and high conductance of electrical and thermal energy. The reason for high conductance is the sp^2 hybridization state of carbon, e.g., fullerene, an allotrope of carbon. Carbon nanotubes have been implicated in combination with other gold nanoparticles and silicon nanowires for the detection of oral cancer and lung cancer (Beishon 2013; Shehada et al. 2015).

19.2.1.2 Nanocrystal

They are crystalline substances having at least one dimension less than 1 μm. Electrical and thermodynamic properties of nanocrystal vary with size. In the range of 10 nm, nanocrystals in the space between them show loose microstructures – nanopores. An Ireland-based pharmaceutical company – Elan Pharma International Limited – developed nanocrystals for nanoparticle drug formulation. Nanocrystals of 2–9.5 nm size have been used to improve solubility of poorly soluble drugs and in labeling of breast cancer markers (Kumar and Vijayalakshmi 2006; Roco et al. 2017).

19.2.1.3 Nanobots

Nanobots are also known as nanorobots. They also fall in the category of nanosized materials. They have been employed for early diagnosis of the disease as well as targeted drug delivery, diabetes monitoring, and in other applications of healthcare mostly in cancer. Nanobot dentifrices are one of the types of nanobots (dentifrobots), used in the identification of pathogenic bacteria in the mouth. When used in the form of mouthwash or toothpaste, they spread over the subgingival surfaces and metabolize trapped bacteria or organic matter into odorless vapors (Shetty et al. 2013).

19.2.1.4 Nanowires

Nanowires (NW) are of nm size and made up of carbon nanotubes or silicon. Being smaller in size, slight change in the electrical properties of it due to binding of additional molecule could be detected. Antibodies could be loaded over its surface as detectors. When the antibody binds to target biomolecules, it undergoes conformational changes which can be detected as signals. When several nanowires are loaded with different antibodies over the surface assembled in a single device, they can work as detectors for cancer (silicon nanowires in field effect transistors (SNW-FET)) (Reimhult and Höök 2015; Lyberopoulou et al. 2015; Takahashi et al. 2015).

19.2.1.5 Quantum Dots

As the name indicates, quantum dots are crystals in the form of small dots of 2–9.5 nm in size. They are constituted with inorganic materials having fluorescent properties. When low energy light falls on it, quantum dots emit fluorescent light. The color of the emitted light depends on the size of dot. On excitation, quantum dots of different size when embedded in a given microbead show spectra of distinct pattern of colors. Quantum dots are very sensitive as even in a general excitation they can produce broad spectra and could be used in image-guided surgery, molecular diagnostics, and genotype determination (Rajasundari and Hamurugu 2011).

19.2.1.6 Nanocapacitor

A capacitor is an electrical device which is formed by the combination of two electrode plates and an insulating material sandwiched between the gaps of plates. If the gap between plates is a few nanometers, nanocapacitors are formed. The capacitance can be measured from the area of plate, distance between plates, and the dielectric constant value of the insulating medium. Nanocapacitor-based devices work on the principle that when a target molecule comes in contact with the dielectric material the value of dielectric constant of the medium changes significantly and thus capacitance will be also changed and the change in capacitance could be measured. When using this device, there is no need for prior labeling of the samples. A limitation of this method is that the quantity of sample should be enough to create appropriate change in dielectric constant of the medium. Transparent nanocapacitors are developed by Kang et al. (2003) to monitor dielectric and optical behavior of biomolecules. Silicon nanolithography was used to create 50 nm gaps between the electrodes (Kang et al. 2003) of a nanocapacitor.

19.2.1.7 Nanoparticles

As the name indicates, nanoparticles are the particles of nm in size. They possess at least one of the diameters <100 nm (Buzea et al. 2007). A number of varieties of nanoparticles are available. The difference in varieties of nanoparticles is due to difference in their shapes, materials, and sizes. The nanoparticles occur in different shapes such as dot, sphere, star, prism, and rod. Nanoparticles are synthesized using metals or polymers. Gold nanoparticles (GNPs) are the best example of metallic nanoparticles (Boisselier and Astruc 2009; Fu et al. 2010). GNPs can be detected by Raman scattering, fluorescence, optical absorption, magnetic/atomic force microscopy, and electrical resistance measurements (Marzán 2006).

19.2.1.8 Nanotechnology-Based Chips: "Nanobiochips"

Nanobiochips are the most advanced application of nanobiotechnology. It is also called Lab-on a-chip. The size of the nanobiochip is much smaller than the size of a cell. The chip is made up of biologically active artificial structures. The microarrays are spread in a very small area over the solid surface of chip which enables to perform several biochemical tests simultaneously. Therefore, identification of novel biomarkers can be verified using multiple tests. Nanobiochips are generally used in analytical separations and identification of biomolecules such as nucleic acids and proteins. Implementation of lab-on a-chip on PCR and other in vitro diagnostics shows its substantial impact in the field of biotechnology (Bahadorimehr et al. 2010; Sharma and Hashim 2013). Lab-on a-chip establishes a very simple method of in vitro diagnostics in which a number of cantilever biosensors can be used together on a single array. The device can be developed in the most advanced manner by the integration of analytical technique and signal extraction system. Surface plasmon resonance (SPR) nanobiochip expresses the interaction of biomolecules in terms of affinity. It is useful in the study of association or dissociation affinity of ligands with its binding partners. The device is very simple, rapid, and advanced than earlier available affinity measurement methods. Very low sample size is required, and there is no need of tracer for labeling. Advanced application of SPR nanobiosensor device has been seen in the field of proteomics, where the sensor is combined with matrix-assisted laser desorption/Ionization time of flight mass spectrometry (MALDI-TOF) MS. This combination is very useful for the interaction study of biomolecules if one molecule is known. If there are varieties of proteins, then it will be very difficult to identify the interacting molecule. However, this technique permits to determine the interaction of molecule from mixture with sensor (Zhang et al. 2005; Marzán 2006).

In future, if a combination is prepared by the integration of nanofluids, nanobiosensors with biochip, the device will be more efficient to identify a number of biomolecules and their concentration. Therefore, nanobiochip offers new opportunities for innovative point-of-care diagnosis.

19.2.1.9 Biosensors and Nanobiosensors

Biosensor is an analytical device which detects the presence or concentration of biological product. It is a combination of a biorecognition element such as enzymes, antibodies, nucleic acids, or whole cell with a physiochemical transducer and an electronic reader device.

Principle of Biosensor

When a biorecognition element binds with a biological analyte, a biological event is performed on the interface and a signal is generated. The generated signal is sensed by transducer and converted into electrical signals which undergo amplification by a detector circuit. The intensity of signal directly denotes the concentration of chemical species (analyte). Signal processing by computer software into meaningful physical parameters describes the process being investigated (Barone et al. 2005; Huang et al. 2009; Lueke and Moussa 2011). With the progress of nanobiotechnology, biosensors have also been modified and fabricated at nanoscale for construction of nanobiosensors. Due to their wider applications, nanobiosensors have created the most important place in nanodiagnostic world. Nanobiosensors are very useful due to its portability, sensitivity, reproducibility, and low sample requirement for the detection of analyte. It is very simple to use and generate results quickly. The main aim of biosensor is to detect any biophysical or biochemical signal associated with the specific disease or disease-causing organism at the level of a single cell or molecule. In molecular diagnostics, integration of nanobiosensor with lab-on-a-chip has great potential for detection of disease-specific metabolites, pathogens, and cancers from a variety of samples. Detection of circulating tumor cells, pathogens, nucleic acids, and proteins not only just facilitates early diagnosis of disease but also creates new horizon for drug discovery and development. In the field of medical diagnosis, integration of nanobiosensors with other medical instruments creates new dimensions for emerging nanodiagnostics. Examples of nanobiosensors are nanowire-based sensors, cantilever biosensor, and ion channel sensing.

Nanowire-Based Sensors

Nanowires are the key component for the fabrication of nanowire-based sensors. The surface of nanowires is coated by biomolecules such as nucleic acids, proteins, or bacteriophages. Nanowires are fibril-like structures of definite length and diameter 10 nm or less. Surface properties of nanowires can be easily modified virtually by coating biomolecules as biorecognition elements over its surface which make them analyte independent. Nanowires are very sensitive for signal detection as chemical binding events change its conductance in a real time and quantitative fashion.

One-dimensional nanowires such as nanotubes and nanosprings can transport electrons efficiently and undergo the process of optical excitation which makes them appropriate nanomaterial for fabrication of ultrasensitive high-density nanoscale devices. Because of their unique properties, minor perturbations influence the electrical properties of nanowires. Carbon nanotubes are equipped with unique properties such as higher surface to unit volume ratio, high ductility, efficient electron transporter, and electrocatalytic properties. Therefore, carbon nanotubes have been found to be excellent material for the development of enzymatic nanobiosensors. Wang et al. (2003) used carbon nanotubes/Nafion electrodes for the immobilization of enzyme glucose oxidase for the detection of glucose. The combination of carbon nanotube/ Nafion was prepared over the surface of an electrode by dispersing the solubilized form of carbon nanotubes in a solution of Nafion. The constructed nanobiosensor was efficient enough to improve the intensity and quality of signals even at a very low potential with fast response time. In the future, multiplexed nanobiosensors can be developed by using an array of highly ordered nanowires and their combination with multiple biorecognition elements. Small size and robustness make nanowire biosensors very efficient in in vivo as well as in vitro sensing and hold great promises in healthcare diagnostics (Cui et al. 2001; Wang et al. 2003).

Cantilever Biosensors

At nanoscale, micromechanically generated cantilevers are found to be an excellent transducer for the fabrication of cantilever-based nanobiosensors. Bending of cantilevers of more than 100 nm due to sensing interaction on surface is detected by a laser. The high sensitivity of microcantilever-based sensor created a very good platform in label-free and less time detection. Microcantilever biosensors can detect processes in both static and dynamic modes. In the static mode, asymmetric chemical absorption on one surface of cantilever results in bending due to stress in the chemical adsorption, and the bending can be measured. In the dynamic mode, the frequency of microcantilever is changed due to mass adsorption over its surface, and the shift in resonating frequency can be measured. Cantilever biosensors have been employed widely for wider detection of ions, vapors, antibiotics, and biomolecules such as DNA, proteins, and a number of disease-causing pathogens. The change in mechanical sensitivity of cantilever nanobiosensors is due to change in molecular interactions with surface in static mode and lateral interactions such as steric hindrance, electrostatic forces, structural change within the layer of molecules. In the static mode, high sensitivity to structural changes is very useful to measure binding of biomolecules attached to the surface of microcantilever. Bergar et al. proposed a model in which a cantilever biosensor is combined with an atomic force microscopy (AFM). In this model, the cantilever biosensor with a tip can detect and analyze at nanoscale. Therefore, cantilever biosensing system will be more efficient for pathogen detection, cancer detection, genotyping, and SNPs (Ziegler 2004; Fritz 2008).

It has been observed by several research groups that microcantilevers can act as signal transducers for different domains such as heat, temperature, stress, and electromagnetic field. At present, because of their high flexibility and low sample size requirement without any previous treatment, cantilever-based nanobiosensors are employed in the field of genomics and proteomics extensively.

Ion Channel-Based Sensing

In biological system, ion channels are formed by the interaction of membranous protein molecules of cells. The size of the ion channels may be in the range of nm or lesser. They are filled with water molecules. Ion channels are selectively permeable, i.e., control the in- and outflow of ions. Therefore, they play a very crucial role in the regulation of electrical and biochemical activities of cells. The novel nanobiosensors are developed which are based on the approach that on binding of biorecognition elements such as enzymes, nucleic acids, and proteins to a population of ion channels, their conductance changes. Nanobiosensors fabricated on this approach can be used for most of the receptors, antibodies, and nucleic acids. The device is very flexible and can sense proteins even at a picomolar concentrations. After reducing its dimensions, the sensor which is actually an impedance element can be fitted into a microelectronic circuit as an integral part. The switch of the ion channels contains a gold electrode which is linked to the membrane lipid which harbors gramicidin ion channels attached to antibodies. The structure of the lipid membrane creates a reservoir between the gold electrode and membrane. The movement of ions results in the dimer formation, and conductance is changed. The change in conductance depends on the number of dimer formation. So, it is very simple to calculate the number of dimers formed as a result of change in conductance. The generated signals can be measured using a reader device. The wider application of the device includes detection of blood, media, nucleic acids, proteins, viruses, and compounds of low molecular weight (Krishnamurthy 2010; Holzinger et al. 2014).

19.2.2 Nanobiosensors for Glucose Detection

Globally, diabetes is a major concern. There are a number of advanced versions of insulin, though patients are suffering from worst outcomes of diabetes. In diabetic patients, blood samples are used for the detection of glucose level. In this situation, there is a need for a diagnostic method which is simple, easy to perform, noninvasive, and sensitive enough to meet the requirements of point-of-care diagnosis. Clark and Lyons developed the first glucose biosensor in 1962. The developed sensor was originally an amperometric glucose biosensor. The sensor was able to detect oxygen pressure. A number of biosensors have been developed for the detection of glucose molecules (Cash and Clark 2010).

19.2.2.1 Electrochemical Biosensors for the Detection of Glucose

In electrochemical biosensor, glucose oxidase is the enzyme which breaks glucose into gluconic acid and hydrogen peroxide (H_2O_2). Under the influence of external oxidation potential, H_2O_2 emit protons, O_2, and electrons. The monitoring signal is directly proportional to the concentration of glucose (Taguchi and Ptitsyn 2014). This approach is not very successful as few electrodes damage the cells and sensitivity is restricted to the quantity of active immobilized enzyme. The sensor was improved using entrapment of enzyme in polymers and covalent bonding of several nanostructures with Glucose oxidase enzyme. The quality of generated signal is improved by the integration of carbon nanotubes, graphene, and metallic nanoparticles in the original design of biosensor (Chen et al. 2007).

19.2.2.2 Optical Biosensors for the Detection of Glucose

Optical glucose biosensors are fabricated to nanoscale by using ultrasensitive nanostructures. Fluorescence resonance energy transfer (FRET) is the technique on which the device works. When a radiation of particular energy falls on a photoluminescent material, its molecules become excited to a higher energy state. When these molecules return to their ground state, they emit the radiation of low energy. This phenomenon is known as fluorescence which creates the platform for fabrication of fluorescent glucose biosensors. This sensor is based on the principle that the signal generated by the detection of glucose is converted into correlated fluorescent signal. Nanomaterials are very good transducer for the conversion of signal. FRET is a sensitive analytical technique and nowadays frequently used in fluorescence biosensing systems (Chen et al. 2017).

19.2.2.3 Nanoparticle-Based Biosensors for the Detection of Glucose

Nanoparticles are extensively used for the fabrication of nanobiosensors due to their small size, catalytic property, high unit to volume surface ratio, and high degree of sensitivity. They are frequently used for labeling of biomolecules and adsorption. They improve the performance of electrochemical and enzymatic biosensors by rapid transfer of electrons and make the process fast by shortening distance between enzyme and electrodes. Nanobiosensors based on surface-enhanced Raman spectroscopy (SERS) are widely used for glucose sensing. Direct concentration of glucose was detected for the first time by integration of silver nanoparticles in biosensor design. In many varieties of glucose nanobiosensors, nanoparticles are immobilized on electrodes. Metallic nanoparticles and quantum dots were used for this design. On the surface of single electrode, arrays of electrodes can be created by arranging the nanoparticles in ordered structures. These arrays improve the signal to noise ratio and detection limits (Garcia and Merkoci 2016).

19.3 Nanomaterials and Nanodevices for the Diagnosis of Infectious Diseases

Nowadays, PCR, ELISA, and sequencing are the main techniques used for the detection of disease-causing pathogens. These techniques are expensive and time taking and require large sample size, and there is a need for the expert to perform the PCR and sequencing techniques. Contrary to this, nanodiagnostics are found to be a very useful approach in pathogen detection. The technique is very simple to use, rapid, and cost-effective and provides reproducible results, and there is no need for expert to perform the test. Nanodiagnostics are well suited to meet the requirement of point-of-care diagnosis. Nanotechnology uses nanoparticle-conjugated antibodies as targeted ligands for binding with the molecules expressed on microbial surface such as proteins and lipids. This is the basis for the development of nanotechnology-based immunoassays.

Gold and silver nanoparticles are widely applied in conjugation with affinity ligands for microbial detection. To develop colorimetric assays, nanoparticles have been also used in conjugation with short nucleic acid sequences to bind complementary sequences present in pathogens. Quantum dots and carbon nanotubes have been also used for the DNA detection of bacteria and viruses. Several immunoassays have been developed using nanoparticles such as fluorescent nanoparticles, metallic nanoparticles, and magnetic nanoparticles. High sensitivity and photostability of fluorescent nanoparticles make them suitable for use as a probe to label different biological targets (Thévenot et al. 2001).

Fluorescent nanoparticles have been evolved as a new strategy for real-time diagnosis of the diseases. They are mainly used for imaging and sensing. Quantum dots (QDs) are fluorescent nanostructures of 1–10 nm in size. They fall in the category of semiconductor nanomaterials (NMs).The unique properties of QDs such as photostability and bright fluorescence can be employed successfully in NDs. The inherent properties of QDs can be easily configured through the change in their size and composition (Peeling and Mabey 2010).

Fluorescent signals of QDs can be amplified by the packing of several QDs inside an individual NP. The arrangement is called quantum dots nano beads (QDNBs). QDNBs were used in a dot-based immunoassay for the detection of HBV. In this assay, surface antigen of HBV (HBsAg) was immobilized on a porous surface of polyvinylidene fluoride (PVDF). QDNB-conjugated antibodies were used as probe. Probes were allowed to incubate with antigen. When a UV light emitted through UV lamp falls on the PVDF surface, combination of conjugated antibody-antigen flourishes in the form of dots. The device is simple to use and sensitive enough to detect picogram concentrations of HBsAg (Li et al. 2018; Yu et al. 2015).

Generally, gold and silver NPs are used in nanodiagnostics. Gold nanoparticles (AuNPs) were the first nanomaterial used in 1996 for DNA detection. AuNPs have the unique property of color change (on excitation through electromagnetic radiation) which makes them a suitable candidate for the fabrication of nanodevices and/or use in NDs (Fuente and Jesus 2006).

Gold nanorods are elongated nanoparticles and received a lot of attention due to their tunable optic properties. Gold nanorods have been used in the diagnosis of human immunodeficiency virus (HIV) with the help of hyper-Rayleigh scattering (HRS) spectroscopy. Oligonucleotide probe of 145 mer was used as a probe to detect 100 picomolar of target DNA with single base mismatch. The detection was based on the change in the intensity of HRS. HRS spectroscopy was also used for the detection of Hepatitis C virus (HCV) using gold nanoparticles. To detect the HCV, gold nanoparticles were conjugated with single-stranded RNA of HCV tagged with Rhodamine 6G. These particles are very sensitive for the detection of up to 80 picomolar concentration of HCV RNA with single base mismatch (Draz and Shafiee 2018; Fazio et al. 2016).

Application of magnetic nanoparticles in nanodiagnostics is based on the fact that when an external magnetic field is applied, separation and detection of magnetic nanoparticles bonded with target is enhanced. Detection of malaria in early stage using conventional approaches such as real-time PCR is very challenging due to the presence of very low concentration of parasite in the blood. The application of magnetic nanoparticles is seen in the early diagnosis of malaria through the magnetic-enriched SERS. In this approach, β-hematin crystals (an equivalent of hemozoin granule biomarker in malaria) were detected through magnetic nanoparticles. In the early diagnosis of malaria, by applying external magnetic field, the detection limit can be improved as low as 5 nM of β-hematin crystals $\approx$ 30 parasites/μL of blood samples (Yuen and Liu 2012; Kumvongpin et al. 2016; Chen et al. 2017).

In another approach, to improve the sensitivity of assay, microfluidics-integrated nanodevices have been used for the enrichment of pathogens or their DNA. The method of enrichment significantly improved the sensitivity of the assay. Therefore, very low concentration of pathogens could be detected from blood samples using conventional diagnostics (Warkiani et al. 2015).

To prepare an advanced nanodevice-based diagnostic platform, many techniques are integrated with nanotechnology, i.e., integration of microfluidics with lab-on-a-chip. The device is very useful in the diagnosis when a patient is infected with more than one strain/parasite (Cabibbe et al. 2015; Dixon et al. 2016). Lab-on-a-chip in integration with other techniques can be used for the detection of multiple tests simultaneously. It reduces cost, time, and sample size.

To detect the blood-borne infections caused by HBV, HCV, and HIV, a high-throughput and multiplexed nanodevice was developed by the integration of quantum dots and microfluidics. The device can precisely detect multiple infecting agents from serum volume as low as 100 μL. The assay can be performed in less than an hour with the sensitivity more than 50 times from the currently available US Food and Drug Administration (USFDA)-approved platforms (Klostranec and Chan 2006). Another nanodevice based on silica membrane was developed to detect envelop glycoprotein gp120 of HIV. Human cell surface receptor CD4 was coated on the surface of membrane. The system is useful to study the real-time interaction between host cell surface receptor CD4 and gp120 of HIV (Cheng et al. 2012).

Immunofluorescence-based integrated nanodevices have been successfully used in the detection of mycobacteria in the sputum. In this device, a microtip sensor in combination with genus-specific antibodies was used for the concentration of the pathogen using electric field and streaming flow technique. The sensitivity of the assay was equivalent to the PCR, but there was no need for additional steps like culture or amplification because it can be finished in 25 min (Yeo et al. 2009; Kim et al. 2012).

When a charged or uncharged dielectric particle is subjected to a nonuniform electric field, an external force is exerted on the particle. In the presence of an electric field, all particles exhibit dielectrophoretic activity irrespective of their charge. The property of dielectrophoresis was used in combination with capillary action to enrich the extracellular DNA on nanostructured tips. In the first step, dielectrophoresis was applied for attraction of DNA and other molecules near the nanostructured tips followed by enrichment of DNA onto the nanotips through capillary action. The detection limit of this method is nearly 6.7 pg/mL of DNA; however, the approach is not very useful for pathogen detection in small volume of samples (Yeo et al. 2009).

19.3.1 Nanostructures in Cancer Diagnosis

Cancer is a malignant tumor and responsible for high mortality rate worldwide. Lung cancer, liver cancer, colorectal cancer, stomach cancer, and breast cancer are placed in the list of the World Health Organization (WHO) in 2015 as top 5 cancer types responsible for maximum deaths across the world. To control the stage transition from curable benign form to advanced metastasis stage, diagnosis of cancer in early stage is required. Early detection of cancers is very important for the onset of treatment and prevents further disease-associated complications in the patient's body. In short, early detection leads to better prognosis. Application of nanotechnology in cancer diagnosis reduces the cost and time of diagnosis (Prasad et al. 2017b). A variety of nanomaterials such as gold nanoparticles, silicon nanowires, quantum dots, carbon nanotubes, and graphene have been successfully used to detect different cancer types. Silicon nanowire (SNW) alone or in combination with field-effect transistors (FETs) has been used for the detection of prostate-specific antigen (PSA) and 8-hydroxydeoxyguanosine (8-OHdG) biomarkers in prostate cancer (Gao et al. 2014). FET-SNW and zinc oxide nanowires (ZnONWs) have been used for the detection of single-stranded DNA (ss-DNA) and micro-RNA (mi-RNA) involved in the progression of different cancer types (Lu et al. 2014). Electrochemical nanotubes have also been used for the detection of other prostate cancer biomarkers such as platelet factor-4 and interleukin-6 (Azmi et al. 2014; Chikkaveeraiah et al. 2009). In case of aggressive prostate cancer, the expression level of cancer-testis antigen (CTA) RNA was measured using nano counter analysis system – a nanowire technology. This technology can be fabricated into sensor chip for a simultaneous detection of panel of biomarkers in specific cancer type (Takahashi et al. 2015). A single nanoparticle dual-mode MRI probe was developed by using unique property of quantum dots. Coating of paramagnetic lipid and silica nanoparticles on quantum dots can provide information about the molecules involved in tumorigenesis (Swierczewska et al. 2011). Detection of some volatile organic compounds in breath samples may be an indicator of cancer. In this concept, carbon nanotubes and silicon wires were used to detect these compounds in lung and gastric cancers (Shehada et al. 2015). Gold nanoparticle-modified graphene oxide-based DNA biosensor was used to detect two breast cancer biomarkers – human epidermal growth factor receptor-2 (HER2) and the cell surface protein CD24 (Saeed et al. 2017). HER2 expression was also measured by immunohistochemistry (IHC) technique by using quantum dots-conjugated trastuzumab (a monoclonal antibody against HER2 expression). Hybridization chain reaction (HCR) is a method of signal amplification. The method has been used with quantum dots-Ru complex dyads to enhance the sensitivity of detection and cellular imaging (Zhang et al. 2018). Scientists reported that micro-RNA in circulation may have potential to be used as a noninvasive cancer biomarker. The study of Li et al. justifies the report. He used the quantum dots-based microarray to test the lung cancer patients and found the presence of significantly different micro-RNA such as smiR-16-5p and miR-17b-5p in cancer patients than in controls (Fan et al. 2016). Besides the nanoparticles, nano-based contrasting agents have also been used for detection of several cancers through imaging modalities such as MRI, PET, and so on. Nanoparticles are mainly used in imaging and point-of-care technology.

19.3.2 Cancer Detection Through Imaging

In imaging, several contrast agents are used for the detection of tumors. Bonding of contrast agents to the surface ligands of nanoparticles may be used as indicator biomarkers of tumors. However, screening of appropriate tumor binding ligand is very challenging. Nanoparticles as contrasting agents have several advantages over the conventional contrasting agents such as elimination of nanoparticles through biological clearance pathways in a controlled manner. They can bind with the targets in a very specific manner (Shilo et al. 2012) and remain in circulation for longer time, thus providing more time for imaging.

19.3.2.1 Magnetic Resonance Imaging (MRI)

MRI is based on the principle of nuclear magnetic resonance (NMR). NMR uses the magnetic properties of atomic nuclei for creation of medical images through MRI. When an atomic nucleus is placed in a magnetic field, it absorbs and emits the electromagnetic radiation with the resonance frequency of substance and creates the image. The resonance frequency is directly proportional to the strength of magnetic field and depends on location and thus can create images of different body organs. The contrasting agents create the image by the alteration of proton relaxation time in longitudinal (T_1) and transverse (T_2) dimensions and dismantle the local magnetic field. The difference in relaxation times of T_1 and T_2 discriminates between tissues, air, and biopsy (Chen et al. 2016). A superparamagnetic particle when used in MRI can perturb the magnetic field 50 times of its diameter and therefore can impact water protons to the deep layer of cells from original location (Shilo et al. 2012). To be used as contrasting agent in vivo, nanoparticles should be stable and nontoxic and remain longer in body circulation. These features can be modulated by change in size and coatings of nanoparticles (Cole et al. 2015). Iron oxide nanoparticles, gold nanoparticles, and gadolinium are frequently used as contrasting agents in MRI. Gadolinium is generally used as contrasting agent in T_1-weighted protocol, but due to toxicity generated by it, it is not suitable for use in patients with renal failure. Therefore, iron oxide nanoparticles can be used as a substitute of gadolinium in these patients. However, iron oxide nanoparticles have limited relaxivity. In general, paramagnetic nanoparticles and superparamagnetic particles are used as T_1 and T_2 contrasting agents, respectively (Lawaczeck et al. 2016).

Gadolinium (Gd^{+3}) is the most commonly used contrasting agent in MRI. Gd^{3+} has huge potential to catalase water signals. The presence of seven unpaired electrons and long electron spin relaxation time make it suitable as a contrasting agent to create positive contrast in MRI. FDA-approved five contrasting agents are based on the use of Gd^{3+} (Tian et al. 2015; Coughlin et al. 2014).

To get the images of cancer cells with higher spatial resolution and sensitivity, two imaging modalities (MRI and optical imaging) were combined to create dual modal imaging (DMI). To be used as contrast agent in DMI, they have luminescent and magnetic properties. For this purpose, Europium-doped gadolinium oxide (Eu-doped Gd_2O_3) nanorods were synthesized and coated with silica (Gayathri et al. 2017).

19.3.2.2 Superparamagnetic Iron Oxide Nanoparticle (SPION)-Enhanced MRI Imaging

Superparamagnetic magnetism occurs due to interaction of a permanent magnet and paramagnetic substance and lies between both of them. Iron oxide nanoparticles and other ferrous materials of diameter 1–100 nm come into the category of superparamagnets. Generally, iron oxide nanoparticles occur in two main forms – magnetite (Fe_3O_4) and maghemite (γ-Fe_2O_3). Maghemite is the oxidized form of magnetite. In MRI, they can be used to identify infection and inflammation. In contrast to the conventional contrasting agents, these particles provide image of high quality and resolution. For better image quality, use of larger SPION for T_2 relaxation imaging and smaller SPION for T_1 relaxation imaging is recommended (Warlin 2013). SPIONs have also great potential for discrimination between benign and cancerous tumors. Seyfer et al. (2014) used SPION in T2-weighted MRI protocol and reported low contrast-to-noise ratio in abscesses than neoplasia.

19.3.2.3 Positron-Emission Topography (PET) Scanning

PET imaging is used in nuclear medicine to get the clinical information about the patients using radiotracers. PET works on detection of radiation produced by decay of tracer inside the body. PET-CT is an advanced form of PET scanning. It provides the information about the onset of disease much earlier than other methods used in imaging. In PET imaging, metal oxide nanoparticles were used as probes. They were used in a nuclear reaction-involved conversion of ^{18}O-enriched aluminum oxide to ^{18}F-labeled nanoparticles by the action of photons on ^{18}O-enriched aluminum oxide (Al_2O_3). Radiolabeled isotopes of nanoparticles as positron emitters have been used to monitor the level of tumors and activity of enzymes involved in tumorigenesis (Perez-Campana et al. 2013).

19.3.2.4 Ultrasound

Ultrasound is another modality of imaging. It is a type of mechanical sound called ultrasonic waves with frequency more than the hearing frequency of human ear (>20 kHz). It is an important tool in diagnosis and imaging in sonography. In imaging as per the point of diagnosis, these waves are focused at a particular depth. A scattered signal is produced due to different acoustic resistance of tissues. The signals are recovered and used to create the reconstructed images of tissues. Mattrey et al. developed perfluorooctylbromide nanoparticles (PFOB-NPs) encapsulated within a pluronic F-68 shells. They have significantly improved echogenicity of the liver than that of kidneys. Therefore, they can be used as a promising contrasting agent to improve the sensitivity of detection in the liver and tumors (Mattrey et al. 1982).

Microbubbles were used as contrasting agents to improve the echogenicity and signaling of tissues. Microbubbles consist of different gases enclosed in a lipid/protein/polymers shell and can create two-/three-dimensional images of tissues and organs (Nie et al. 2014; Daraee et al. 2016). Microbubbles were developed with the intention for imaging of blood flow and tissue perfusions. Perfluorooctyl bromide (PFOB)-gold core-shell complex as a contrasting agent was used in imaging of the kidney and liver of mouse (Ke et al. 2014). Gold and graphene oxide nanoparticles containing poly (lactic acid) microcapsules (PLA microcapsules) developed by Jin can be used to deliver payload in tissues (Jin et al. 2013). Another group of scientists used PLA microbubbles containing gold nanoparticles to deliver payload. Signal with 50% decrease was attained in this study (Teraphongphom et al. 2015).

Although microbubbles offer very sensitive detection due to their strong nonlinear response, their applications have several limitations such as in vivo short half-life span and high background signal. Nanoparticles overcome these limitations as they have a long circulation half-life and are able to accumulate in the interstitial space of tumors through enhanced electron paramagnetic imaging (EPR) (Zhang et al. 2014).

Ultrasound imaging offers several advantages such as real-time imaging, affordability than other imaging modalities, biocompatibility, and portability. The contrasting agents used in ultrasound imaging can be also used as theranostic purpose, i.e., therapy and diagnosis in major diseases like cancer. In theranostic applications, contrasting agents are conjugated with superparamagnetic iron oxide nanoparticles (SPIONs), CuS nanoparticles, DNA, siRNA, gold nanoparticles (GNPs), gold nanorods (GNRs), gold nanoshells (GNS), graphene oxides (GOs), polypyrrole (PPy) nanocapsules, Prussian blue (PB) nanoparticles, and so on to different types of UCAs (Fu and Ke 2016).

19.3.2.5 X-Ray/Computed Tomography

X-ray imaging is used to generate high-resolution images of internal structures of body through the use of X-rays – a high-energy electromagnetic radiation. It is the most popular method of imaging and accounts for 50–70% of all medical imaging done. It is a safe and cost-effective method of imaging and can be used by taking care of patient safety and limited exposure of radiation (Alric et al. 2008). When X-rays pass through the body, there is loss in beam intensity. This is due to the photoelectric absorption or scattering. The loss of beam intensity is called attenuation. Repeated use of iodine-based contrast agents offers adverse side effects such as allergies and nephrotoxicity. Previously, use of gold nanoparticles as contrast agent have been reported for X-ray imaging. Gold nanoparticles are safer and less toxic than other contrast agents and have higher absorption coefficient value at low energy for X-rays (Kim et al. 2007; Kojima et al. 2010). Hainfeld et al. (2006) reported the detection of intravenously injected gold nanoparticles through X-ray imaging.

The technique of Computed Tomography came into existence in 1973. The technique was able to create the three-dimensional X-ray images through the rotation of detector and X-ray source around the body (Hounsfield 1973). Iodine-based contrast agents are rapidly cleared from the body and thus provide very less time for imaging. They are also distributed unevenly in intracellular and extracellular vasculature; therefore, they are unable to create high-quality and high-resolution images of CT. To overcome these limitations, iodine-based contrast agents are developed in nm range and in the form of micelle, liposomes, and polymers.

Prolonged exposure of nanoparticles in circulation can lead to the high-quality contrast CT images. On the other hand, clinically approved contrast agents are smaller in size and therefore rapidly clear from circulation via excretion through the kidney. To remain longer in circulation, nanoparticles can be synthesized bigger than the size of fenestra (Choi et al. 2014). Nanoparticle-based contrast agent for longer duration in circulation was synthesized by Torchilin (Torchilin 2001; Trubetskoy et al. 1997). The use of gold nanoparticles as contrast agent showed their existence for up to 12 h in blood vessels (Kim et al. 2007; Cai et al. 2007). Nanoparticle-based contrast agents can be synthesized in the form of a core containing atoms for contrast generation and coated by lipid, proteins, silica, or polymers. The coating layer can be easily modified for insertion of antibodies, nucleic acids, drugs, or other contrast-generating moieties for multimodal imaging (Lee et al. 2012; Jia et al. 2013). Therefore, synthesis of nanoparticles equipped with optical and/or magnetic properties potentiates them for multifunctional use (Xu et al. 2011). Nanoparticles in the form of lipid formulations such as liposomes, micelles, and lipoproteins and solid core nanoparticles like salt and alloy of metals are frequently used in CT applications (Ghaghada et al. 2016).

19.3.3 Nanocarriers as a Weapon for Drug Dissolution and Drug Discovery

Nanocarriers in drug delivery offer several advantages: (1) enhance solubility of very low-soluble drugs, (2) enhance targeted drug delivery specificity, (3) can move across epithelial and endothelial barriers, (4) improve direct delivery of large molecular drugs into cells, (5) can be used in single or combination delivery of drugs, (6) can be used in drug delivery site monitoring by tagging of therapeutic agent with molecules having imaging properties, and (7) can be used for real-time monitoring of drug delivery and high in vivo efficacy (Farokhzad and Langer 2009; Prasad et al. 2017b).

For efficient working, nanocarriers should be in circulation for a longer time for drug accumulation, but practically it is difficult. Immune system treats nanocarriers as foreign particles and rapidly clears it through opsonization (Torchilin 2012). Premature release of entrapped drugs at other sites than drug targets using nanocarriers leads to systemic toxicity (Jin et al. 2014).

To maintain the concentration of drugs across the target by means of active or passive transport, nanocarriers undergo conjugation with polyethylene glycol (PEG) molecules on its surface, resulting in low interaction of nanocarriers with blood, improved water solubility, and colloidal stability. This combination significantly improved the circulation time of nanocarriers. PEG is not suitable in all conditions; therefore, different approaches are being explored. It includes nonfouling material and zwitterions such as phosphorylcholine (PC), carboxybetaine (CB), and sulfobetaine (SB).

The efficacy of drug and bioavailability mainly depends on the physical and chemical properties of drugs such as aqueous solubility, pK_a, and partition coefficient-Log P. Solubility, permeability, and oral absorption of drugs are significantly affected by pK_a. One of the challenging problem of pharmaceutical industry is the aqueous insolubility of drugs (Cheng et al. 2006).

In case of drug delivery through intravenous route, aqueous-based formulations are highly desirable. To prepare such type of formulations, water solubility of drug is the prerequisite (Yeh et al. 2009). For oral drug administration, before the absorption, drug should be dissolved properly in aqueous environment of the gut (Aulton

2007). Therefore, the absorption rate of drugs with high permeability but low solubility taken orally depends on the rate of drug dissolution.

Growing demand to fabricate novel drug delivery systems for delivering high content of active ingredient with high drug-targeting capacities led to the use of nanostructures as drug delivery agents – "nanocarriers." Nanostructures such as nanocrystals, liposomes, nano-emulsions, solid lipid nanoparticles (SLN), polymeric nanoparticles, and polymeric self-assemblies have been successfully used to improve solubility of drugs.

Being smaller in size nanocarriers can easily cross the barrier of gut epithelium results in enhanced drug absorption and bioavailability (Ramesan and Sharma 2009). In nanocarrier delivery system, carrier matrix harbors the attached or adsorbed active ingredient either in dissolved state or in an encapsulated form (Caban et al. 2014).

Advantages of Nanocarriers to Improve the Solubility of Drugs
Nanocarriers are the materials of nanoscale. The miniaturization of nanocarriers offers several advantages:

(i) Nanocarriers can be used to enhance drug pharmacokinetics and reduce systemic toxicity of bioactive material by moving to the particular target site.
(ii) Nanocarriers can also be used to enhance solubility of hydrophobic drugs.
(iii) Nanocarriers can increases the stability of drugs and can be used for sustained or controlled release of drugs
(iv) Nanocarriers can easily cross the blood-brain barrier and tight epithelial junctions. Therefore, they can deliver drug across the barrier in a much better manner (Caban et al. 2014).

19.3.3.1 Types of Nanocarriers

Nanocarriers include the nanosized structures such as nanoparticles, nanocapsules, lipid complexes, polymeric micelles, and dendrimers (Sahoo and Labhasetwar 2003).

Nanosuspensions

Nanosuspensions are crystals in which 100–1000 molecules are arranged in the form of aggregates. Nanosuspensions are prepared by using a thin-coated drug containing one or mixture of surfactants. Nanonization is the method through which nanosuspensions are formulated (Rabinow 2004). In the first step, surfactants are dissolved in the water. Thereafter, macrosuspension of drug is prepared by dispersal of drug powder in aqueous solution through high-speed agitation. In the next step, macrosuspensions are homogenized using techniques such as wet milling (Liversidge et al. 2003), high-pressure homogenization (Müller et al. 2001), spray-drying, and nanocrystallization.

Nanosuspensions are capable of overcoming the problems associated with drugs such as poor solubility, poor bioavailability, difficulty in preparation in parenteral dosage form, and poor absorption pattern. In nanosuspensions for stabilization, mild quantity of surfactants or stabilizer is used. As per the dissolution property of the drugs, aqueous (water/buffer) or nonaqueous (lipid solvents) media can be used in nanosuspensions (Shegokar and Müller 2010; Junghanns and Müller 2008).

Advantages of Nanosuspensions

(i) Nanosuspensions are smaller in size; therefore they provide larger surface area for increased dissolution and absorption and faster onset of action.
(ii) Nanosuspensions increase the dissolution of drugs.
(iii) Nanosuspensions increase the oral bioavailability.
(iv) Nanosuspensions can be used to decrease the dose of drugs.
(v) Nanosuspensions can be used to decrease side effects of drugs (Savjani et al. 2012).

19.3.3.2 Liposomes

Liposomes are spherical structures in which an aqueous reservoir is completely surrounded by phospholipid bilayer. The phospholipid membrane of liposome contains a hydrophilic (attract water) head and a hydrophobic (repel water) tail group (Rawat et al. 2006). The property of liposomes varies with size, charge, lipid content, and the method of preparation. Liposomes can prevent the degradation of encapsulated drugs and are able to reduce systemic toxicity. As per the need, availability of liposomes can be increased by the incorporation of polyethylene glycol (PEG) units to the lipid bilayer. Conjugation of liposomes with antibodies or ligands improves the precision of targeted drug delivery (Sahoo and Labhasetwar 2003). Biphasic nature of liposomes makes them suitable as a carrier for hydrophilic as well as for hydrophobic drugs (Farokhzad and Langer 2009). Liposomes have been successfully used as a carrier in the field of drug delivery, cosmetics, and diagnostics (Akbarzadeh et al. 2013).

Liposomes can be used as a potent carrier for ocular drug delivery due to the presence of natural phospholipid membrane and biocompatibility. In case of topical application, liposomes can attach with the epithelial cells of cornea and deliver the bound drug and therefore improve the pharmacokinetics and decrease the toxicity-related side effects of the drug (Chetoni et al. 2007). Nanosized version of liposome is called nanoliposome. Common laboratory methods of nanoliposome production include sonication, freeze-thawing, extrusion, micro-fluidization, and ether injection (Mozafari 2010).

Advantages of Liposomes

(i) Liposomes can be used for selective passive targeting of tumors.
(ii) Liposomes enhance stability, efficacy, and therapeutic index of the drug.
(iii) Liposomes also decrease the toxicity and side effects of drug.
(iv) Liposomes enhance the pharmacokinetic effects of drugs (Dua et al. 2012).

19.3.3.3 Solid Lipid Nanoparticles (SLN)

Solid lipid nanoparticles are the colloidal carriers of diameter 50–1000 nm. Solid lipid nanoparticles consist of solid lipids distributed in aqueous solution or surfactant solution in water. Solid core is the drug containing area and are rich in high fat matrix. The fat matrix contains a chain of phospholipids which are hydrophobic in nature. The physical stabilization is attained through the addition of surfactant or emulsifying agent depending on the lipid contents and type (Rawat et al. 2006). SLN does not cause toxicity and can be prepared by several methods such as high-pressure homogenization, precipitation, lipid nano pellets, and so on (Mehnert and Mäder 2001; Müller et al. 1996). Positively charged SLN can be used as a potent non-viral carrier (Olbrich et al. 2001; Pedersen et al. 2006). SLN is an effective vaccine carrier. Chitosan-coated lipid nanoparticles have been employed in delivery of peptide drug through oral route (Fonte et al. 2011).

Currently, the use of lipid-based formulations has been emphasized to enhance the solubility of poor water-soluble drugs for oral bioavailability.

Advantage of Solid Lipid Nanoparticles
In comparison with nano particulate carriers, SLN offers the following advantages:

(i) SLN provides higher tolerability and biocompatibility.
(ii) SLN offers controlled release of drugs.
(iii) SLN also protects the incorporated drugs.
(iv) SLN significantly enhances the oral bioavailability.
(v) SLN can be prepared on a large scale.
(vi) SLN also provides the chemical stability (Varshosaz et al. 2010).

19.3.3.4 Dendrimers

Dendrimers are highly branched treelike structure. They consist of repeated units of monomers, a central core, an internal cavity, and peripheral groups. The macromolecule of dendrimers have large molecular weight and equipped with the entrapment property. They are formed from the monomeric units. Dendrimers have special physiochemical properties due to their components such as organic molecules and polymers. The internal cavity of dendrimers is used to encapsulate hydrophobic drugs. In comparison with traditional macromolecules, dendrimers have a higher functional group density which enhances the solubility of drugs (Svenson 2009). Higher reactivity of dendrimers is due to the presence of functional groups on its outer surface. These groups can be conjugated with other molecules (Yogesh et al. 2011). Therefore, drugs can be loaded in two ways: either conjugated with surface functional groups through the electrostatic bonds or encapsulated in internal cavity.

Dendrimers coming in the category of novel polymeric materials have spherical 3D structures and offer higher surface group functionality.

Dendrimer-based drug delivery includes two mechanisms:

(i) Cleaving of drug-dendrimer conjugation involves the presence of enzymes in vivo.
(ii) Drugs are released in vivo due to change in physiochemical environment such as pH and temperature.

These mechanisms come into play either on the outer surface or in the internal cavity (Liu and Fréchet 1999; Caminade and Turrin 2014).

Dendrimers as nanocarriers can deliver drugs to the pulmonary system, across transdermal route, and eyes and can be used for controlled release and targeted drug delivery.

19.3.3.5 Polymeric micelles (PM)

Polymeric micelles (PMs) were developed by Ringsdorf as potent vehicle for drug delivery to overcome the problem of low drug solubility and systemic toxicity (Williams et al. 2013). Thereafter, PMs have been used as delivery vehicle for anti-cancer drugs, contrast agents, lipids, proteins, plasmids, components of antisense technology, and RNAi-short interfering RNA (siRNA). These formulations are undergoing clinical trials (Kedar et al. 2010).

PM mainly consists of a bio inert material with a core and outer shell. The inner core is composed of a hydrophobic polymer, whereas the outer shell is from the hydrophilic polymer. The inner cores are meant for drug loading, and the outer shell is for the protection of core from bioenzymes and to prevent adsorption of proteins on the surface of PM (Miyata et al. 2011).

Being smaller in size, PM can withstand against the renal excretion and uptake by reticuloendothelial system and it also facilitates the use of PM in targeting tumors and prolonged retention in circulation. Several approaches have been described for the preparation of pH-sensitive PM (Tan et al. 2013). They can be prepared by attaching a titratable group, e.g., –COOH or $–NH_2$, into block copolymer. These groups release protons and also control the micelle formation. Micelles are formed when the concentration of block copolymer exceeds the critical micelle concentration (CMC).

Advantages of Polymeric Micelles

(i) The polymeric micelles impart thermodynamic stability in the living system; therefore, they decrease the in vivo dissolution (Xu et al. 2013).
(ii) The polymeric micelles can be used as a carrier for poor water-soluble drugs due to their core-shell structure.
(iii) The polymeric micelles are a suitable carrier for intravenous administration as the poor soluble drugs present in hydrophobic core and outer hydrophilic layer support dispersal in an aqueous environment (Ding et al. 2012).
(iv) Being smaller in size, RES cannot remove them from circulation.

19.4 Conclusion

Miniaturization of matter at nano scale founded the nano world. We can see one of the best applications of nanotechnology in the field of medicine. Nanotechnology in medicine revolutionized the diagnostic world and expanded the new approach toward better healthcare. Conventional methods of disease diagnosis are expensive and time taking need expertise and large quantity of samples. Nano-based diagnostics diminished these problems through point-of-care diagnosis tests (POCTs). Nanomaterials and fabricated nanodevices are frequently in use not only in the field of diagnosis but also as efficient nanocarriers for drug dissolution and targeted drug therapy. Integration of nanotechnology with other techniques such as microfluidics enhanced the sensitivity of the diagnostic modalities and decreased the time of detection. In a nutshell, application of nanotechnology in medicine is a very promising approach and expanding new horizons in future diagnostics and pharmaceuticals.

References

Akbarzadeh A, Sadabady RR, Davaran S (2013) Liposome: classification, preparation, and applications. Nanoscale Res Lett 8(1):102

Alric C, Taleb J, Le Duc G et al (2008) Gadolinium chelate coated gold nanoparticles as contrast agents for both X-ray computed tomography and magnetic resonance imaging. J Am Chem Soc 130:5908–5915

Aulton ME (2007) Pharmaceutics: the design and manufacture of medicines, 3rd edn. Churchill Livingstone, London

Aziz N, Faraz M, Pandey R, Sakir M, Fatma T, Varma A, Barman I, Prasad R (2015) Facile algae-derived route to biogenic silver nanoparticles: synthesis, antibacterial and photocatalytic properties. Langmuir 31:11605–11612

Aziz N, Pandey R, Barman I, Prasad R (2016) Leveraging the attributes of *Mucor hiemalis*-derived silver nanoparticles for a synergistic broad-spectrum antimicrobial platform. Front Microbiol 7:1984. https://doi.org/10.3389/fmicb.2016.01984

Aziz N, Faraz M, Sherwani MA, Fatma T, Prasad R (2019) Illuminating the anticancerous efficacy of a new fungal chassis for silver nanoparticle synthesis. Front Chem. https://doi.org/10.3389/fchem.2019.00065

Azmi MA, Tehrani Z, Lewis RP (2014) Highly sensitive covalently functionalized integrated silicon nanowire biosensor devices for detection of cancer risk biomarker. Biosens Bioelectron 52:216–224

Bahadorimehr AR, Jumril Y, Majlis BY (2010) Low cost fabrication of microfluidic microchannels for lab-on-a-chip applications. International conference on electronic devices, systems and applications, pp 242–244

Barone PW, Baik S, Heller DA, Strano MS (2005) Near-infrared optical sensors based on single-walled carbon nanotubes. Nat Mater 4:86–92

Beishon M (2013) Exploiting a Nano-Sized Breach in Cancer's Defences. Cancer World :14–21

Boisselier BE, Astruc D (2009) Gold nanoparticles in nanomedicine: preparations, imaging, diagnostics, therapies and toxicity. Chem Soc Rev 38:1759–1782

Buzea C, Pacheco II, Robbie K (2007) Nanomaterials and nanoparticles: sources and toxicity. Biointerphases 4:MR17–MR71

Caban S, Aytekin E, Sahin A, Capan Y (2014) Nanosystems for drug delivery. OA Drug Des Deliv 2(1):2

Cabibbe AM, Miotto P, Moure R (2015) Lab-on-chip-based platform for fast molecular diagnosis of multidrug-resistant tuberculosis. J Clin Microbiol 53(12):3876–3880

Cai QY, Kim SH, Choi KS et al (2007) Colloidal gold nanoparticles as a blood-pool contrast agent for x-ray computed tomography in mice. Investig Radiol 42(12):797–806

Caminade AM, Turrin CO (2014) Dendrimers for drug delivery. J Mater Chem B 2:4055–4066

Cash KJ, Clark HA (2010) Nanosensors and nanomaterials for monitoring glucose in diabetes. Trends Mol Med 16(12):584–593
Challa S, Kumar SR (2007) Nanomaterials for medical diagnosis and therapy (Handbook). The Wiley network USA
Chan HK, Kwok PCL (2011) Production methods for nanodrug particles using the bottom-up approach. Adv Drug Deliv Rev 63:406–416
Chen Z, Meng H, Xing G (2007) Toxicological and biological effects of nanomaterials. Int J Nanotechnol 4:179–196
Chen Y, Ai K, Liu J, Ren X (2016) Polydopamine-based coordination nanocomplex for T1/T2 dual mode magnetic resonance imaging-guided chemo-photothermal synergistic therapy. Biomaterials 77:198–206
Chen K, Yuen C, Aniweh Y (2017) Recent progress in the development of diagnostic tests for malaria. Diagnostics (Basel) 7(3):54
Cheng WP, Gray AI, Tetley L, Hang TLB (2006) Polyelectrolyte nanoparticles with high drug loading enhance the oral uptake of hydrophobic compounds. Biomacromolecules 7:1509–1520
Cheng K, El-Boubbou K, Landry CC (2012) Binding of HIV-1 gp120 glycoprotein to silica nanoparticles modified with CD4 glycoprotein and CD4 peptide fragments. ACS Appl Mater Interfaces 4(1):235–243
Chetoni P, Burgalassi S, Monti D, Najarro M (2007) Liposome-encapsulated mitomycin C for the reduction of corneal healing rate and ocular toxicity. J Drug Deliv Sci Technol 17(1):43–48
Chikkaveeraiah BV, Bhirde A, Malhotra R, Patel V, Gutkind JS (2009) Single-wall carbon nanotube forest arrays for immunoelectrochemical measurement of four protein biomarkers for prostate cancer. Anal Chem 81(21):9129–9134
Choi S, Tripathi A, Singh D (2014) Smart nanomaterials for biomedics. J Biomed Nanotechnol 10(10):3162–3188
Cole LE, Ross RD, Tilley JM, Vargo-Gogola T (2015) Gold nanoparticles as contrast agents in X-ray imaging and computed tomography. Nanomedicine 10:321–341
Coughlin AJ, Ananta JS, Deng N, Larina IV, Decuzzi P, West JL (2014) Gadolinium-conjugated gold nanoshells for multimodal diagnostic imaging and photothermal cancer therapy. Small 10:556–565
Cui Y, Wei Q, Park H, Lieber CM (2001) Nanowire nanosensors for highly sensitive and selective detection of biological and chemical species. Science 293:89–92
Daraee H, Eatemadi A, Abbasi E, Aval SF, Kouhi M (2016) Application of gold nanoparticles in biomedical and drug delivery. Artif Cells Nanomed Biotechnol 44:410–422
Ding H, Wang X, Zhang S, Liu X (2012) Applications of polymeric micelles with tumor targeted in chemotherapy. J Nanopart Res 14:1–13
Dixon C, Ng AHC, Fobel R, Miltenburg MB, Wheeler AR (2016) An inkjet printer, roll-coated digital microfluidic device for inexpensive, miniaturized diagnostic assays. Lab Chip 16(23):4560–4568
Draz MS, Shafiee H (2018) Applications of gold nanoparticles in virus detection. Theranostics 8(7):1985–2017
Dua JS, Rana AC, Bhandari AK (2012) Liposomes methods of preparation and applications. Int J Pharm Stud Res 3:14–20
Fan L, Qi H, Teng J, Su B (2016) Identification of serum miRNAs by nano-quantum dots microarray as diagnostic biomarkers for early detection of non-small cell lung cancer. Tumour Biol 37:7777–7784
Farokhzad OC, Langer R (2009) Impact of nanotechnology on drug delivery. ACS Nano 3:16–20
Fazio B, Andrea CD, Foti A, Messina E (2016) SERS detection of Biomolecules at Physiological pH via aggregation of Gold Nanorods mediated by Optical Forces and Plasmonic Heating. Sci Rep 6:26952
Fonte P, Noguiera T, Gehm C (2011) Chitosan-coated solid lipid nanoparticles enhance the oral absorption of insulin. Drug Deliv Transl Res 1(4):299–308
Fritz J (2008) Cantilever biosensors. Analyst 133:855–863

Fu L, Ke H (2016) Nanomaterials incorporated ultrasound contrast agents for cancer theranostics. Cancer Biol Med 13(3):313–324
Fu Y, Li P, Wang T (2010) Novel polymeric bionanocomposites with catalytic Pt nanoparticles label immobilized for high performance amperometric immunoassay. Biosens Bioelectron 25:1699–1704
Fuente DL, Jesus M (2006) Gold and gold-Iron oxide magnetic glyconanoparticles: synthesis, characterization and magnetic properties. J Phys Chem B 110(26):13021–13028
Gao A, Lu N, Dai P, Fan C, Wang Y, Li T (2014) Direct ultrasensitive electrical detection of prostate cancer biomarkers with CMOS-compatible n- and p-type silicon nanowire sensor arrays. Nanoscale 6(21):13036–13042
Garcia AC, Merkoci A (2016) Nanobiosensors in diagnostics. Nano 3:1–26
Gayathri T, Kumar RA, Panigrahi BS, Devanand B (2017) Silica-coated europium-doped gadolinium oxide nanorods for dual-modal imaging of cancer cells. Nano 12(06):1750073
Geho DH, Jones CD, Petricoin EF, Liotta LA (2006) Nanoparticles: potential biomarker harvesters. Curr Opin Chem Biol 10:56–61
Ghaghada KB, Sato AF, Starosolski ZA, Berg J (2016) Computed tomography imaging of solid tumors using a liposomal-iodine contrast agent in companion dogs with naturally occurring cancer. PLoS One 11(3):e0152718
Hainfeld JF, Slatkin DN, Focella TM, Smilowitz HM (2006) Gold nanoparticles: a new X-ray contrast agent. Br J Radiol 79:248–253
Holzinger M, Goff AL, Cosneir S (2014) Nanomaterials for biosensing applications: a review. Front Chem 2:63
Hounsfield GN (1973) Computerized transverse axial scanning (tomography), description of system. Br J Radiol 46(552):1016–1022
Huang X, Li S, Schultz JS, Wang Q, Lin Q (2009) A MEMS affinity glucose sensor using a biocompatible glucose-responsive polymer. Sensors Actuators B Chem 140(2):603–609
Jackson TC, Patani BO, Ekpa DE (2017) Nanotechnology in diagnosis: a review. Adv Nanopart 6:93–102
Jia F, Liu X, Li L, Mallapragada S, Narasimhan B, Wang Q (2013) Multifunctional nanoparticles for targeted delivery of immune activating and cancer therapeutic agents. J Control Release 172(3):1020–1034
Jin Y, Wang J, Ke H, Wang S, Dai Z (2013) Graphene oxide modified PLA microcapsules containing gold nanoparticles for ultrasonic/CT bimodal imaging guided photothermal tumor therapy. Biomaterials 34:4794–4802
Jin Q, Chen Y, Wang Y, Ji J (2014) Zwitterionic drug nanocarriers: a biomimetic strategy for drug delivery. Colloids Surf B Biointerfaces 124:80–86
Junghanns J, Müller RH (2008) Nanocrystal technology, drug delivery and clinical applications. Int J Nanomedicine 3:295–309
Kang HK, Seo J, Carlo DD, Choi YK (2003) Planar nanogap capacitor arrays on quartz for optical and dielectric bioassays. In: Proceedings of the micro total analysis systems, Squaw Valley, California, USA. pp 697–700
Ke H, Yue X, Wang J (2014) Gold nanoshelled liquid perfluorocarbon nanocapsules for combined dual modal ultrasound/CT imaging and photothermal therapy of cancer. Small 10:1220–1227
Kedar U, Phutane P, Shidhaye S, Kadam V (2010) Advances in polymeric micelles for drug delivery and tumor targeting. Nanomedicine 6:714–729
Kim D, Park S, Lee JH, Jeong YY, Jon S (2007) Antibiofouling polymer-coated gold nanoparticles as a contrast agent for in vivo X-ray computed tomography imaging. J Am Chem Soc 129:7661–7665
Kim JH, Yeo WH, Shu Z (2012) Immunosensor towards low-cost, rapid diagnosis of tuberculosis. Lab Chip 12(8):1437–1440
Klostranec JM, Chan WCW (2006) Quantum dots in biological and biomedical research: recent progress and present challenges. Adv Mater 18:1953–1964
Kojima C, Umeda Y, Ogawa M, Harada A, Magata Y, Kono K (2010) X-ray computed tomography contrast agents prepared by seeded growth of gold nanoparticles in PEGylated dendrimer. Nanotechnology 21:245104

Krishnamurthy V (2010) Ion-channel biosensors-part i: construction, operation, and clinical studies. IEEE Trans Nanotechnol 9(3):303–312
Krukemeyer MG, Krenn V, Huebner F, Wagner W (2015) History and possible uses of nanomedicine based on nanoparticles and nanotechnological progress. Nanomed Nanotechnol 6:336
Kumar SR, Vijayalakshmi R (2006) Nanotechnology in dentistry. Indian J Dent Res 17:62–69
Kumvongpin R, Jearanaikool P, Wilailuckana C, Sae-Ung N (2016) High sensitivity, loop-mediated isothermal amplification combined with colorimetric gold-nanoparticle probes for visual detection of high risk human papillomavirus genotypes 16 and 18. J Virol Methods 234:90–95
Lawaczeck R, Bauer H, Frenzel T (2016) Magnetic Iron oxide particles coated with carboxydextran for parenteral administration and liver contrasting. Acta Radiol 38:584–597
Lee D, Koo H, Sun IC, Ryu JH (2012) Multifunctional nanoparticles for multimodal imaging and theragnosis. Chem Soc Rev 41(7):2656–2672
Li Y, Artés JM, Demir B, Gokce S (2018) Detection and identification of genetic material via single-molecule conductance. Nat Nanotechnol 13(12):1167–1173
Liu M, Fréchet JM (1999) Designing dendrimers for drug delivery. Pharm Sci Technol Today 2:393–401
Liversidge EM, Liversidge GG, Cooper ER (2003) Nanosizing: a formulation approach for poorly-water-soluble compounds. Eur J Pharm Sci 18:113–120
Lu N, Gao A, Dai P (2014) CMOS-compatible silicon nanowire field-effect transistors for ultrasensitive and label-free microRNAs sensing. Small 10(10):2022–2028
Lueke J, Moussa WA (2011) MEMS-based power generation techniques for implantable Biosensing applications. Sensors (Basel) 11(2):1433–1460
Lyberopoulou A, Stathopoulos EP, Gazouli M (2015) Nanodiagnostic and nanotherapeutic molecular platforms for cancer management. J Cancer Res Updat 4:153–162
Marzán LM (2006) Tailoring surface plasmons through the morphology and assembly of metal nanoparticles. Langmuir 22(1):32–41
Mascini M, Tombelli S (2008) Biosensors for biomarkers in medical diagnostics. Biomarkers 13:637–657
Mattrey RF, Scheible FW, Gosink BB, Leopold GR, Long DM, Higgins CB (1982) Perfluoroctylbromide: a liver/spleen-specific and tumorimaging ultrasound contrast material. Radiology 145(3):759–762
Mehnert W, Mäder K (2001) Solid lipid nanoparticles: production, characterization and applications. Adv Drug Deliv Rev 47:165–196
Miyata K, Christie RJ, Kataoka K (2011) Polymeric micelles for nano-scale drug delivery. React Funct Polym 71:227–234
Mozafari M (2010) Liposomes: methods and protocols, pharmaceutical nanocarriers, 1st edn. Humana Press, Totowa, pp 29–50
Müller R, Maaβen S, Weyhers H (1996) Cytotoxicity of magnetite-loaded polylactide, polylactide/glycolide particles and solid lipid nanoparticles. Int J Pharm 138:85–94
Müller R, Jacobs C, Kayser O (2001) Nanosuspensions as particulate drug formulations in therapy. Rationale for development and what we can expect for the future. Adv Drug Deliv Rev 47:3–19
Nasimi P, Haidari M (2013) Medical use of nanoparticles: drug delivery and diagnosis diseases. Int J Green Nanotechnol 1:1–5
Nie L, Liu F, Ma P, Xiao X (2014) Applications of gold nanoparticles in optical biosensors. J Biomed Nanotechnol 10:2700–2721
Olbrich C, Bakowsky U, Lehr CM (2001) Cationic solid-lipid nanoparticles can efficiently bind and transfect plasmid DNA. J Control Release 77(3):345–355
Pedersen N, Hansen S, Heydenreich AV, Kristensen HG (2006) Solid lipid nanoparticles can effectively bind DNA, streptavidin and biotinylated ligands. Eur J Pharm Biopharm 62:155–162
Peeling RW, Mabey D (2010) Point-of-care tests for diagnosing infections in the developing world. Clin Microbiol Infect 16(8):1062–1069
Perez-Campana C, Gomez-Vallejo V, Puigivila M, Martin A, Calvo-Fernandez T, Moya SE, Ziolo RF, Reese T, Llop J (2013) Biodistribution of different sized nanoparticles assessed by positron

emission tomography: a general strategy for direct activation of metal oxide particles. ACS Nano 7(4):3498–3505
Prasad R (2014) Synthesis of silver nanoparticles in photosynthetic plants. J Nanopart 963961. https://doi.org/10.1155/2014/963961
Prasad R, Pandey R, Barman I (2016) Engineering tailored nanoparticles with microbes: quo vadis. Wiley Interdiscip Rev Nanomed Nanobiotechnol 8:316–330. https://doi.org/10.1002/wnan.1363
Prasad R (2017) Fungal nanotechnology: applications in agriculture, industry, and medicine. Springer Nature, Singapore. (ISBN 978-3-319-68423-9)
Prasad R, Bhattacharyya A, Nguyen QD (2017a) Nanotechnology in sustainable agriculture: Recent developments, challenges, and perspectives. Front Microbiol 8:1014. doi: 10.3389/fmicb.2017.01014
Prasad R, Pandey R, Varma A, Barman I (2017b) Polymer based nanoparticles for drug delivery systems and cancer therapeutics. In: Kharkwal H and Janaswamy S (eds.), Natural Polymers for Drug Delivery CAB International, UK, pp 53–70
Prasad R, Jha A, Prasad K (2018) Exploring the realms of nature for nanosynthesis. Springer International Publishing. https://www.springer.com/978-3-319-99570-0
Rabinow BE (2004) Nanosuspensions in drug delivery. Nat Rev Drug Discov 3:785–796
Rajasundari K, Hamurugu K (2011) Nanotechnology and its application in medical diagnosis. J Basic Pure Appl Chem 1:26–32
Ramesan RM, Sharma CP (2009) OMICS: biomedical perspectives and applications. Expert Rev Med Devices 6:665–676
Rawat M, Singh D, Saraf S (2006) Nanocarriers: promising vehicle for bioactive drugs. Biol Pharm Bull 29(9):1790–1798
Reimhult E, Höök F (2015) Design of Surface Modifications for nanoscale sensor applications. Sensors 15. https://doi.org/10.3390/s150101635
Roco MC, Harthorn B, Guston D, Shapira P (2017) Innovative and responsible governance of nanotechnology for societal development. In: Nanotechnology research directions for societal needs in 2020, pp 441–488 Springer USA
Saeed AA, Sanchez JLA, O'Sullivan CK (2017) DNA biosensors based on gold nanoparticles-modified graphene oxide for the detection of breast cancer biomarkers for early diagnosis. Bioelectrochemistry 118:91–99
Sahoo SK, Labhasetwar V (2003) Nanotech approaches to drug delivery and imaging. Drug Discov Today 8:1112–1120
Savjani KT, Anuradha K, Gajjar AK, Savjani JK (2012) Drug solubility: importance and enhancement techniques. ISRN Pharm 2012:195727
Seyfer P, Pagenstecher A, Mandic R, Klose KJ, Heverhagen JT (2014) Cancer and inflammation: differentiation by USPIO-enhanced MR imaging. J Magn Reson Imaging 39:665–672
Sharma RB, Hashim U (2013) Microfluidic photomask design using CAD software for application in lab-on-chip biomedical nano diagnostics. Adv Mater Res 795:388–392
Shegokar R, Müller RH (2010) Nanocrystals: industrially feasible multifunctional formulation technology for poorly soluble actives. Int J Pharm 399:129–139
Shehada N, Brönstrup G, Funka K, Christiansen S, Leja M, Haick H (2015) Ultrasensitive silicon nanowire for real-world gas sensing: noninvasive diagnosis of cancer from breath Volatolome. Nano Lett 15(2):1288–1295
Shetty NJ, Swati P, David K (2013) Nanorobots: future in dentistry. Saudi Dent J 25(2):49–52
Shilo M, Reuveni T, Motiei M (2012) Nanoparticles as computed tomography contrast agents. Nanomedicine 7(2):257–259
Svenson S (2009) Dendrimers as versatile platform in drug delivery applications. Eur J Pharm Biopharm 71:445–462
Swierczewska M, Lee S, Chen X (2011) Inorganic nanoparticles for multimodal molecular imaging. Mol Imaging 10(1):3–16
Taguchi M, Ptitsyn A (2014) Nanomaterial-mediated biosensors for monitoring glucose. J Diabetes Sci Technol 8(2):403–411

Takahashi S, Shiraishi T, Miles N, Trock BJ, Kulkarni P, Getzenberg RH (2015) Nanowire analysis of cancer-testis antigens as biomarkers of aggressive prostate cancer. Urology 85(3):704.e1–704.e7
Tan C, Wang Y, Fan Y (2013) Exploring polymeric micelles for improved delivery of anticancer agents: recent developments in preclinical studies. Pharmaceutics 5:201–219
Teraphongphom N, Chhour P, Eisenbrey JR (2015) Nanoparticle loaded polymeric microbubbles as contrast agents for multimodal imaging. Langmuir 31(43):11858–11867
Thévenot DR, Toth K, Durst RA (2001) Electrochemical biosensors: recommended definitions and classification. Biosens Bioelectron 16(1–2):121–131
Tian C, Zhu L, Lin F, Boyes SG (2015) Poly(acrylic acid) bridged gadolinium metal-organic framework-gold nanoparticle composites as contrast agents for computed tomography and magnetic resonance bimodal imaging. ACS Appl Mater Interfaces 7:17765–17775
Torchilin VP (2001) Structure and design of polymeric surfactant-based drug delivery systems. J Control Release 73:137–172
Torchilin VP (2012) Multifunctional nanocarriers. Adv Drug Deliv Rev 64:302–315
Trubetskoy VS, Gazelle GS, Wolf GL, Torchilin VP (1997) Block-copolymer of polyethylene glycol and polylysine as a carrier of organic iodine: design of long-circulating particulate contrast medium for x-ray computed tomography. J Drug Target 4(6):381–388
Varshosaz J, Minayian M, Moazen E (2010) Formulation and optimization of solid lipid nanoparticles of buspirone HCl for enhancement of its oral bioavailability. J Liposome Res 20:115–123
Wang J, Musameh M, Lin YH (2003) Solubilization of carbon nanotubes by Nafion toward the preparation of amperometric biosensors. J Am Chem Soc 125:2408
Warkiani ME, Tay AK, Khoo BL, Xiaofeng X, Han J, Lim CT (2015) Malaria detection using inertial microfluidics. Lab Chip 15(4):1101–1109
Warlin D (2013) Superparamagnetic iron oxide nanoparticles for MRI: contrast media pharmaceutical company R&D perspective. Wiley Interdiscip Rev Nanomed Nanobiotechnol 5(5):411–422
Williams HD, Trevaskis NL, Charman SA, Shanker RM (2013) Strategies to address low drug solubility in discovery and development. Pharmacol Rev 65:315–499
Xu C, Mu L, Roes I, Miranda-Nieves I, Nahrendorf D (2011) Nanoparticle-based monitoring of cell therapy. Nanotechnology 22(49):494001
Xu W, Ling P, Zhang T (2013) Polymeric micelles, a promising drug delivery system to enhance bioavailability of poorly water-soluble drugs. J Drug Deliv 2013:340315
Yeh MK, Chang LC, Chiou AHJ (2009) Improving tenoxicam solubility and bioavailability by cosolvent system. AAPS PharmSciTech 10:166–171
Yeo WH, Chung JH, Liu Y, Lee KH (2009) Size-specific concentration of DNA to a nanostructured tip using dielectrophoresis and capillary action. J Phys Chem B 113(31):10849–10858
Yogesh ST, Indrajeet DG, Avinash HS (2011) Solubility enhancement techniques: a review on conventional and novel approaches. Int J Pharm Sci Res 2:2501–2513
Yu L, Shi Z, Fang C, Zhang Y, Liu Y (2015) Disposable lateral flow-through strip for smartphone-camera to quantitatively detect alkaline phosphatase activity in milk. Biosens Bioelectron 69:307–315
Yuen C, Liu QJ (2012) Magnetic field enriched surface enhanced resonance Raman spectroscopy for early malaria diagnosis. J Biomed Opt 17(1):017005
Zhang CY, Yeh HC, Kuroki MT, Wang TH (2005) Single-quantum-dot-based DNA nanosensor. Nat Mater 4(11):826–831
Zhang B, Wang K, Si J, Sui M, Shen Y (2014) Charge-reversal polymers for biodelivery. In: Gu Z (ed) Bioinspired and biomimetic polymer systems for drug and gene delivery. Wiley-VCH Verlag GmbH & Co. KGaA, Weinheim, pp 223–242
Zhang Z, Xia X, Xiang X, Huang F (2018) Quantum dots-ru complex assembling dyads for cancer cell detection and cellular imaging based on hybridization chain reaction. Sensors Actuators B Chem 257:1–8
Ziegler C (2004) Cantilever-based biosensors. Anal Bioanal Chem 379(7–8):946–959

Index

R. Prasad et al. (eds.), *Nanobiotechnology in Bioformulations*, Nanotechnology in the Life Sciences, https://doi.org/10.1007/978-3-030-17061-5

O

MIX
Papier aus verantwortungsvollen Quellen
Paper from responsible sources
FSC® C105338

If you have any concerns about our products,
you can contact us on
ProductSafety@springernature.com

In case Publisher is established outside the EU,
the EU authorized representative is:
Springer Nature Customer Service Center GmbH
Europaplatz 3, 69115 Heidelberg, Germany

Printed by Libri Plureos GmbH
in Hamburg, Germany